Lecture Notes in Computer Science 5580

Commenced Publication in 1973
Founding and Former Series Editors:
Gerhard Goos, Juris Hartmanis, and Jan van Leeuwen

Bart Preneel (Ed.)

Progress in Cryptology – AFRICACRYPT 2009

Second International Conference on Cryptology in Africa
Gammarth, Tunisia, June 21-25, 2009
Proceedings

 Springer

Volume Editor

Bart Preneel
Katholieke Universiteit Leuven
Dept. Electrical Engineering-ESAT/COSIC
Kasteelpark Arenberg 10, Bus 2446, 3001 Leuven, Belgium
E-mail: bart.preneel@esat.kuleuven.be

Library of Congress Control Number: Applied for

CR Subject Classification (1998): E.3, F.2.1-2, G.2.1, D.4.6, K.6.5, C.2, J.1

LNCS Sublibrary: SL 4 – Security and Cryptology

ISSN   0302-9743
ISBN-10   3-642-02383-5 Springer Berlin Heidelberg New York
ISBN-13   978-3-642-02383-5 Springer Berlin Heidelberg New York

springer.com

© Springer-Verlag Berlin Heidelberg 2009
Printed in Germany

Typesetting: Camera-ready by author, data conversion by Scientific Publishing Services, Chennai, India
Printed on acid-free paper   SPIN: 12693764   06/3180   5 4 3 2 1 0

Preface

AFRICACRYPT 2009 was held during June 21–25, 2009 in Gammarth, Tunisia. After AFRICACRYPT 2008 in Casablanca, Morocco, it was the second international research conference in Africa dedicated to cryptography.

The conference received 70 submissions; four of these were identified as irregular submissions. The remaining papers went through a careful doubly anonymous review process. Every paper received at least three reports; papers with a Program Committee member as co-author received five reports. After the review period, 25 papers were accepted for presentation. The authors were requested to revise their papers based on the comments received. The program was completed with invited talks by Antoine Joux, Ueli Maurer and Nigel Smart.

First and foremost we would like to thank the members of the Program Committee for the many hours spent on reviewing and discussing the papers, thereby producing more than 600 Kb of comments. They did an outstanding job. We would also like to thank the numerous external reviewers for their assistance. We are also indebted to Shai Halevi for the support provided for his excellent Web-Submission-and-Review software package. We also wish to heartily thank Sami Ghazali, the General Chair, and Sami Omar, the General Co-chair, for their efforts in the organization of the conference. Special thanks go to the Tunisian Ministry of Communication Technologies, the National Digital Certification Agency, and the Tunisian Internet Agency for their support of the organization. Finally, we would like to thank the participants, submitters, authors and presenters who all together made AFRICACRYPT 2009 a great success.

I hope that the AFRICACRYPT conference tradition has now taken firm root and that we will witness a fruitful development of academic research in cryptology in Africa.

April 2009   Bart Preneel

Organization

AFRICACRYPT 2009 was organized under the patronage of the Tunisian Ministry of Communication Technologies by the National Digital Certification Agency, and the Tunisian Internet Agency. AFRIACRYPT 2009 was organized in cooperation with the International Association for Cryptologic Research (IACR).

Executive Committee

Conference Chair   Sami Ghazali (National Digital Certification
                        Agency, Tunisia)
Conference Co-chair   Sami Omar (University of Tunis, Tunisia)
Program Chair   Bart Preneel (Katholieke Universiteit Leuven,
                        Belgium)

Program Committee

Michel Abdalla   ENS, Paris, France
Paulo Barreto   University of São Paulo (USP), Brazil
Tom Berson   Anagram Laboratories, USA
Anne Canteaut   INRIA Paris-Rocquencourt, France
Dario Catalano   University of Catania, Italy
Hervé Chabanne   Sagem Sécurité and Télécom Paristech, France
Jean-Marc Couveignes   Université Toulouse 2, France
Kris Gaj   George Mason University, USA
Henri Gilbert   Orange Labs, France
Helena Handschuh   Spansion, France
Martin Hirt   ETH Zurich, Switzerland
Seokhie Hong   Korea University, Korea
Tetsu Iwata   Nagoya University, Japan
Yassine Lakhnech   University Joseph Fourier-Grenoble, France
Tanja Lange   Technische Universiteit Eindhoven, The Netherlands
Helger Lipmaa   Cybernetica AS, Estonia
Keith Martin   Royal Holloway University of London, UK
Mitsuru Matsui   Mitsubishi Electric, Japan
Alexander May   R.U. Bochum, Germany
Sihem Mesnager   Université Paris 8, France
Kaisa Nyberg   Helsinki University of Technology and Nokia, Finland
Sami Omar   University of Tunis, Tunisia
Elisabeth Oswald   University of Bristol, UK
Reihaneh Safavi-Naini   University of Calgary, Canada
Kazue Sako   NEC, Japan
Ali Aydın Selçuk   Bilkent University, Turkey

Christine Swart   University of Cape Town, South Africa
Serge Vaudenay   EPFL, Switzerland
Frederik Vercauteren   Katholieke Universiteit Leuven, Belgium
Michael Wiener   Cryptographic Clarity, Canada
Yiqun Lisa Yin   Independent Security Consultant, USA

External Reviewers

Divesh Aggarwal   Benedikt Gierlichs   Sylvain Pasini
Hadi Ahmadi   Malakondayya Gorantla   Maura Paterson
Murat Ak   Nicolas Guillermin   Ludovic Perret
Toshinori Araki   Tim Güneysu   Duong Hieu Phan
Roberto Avanzi   Risto Hakala   Dominik Raub
Lejla Batina   Jens Hermans   Yu Sasaki
Aurelie Bauer   Miia Hermelin   Berry Schoenmakers
Zuzana Beerliova   Thomas Icart   Nicolas Sendrier
Daniel J. Bernstein   Toshiyuki Isshiki   Hongsong Shi
Arnaud Boscher   Takashi Ito   Igor Shparlinski
Julien Bringer   Orhun Kara   Francesco Sica
Renier Broker   Kamer Kaya   Michal Sramka
Bogdan Carbunar   Bruno Kindarji   Drew Sutherland
Claude Carlet   Miroslav Knezevic   Daisuke Suzuki
Rafik Chaabouni   Patrick Lacharme   Björn Tackmann
Donghoon Chang   Fabien Laguillaumie   Katsuyuki Takashima
Claude Crépeau   Thanh Ha Le   Isamu Teranishi
Hüseyin Demirci   Jesang Lee   Stefano Tessaro
Alex Dent   Benoît Libert   Mike Tunstall
Mario Di Raimondo   Moses Liskov   Damien Vergnaud
Orr Dunkelman   Christoph Lucas   Ivan Visconti
Junfeng Fan   Andrew Moss   Bogdan Warinschi
Dario Fiore   María Naya-Plasencia   Qianhong Wu
Jun Furukawa   Gregory Neven   Vassilis Zikas
Martin Gagne   Satoshi Obana
Pierrick Gaudry   Khaled Ouafi

Table of Contents

Asymmetric Encryption and Anonymity

Key Agreement Protocols

Cryptographic Protocols

Efficient Implementations

Implementation Attacks

Second Preimage Attack on 5-Pass HAVAL and Partial Key-Recovery Attack on HMAC/NMAC-5-Pass HAVAL

Gaoli Wang[1,*] and Shaohui Wang[2]

[1] School of Computer Science and Technology, Donghua University,
Shanghai 201620, China
wanggaoli@dhu.edu.cn
[2] Nanjing University of Posts and Telecommunications,
Nanjing 210046, China
wangshaohui@njupt.edu.cn

Abstract. HAVAL is a cryptographic hash function with variable hash value sizes proposed by Zheng, Pieprzyk and Seberry in 1992. It has 3, 4, or 5 passes, and each pass contains 32 steps. There was a collision attack on 5-pass HAVAL, but no second preimage attack. In this paper, we present a second preimage differential path for 5-pass HAVAL with probability 2^{-227} and exploit it to devise a second preimage attack on 5-pass HAVAL. Furthermore, we utilize the path to recover the partial key of HMAC/NMAC-5-pass HAVAL with 2^{235} oracle queries and 2^{35} memory bytes.

Keywords: HMAC, NMAC, second preimage attack, key-recovery, HAVAL.

1 Introduction

A cryptographic hash function compresses a message of arbitrary length to a fixed-length hash value. It is defined as a mapping $h : \{0,1\}^* \longrightarrow \{0,1\}^n$, where $\{0,1\}^*$ denotes the set of strings of arbitrary length, and $\{0,1\}^n$ denotes the set of strings of n-bit length. Hash function is a fundamental primitive in many cryptographic schemes and protocols.

From the security perspective, a hash function h with inputs x, x' and outputs y, y' should satisfy some or all of the following properties. The first property is preimage resistance, i.e., for any output y, it is computationally infeasible to get an input x such that $h(x) = y$. The second property is second preimage resistance, i.e., for any given x, it is computationally infeasible to get another input $x' (\neq x)$ such that $h(x) = h(x')$. The third property is collision resistance, i.e., it is computationally infeasible to get any two distinct inputs x, x' such that $h(x) = h(x')$.

---

* Supported by Foundation of State Key Laboratory of Information Security (Institute of Software, Chinese Academy of Sciences).

B. Preneel (Ed.): AFRICACRYPT 2009, LNCS 5580, pp. 1–13, 2009.

The hash function HAVAL [2] was proposed by Zheng, Pieprzyk and Seberry at Auscrypt'92. It is a Merkle-Damgård hash function, which uses a compression function to digest messages. In recent years, there has been a great progress in finding collisions of hash functions, and many hash functions based on MD4, such as HAVAL, MD5, RIPEMD, SHA-0 and SHA-1 etc. are not collision resistant [4,5,6,7,8,9,10,12]. [11] presents a second preimage attack on MD4 which finds a second preimage for a random message with probability 2^{-56}.

Several articles that focus on attacks on NMAC and HMAC based on the MD4 family were presented. Kim et al. [13] proposed distinguishing and forgery attacks on NMAC and HMAC based on the full or reduced HAVAL, MD4, MD5, SHA-0 and SHA-1. Contini and Yin [14] proposed forgery and partial key-recovery attacks on NMAC-MD5 and HMAC/NMAC based on MD4, SHA-0 and 34-round reduced SHA-1. Full key-recovery attacks on HMAC/NMAC-MD4, 61-round reduced SHA-1 and NMAC-MD5 were proposed in [16], and these attacks were independently presented in [17]. More recently, [18] presented new key-recovery attacks on HMAC/NMAC-MD4 and NMAC-MD5 and [19] presented a second preimage attack on 3-pass HAVAL and partial key-recovery attacks on HMAC/NMAC-3-Pass HAVAL. Yu [15] gave partial key-recovery attacks on HMAC-3-pass, -4-pass HAVAL and second preimage attack on 3-pass and 4-pass HAVAL in her doctoral dissertation.

In this paper, we present a second preimage differential path for 5-pass HAVAL with probability 2^{-227}. Then we give a partial key-recovery attack on HMAC/NMAC-5-Pass HAVAL with the complexity of 2^{235} oracle queries and 2^{35} memory bytes.

The rest of the paper is organized as follows. We describe HMAC, NMAC and HAVAL algorithms in Section 2. In Section 3, we give a second preimage attack on 5-pass HAVAL. Section 4 shows how to recover a partial key of HMAC/NMAC-5-pass HAVAL. Finally, we summarize the paper in section 5.

2   Preliminaries

In this section, we will describe the HMAC, NMAC algorithms, the 5-pass HAVAL hash function, and the notations used in this paper.

2.1   Description of HMAC and NMAC

HMAC and NMAC are hash-based message authentication codes proposed by Bellare, Canetti and Krawczyk [3]. The construction of HMAC/NMAC is based on a keyed hash function. Let H be an iterated Merkle-Damgård hash function, which defines a keyed hash function H_k by replacing the IV with the key k. Then HMAC and NMAC are defined as:

$$\text{HMAC}_k(M) = H_{IV}(\bar{k} \oplus opad || H_{IV}(\bar{k} \oplus ipad || M));$$

$$\text{NMAC}_{k_1,k_2}(M) = H_{k_1}(H_{k_2}(M)),$$

where M is the input message, k and (k_1, k_2) are the secret keys of HMAC and NMAC respectively, $\bar{k}$ means k padded to a single block, $\|$ means concatenation, and opad and ipad are two one-block length constants. NMAC is the theoretical foundation of HMAC: HMAC_k is essentially the same as $\text{NMAC}_{H_{IV}(\bar{k} \oplus opad), H_{IV}(\bar{k} \oplus ipad)}$, except with a change in the length value included in the padding. k_1 and k_2 (for HMAC: $H_{IV}(\bar{k} \oplus opad)$ and $H_{IV}(\bar{k} \oplus ipad)$ with the appropriate changes in the padding) are referred to as the outer key and the inner key respectively.

2.2　Description of 5-Pass HAVAL

HAVAL allows hash values in different sizes of 128, 160, 192, 224, and 256 bits. The main algorithm produces 256-bit hash values, and the other sizes are obtained by post-processing the 256-bit hash value. Therefore for our purposes we may consider HAVAL as a hash function with the hash value size of 256 bits.

HAVAL has a similar structure as the well-known hash function MD4 [1]. Firstly HAVAL pads any given message into a message with the length of 1024 bits multiple, then the compression function takes a 1024-bit message block and a 256-bit chaining value as inputs and produces another 256-bit chaining value as output.

The initial value of HAVAL is: $(0x243f6a88, 0x85a308d3, 0x13198a2e, 0x03707344, 0xa4093822, 0x299f31d0, 0x082efa98, 0xec4e6c89)$.

The boolean functions of 5-pass HAVAL are presented in Table 1. Here x_i (i=0, ..., 6) are 32-bit words and the operations of the four functions are all bitwise.

Table 1. Boolean functions

Pass	Function
1	$f_1(x_6, x_5, x_4, x_3, x_2, x_1, x_0) = x_0x_3 \oplus x_1x_2 \oplus x_2x_6 \oplus x_4x_5 \oplus x_6$
2	$f_2(x_6, x_5, x_4, x_3, x_2, x_1, x_0) = x_0x_2 \oplus x_0x_3x_4 \oplus x_1x_2 \oplus x_1x_2x_3 \oplus x_1x_4 \oplus x_3x_4 \oplus$ $x_3x_5 \oplus x_5 \oplus x_3x_6$
3	$f_3(x_6, x_5, x_4, x_3, x_2, x_1, x_0) = x_0x_1 \oplus x_1x_3x_4 \oplus x_2x_4 \oplus x_3x_6 \oplus x_4x_5 \oplus x_5$
4	$f_4(x_6, x_5, x_4, x_3, x_2, x_1, x_0) = x_0x_1 \oplus x_0x_2x_4 \oplus x_0x_3x_5 \oplus x_1x_2 \oplus x_1x_2x_3 \oplus x_1x_3 \oplus$ $x_1x_3 \oplus x_2x_3 \oplus x_2x_5 \oplus x_3x_4 \oplus x_3x_6 \oplus x_3x_5 \oplus x_6$
5	$f_5(x_6, x_5, x_4, x_3, x_2, x_1, x_0) = x_0x_3 \oplus x_1 \oplus x_1x_3x_4x_6 \oplus x_1x_5 \oplus x_2x_6 \oplus x_4x_5$

5-Pass HAVAL Compression Function For a 1024-bit block M, $M = (m_0, m_1, \ldots, m_{31})$, the compressing process is as follows:

1. Let $(a_{-1}, a_{-2}, a_{-3}, a_{-4}, a_{-5}, a_{-6}, a_{-7}, a_{-8})$ be the input of compressing process for M. If M is the first block to be hashed, $(a_{-1}, a_{-2}, a_{-3}, a_{-4}, a_{-5}, a_{-6}, a_{-7}, a_{-8})$ is the initial value. Otherwise it is the output of the previous block compressing.

2. Perform the following 160 steps:

For $j = 1, 2, 3, 4, 5$

For $i = 32(j-1), 32(j-1) + 1, \ldots, 32(j-1) + 31$

$a_i = ((f_j(a_{i-7}, a_{i-6}, a_{i-5}, a_{i-4}, a_{i-3}, a_{i-2}, a_{i-1})) \ggg 7) + (a_{i-8} \ggg 11) + m_{ord(j,i)} + k_{j,i}$.

The operation in each step employs a constant $k_{j,i}$ (See [2]). The orders of the message words in each pass can refer to Table 2. Here and in the follows, + denotes addition modulo 2^{32}, - denotes abstract modulo 2^{32}.

3. Add $a_{159}, a_{158}, a_{157}, a_{156}, a_{155}, a_{154}, a_{153}, a_{152}$ respectively to the input value, i.e., $aa = a_{159} + a_{-1}$, $bb = a_{158} + a_{-2}$, ..., $hh = a_{152} + a_{-8}$.

4. $H(M) = hh\|gg\|ff\|ee\|dd\|cc\|bb\|aa$, where $\|$ denotes the bit concatenation.

If M is the last message block of the message MM, then $H(M) = hh\|gg\|ff\|ee\|dd\|cc\|bb\|aa$ is the hash value for the message MM. Otherwise take $(aa, bb, cc, dd, ee, ff, gg, hh)$ as input, and repeat the compression process for the next 1024-bit message block.

Table 2. Word Processing Orders

$pass_1$	0	1	2	3	4	5	6	7	8	9	10	11	12	13	14	15
	16	17	18	19	20	21	22	23	24	25	26	27	28	29	30	31
$pass_2$	5	14	26	18	11	28	7	16	0	23	20	22	1	10	4	8
	30	3	21	9	17	24	29	6	19	12	15	13	2	25	31	27
$pass_3$	19	9	4	20	28	17	8	22	29	14	25	12	24	30	16	26
	31	15	7	3	1	0	18	27	13	6	21	10	23	11	5	2
$pass_4$	24	4	0	14	2	7	28	23	26	6	30	20	18	25	19	3
	22	11	31	21	8	27	12	9	1	29	5	15	17	10	16	13
$pass_5$	27	3	21	26	17	11	20	29	19	0	12	7	13	8	31	10
	5	9	14	30	18	6	28	24	2	23	16	22	4	1	25	15

2.3 Notations

$M = (m_0\|m_1\|\ldots\|m_{31})$ and $M' = (m'_0\|m'_1\|\ldots\|m'_{31})$ denote two 1024-bit message blocks where m_i $(i=0, \ldots, 31)$ and m_j $(j=0, \ldots, 31)$ are 32-bit words. $a_i, b_i, c_i, d_i, e_i, f_i, g_i, h_i$ and $a'_i, b'_i, c'_i, d'_i, e'_i, f'_i, g'_i, h'_i$ $(i = 0, 1, \ldots, 159)$ denote the chaining variables after the i-th step corresponding to the message blocks M and M' respectively. The updated values of the i-th step are a_i and a'_i. $\Delta m_i = m'_i - m_i$, $\Delta a_i = a'_i - a_i$, ..., $\Delta h_i = h'_i - h_i$ denote the modular differences of two variables. For $1 \leq j \leq 32$, $a_{i,j}$ denotes the j-th bit of 32-bit word a_i, $a_i[j]$ is the resulting value by only changing the j-th bit of a_i from 0 to 1, $a_i[-j]$ is the resulting value by only changing the j-th bit of a_i from 1 to 0, $\Delta a_i = [j]$ denotes $a_{i,j} = 0$ and $a'_{i,j} = 1$, $\Delta a_i = [-j]$ denotes $a_{i,j} = 1$ and $a'_{i,j} = 0$.

3   Second Preimage Attack on 5-Pass HAVAL

In this section, first we will recall some properties of the boolean functions, then we will construct a second preimage differential path for 5-pass HAVAL with probability 2^{-227}.

3.1   Some Basic Propositions

Here, we only recall some popular used properties of the nonlinear function f_1 in our attack. We omit describing the properties of the other four functions f_2, f_3, f_4 and f_5 because of the space constraint.

Proposition. For the nonlinear function $f_1(x_6, x_5, x_4, x_3, x_2, x_1, x_0) = x_0 x_3 \oplus x_1 x_2 \oplus x_2 x_6 \oplus x_4 x_5 \oplus x_6$, the following properties hold:

1. $f_1(x_6, x_5, x_4, x_3, x_2, x_1, x_0) = f_1(x_6, x_5, x_4, x_3, x_2, x_1, \neg x_0)$ if and only if $x_3 = 0$.
2. $f_1(x_6, x_5, x_4, x_3, x_2, x_1, x_0) = f_1(x_6, x_5, x_4, x_3, x_2, \neg x_1, x_0)$ if and only if $x_2 = 0$.
3. $f_1(x_6, x_5, x_4, x_3, x_2, x_1, x_0) = f_1(x_6, x_5, x_4, x_3, \neg x_2, x_1, x_0)$ if and only if $x_1 = x_6$.
4. $f_1(x_6, x_5, x_4, x_3, x_2, x_1, x_0) = f_1(x_6, x_5, x_4, \neg x_3, x_2, x_1, x_0)$ if and only if $x_0 = 0$.
5. $f_1(x_6, x_5, x_4, x_3, x_2, x_1, x_0) = f_1(x_6, x_5, \neg x_4, x_3, x_2, x_1, x_0)$ if and only if $x_5 = 0$.
6. $f_1(x_6, x_5, x_4, x_3, x_2, x_1, x_0) = f_1(x_6, \neg x_5, x_4, x_3, x_2, x_1, x_0)$ if and only if $x_4 = 0$.
7. $f_1(x_6, x_5, x_4, x_3, x_2, x_1, x_0) = f_1(\neg x_6, x_5, x_4, x_3, x_2, x_1, x_0)$ if and only if $x_2 = 1$.

Here, $x_i \in \{0, 1\}$ $(0 \leq i \leq 6)$, and $\neg x_i$ denotes the complement of the bit x_i.

3.2   Description of the Attack

Our second preimage attack on 5-pass HAVAL contains two phases. Firstly, we choose an appropriate message difference and deduce the differential path according to the specified message difference. Secondly, we determine the corresponding chaining variable conditions.

Choosing an appropriate message difference. We select two 1024-bit message blocks $M = (m_0, m_1, \ldots, m_{31})$ and $M' = M + \Delta M$, where $\Delta M = (\Delta m_0, \ldots, \Delta m_{31})$ such that $\Delta m_i = 0$ $(i \neq 11)$ and $\Delta m_{11} = -1$. Then we can construct a second preimage differential path with probability 2^{-227}.

Deducing the differential path. The first inner collision is from step 11 to step 36. The message word m_{11} is first used at step 11, and the difference introduced at step 11 is propagated until step 36, where m_{11} is again used. The message word m_{11} is again used at step 93, near the end of the pass 3, and the second inner collision is from step 93 to step 133. Table 3 and Table 4 shows the differential path.

Deriving the sufficient conditions for the differential path. We will deduce a set of sufficient conditions which ensure the second preimage differential path to hold. We give an example to explain how to derive the set of sufficient conditions.

The message difference $\Delta m_{11} = -1$ produces the output difference of the 11-th step $(\Delta a_{11}, \Delta a_{10}, \Delta a_9, \Delta a_8, \Delta a_7, \Delta a_6, \Delta a_5, \Delta a_4) = (a_{11}[1, 2, -3], 0, 0, 0, 0, 0, 0, 0)$ if and only if the conditions $a_{11,1} = 0$, $a_{11,2} = 0$ and $a_{11,3} = 1$ hold.

The difference $a_{11}[1, -3]$ doesn't produce any more bit differences between step 11 and step 18, and the difference $a_{11}[2]$ is used to produce the differences $a_{13}[-27, -28, -29, 30]$ in step 13 and $a_{16}[27]$ in step 16.

1. In step 11, $a'_{11} = a_{11}[1, 2, -3] \Leftrightarrow a_{11,1} = 0$, $a_{11,2} = 0$ and $a_{11,3} = 1$.
2. In step 12, $(a_{11}[1, 2, -3], a_{10}, a_9, a_8, a_7, a_6, a_5, a_4) \rightarrow (a_{12}, a_{11}[1, 2, -3], a_{10}, a_9, a_8, a_7, a_6, a_5)$. According to 1 of Proposition, we know that $a'_{12} = a_{12} \Leftrightarrow a_{8,1} = 0$, $a_{8,2} = 0$ and $a_{8,3} = 0$.
3. In step 13, $(a_{12}, a_{11}[1, 2, -3], a_{10}, a_9, a_8, a_7, a_6, a_5) \rightarrow (a_{13}[-27, -28, -29, 30], a_{12}, a_{11}[1, 2, -3], a_{10}, a_9, a_8, a_7, a_6)$. According to the proposition of the nonlinear function f_1, we know that $a'_{13} = a_{13}[-27, -28, -29, 30] \Leftrightarrow a_{10,1} = 0$, $a_{10,2} = 1$, $a_{10,3} = 0$, $a_{12,2}a_{9,2} \oplus a_{10,2}a_{6,2} \oplus a_{8,2}a_{7,2} \oplus a_{6,2} = 0$, $a_{13,27} = 1$, $a_{13,28} = 1$, $a_{13,29} = 1$ and $a_{13,30} = 0$.
4. In step 14, $(a_{13}[-27, -28, -29, 30], a_{12}, a_{11}[1, 2, -3], a_{10}, a_9, a_8, a_7, a_6) \rightarrow (a_{14}, a_{13}[-27, -28, -29, 30], a_{12}, a_{11}[1, 2, -3], a_{10}, a_9, a_8, a_7)$. According to 1 and 3 of Proposition, we know that $a'_{14} = a_{14} \Leftrightarrow a_{12,1} = a_{7,1}$, $a_{12,2} = a_{7,2}$, $a_{12,3} = a_{7,3}$, $a_{10,27} = 0$, $a_{10,28} = 0$, $a_{10,29} = 0$ and $a_{10,30} = 0$.
5. In step 15, $(a_{14}, a_{13}[-27, -28, -29, 30], a_{12}, a_{11}[1, 2, -3], a_{10}, a_9, a_8, a_7) \rightarrow (a_{15}, a_{14}, a_{13}[-27, -28, -29, 30], a_{12}, a_{11}[1, 2, -3], a_{10}, a_9, a_8)$. According to 2 and 4 of Proposition, we know that $a'_{15} = a_{15} \Leftrightarrow a_{14,1} = 0$, $a_{14,2} = 0$, $a_{14,3} = 0$, $a_{12,27} = 0$, $a_{12,28} = 0$, $a_{12,29} = 0$ and $a_{12,30} = 0$.
6. In step 16, $(a_{15}, a_{14}, a_{13}[-27, -28, -29, 30], a_{12}, a_{11}[1, 2, -3], a_{10}, a_9, a_8) \rightarrow (a_{16}[27], a_{15}, a_{14}, a_{13}[-27, -28, -29, 30], a_{12}, a_{11}[1, 2, -3], a_{10}, a_9)$. According to the proposition of the nonlinear function f_1, we know that $a'_{16} = a_{16}[27] \Leftrightarrow a_{10,1} = 0$, $a_{10,2} = 1$, $a_{10,3} = 0$, $a_{15,2}a_{12,2} \oplus a_{14,2}a_{13,2} \oplus a_{13,2}a_{9,2} \oplus a_{9,2} = 0$, $a_{14,27} = a_{9,27}$, $a_{14,28} = a_{9,28}$, $a_{14,29} = a_{9,29}$, $a_{14,30} = a_{9,30}$ and $a_{16,27} = 0$.
7. In step 17, $(a_{16}[27], a_{15}, a_{14}, a_{13}[-27, -28, -29, 30], a_{12}, a_{11}[1, 2, -3], a_{10}, a_9) \rightarrow (a_{17}, a_{16}[27], a_{15}, a_{14}, a_{13}[-27, -28, -29, 30], a_{12}, a_{11}[1, 2, -3], a_{10})$. According to the proposition of the nonlinear function f_1, we know that $a'_{17} = a_{17} \Leftrightarrow a_{12,1} = 0$, $a_{12,2} = 0$, $a_{12,3} = 0$, $a_{16,28} = 0$, $a_{16,29} = 0$ and $a_{16,30} = 0$.
8. In step 18, $(a_{17}, a_{16}[27], a_{15}, a_{14}, a_{13}[-27, -28, -29, 30], a_{12}, a_{11}[1, 2, -3], a_{10}) \rightarrow (a_{18}, a_{17}, a_{16}[27], a_{15}, a_{14}, a_{13}[-27, -28, -29, 30], a_{12}, a_{11}[1, 2, -3])$. According to the proposition of the nonlinear function f_1, we know that $a'_{18} = a_{18} \Leftrightarrow a_{15,1} = 1$, $a_{15,2} = 1$, $a_{15,3} = 1$, $a_{15,27} = a_{12,27}$, $a_{12,28} = 0$, $a_{12,29} = 0$ and $a_{12,30} = 0$.

Table 3. Second preimage differential path for 5-pass HAVAL

Step	m_i	Δm_i	Δa_i	Δa_{i-1}	Δa_{i-2}	Δa_{i-3}	Δa_{i-4}	Δa_{i-5}	Δa_{i-6}	Δa_{i-7}
0	m_0	0	0	0	0	0	0	0	0	0
...	...	...	...	...	...	...	...	...	...	...
11	m_{11}	−1	[1,2,−3]	0	0	0	0	0	0	0
12	m_{12}	0	0	[1,2,−3]	0	0	0	0	0	0
13	m_{13}	0	[−27,−28,−29,30]	0	[1,2,−3]	0	0	0	0	0
14	m_{14}	0	0	[−27,−28,−29,30]	0	[1,2,−3]	0	0	0	0
15	m_{15}	0	0	0	[−27,−28,−29,30]	0	[1,2,−3]	0	0	0
16	m_{16}	0	[27]	0	0	[−27,−28,−29,30]	0	[1,2,−3]	0	0
17	m_{17}	0	0	[27]	0	0	[−27,−28,−29,30]	0	[1,2,−3]	0
18	m_{18}	0	0	0	[27]	0	0	[−27,−28,−29,30]	0	[1,2,−3]
19	m_{19}	0	0	0	0	[27]	0	0	[−27,−28,−29,30]	0
20	m_{20}	0	[23]	0	0	0	[27]	0	0	[−27,−28,−29,30]
21	m_{21}	0	0	[23]	0	0	0	[27]	0	0
22	m_{22}	0	0	0	[23]	0	0	0	[27]	0
23	m_{23}	0	0	0	0	[23]	0	0	0	[27]
24	m_{24}	0	0	0	0	0	[23]	0	0	0
25	m_{25}	0	0	0	0	0	0	[23]	0	0
26	m_{26}	0	0	0	0	0	0	0	[23]	0
27	m_{27}	0	0	0	0	0	0	0	0	[23]
28	m_{28}	0	[12]	0	0	0	0	0	0	0
29	m_{29}	0	0	[12]	0	0	0	0	0	0
30	m_{30}	0	0	0	[12]	0	0	0	0	0
31	m_{31}	0	0	0	0	[12]	0	0	0	0
32	m_5	0	0	0	0	0	[12]	0	0	0
33	m_{14}	0	0	0	0	0	0	[12]	0	0
34	m_{26}	0	0	0	0	0	0	0	[12]	0
35	m_{18}	0	0	0	0	0	0	0	0	[12]
36	m_{11}	−1	0	0	0	0	0	0	0	0
...	...	...	...	...	...	...	...	...	...	...
93	m_{11}	−1	[−1]	0	0	0	0	0	0	0
94	m_5	0	[−26,−27,−28,29]	[−1]	0	0	0	0	0	0
95	m_2	0	0	[−26,−27,−28,29]	[−1]	0	0	0	0	0
96	m_{24}	0	0	0	[−26,−27,−28,29]	[−1]	0	0	0	0
97	m_4	0	0	0	0	[−26,−27,−28,29]	[−1]	0	0	0

The conditions $a_{12,2} = 0$, $a_{10,2} = 1$ and $a_{7,2} = 0$ make the condition $a_{12,2}a_{9,2} \oplus a_{10,2}a_{6,2} \oplus a_{8,2}a_{7,2} \oplus a_{6,2} = 0$ in step 13 always hold.

The condition $a_{15,2}a_{12,2} \oplus a_{14,2}a_{13,2} \oplus a_{13,2}a_{9,2} \oplus a_{9,2} = 0$ in step 16 is simplified to $a_{13,2}a_{9,2} \oplus a_{9,2} = 0$ due to the conditions $a_{12,2} = 0$ and $a_{14,2} = 0$.

From the conditions $a_{12,i} = a_{7,i}$ and $a_{12,i} = 0$, we can obtain $a_{7,i} = 0$ (i=1,2,3). The condition $a_{15,27} = 0$ is derived from the conditions $a_{15,27} = a_{12,27}$ and $a_{12,27} = 0$.

Table 4. Second preimage differential path for 5-pass HAVAL (Continued from Table 3)

98	m_0	0	[19, 20, 21, −22]	0	0	0	[−26, −27, −28, 29]	[−1]	0	0
99	m_{14}	0	0	[19, 20, 21, −22]	0	0	0	[−26, −27, −28, 29]	[−1]	0
100	m_2	0	0	0	[19, 20, 21, −22]	0	0	0	[−26, −27, −28, 29]	[−1]
101	m_7	0	0	0	0	[19, 20, 21, −22]	0	0	0	[−26, −27, −28, 29]
102	m_{28}	0	0	0	0	0	[19, 20, 21, −22]	0	0	0
103	m_{23}	0	0	0	0	0		[19, 20, 21, −22]	0	0
104	m_{26}	0	0	0	0	0		0	[19, 20, 21, −22]	0
105	m_6	0	0	0	0	0		0		[19, 20, 21, −22]
106	m_{30}	0	[8, −9]	0	0	0	0	0	0	0
107	m_{20}	0	0	[8, −9]	0	0	0	0	0	0
108	m_{18}	0	0	0	[8, −9]	0	0	0	0	0
109	m_{25}	0	[−2, −3, 4]	0	0	[8, −9]	0	0	0	0
110	m_{19}	0	0	[−2, −3, 4]	0	0	[8, −9]	0	0	0
111	m_3	0	0	0	[−2, −3, 4]	0	0	[8, −9]	0	0
112	m_{22}	0	0	0	0	[−2, −3, 4]	0	0	[8, −9]	0
113	m_{11}	−1	0	0	0	0	[−2, −3, 4]	0	0	[8, −9]
114	m_{31}	0	0	0	0	0	0	[−2, −3, 4]	0	0
115	m_{21}	0	0	0	0	0	0	0	[−2, −3, 4]	0
116	m_8	0	0	0	0	0	0	0	0	[−2, −3, 4]
117	m_{27}	0	[23]	0	0	0	0	0	0	0
118	m_{12}	0	0	[23]	0	0	0	0	0	0
119	m_9	0	0	0	[23]	0	0	0	0	0
120	m_1	0	0	0	0	[23]	0	0	0	0
121	m_{29}	0	0	0	0	0	[23]	0	0	0
122	m_5	0	0	0	0	0	0	[23]	0	0
123	m_{15}	0	0	0	0	0	0	0	[23]	0
124	m_{17}	0	0	0	0	0	0	0	0	[23]
125	m_{10}	0	[12]	0	0	0	0	0	0	0
126	m_{16}	0	0	[12]	0	0	0	0	0	0
127	m_{13}	0	0	0	[12]	0	0	0	0	0
128	m_{27}	0	0	0	0	[12]	0	0	0	0
129	m_3	0	0	0	0	0	[12]	0	0	0
130	m_{21}	0	0	0	0	0	0	[12]	0	0
131	m_{26}	0	0	0	0	0	0	0	[12]	0
132	m_{17}	0	0	0	0	0	0	0	0	[12]
133	m_{11}	−1	0	0	0	0	0	0	0	0
...	...	...	...	...	...	...	...	...	...	...
159	m_{15}	0	0	0	0	0	0	0	0	0

There are 43 equations in steps 11-18.

$a_{7,1} = 0$, $a_{7,2} = 0$, $a_{7,3} = 0$, $a_{8,1} = 0$, $a_{8,2} = 0$, $a_{8,3} = 0$, $a_{10,1} = 0$, $a_{10,2} = 1$, $a_{10,3} = 0$, $a_{11,1} = 0$, $a_{11,2} = 0$, $a_{11,3} = 1$, $a_{12,1} = 0$, $a_{12,2} = 0$, $a_{12,3} = 0$, $a_{14,1} = 0$, $a_{14,2} = 0$, $a_{14,3} = 0$, $a_{15,1} = 1$, $a_{15,2} = 1$, $a_{15,3} = 1$, $a_{10,27} = 0$, $a_{10,28} = 0$, $a_{10,29} = 0$, $a_{10,30} = 0$, $a_{12,27} = 0$, $a_{12,28} = 0$, $a_{12,29} = 0$, $a_{12,30} = 0$, $a_{13,27} = 1$, $a_{13,28} = 1$, $a_{13,29} = 1$, $a_{13,30} = 0$, $a_{14,27} = a_{9,27}$, $a_{14,28} = a_{9,28}$,

Table 5. A set of sufficient conditions for the differential path given in Table 3 and 4

Step	Sufficient conditions of the chaining variable in each step
7	$a_{7,1} = 0, a_{7,2} = 0, a_{7,3} = 0$
8	$a_{8,1} = 0, a_{8,2} = 0, a_{8,3} = 0$
9	$a_{9,2} = 0, a_{9,27} = 0, a_{9,28} = 0, a_{9,29} = 1, a_{9,30} = 0$
10	$a_{10,1} = 0, a_{10,2} = 1, a_{10,3} = 0, a_{10,27} = 0, a_{10,28} = 0, a_{10,29} = 0, a_{10,30} = 0$
11	$a_{11,1} = 0, a_{11,2} = 0, a_{11,3} = 1$
12	$a_{12,1} = 0, a_{12,2} = 0, a_{12,3} = 0, a_{12,27} = 0, a_{12,28} = 0, a_{12,29} = 0, a_{12,30} = 0$
13	$a_{13,2} = 0, a_{13,27} = 1, a_{13,28} = 1, a_{13,29} = 1, a_{13,30} = 0$
14	$a_{14,1} = 0, a_{14,2} = 0, a_{14,3} = 0, a_{14,23} = 1, a_{14,27} = 0, a_{14,28} = 0, a_{14,29} = 1, a_{14,30} = 0$
15	$a_{15,1} = 1, a_{15,2} = 1, a_{15,3} = 1, a_{15,27} = 0, a_{15,29} = 1$
16	$a_{16,2} = 0, a_{16,23} = 0, a_{16,27} = 0, a_{16,28} = 0, a_{16,29} = 0, a_{16,30} = 0$
17	$a_{17,23} = 1, a_{17,27} = 0, a_{17,28} = 1, a_{17,29} = 1, a_{17,30} = 0$
18	$a_{18,23} = 0, a_{18,29} = 1$
19	$a_{19,23} = 0, a_{19,27} = 1$
20	$a_{20,23} = 0, a_{20,27} = 1$
21	$a_{21,23} = 0,$
23	$a_{23,23} = 1$
24	$a_{24,12} = 0, a_{24,23} = 1$
25	$a_{25,12} = 0$
27	$a_{27,12} = 0$
28	$a_{28,12} = 0$
29	$a_{29,12} = 0$
31	$a_{31,12} = 0$
32	$a_{32,12} = 1$
89	$a_{89,1} = 1$
90	$a_{90,1} = 0$
91	$a_{91,1} = 1, a_{91,26} = 0, a_{91,27} = 0, a_{91,28} = 0, a_{91,29} = 0$
92	$a_{92,1} = 1, a_{92,26} = 0, a_{92,27} = 0, a_{92,28} = 0, a_{92,29} = 0$
93	$a_{93,1} = 1, a_{93,26} = 0, a_{93,27} = 0, a_{93,28} = 0, a_{93,29} = 0$
94	$a_{94,1} = 0, a_{94,19} = 1, a_{94,20} = 1, a_{94,21} = 1, a_{94,26} = 1, a_{94,27} = 1, a_{94,28} = 1, a_{94,29} = 0$
95	$a_{95,1} = 1, a_{95,19} = 0, a_{95,20} = 0, a_{95,21} = 0, a_{95,22} = 0, a_{95,26} = 0, a_{95,27} = 0, a_{95,28} = 0, a_{95,29} = 0$
96	$a_{96,1} = 1, a_{96,19} = 1, a_{96,20} = 1, a_{96,21} = 1, a_{96,22} = 0, a_{96,26} = 1, a_{96,27} = 1, a_{96,28} = 1, a_{96,29} = 1$
97	$a_{97,1} = 0, a_{97,19} = 1, a_{97,20} = 1, a_{97,21} = 1, a_{97,22} = 0, a_{97,26} = 1, a_{97,27} = 1, a_{97,28} = 1, a_{97,29} = 0$
98	$a_{98,1} = 0, a_{98,19} = 0, a_{98,20} = 0, a_{98,21} = 0, a_{98,22} = 1, a_{98,26} = 0, a_{98,27} = 0, a_{98,28} = 0, a_{98,29} = 0$
99	$a_{99,19} = 1, a_{99,20} = 1, a_{99,21} = 1, a_{99,22} = 0, a_{99,26} = 0, a_{99,27} = 0, a_{99,28} = 0, a_{99,29} = 1$
100	$a_{100,19} = 1, a_{100,20} = 1, a_{100,21} = 1, a_{100,22} = 1$
101	$a_{101,9} = 0, a_{101,19} = 1, a_{101,20} = 1, a_{101,21} = 1, a_{101,22} = 1$
102	$a_{102,8} = 1, a_{102,9} = 1, a_{102,19} = 1, a_{102,20} = 1, a_{102,21} = 1, a_{102,22} = 0$

$a_{14,29} = a_{9,29}$, $a_{14,30} = a_{9,30}$, $a_{15,27} = 0$, $a_{16,27} = 0$, $a_{16,28} = 0$, $a_{16,29} = 0$, $a_{16,30} = 0$ and $a_{13,2}a_{9,2} \oplus a_{9,2} = 0$.

Each equation of the first 42 equations holds with probability $\frac{1}{2}$, and the last equation holds with probability $\frac{3}{4}$.

Similarly, according to the propositions of the nonlinear functions f_1, f_2, f_3, f_4 and f_5, we can deduce all the other conditions which ensure the differential path in Table 3 and Table 4 hold. All these sufficient conditions are collected in Table 5 and Table 6.

From the conditions in Table 5 and Table 6, we know that for a given message M, M holds the sufficient conditions with probability 2^{-227}, which is larger than 2^{-256}. If the message M satisfies all the 227 conditions, then $M' = M + \Delta M$ is the second preimage of the hash value $h(M)$.

Table 6. A set of sufficient conditions for the differential path given in Table 3 and 4 (Continued from Table 5)

Step	Sufficient conditions of the chaining variable in each step
103	$a_{103,8} = 0, a_{103,9} = 1, a_{103,19} = 0, a_{103,20} = 0, a_{103,21} = 0, a_{103,22} = 0$
104	$a_{104,8} = 1, a_{104,9} = 0$
105	$a_{105,2} = 1, a_{105,3} = 1, a_{105,4} = 1, a_{105,8} = 1, a_{105,9} = 0$
106	$a_{106,2} = 0, a_{106,3} = 0, a_{106,4} = 0, a_{106,8} = 0, a_{106,9} = 1$
107	$a_{107,2} = 1, a_{107,3} = 1, a_{107,4} = 1, a_{107,8} = 1, a_{107,9} = 0$
108	$a_{108,2} = 1, a_{108,3} = 1, a_{108,4} = 1, a_{108,8} = 1, a_{108,9} = 1$
109	$a_{109,2} = 1, a_{109,3} = 1, a_{109,4} = 0, a_{109,8} = 1, a_{109,9} = 1$
110	$a_{110,2} = 1, a_{110,3} = 1, a_{110,4} = 1, a_{110,8} = 1, a_{110,9} = 0$
111	$a_{111,2} = 1, a_{111,3} = 1, a_{111,4} = 1, a_{111,8} = 0, a_{111,9} = 0$
112	$a_{112,2} = 1, a_{112,3} = 1, a_{112,4} = 1$
113	$a_{113,2} = 1, a_{113,3} = 1, a_{113,4} = 0, a_{113,23} = 1$
114	$a_{114,2} = 0, a_{114,3} = 0, a_{114,4} = 0, a_{114,23} = 0$
115	$a_{115,23} = 1$
116	$a_{116,23} = 1$
117	$a_{117,23} = 0$
118	$a_{118,23} = 1$
119	$a_{119,23} = 1$
120	$a_{120,23} = 1$
121	$a_{121,12} = 0, a_{121,23} = 1$
122	$a_{122,12} = 0, a_{122,23} = 0$
123	$a_{123,12} = 0$
124	$a_{124,12} = 0$
125	$a_{125,12} = 0$
126	$a_{126,12} = 0$
128	$a_{128,12} = 0$
129	$a_{129,12} = 0$

4   Partial Key-Recovery Attack on HMAC/NMAC-5-Pass HAVAL

In this section, based on our differential path described in Section 3, we present a partial key-recovery attack on HMAC/NMAC-5-pass HAVAL. If $H_{IV}(\bar{k} \oplus ipad) = k_2$ and $H_{IV}(\bar{k} \oplus opad) = k_1$, then HMAC is equivalent to NMAC, so we focus on recovering the partial key K_2 of NMAC-5-pass HAVAL. Recovering the partial key K_2 of NMAC-5-pass HAVAL is equivalent to getting the initial value of 5-pass HAVAL because K_2 is put as the initial value in NMAC.

In [19], the attacker has access to the oracle $\mathcal{O}$ (=NMAC-3-pass HAVAL) and uses the idea that the attacker can recover the initial state of NMAC-3-pass HAVAL if he knows a 256-bit chaining variables at any step of 3-pass HAVAL. This idea was firstly introduced in [14]. In our attack, we also use the similar idea and have access to an oracle NMAC-5-pass HAVAL to recover the partial key K_2.

We use the method depicted in [19] to recover the variables a_7, a_8, a_9, a_{10}, a_{11}, a_{12}, a_{13} and a_{14}.

In order to recover the value a_7, we use the condition $a_{7,1} = 0$ depicted in Table 5. Let $\beta_i = ((f_j(a_{i-7}, a_{i-6}, a_{i-5}, a_{i-4}, a_{i-3}, a_{i-2}, a_{i-1})) \ggg 7) + (a_{i-8} \ggg 11) + k_{j,i}$, then $a_i = \beta_i + m_{ord(j,i)}$. The attack procedure is as follows:

1. Choose 2^{233} message pairs $M = (m_0, \ldots, m_{31})$ and $M' = (m'_0, \ldots, m'_{31})$ which satisfy $M' - M = \Delta M$ (ΔM is shown in Section 3) and make 2^{234} queries to the oracle NMAC-5-pass HAVAL. The 2^{233} message pairs are chosen as follows: $m_0, \ldots, m_6$ and $m'_0, \ldots, m'_6$ are all identically fixed, m_7 and m'_7 vary in all 2^{32} possible values, $m_8, \ldots, m_{31}$ and $m'_8, \ldots, m'_{31}$ are randomly chosen to construct 2^{201} message pairs. The message words m_0, $\ldots$, m_6 are fixed, so β_7 is identically fixed for all the 2^{233} message pairs even though the attack doesn't know the actual value.
2. (a) Select the message pairs (M, M') that make collisions for the corresponding MAC pairs.
   (b) Guess the 32-bit word β_7, and for each of the guessed β_7, the attacker counts the number of the message pairs (M, M') chosen in Step 2(a) that satisfy the 1-st bit of $\beta_7 + m_7$ is equal to 1, i.e., $(\beta_7 + m_7)_1 = 1$.
3. Output $\beta_7 + m_7$ as a_7, where β_7 has the least count number in Step 2(b).

Our differential path holds with probability 2^{-227}. If the message pair (M, M') makes the first bits of a_7 and a'_7 to be 1, then the probability that the message pair (M, M') makes a collision is 2^{-256}. Therefore, the expected number of the message pairs (M, M') that make collisions in Step 2(a) is $2^6 + 2^{-23} (= 2^{233} \times (2^{-227} + 2^{-256}))$.

If the right β_7 is guessed, the number of the message pairs (M, M') that satisfy $(\beta_7 + m_7)_1 = 1$ in Step 2(b) is 2^{-23}. On the other hand, if the wrong β_7 is guessed, the number of the message pairs (M, M') that satisfy $(\beta_7 + m_7)_1 = 1$ in Step 2(b) is $2^5 + 2^{-24}$, because if β_7 is wrong guessed there are on average half message pairs satisfying $(\beta_7 + m_7)_1 = 1$ among the message pairs that make collisions. The probability that there is no collision pair among the 2^{233} chosen message pairs is lower than $e^{-64} (\approx (1 - 2^{-227})^{233} \times (1 - 2^{-256})^{233})$. Therefore, we can get the right β_7. The attack requires 2^{234} oracle queries and 2^{32} memory bytes.

Use the similar idea, we can recover a_8 by using the condition $a_{8,1} = 0$. We chooses 2^{232} message pairs (M, M') which satisfy $M' - M = \Delta M$. $m_0, \ldots, m_7$ and $m'_0, \ldots, m'_7$ are selected the same as those selected in the attack to recover a_7 which leads to $a_{7,1} = 0$. This improves the probability of our differential path by twice. Therefore, compared with the procedure to recover a_7, recovering a_8 needs half of the message pairs. The remaining analysis is the same as the procedure when recovering a_7 and the attack requires 2^{233} oracle queries and 2^{32} memory bytes.

Recovering $a_9, \ldots, a_{14}$ are similar to recovering a_7 and a_8, which require 2^{232}, 2^{231}, 2^{230}, 2^{229}, 2^{228} and 2^{227} oracle queries respectively.

For the message pair (M, M') selected from the above procedures which makes a collision and $a_7, \ldots, a_{14}$, recover the initial value k_2 by computing from the inverse direction.

The total cost for finding our partial key k_2 is $2^{234} + 2^{233} + \ldots + 2^{227} \approx 2^{235}$ oracle queries with a required storage of $2^{35} = 8 \times 2^{32}$ bytes.

5   Conclusion

In this paper, we first design a second preimage differential path for 5-pass HAVAL with probability 2^{-227}. Based on the differential path, we present a second preimage attack on 5-pass HAVAL and a partial key-recovery attack on HMAC/NMAC-5-pass HAVAL with about 2^{235} oracle queries and 2^{35} memory bytes. Even though our attacks are not practical, they show that the security margin of the second preimage attack on 5-pass HAVAL and the partial key-recovery attack on HMAC/NMAC-5-pass HAVAL are not as high as expected.

References

1. Rivest, R.L.: The MD4 message digest algorithm. In: Menezes, A., Vanstone, S.A. (eds.) CRYPTO 1990. LNCS, vol. 537, pp. 303–311. Springer, Heidelberg (1991)
2. Zheng, Y., Pieprzyk, J., Seberry, J.: HAVAL - A One-way Hashing Algorithm with Variable Length of Output. In: Zheng, Y., Seberry, J. (eds.) AUSCRYPT 1992. LNCS, vol. 718, pp. 83–104. Springer, Heidelberg (1993)
3. Bellare, M., Canetti, R., Krawczyk, H.: Keying hash functions for message authentication. In: Koblitz, N. (ed.) CRYPTO 1996. LNCS, vol. 1109, pp. 1–15. Springer, Heidelberg (1996)
4. Biham, E., Chen, R.: Near-Collisions of SHA-0. In: Franklin, M. (ed.) CRYPTO 2004. LNCS, vol. 3152, pp. 290–305. Springer, Heidelberg (2004)
5. Wang, X.Y., Lai, X.J., Feng, D.G., Chen, H., Yu, X.: Cryptanalysis for Hash Functions MD4 and RIPEMD. In: Cramer, R. (ed.) EUROCRYPT 2005. LNCS, vol. 3494, pp. 1–18. Springer, Heidelberg (2005)
6. Wang, X.Y., Yu, H.B.: How to Break MD5 and Other Hash Functions. In: Cramer, R. (ed.) EUROCRYPT 2005. LNCS, vol. 3494, pp. 19–35. Springer, Heidelberg (2005)
7. Biham, E., Chen, R., Joux, A., Carribault, P., Lemuet, C., Jalby, W.: Collisions of SHA-0 and Reduced SHA-1. In: Cramer, R. (ed.) EUROCRYPT 2005. LNCS, vol. 3494, pp. 36–57. Springer, Heidelberg (2005)
8. Wang, X.Y., Yu, H.B., Lisa, Y.: Efficient Collision Search Attacks on SHA-0. In: Shoup, V. (ed.) CRYPTO 2005. LNCS, vol. 3621, pp. 1–16. Springer, Heidelberg (2005)
9. Wang, X.Y., Lisa, Y., Yu, H.B.: Finding collisions on the Full SHA-1. In: Shoup, V. (ed.) CRYPTO 2005. LNCS, vol. 3621, pp. 17–36. Springer, Heidelberg (2005)
10. Wang, X.Y., Feng, F.D., Yu, X.: An attack on HAVAL function HAVAL-128. Science in China Ser. F Information Sciences 48(5), 1–12 (2005)
11. Yu, H.B., Wang, G.L., Zhang, G.Y., Wang, X.Y.: The Second-Preimage Attack on MD4. In: Desmedt, Y.G., Wang, H., Mu, Y., Li, Y. (eds.) CANS 2005. LNCS, vol. 3810, pp. 1–12. Springer, Heidelberg (2005)

12. Yu, H.B., Wang, X.Y., Yun, A., Park, S.: Cryptanalysis of the Full HAVAL with 4 and 5 Passes. In: Robshaw, M.J.B. (ed.) FSE 2006. LNCS, vol. 4047, pp. 89–110. Springer, Heidelberg (2006)
13. Kim, J., Biryukov, A., Preneel, B., Hong, S.: On the Security of HMAC and NMAC Based on HAVAL, MD4, MD5, SHA-0 and SHA-1. In: De Prisco, R., Yung, M. (eds.) SCN 2006. LNCS, vol. 4116, pp. 242–256. Springer, Heidelberg (2006)
14. Contini, S., Lisa, Y.: Forgery and Partial Key-Recovery Attacks on HMAC and NMAC Using Hash Collisions. In: Lai, X., Chen, K. (eds.) ASIACRYPT 2006. LNCS, vol. 4284, pp. 37–53. Springer, Heidelberg (2006)
15. Yu, H.B.: Cryptanalysis of Hash Functions and HMAC/NMAC, Doctoral dissertation, Shandong University (2007)
16. Rechberger, C., Rijmen, V.: On Authentication With HMAC and Non-Rondom Properties. In: Dietrich, S., Dhamija, R. (eds.) FC 2007 and USEC 2007. LNCS, vol. 4886, pp. 119–133. Springer, Heidelberg (2007)
17. Fouque, P., Leurent, G., Nguyen, P.Q.: Full Key-Recovery Attacks on HMAC/NMAC-MD4 and NMAC-MD5. In: Menezes, A. (ed.) CRYPTO 2007. LNCS, vol. 4622, pp. 13–30. Springer, Heidelberg (2007)
18. Wang, L., Ohta, K., Kunihiro, N.: New Key-Recovery Attacks on HMAC/NMAC-MD4 and NMAC-MD5. In: Smart, N.P. (ed.) EUROCRYPT 2008. LNCS, vol. 4965, pp. 237–253. Springer, Heidelberg (2008)
19. Lee, E., Chang, D., Kim, J., Sung, J., Hong, S.: Second Preimage Attack on 3-Pass HAVAL and Partial Key-Recovery Attacks on HMAC/NMAC-3-Pass HAVAL. In: Nyberg, K. (ed.) FSE 2008. LNCS, vol. 5086, pp. 189–206. Springer, Heidelberg (2008)

Cryptanalysis of Vortex*

Jean-Philippe Aumasson[1,**], Orr Dunkelman[2,***], Florian Mendel[3],
Christian Rechberger[3], and Søren S. Thomsen[4]

[1] FHNW, Windisch, Switzerland
[2] École Normale Supérieure, Paris, France
[3] IAIK, Graz University of Technology, Austria
[4] DTU Mathematics, Technical University of Denmark

Abstract. Vortex is a hash function that was first presented at ISC'2008,
then submitted to the NIST SHA-3 competition after some modifications. This paper describes several attacks on both versions of Vortex, including collisions, second preimages, preimages, and distinguishers. Our
attacks exploit flaws both in the high-level design and in the lower-level
algorithms.

1 Introduction

Vortex is the name of an AES-based hash function proposed by Gueron and
Kounavis [5] at ISC'2008, and is also the name of the modified version [8] submitted to the NIST Hash Competition[1]. To distinguish between the two, we call
them Vortex-0 and Vortex-1, respectively. We present attacks on both, making
the latter unsuitable for selection as SHA-3. Table 1 summarizes our results.

The paper is structured as follows: §2 briefly introduces the hash functions
Vortex-0 and Vortex-1; §3 shows that many digests *cannot* be produced by Vortex
(both versions); §4 and §5 present collision attacks for Vortex-0 and Vortex-1,
respectively; second preimage attacks on Vortex-1 are given in §6, and preimage
attacks in §7. Finally, §8 concludes, and Appendix A provides details on the
collision attack on Vortex-0.

2 Vortex-0 and Vortex-1

2.1 Vortex-0

Vortex-0 is a Merkle-Damgård iterated hash function with 256-bit chaining value
and 256-bit digest. Given a 2×128-bit chaining value $A \| B$ and a 4×128-bit

---

* The work in this paper has been supported in part by the European Commission
under contract ICT-2007-216646 (ECRYPT II) and by the IAP Programme P6/26
BCRYPT of the Belgian State (Belgian Science Policy).
** Supported by the Swiss National Science Foundation, project no. 113329.
*** The second author was supported by the France Telecom chaire.
[1] See http://www.nist.gov/hash-competition.

B. Preneel (Ed.): AFRICACRYPT 2009, LNCS 5580, pp. 14–28, 2009.
© Springer-Verlag Berlin Heidelberg 2009

Table 1. Summary of our results on Vortex-0 and Vortex-1 (256-bit digest)

Target	Type	Time	Memory	Section
Vortex-0	distinguisher	2^{97}	negl.	3
Vortex-1	distinguisher	2^{97}	negl.	3
Vortex-0	collision	2^{62}	negl.	4
Vortex-1	pseudo-collision	2^{64}	negl.	5.1
Vortex-1	free-start collision	2^{64}	negl.	5.2
Vortex-1	collision	$2^{124.5}$	$2^{124.5}$	5.3
Vortex-1	second-preimage*	2^{129}	negl.	6.1
Vortex-1	second-preimage*◇	2^{33}	2^{135}	6.1
Vortex-1	second-preimage	2^{192}	2^{64}	6.3
Vortex-1	preimage	2^{195}	2^{64}	7

*: for a small class of weak messages
◇: with 2^{135} precomputation time

message block $W_0\|W_1\|W_2\|W_3$, the compression function of Vortex-0 sequentially computes

$$A\|B \leftarrow (A\|B) \oplus \mathsf{subblock}(A, B, W_0, W_1)$$
$$A\|B \leftarrow (A\|B) \oplus \mathsf{subblock}(A, B, W_2, W_3)$$

and returns the new $A\|B$ as the new chaining value (or as the digest, if the message block is the last one). The function $\mathsf{subblock}(A, B, W_i, W_j)$ returns the 256-bit value

$$V\left(\mathsf{C}_{W_i}(A), \mathsf{C}_{W_j}(B)\right) .$$

The block cipher C is a reduced version of AES with three rounds and a simplified key schedule.

The merging function $V : \{0,1\}^{256} \mapsto \{0,1\}^{256}$ takes two 128-bit inputs A and B, which are parsed as four 64-bit words as $A_1\|A_0 \leftarrow A$ and $B_1\|B_0 \leftarrow B$. The function V updates these words as follows ("$\otimes$" denotes carryless multiplication[2], and integer addition is modulo 2^{64}):

- $L_1\|L_0 \leftarrow A_1 \otimes B_0$
- $O_1\|O_0 \leftarrow A_0 \otimes B_1$
- $A_0 \leftarrow A_0 \oplus L_0$
- $A_1 \leftarrow A_1 \oplus O_1$
- $B_0 \leftarrow B_0 + O_0$
- $B_1 \leftarrow B_1 + L_1$

Note that the new A_1 has the same most significant bit (MSB) as the original A_1. This is because $O_1\|O_0$ is the 128-bit representation of a polynomial of degree

---

[2] The carryless multiplication used in Vortex corresponds to the new PCLMULQDQ instruction in Intel processors, which multiplies two polynomials over GF(2) of degree at most 63 and return a polynomial of degree at most 126.

at most 126 (A_1 and B_1 respresent polynomials of degree at most 63). Hence, the MSB of O_1 will always be zero since it encodes the coefficient of the term of degree 127.

Another observation on the structure of V was given by Ferguson [4]: he observed that the least significant bit (LSB) of the new A_0 and of the new B_0 are equal with probability 5/8, leading to a distinguisher for both versions of Vortex.

2.2   Vortex-1

Vortex-1 is very similar to Vortex-0 but it has a different compression function, which computes

$$A\|B \leftarrow \mathsf{subblock}(A, B, W_0, W_1)$$
$$A\|B \leftarrow \mathsf{subblock}(A, B, W_2, W_3)$$

Note that, compared to the original version, Vortex-1 omits the feedforward of the chaining value $A\|B$. Furthermore, $\mathsf{subblock}(A, B, W_i, W_j)$ now computes

$$A\|B \leftarrow V(\mathsf{C}_A(W_i) \oplus W_i, \mathsf{C}_B(W_i) \oplus W_i)$$
$$A\|B \leftarrow V(\mathsf{C}_A(W_j) \oplus W_j, \mathsf{C}_B(W_j) \oplus W_j) \,,$$

where, in the 256-bit version, A, B, W_i, and W_j are 128-bit words (see also Fig. 1). The AES-like cipher C still makes three rounds, and the V function is the same as in Vortex-0. Note that the compression function of Vortex-1 is similar to MDC-2 [10], except that the final transform V is not a permutation (see §3).

The iteration mode of Vortex-1 slightly differs from the classical Merkle-Damgård: the last message block is 256-bit, instead of 512-bit for the previous blocks, and is processed differently. A detailed description of this mode is not necessary to the understanding of our attacks.

A 512-bit version of Vortex-1 is described in [8]; instead of 128-bit Rijndael rounds, 512-bit Vortex-1 uses 256-bit Rijndael rounds. The merging function V is similar but with words which are twice as large, and the message blocks are twice as large as well. The attacks in this paper are mainly described on the 256-bit version, but apply to the 512-bit version as well (with higher complexities).

3   On Impossible Images of V

We show that both versions of Vortex have *impossible images*. That is, the range of Vortex-0 and Vortex-1 does not span their codomain $\{0, 1\}^{256}$. This observation allows slightly faster preimage and collision search, and can be used to mount distinguishers on function ensembles based on Vortex (e.g. HMAC [2]).

Both Vortex-0 and Vortex-1 use the V function after each evaluation of C. In particular, their final output is an output of V. But V is non-surjective. Hence, there exist impossible images by the Vortex hash functions. Experiments on reduced versions suggest that V behaves more like a random function than

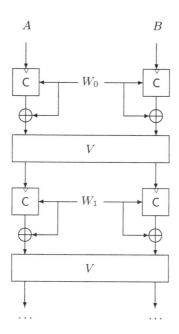

Fig. 1. Schematical view of Vortex-1's subblock function (a hatch marks the key input)

like a random permutation: for example, with 12-bit A and B, about 66% of the outputs are reachable (against $1 - 1/e \approx 63\%$ for a random function)[3]. It appears that the longer A and B, the closer V is to a random permutation.

We let hereafter $\mathcal{R}_V$ be the range of V; we thus have $\mathcal{R}_V \subsetneq \{0,1\}^{256}$.

3.1 Fixed Points

Any input of the form $(A_1\|A_0, B_1\|B_0) = (0\|x, 0\|y)$ or $(z\|0, w\|0)$ is a fixed point for V. Indeed, we then get

$$L_1\|L_0 = O_1\|O_0 = 0\|0$$

and thus the input is unchanged by the V transform. There are $2^{129} - 1$ such fixed points.

3.2 Multicollisions for V

Recall that a *multicollision* for a hash function is a set of distinct messages that map to the same digest; when there are r messages, we talk of an r-collision (a collision is thus a 2-collision).

---

[3] Note that a random function $\{0,1\}^n \mapsto \{0,1\}^n$ has in average about 63% of its outputs reachable, but a random function $\{0,1\}^m \mapsto \{0,1\}^n$, $m > n + \log n$, has the space $\{0,1\}^n$ as range with high probability.

We present a simple method to find multicollisions for V: set $A_1 = B_1 = 1$, and choose a A_0 and a B_0 that have no bit "1" at the same position; that is, $A_0 \wedge B_0 = 0$, where $\wedge$ is logical AND (for example, $A_0 = \texttt{FF00}\ldots\texttt{00}$ and $B_0 = \texttt{00FF}\ldots$). The V function then sets:

- $L_1 \| L_0 \leftarrow 0 \| B_0$
- $O_1 \| O_0 \leftarrow 0 \| A_0$
- $A_0 \leftarrow A_0 \oplus L_0 = A_0 \oplus B_0$
- $A_1 \leftarrow A_1 \oplus O_1 = 1$
- $B_0 \leftarrow B_0 + O_0 = B_0 + A_0 = B_0 \oplus A_0$
- $B_1 \leftarrow B_1 + L_1 = 1$

The equality $B_0 + A_0 = B_0 \oplus A_0$ holds because A_0 and B_0 were chosen with no bit "1" at the same position, thus avoiding carries in the modulo 2^{64} addition.

Now one can modify the initial A_0 and B_0 such that $A_0 \oplus B_0$ remains unchanged (and still have no bit "1" at the same position), which does not affect the output. Note that this even allows multicollisions, since for a given pair (A_0, B_0) there exist many colliding modified pairs.

One can easily derive an upper bound on $|\mathcal{R}_V|$ from the above technique: all images obtained have the form $(A_1 \| A_0, B_1 \| B_0) = (1 \| x, 1 \| x)$, and we can find preimages for any choice of x. In particular, for a x of weight i, $0 \le i \le 64$, we can find 2^i preimages, and deduce that there are $2^i - 1$ 256-bit values that are impossible images by V. In total there are $\binom{64}{i}$ weight-i images, so we have about

$$\sum_{i=0}^{64} \binom{64}{i} \cdot 2^i \approx 2^{101}$$

impossible images, i.e. $|\mathcal{R}_V| < 2^{256} - 2^{101}$. It follows that search for preimages and collisions is slightly faster than expected. Note that this remark also applies to the new (third) version of Vortex, as presented at the First SHA-3 Conference [7].

3.3 Inverting V

Given a random element y of $\{0,1\}^{256}$, the best generic algorithm to decide whether y lies in $\mathcal{R}_V$ is to try all the 2^{256} inputs. Below we describe an algorithm that solves this problem much faster, and finds a preimage of y when $y \in \mathcal{R}_V$.

- Let $y = C \| D = C_1 \| C_0 \| D_1 \| D_0$ be the given 256-bit output value.
- Guess the 43 LSBs of A_0 and B_1 (2^{86} choices).
- From the 43 LSBs of A_0 and C_0, deduce the 43 LSBs of L_0.
- From the 43 LSBs of B_1 and D_1, deduce the 43 LSBs of L_1.
- From the 43 LSBs of A_0 and of B_1, deduce the 43 LSBs of O_0.
- From the 43 LSBs of O_0 and D_0, deduce the 43 LSBs of B_0.
- From the 43 LSBs of B_0 and of L_0, deduce the 43 LSBs of A_1.
- From the 43 LSBs of L_1, of A_1, and of B_0, find the $20 + 21$ unknown bits of A_1 and B_0 by solving a linear system of equations.
- From the 43 LSBs of A_1 and C_1, deduce the 43 LSBs of O_1.

- From the 43 LSBs of O_1, of A_0, and of B_1, find the $21 + 21$ unknown bits of A_0 and B_1 by solving a linear system of equations.
- If no solution for A and B is found (which happens when the systems of equations contain contradictions), return "$y \notin \mathcal{R}_V$", else return "$y \in \mathcal{R}_V$" and the preimage found.

The running time of the algorithm is dominated by the solving of two sets of 42 GF(2) equations for each guess, i.e., in total finding 2^{86} solutions (we guess 86 bits) to a set of 42 equations in 42 unknowns over GF(2). These systems have to be solve only once, thus a rough estimate yields complexity $42^2 \times 2^{86} \approx 2^{97}$. Note that parallelism provides a linear speedup to this algorithm.

We can now distinguish the output of Vortex from a random string by running the above algorithm (if the algorithm fails to find a preimage of the output, then the string was not produced by Vortex).

Note that similar claims may be made about a Merkle-Damgård hash function based on a Davies-Meyer compression function (e.g. SHA-1), when the last block is a full padding block. In such a case, the output space is indeed only 63% of the possible values. However, unlike the case of Vortex, this space changes when the padding changes (i.e., a different number of bits is hashed). Moreover, in the case of a Merkle-Damgård construction with a Davies-Meyer compression function, the adversary has to try all possible input chaining values before deducing that the output is indeed not in the range of the specific case of the function, which is clearly not the case for Vortex.

4   Collision Attack on Vortex-0

We present a collision attack on Vortex-0 that exploits structural properties of the subblock function, assuming that the block cipher C is ideal. In Appendix A, we prove that the actual C cipher in Vortex-0 is close enough to an ideal cipher to be vulnerable to our attack.

The attack goes as follows: given the IV $A\|B$, choose arbitrary W_1, W_2, W_3, and compute $\mathsf{C}_{W_0}(A)$ for 2^{64} distinct values of W_0; in the ideal cipher model, one thus gets 2^{64} random values, each uniformly distributed over $\{0,1\}^{128}$, hence a collision

$$\mathsf{C}_{W_0}(A) = \mathsf{C}_{W_0'}(A)$$

occurs with probability $1 - 1/e^2 \approx 0.39$ (by the birthday paradox), which directly gives a collision for the compression function. The cost of this attack is 2^{64} evaluations of C (which is equivalent to 2^{62} evaluations of the compression function of Vortex-0), whereas 2^{128} compressions was conjectured in [5] to be a minimum.

The attack would not work if the map key-to-ciphertext induced by C were significantly more injective than a random function. In Appendix A, we prove that, under reasonable assumptions, we have, for any $x \in \{0,1\}^{128}$

$$\Pr_{K,K'}[\mathsf{C}_K(x) = \mathsf{C}_{K'}(x)] \approx \frac{1}{2^{128}}.$$

More precisely, we show that for C with two rounds (denoted C^2), instead of three, we have

$$\Pr_{K,K'}[C_K^2(x) = C_{K'}^2(x)] = \frac{2^{128} - 2}{(2^{128} - 1)^2} \approx \frac{1}{2^{128}},$$

which means that our collision attack works with the actual C.

5 Collision Attacks on Vortex-1

5.1 Pseudo-collisions

We now show how to find a pair of colliding messages for Vortex-1 with two distinct IV's, of the form $A\|B$ and $A'\|B$, respectively, for any fixed B and random A and A'. Observe that for an IV $A\|B$, the 128-bit A is used only once in the compression function, to compute $C_A(W_0)$. One can thus find a collision on the compression function by finding a collision for $C_A(W_0) \oplus W_0$: fix W_0 and cycle through 2^{64} distinct A's to find a collision with high probability. One can thus find a pseudo-collision for two IV's $A\|B$ and $A'\|B$ in 2^{65} evaluations of C (2^{64} for A and 2^{64} for A'), instead of 2^{128} compressions ideally.

5.2 Free-Start Collisions

We show how to find a pair of colliding messages for Vortex-1 with any IV of the form $A\|B = A\|A$, which we call a *symmetric* IV. It suffices to find W_0, W_0' such that

$$C_A(W_0) \oplus W_0 = C_A(W_0') \oplus W_0'$$

to get two colliding messages with the same random IV. When the IV is fixed (and not symmetric) one can precompute a message block that leads to an IV $A\|A$ within 2^{128} trials, and then find collisions in 2^{64}.

5.3 Full Collisions

As mentioned in §2.2, Vortex-1's construction is very similar to MDC-2 [10]; a recently discovered collision attack on MDC-2 [6] applies to Vortex-1 as well. In the following, n denotes the size of C's block (128 and 256 for the 256- and 512-bit versions of Vortex-1, respectively).

For ease of exposition, we introduce the notation halfsubblock$(h\|\tilde{h}, W_i)$ to denote, in subblock, the result of the operations

1. $a \leftarrow C_h(W_i) \oplus W_i$
2. $\tilde{a} \leftarrow C_{\tilde{h}}(W_i) \oplus W_i$
3. $h\|\tilde{h} \leftarrow V(a\|\tilde{a})$

We write $\mathsf{mmo}(x, y) = \mathsf{C}_x(y) \oplus y$. The function $\mathsf{halfsubblock}$ can then be written:

$$\mathsf{halfsubblock}(h\|\tilde{h}, m) = V\left(\mathsf{mmo}(h, m)\|\mathsf{mmo}(\tilde{h}, m)\right).$$

Let $A\|\tilde{A} = \mathsf{halfsubblock}(h\|\tilde{h}, m)$, where $|A| = |\tilde{A}| = n$. The idea of the collision attack is to first find (by brute force) an r-collision in the variable A, resulting in r different values $\tilde{A}_1, \ldots, \tilde{A}_r$ of $\tilde{A}$. The n-bit message sub-blocks producing the r-collision are denoted by W_0^i, $i = 1, \ldots, r$. Then, one finds a message sub-block W_1 such that $\mathsf{mmo}(\tilde{A}_i, W_1) = \mathsf{mmo}(\tilde{A}_j, W_1)$ for some $i \neq j$. Since $A_i = A_j$, we know that $\mathsf{mmo}(A_i, W_1) = \mathsf{mmo}(A_j, W_1)$, and hence we have found a two-block collision.

More precisely, given an arbitrary chaining value $h\|\tilde{h}$, one proceeds as follows (see also Fig. 2).

1. Find r different message blocks W_0^i, $i = 1, \ldots, r$, such that

$$A_i\|\tilde{A}_i = V(\mathsf{mmo}(h, W_0^i)\|\mathsf{mmo}(\tilde{h}, W_0^i)),$$

   for $1 \leq i \leq r$, and $A_i = A_j$ for all i, j.
2. Choose an arbitrary message block W_1, and compute $\tilde{B}_i = \mathsf{mmo}(\tilde{A}_i, W_1)$ for all i, $1 \leq i \leq r$. If $\tilde{B}_i = \tilde{B}_j$ for some i, j, $i \neq j$, then the two messages $W_0^i\|W_1$ and $W_0^j\|W_1$, collide, since $\mathsf{mmo}(A_i, W_1) = \mathsf{mmo}(A_j, W_1)$. If no such pair exists, repeat this step with a different choice of W_1.

Finding the r-collision in step 1 requires the equivalent of about $(r! \times 2^{(r-1)n})^{1/r}$ evaluations of $\mathsf{halfsubblock}$ [12]. In the second step, mmo is evaluated r times.

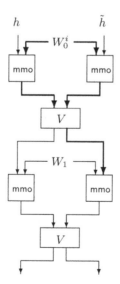

Fig. 2. The collision attack on Vortex-1. Thick lines mean that there are r different values of this variable. Thin lines mean that there is only one.

Assuming that V takes negligible time compared to mmo, the r evaluations of mmo correspond to about $r/2$ applications of halfsubblock. If evaluating V is less efficient than assumed here, then the complexity of step 2 in terms of evaluations of halfsubblock is lower. For each choice of W_1 in step 2, the probability that a collision is found is about $\binom{r}{2} \times 2^{-n} = r(r-1)/2 \times 2^{-n}$. Hence, the expected time spent in step 2 is about $2^n/(r-1)$.

For the optimal r (14), our attack on 256-bit Vortex-1 runs in $2^{124.5}$, and requires as much memory. For the 512-bit version, the optimal r is 24, and the attack runs in time $2^{251.7}$. These attacks work as well on the latest version of Vortex, as presented in [7].

6   Second-Preimage Attacks on Vortex-1

6.1   Second Preimages for Weak Messages

We now show how to find second preimages of messages that produce a symmetric chaining value (that is, of the form $A\|A$) during the digest computation. A key observation is that if $A = B$ and is of the form $(x\|0)$ or $(0\|y)$, then $V(A, B)$ maintains the equality of A and B.

The attack works as follows: given a message that produces the chaining value $\tilde{A}\|\tilde{A}$, find a message that leads after V to a symmetric chaining value $A\|B = A\|A$. Then find message blocks that preserve the symmetry and that eventually give $A = \tilde{A}$ (after as many blocks as in the original message). One then fills the new message with the blocks of the original message to get a second preimage of it. Reaching the first symmetric chaining value costs about 2^{128}, preserving the property for each step costs 2^{64}, and the connection costs 2^{128}. The total complexity is thus about 2^{129}. Note that the computation of a message that leads to a symmetric chaining value is message-independent, hence can be precomputed.

This attack, however, applies with low probability to a random message of reasonable size: for a random m-bit message, there are about $\lfloor m/128 \rfloor - 1$ "chaining values" (note that we can connect inside the compression function as well), thus the probability that a random message is weak is

$$1 - (1 - 2^{-128})^{\lfloor m/128 - 1 \rfloor},$$

which approximately equals $2^{-128} \times \lfloor m/128 - 1 \rfloor$ for short messages.

6.2   Time-Memory Tradeoff

We show that a variant of the above attack with precomputation 2^{135} and memory 2^{135} for bits runs in only 2^{33} trials, using the tree-construction technique in [1].

Consider a set of *special chaining values*

$$\mathcal{S} = \left\{ (x\|0)\|(x\|0), x \in \{0,1\}^{64} \right\} \cup \left\{ (0\|y)\|(0\|y), y \in \{0,1\}^{64} \right\}.$$

As noted earlier, these chaining values are maintained under the V transformation. The preprocessing phase is composed of three phases:

1. Find a message block W such that $A\|A \leftarrow V(C_{IV_0}(W) \oplus W, C_{IV_1}(W) \oplus W)$ where the IV is treated as $IV = IV_0\|IV_1$.
2. For each B, find a special chaining value s and a message word W' such that

$$C_s(W') \oplus W' = B ,$$

and store it in a table for each possible B.
3. For each $s \in \mathcal{S}$, find two message blocks W_1 and W_2 such that

$$C_s(W_1) \oplus W_1, C_s(W_2) \oplus W_2 \in \mathcal{S} .$$

It is easy to see that the first precomputation phase takes 2^{128} calls to $V(C(\cdot)\|C(\cdot))$, and outputs one message word of 128 bits to memorize. The second phase can be done by picking $s \in \mathcal{S}$ and message words W at random, until all outputs B are covered. Assuming that the process is random (i.e., by picking s and W randomly and independently from previous calls) we can model the problem as the coupon collector (see e.g. [11, p.57]), which means that about $\ln(2^{128}) \cdot 2^{128} < 2^{135}$ computations are performed, and about 2^{128} memory cells are needed. Finally, the third phase can be done in means of exhaustive search for each special chaining value s, and we expect about 2^{64} computations for each of the $2^{65} - 1$ special values. The memory needed for the output of the last precomputation is about 2^{66} memory cells. With respect to the precomputation we note that it is entirely parallelizable, and can enjoy a speed up of a factor N given N processors.

The online phase of the attack is as follows. Given the weak message that has a chaining value $\tilde{A}\|\tilde{A}$, we find in the first table the special chaining value $s \in \mathcal{S}$ and the message block W that lead to $\tilde{A}\|\tilde{A}$. We then start from the IV, and using the precomputed message blocks, reach a state $s' \in \mathcal{S}$. Now we have to find a path of message blocks from s' to s. This is done by randomly picking message blocks from s' which maintain the chaining value in the special set, until the distance between the reached state s'' and s is 65 message blocks.

To connect s'' and s we use the tree-construction technique described in [1]: from s'' one constructs a tree with all the 2^{33} possible special chaining values reachable after 33 blocks; similarly, one constructs a tree with the (expected) 2^{32} possible chaining values that may arrive to s after 32 blocks. As the size of the space is about 2^{65}, we expect a collision, and a path from s'' to s.

The preprocessing of this phase costs 2^{128} trials, storage is 2^{64}, and the on-line complexity is composed of performing a birthday on space of about 2^{65} values—which we expect to take about 2^{33} operations. So given about 2^{128} pre-computation, storage for 2^{135} bits that needs to be accessed once (store it on DVDs and put them in the closet), storage for 2^{71} bits that is going to be accessed randomly, the online complexity of the attack is only 2^{33}.

6.3   A Second Preimage Attack

This attack is based on a partial meet-in-the-middle attack, and finds a second preimage for any message. The attack applies to messages of three partial blocks or more (i.e., 384 bits or more), and replaces the first three blocks. We denote the consecutive chaining values of these partial blocks by $IV = A_0\|B_0$, $A_1\|B_1$, $A_2\|B_2$, $A_3\|B_3$, etc., and write

$$W \oplus \mathsf{C}_W(A_2)\|W \oplus \mathsf{C}_W(B_2) = X_2\|Y_2 \ ,$$

so that $A_3\|B_3 = V(X_2, Y_2)$.

The attack goes as follows:

1. For every $A_2 = 0\|x$ (where x is 64-bit), the attacker tries all W's, until he finds W_x such that $W_x \oplus \mathsf{C}_{W_x}(0\|x) = X_2$. On average, there is one such W_x for each x. The attacker stores the pairs (x, W_x) in a table.
2. The attacker takes 2^{192} two partial block messages, and computes for them $A_2\|B_2$. If A_2 is not of the form $0\|x$, the attacker discards the message; otherwise (i.e., $A_2 = 0\|y$), the attacker retrieves W_y from the table, and checks whether Y_2 equals $\mathsf{C}_{W_y}(B_2) \oplus W_y$. If yes, the two partial blocks along with W_y, can replace the first three message blocks of the original message.

As we start with 2^{192} messages, we expect about 2^{128} messages which generate the required pattern for A_2. For each of these messages, the probability that indeed $Y_2 = \mathsf{C}_{W_y}(B_2) \oplus W_y$, is 2^{-128}, and thus we expect one second preimage to be found.

We note that if multiple computing devices are available, they can be used efficiently. By picking the special structure of A_2, it is possible to "discard" many wrong trials, and access the memory very rarely. It is also possible to route the queries in the second phase between the various devices if each device is allocated a different segment of the special A_2's. Once one of the devices finds a message block which leads to a special A_2, it can send the message block to the computing device.

Finally, note that the attacks presented in this section apply as well to the most recent version of Vortex, as presented in [7].

7   Preimage Attacks on Vortex-1

A preimage attack on MDC-2 having complexity below the brute force complexity of 2^{2n} (where, again, n is the size of the underlying block cipher) has been known for many years [9]. The attack has time complexity about $2^{3n/2}$. The attack applies to Vortex-1 as well, but is slightly more complicated due to the finalization process, and due to V not being efficiently invertible.

Consider first a second preimage attack, where the chaining value after processing the first $t-1$ message blocks of the first preimage is known. That is, we can ignore the EMD extension [3] for now. Let this chaining value be $h_T\|\tilde{h}_T$. We may find a second preimage by the following meet-in-the-middle method, similar to the attack described in [9] (mmo is defined as in the collision attack described in §5.3).

1. Compute $Z\|\tilde{Z} = V^{-1}(h_T\|\tilde{h}_T)$ by inverting V as described in §3.3.
2. Pick a W_3 and search for a such that $\mathsf{mmo}(a, W_3) = Z$.

3. Likewise, compute $\mathsf{mmo}(\tilde{a}, W_3)$ for many different values of $\tilde{a}$, until $\mathsf{mmo}(\tilde{a}, W_3) = \tilde{Z}$.
4. Repeat $2^{n/2}$ times the above two steps with different choices of W_3. This yields $2^{n/2}$ values of $a\|\tilde{a}$ and W_3 such that $\mathsf{halfsubblock}(a\|\tilde{a}, W_3) = h_{\mathrm{T}}\|\tilde{h}_{\mathrm{T}}$.
5. Compute

$$\mathsf{halfsubblock}(\mathsf{halfsubblock}(\mathsf{halfsubblock}(h_0\|\tilde{h}_0, W_0), W_1), W_2)$$

for different choices of W_0, W_1, W_2, until a triple is found such that

$$\mathsf{halfsubblock}(\mathsf{halfsubblock}(\mathsf{halfsubblock}(h_0\|\tilde{h}_0, W_0), W_1), W_2) = a\|\tilde{a}$$

for some $a\|\tilde{a}$ computed in the previous step.

Here we produce a preimage $W_0\|W_1\|W_2\|W_3$ of $h_{\mathrm{T}}\|\tilde{h}_{\mathrm{T}}$, ignoring padding and the EMD extension. Step 1 takes expected time 2^{97} (see §3.3), and Steps 2 and 3 can be done combined in time about 2^n each, which means that when repeated $2^{n/2}$ times, the time complexity is about $2^{3n/2}$. Step 5 takes expected time about $2^{3n/2}$ in terms of evaluations of halfsubblock. Hence, the total time complexity is roughly $2^{3n/2+1}$. Taking length padding into account is not a problem in Step 5. One may simply partially hash a message of the appropriate length, and carry out Step 5 starting from the resulting intermediate hash value.

In a preimage attack we do not know the chaining value before the EMD extension. However, we can invert the hash function through the EMD extension as follows. Let the target image be $h_{\mathrm{T}}\|\tilde{h}_{\mathrm{T}}$. First, we note that the EMD extension can be seen as a short Merkle-Damgård iteration by itself, using a single 512-bit message block (in the 256-bit case), or equivalently, four 128-bit message blocks W_0, W_1, W_2, W_3. The initial value of this Merkle-Damgård iteration is $T\|\tilde{T}$; the first two sub-blocks, W_0 and W_1, form the chaining value from the processing of the first $t-1$ message blocks, and the last two sub-blocks, W_2 and W_3, contain at least 65 bits of padding. W_2 and W_3 are treated specially, since the subblock function is applied to them five times.

1. Choose a final message length, and construct $2^{n/2}$ different versions of $W_2\|W_3$, e.g., varying bits in W_2 only (at least 65 bits in W_3 are fixed by the choice of message length and one bit of padding).
2. For each version of $W_2\|W_3$, invert five times the function subblock using the same technique as in steps 1–3 above. Now we have $2^{n/2}$ values of $a\|\tilde{a}$ and $W_2\|W_3$ such that $\mathsf{subblock}^5(a\|\tilde{a}, W_2, W_3) = h_{\mathrm{T}}\|\tilde{h}_{\mathrm{T}}$.
3. Compute $\mathsf{halfsubblock}(\mathsf{halfsubblock}(T\|\tilde{T}, W_0), W_1)$ for different choices of W_0, W_1, until a pair is found such that

$$\mathsf{halfsubblock}(\mathsf{halfsubblock}(T\|\tilde{T}, W_0), W_1) = a\|\tilde{a}$$

for some $a\|\tilde{a}$ computed in the previous step.

The attack yields a chaining value $W_0\|W_1$ that may be used in place of $h_{\mathrm{T}}\|\tilde{h}_{\mathrm{T}}$ in the second preimage attack described above. Hence, one may now carry out this attack, keeping in mind that the message length has been fixed.

The time required to invert through the EMD extension is about $21 \times 2^{3n/2}$ (two inversions of each of mmo and V are needed per application of the subblock function). The different phases of the attack can be scaled differently to reduce the time complexity by a factor about four. Of course, the attack also fixes the padded version of the t-th message block $W_2\|W_3$.

Our attack runs in 2^{195} on the 256-bit version, and in 2^{387} on the 512-bit version, with memory requirements about 2^{64} and 2^{128}, respectively.

8   Conclusion

We presented several attacks on the hash function Vortex as submitted to NIST, and on its original version. The new version of Vortex appears to be stronger than the original, but fails to provide ideal security against collision attacks and (second) preimage attacks, and suffers from impossible images, which slightly reduces the entropy of a digest. These results seem to make Vortex unsuitable as a new hash standard. As a response to our attacks, another new version of Vortex was presented at the First SHA-3 Conference [7].

Acknowledgments

The authors would like to thank Sebastiaan Indesteege and Adi Shamir for stimulating discussions.

References

1. Andreeva, E., Bouillaguet, C., Fouque, P.-A., Hoch, J.J., Kelsey, J., Shamir, A., Zimmer, S.: Second preimage attacks on dithered hash functions. In: Smart, N.P. (ed.) EUROCRYPT 2008. LNCS, vol. 4965, pp. 270–288. Springer, Heidelberg (2008)
2. Bellare, M., Canetti, R., Krawczyk, H.: Keying hash functions for message authentication. In: Koblitz, N. (ed.) CRYPTO 1996. LNCS, vol. 1109, pp. 1–15. Springer, Heidelberg (1996)
3. Bellare, M., Ristenpart, T.: Multi-property-preserving hash domain extension and the EMD transform. In: Lai, X., Chen, K. (eds.) ASIACRYPT 2006. LNCS, vol. 4284, pp. 299–314. Springer, Heidelberg (2006)
4. Ferguson, N.: Simple correlation on some of the output bits of Vortex. OFFICIAL COMMENT (local link) (2008), http://ehash.iaik.tugraz.at/uploads/6/6d/Vortex_correlation.txt
5. Gueron, S., Kounavis, M.E.: Vortex: A new family of one-way hash functions based on AES rounds and carry-less multiplication. In: Wu, T.-C., Lei, C.-L., Rijmen, V., Lee, D.-T. (eds.) ISC 2008. LNCS, vol. 5222, pp. 331–340. Springer, Heidelberg (2008)
6. Knudsen, L.R., Mendel, F., Rechberger, C., Thomsen, S.S.: Cryptanalysis of MDC-2. In: Joux, A. (ed.) EUROCRYPT 2009. LNCS, pp. 106–120. Springer, Heidelberg (2009)
7. Kounavis, M.: Vortex – a new family of one way hash functions based on Rijndael rounds and carry-less multiplication. In: Candidate presentation at the First SHA-3 Conference (February 2009), http://csrc.nist.gov/groups/ST/hash/sha-3/Round1/Feb2009/program.html

8. Kounavis, M., Gueron, S.: Vortex: A new family of one way hash functions based on Rijndael rounds and carry-less multiplication. In: NIST (2008) (submission), http://eprint.iacr.org/2008/464.pdf
9. Lai, X., Massey, J.L.: Hash function based on block ciphers. In: Rueppel, R.A. (ed.) EUROCRYPT 1992. LNCS, vol. 658, pp. 55–70. Springer, Heidelberg (1993)
10. Meyer, C.H., Schilling, M.: Secure program load with manipulation detection code. In: SECURICOM 1988, pp. 111–130 (1988)
11. Motwani, R., Raghavan, P.: Randomized Algorithms. Cambridge University Press, Cambridge (1995)
12. Suzuki, K., Tonien, D., Kurosawa, K., Toyota, K.: Birthday paradox for multi-collisions. In: Rhee, M.S., Lee, B. (eds.) ICISC 2006. LNCS, vol. 4296, pp. 29–40. Springer, Heidelberg (2006)

A   Why the Collision Attack on Vortex-0 Works

In Vortex-0 a *round* consists of the AddRoundKey operation (which xors the 128-bit round key with the 128-bit state), followed by a permutation defined by the sequence SubBytes, ShiftRows, and MixColumns. The key schedule of C is much simpler than that of Rijndael: given a 128-bit key K, it computes the 128-bit rounds keys

$$RK_1 \leftarrow \pi_1(K)$$
$$RK_2 \leftarrow \pi_2(RK_1)$$
$$RK_3 \leftarrow \pi_3(RK_2)$$

where the π_i's are permutations defined by S-boxes, bit permutations and addition with constants. We denote RK_i^K a round key derived from K.

Denote Π_i^K the permutation corresponding to the i-th round of C; Π_i depends of the RK_i derived from K. Observe that for any state x and any distinct keys K, K', we have $K \oplus x \neq K' \oplus x$, therefore for any x

$$\Pi_i^K(x) \neq \Pi_i^{K'}(x)\cdot$$

In other words, a 1-round C mapping a key to a ciphertext, for any fixed plaintext, is a *permutation*. In the following we show that for 2 rounds it is *not* a permutation.

From the above observation, we have, for any $K \neq K'$, and for any x_1, x_2,

$$\Pi_1^K(x_1) \neq \Pi_1^{K'}(x_1)$$
$$\Pi_2^K(x_2) \neq \Pi_2^{K'}(x_2)\cdot$$

We show that, however, the probability over K, K', x that

$$\Pi_2^K \circ \Pi_1^K(x) = \Pi_2^{K'} \circ \Pi_1^{K'}(x)$$

is nonzero, and is even close to what one would expect if $\Pi_2 \circ \Pi_1$ were a random permutation; in the latter, for clarity, we write $\Pi_i = \Pi_i^K$, $\Pi_i' = \Pi_i^{K'}$, and

$\Pi = \{\Pi_1, \Pi_2\}$, $\Pi' = \{\Pi'_1, \Pi'_2\}$. Recall that Π_i, Π'_i are permutations such that: $\nexists x, \Pi_i(x) = \Pi'_i(x)$, for $i = 1, 2$.

We now compute the probability of a collision after two rounds. First, observe that

$$\Pr_{\Pi, \Pi', x}[\Pi_1 \circ \Pi_2(x) = \Pi'_1 \circ \Pi'_2(x)] = \Pr_{y \neq y', \Pi_1, \Pi'_1}[\Pi_1(y) = \Pi'_1(y')] .$$

The probability holds over random *distinct* 128-bit y and y'. We have (with $N = 2^{128}$):

$$\Pr_{y \neq y', \Pi_1, \Pi'_1}[\Pi_1(y) = \Pi'_1(y')] = \frac{1}{N} \sum_{y=0}^{N-1} \Pr_{y' \neq y, \Pi_1, \Pi'_1}[\Pi_1(y) = \Pi'_1(y')]$$

$$= \frac{1}{N} \sum_{y=0}^{N-1} \frac{1}{N-1} \sum_{\Pi_1(y)=0, \Pi_1(y) \neq y}^{N-1} \Pr[\Pi_1(y) = \Pi'_1(y')]$$

$$= \frac{1}{N} \sum_{y=0}^{N-1} \frac{1}{N-1} \sum_{\Pi_1(y)=0, \Pi_1(y) \neq y}^{N-1} (0 + (N-2) \times \frac{1}{N-1})$$

$$= \frac{1}{N} \sum_{y=0}^{N-1} \frac{N-2}{(N-1)^2}$$

$$= \frac{N-2}{(N-1)^2} = \frac{2^{128} - 2}{(2^{128} - 1)^2} \approx \frac{1}{2^{128}} .$$

The above result suggests that the 2-round C seen as a key-to-ciphertext mapping, for any fixed plaintext, has a distribution close to that of a random function. With three rounds, the distribution is much closer to that of a random function. Therefore, the birthday paradox is applicable, and so our attack works on the real Vortex-0 algorithm.

Two Passes of Tiger Are Not One-Way

Florian Mendel

Institute for Applied Information Processing and Communications (IAIK),
Graz University of Technology, Inffeldgasse 16a, A-8010 Graz, Austria
Florian.Mendel@iaik.tugraz.at

Abstract. Tiger is a cryptographic hash function proposed by Anderson and Biham in 1996. and produces a 192-bit hash value. Recently, weaknesses have been shown in round-reduced variants of the Tiger hash function. Collision attacks have been presented for Tiger reduced to 16 and 19 (out of 24) rounds at FSE 2006 and Indocrypt 2006. Furthermore, Mendel and Rijmen presented a 1-bit pseudo-near-collision for the full Tiger hash function at ASIACRYPT 2007. The attack has a complexity of about 2^{47} compression function evaluations. While there exist several collision-style attacks for Tiger, the picture is different for preimage attacks. At WEWoRC 2007, Indesteege and Preneel presented a preimage attack on Tiger reduced to 12 and 13 rounds with a complexity of $2^{64.5}$ and $2^{128.5}$, respectively.

In this article, we show a preimage attack on Tiger with two passes (16 rounds) with a complexity of about 2^{174} compression function evaluations. Furthermore, we show how the attack can be extended to 17 rounds with a complexity of about 2^{185}. Even though the attacks are only slightly faster than brute force search, they present a step forward in the cryptanalysis of Tiger.

1   Introduction

A cryptographic hash function H maps a message M of arbitrary length to a fixed-length hash value h. A cryptographic hash function has to fulfill the following security requirements:

- *Collision resistance:* it is practically infeasible to find two messages M and M^*, with $M^* \neq M$, such that $H(M) = H(M^*)$.
- *Second preimage resistance:* for a given message M, it is practically infeasible to find a second message $M^* \neq M$ such that $H(M) = H(M^*)$.
- *Preimage resistance:* for a given hash value h, it is infeasible to find a message M such that $H(M) = h$.

The resistance of a hash function to collision and (second) preimage attacks depends in the first place on the length n of the hash value. Regardless of how a hash function is designed, an adversary will always be able to find preimages or second preimages after trying out about 2^n different messages. Finding collisions requires a much smaller number of trials: about $2^{n/2}$ due to the birthday paradox. A function is said to achieve *ideal security* if these bounds are guaranteed.

B. Preneel (Ed.): AFRICACRYPT 2009, LNCS 5580, pp. 29–40, 2009.

Tiger is a cryptographic iterated hash function that processes 512-bit blocks and produces a 192-bit hash value. It was proposed by Anderson and Biham in 1996. Recent cryptanalytic results on the hash function Tiger mainly focus on collision attacks. At FSE 2006, Kelsey and Lucks presented a collision attack on 16 and 17 (out of 24) rounds of Tiger [6]. Both attacks have a complexity of about 2^{44} evaluations of the compression function. These results were later improved by Mendel *et al.* in [9]. They showed that a collision can be found for Tiger reduced to 19 rounds with a complexity of about 2^{62} evaluations of the compression function. At Asiacrypt 2007, Mendel and Rijmen presented the first attack on the full Tiger hash function [10]. They showed that a 1-bit pseudo-near-collision for Tiger can be constructed with a complexity of about 2^{47} compression function evaluations.

While several results have been published regarding the collision-resistance of Tiger, this picture is different for preimage attacks. At WEWoRC 2007, Indesteege and Preneel [4] presented a preimage attack on Tiger reduced to 12 and 13 rounds with a complexity of $2^{64.5}$ and $2^{128.5}$, respectively.

In this article, we will present a security analysis with respect to preimage resistance for the hash function Tiger. We show a preimage attack on Tiger reduced to 2 passes (16 rounds). It has a complexity of about 2^{174} compression function evaluations and memory requirements of 2^{39}. Very recently Isobe and Shibutani presented a preimage attack on 2 passes of Tiger with complexity of about 2^{161} and memory requirements of 2^{32} [5]. This is slightly more efficient than the attack presented in this paper. However, their attack method seems to be limited to 2 passes, while our attack can be extended to 17 rounds. In detail, we show how the attack can be extended to 17 rounds with a complexity of about 2^{185} and memory requirements of 2^{160}

In the preimage attack on Tiger, we combine weaknesses in the key schedule of Tiger with a generic meet-in-the-middle approach to construct a preimage for the compression function faster than brute force search. A similar attack strategy was use to construct preimages for the compression function of round-reduced MD5 in [2,12]. Once we have found a preimage for the compression function of round-reduced Tiger, we use a meet-in-the-middle attack respectively a tree based approach, to turn it into a preimage attack for the hash function.

The remainder of this article is structured as follows. A description of the Tiger hash function is given in Section 2. In Section 3, we present preimages for the compression function of Tiger reduced to 16 rounds (2 passes) and 17 rounds. We show how to extend these attacks for the compression function to the hash function in Section 4. Finally, we present conclusions in Section 5.

2   Description of the Hash Function Tiger

Tiger is an iterated hash function based on the Merkle-Damgård construction. It processes 512-bit input message blocks, maintains a 192-bit state and produces a 192-bit hash value. In the following, we briefly describe the hash function. It basically consists of two parts: the key schedule and the state update transformation.

Table 1. Notation

Notation	Meaning
$A \boxplus B$	addition of A and B modulo 2^{64}
$A \boxminus B$	subtraction of A and B modulo 2^{64}
$A \boxtimes B$	multiplication of A and B modulo 2^{64}
$A \oplus B$	bit-wise XOR-operation of A and B
$\neg A$	bit-wise NOT-operation of A
$A \ll n$	bit-shift of A by n positions to the left
$A \gg n$	bit-shift of A by n positions to the right
$A[i]$	the i-th bit of the word A (64 bits)
X_i	message word i (64 bits)
round	single execution of the round function
pass	set of consecutive *round*, has a size of 8 (1 *pass* = 8 *rounds*)

A detailed description of the hash function is given in [1]. For the remainder of this article, we will follow the notation given in Table 1.

2.1   State Update Transformation

The state update transformation of Tiger starts from a (fixed) initial value IV of three 64-bit words and updates them in three passes of eight rounds each. In each round one 64-bit word X introduced by the key schedule is used to update the three state variables A, B and C as follows:

$$C = C \oplus X$$
$$A = A \boxminus \mathbf{even}(C)$$
$$B = B \boxplus \mathbf{odd}(C)$$
$$B = B \boxtimes \mathtt{mult}$$

The results are then shifted such that A, B, C will be B, C, A in the next iteration. Fig. 1 shows one round of the state update transformation of Tiger.

The non-linear functions **even** and **odd** used in each round are defined as follows:

$$\mathbf{even}(C) = T_1[c_0] \oplus T_2[c_2] \oplus T_3[c_4] \oplus T_4[c_6]$$
$$\mathbf{odd}(C) = T_4[c_1] \oplus T_3[c_3] \oplus T_2[c_5] \oplus T_1[c_7]$$

where state variable C is split into eight bytes $c_7, \ldots, c_0$ with c_7 is the most significant byte and c_0 is the least significant byte. Four S-boxes $T_1, \ldots, T_4 :$ $\{0,1\}^8 \rightarrow \{0,1\}^{64}$ are used to compute the output of the non-linear functions **even** and **odd**. For the definition of the S-boxes we refer to [1]. Note that state variable B is multiplied with the constant $\mathtt{mult} \in \{5, 7, 9\}$ at the end of each round. The value of the constant is different in each pass of the Tiger hash function.

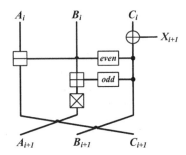

Fig. 1. The round function of Tiger

After the last round of the state update transformation, the initial values A_{-1}, B_{-1}, C_{-1} and the output values of the last round A_{23}, B_{23}, C_{23} are combined, resulting in the final value of one iteration (feed forward). The result is the final hash value or the initial value for the next message block.

$$A_{24} = A_{-1} \oplus A_{23}$$
$$B_{24} = B_{-1} \boxminus B_{23}$$
$$C_{24} = C_{-1} \boxplus C_{23}$$

2.2 Key Schedule

The key schedule takes a 512-bit message block $X_0, \ldots, X_7$ and produces 24 64-bit words $X_0, \ldots, X_{23}$. It is an invertible function which ensures that changing a small number of bits in the message will affect a lot of bits in the next pass. While the message words $X_0, \ldots, X_7$ are used in the first pass to update the state variables, the remaining 16 message words, 8 for the second pass and 8 for the third pass, are generated by applying the key schedule as follows:

$$(X_8, \ldots, X_{15}) = \text{KeySchedule}(X_0, \ldots, X_7)$$
$$(X_{16}, \ldots, X_{23}) = \text{KeySchedule}(X_8, \ldots, X_{15})$$

The key schedule modifies the inputs $(I_0, \ldots, I_7)$ in two steps:

first step

$T_0 = I_0 \boxminus (I_7 \oplus \texttt{A5A5A5A5A5A5A5A5})$
$T_1 = I_1 \oplus T_0$
$T_2 = I_2 \boxplus T_1$
$T_3 = I_3 \boxminus (T_2 \oplus ((\neg T_1) \lll 19))$
$T_4 = I_4 \oplus T_3$
$T_5 = I_5 \boxplus T_4$
$T_6 = I_6 \boxminus (T_5 \oplus ((\neg T_4) \ggg 23))$
$T_7 = I_7 \oplus T_6$

second step

$O_0 = T_0 \boxplus T_7$
$O_1 = T_1 \boxminus (O_0 \oplus ((\neg T_7) \lll 19))$
$O_2 = T_2 \oplus O_1$
$O_3 = T_3 \boxplus O_2$
$O_4 = T_4 \boxminus (O_3 \oplus ((\neg O_2) \ggg 23))$
$O_5 = T_5 \oplus O_4$
$O_6 = T_6 \boxplus O_5$
$O_7 = T_7 \boxminus (O_6 \oplus \texttt{0123456789ABCDEF})$

The final values $(O_0, \ldots, O_7)$ are the output of the key schedule.

3   Preimage Attacks on the Compression Function

In this section, we will present two preimage attacks on the compression function of Tiger – one for Tiger with 2 passes (16 rounds) and one for 17 rounds. Both attacks are based on structural weaknesses in the key schedule of Tiger. By combining these weaknesses with a generic meet-in-the-middle approach we can construct a preimage for the compression function of Tiger reduced to 16 rounds (2 passes) with a complexity of about 2^{173} compression function evaluations and memory requirements of 2^{38}. The attack can be extended to 17 rounds of Tiger at the cost of about 2^{184} compression function evaluations and memory requirements of 2^{159}. In the following, we will describe both attacks in more detail.

3.1   Preimage Attack on Two Passes of Tiger

Before describing the preimage attack on the compression function of Tiger reduced to 2 passes (16 rounds), we first have a closer look at the key schedule of Tiger. In the following, we present a differential characteristic for the key schedule of Tiger which we can use to construct preimages for the compression function faster than brute force search. Consider the differential

$$(\delta_1, 0, \delta_2, 0, 0, 0, 0, 0) \rightarrow (\delta_1, 0, 0, 0, 0, 0, 0, 0), \tag{1}$$

with $\delta_1 \boxplus \delta_2 = 0$, where δ_1 and δ_2 denote modular differences in the 19 most significant bits of the message words X_0, X_2 and X_8. In order to guarantee that this characteristic holds in the key schedule of Tiger, several conditions have to be fulfilled.

Due to the design of the key schedule of Tiger, the difference δ_1 in X_0 will lead to the same difference δ_1 in $T_0 = X_0 \boxminus (X_7 \oplus \mathsf{A5A5A5A5A5A5A5A5})$. Furthermore, by choosing $X_1 = 0$, we get $T_1 = T_0$ and hence $\Delta T_1 = \Delta T_0 = \delta_1$. Since $\Delta T_1 = \delta_1$, $\Delta X_2 = \delta_2$ and $\delta_1 \boxplus \delta_2 = 0$, there will be no difference in $T_2 = X_2 \boxplus T_1$. Note that by restricting the choice of δ_1 and hence δ_2 to differences in the 19 most significant bits we can ensure that there will be no differences in $T_3 = X_3 \boxminus (T_2 \oplus ((\neg T_1) \ll 19))$. It is easy to see, that due to the left shift of T_1 by 19 bits these differences will be canceled. Since there are no difference in T_2 and T_3, there will be no differences in $T_4, \ldots, T_7$. To ensure that there will be only a difference in $X_8 = T_0 \boxplus T_7$, namely δ_1 after the second step of the key schedule of Tiger, we need that $T_7 = 0$. This can be achieved by adjusting X_6 accordingly, such that $T_6 \oplus X_7 = 0$. It is easy to see that if $T_7 = 0$ then $X_8 = T_0$ and hence $\Delta X_8 = \Delta T_0 = \delta_1$. Furthermore, $X_9 = T_1 \boxminus X_8$ and hence $\Delta X_9 = \delta_1 \boxminus \delta_1 = 0$. Since $\Delta X_9 = 0$ and there are no differences in $T_2, \ldots, T_7$ there will be no differences in $X_{10}, \ldots, X_{15}$. By fulfilling all these conditions on the message words and restricting the differences of δ_1 and hence δ_2 to the 19 most significant bits, this characteristic for the key schedule of Tiger will always hold.

We will use this characteristic for the key schedule of Tiger to show a preimage attack on Tiger reduced to 16 rounds (2 passes). We combine the characteristic for the key schedule of Tiger with a generic meet-in-the-middle approach, to construct a preimage for the compression function of Tiger with 2 passes. The attack has a complexity of about 2^{173} compression function evaluations and memory requirements of 2^{38}. It can be summarized as follows.

1. Suppose we seek a preimage of $h = AA\|BB\|CC$, then we chose $A_{-1} = AA$, $B_{-1} = BB$, and $C_{-1} = CC$. To guarantee that the output after the feed forward is correct, we need that $A_{15} = 0$, $B_{15} = 0$, and $C_{15} = 0$.
2. In order to guarantee that the characteristic for the key schedule of Tiger holds, we choose random values for the message words $X_0, X_2, \ldots, X_7$ and set $X_1 = 0$. Furthermore, we adjust X_6 accordingly, such that $T_7 = 0$.
3. Next we compute A_7, B_7, and C_7 for all 2^{38} choices of $B_{-1}[63 - 45]$ and $C_{-1}[63 - 45]$ and save the result in a list L. In other words, we get 2^{38} entries in the list L by modifying the 19 most significant bits of B_{-1} and the 19 most significant bits of C_{-1}.
4. For all 2^{38} choices of the 19 most significant bits of B_{15} and the 19 most significant bits of C_{15} we compute A'_7, B'_7, C'_7 (by going backward) and check if there is an entry in the list L such that the following conditions are fulfilled:

$$A_7[i] = A'_7[i] \quad \text{for } 0 \le i \le 63$$
$$B_7[i] = B'_7[i] \quad \text{for } 0 \le i \le 63$$
$$C_7[i] = C'_7[i] \quad \text{for } 0 \le i \le 44$$

These conditions will hold with probability of 2^{-173}. Note that we can always adjust the 19 most significant bits of X_8 such that the 19 most significant bits of C_7 and C'_7 match.

Since there are 2^{38} entries in the list L and we test 2^{38} candidates, we expect to find a matching entry with probability of $2^{-173} \cdot 2^{76} = 2^{-97}$. Hence, finishing this step of the attack has a complexity of about $2^{38} \cdot 2^{97} = 2^{135}$ evaluations of the compression function of Tiger and memory requirements of 2^{38}.

5. Once we have found a solution, we have to modify the 19 most significant bits of X_0 and X_2 such that the characteristic in the key schedule of Tiger holds. To cancel the differences in X_0 and X_2, we have to adjust the 19 most significant bits of B_{-1} and C_{-1} accordingly. Thus, after applying the feed-forward we get a partial pseudo preimage for 154 (out of 192) bits of the compression function of Tiger reduced to 16 rounds.

Hence, we will find a partial pseudo preimage (154 out of 192 bits) with a complexity of 2^{135} and memory requirements of 2^{38}. By repeating the attack 2^{38} times we will find a preimage for the compression function with a complexity of about 2^{173} instead of the expected 2^{192} compression function evaluations. Note that the partial pseudo preimage (154 out of 192 bits) is also a fixed-point in 154 bits for the compression function f. We will need this later to turn the attack on the compression function into an attack on the hash function.

3.2   Going beyond Two Passes

In a similar way as we can construct a preimage for the compression function of Tiger reduced to 16 rounds, we can also construct a preimage for the compression function of Tiger reduced to 17 rounds. The attack has a complexity of about 2^{184} compression function evaluations and has memory requirements of 2^{159}.

For the attack on 17 rounds we use a slightly different characteristic for the key schedule of Tiger. It is shown below.

$$(0, \delta_1, 0, 0, 0, 0, 0, \delta_2) \rightarrow (0, 0, 0, 0, 0, 0, 0, \delta_3) \rightarrow (\delta_4, ?, ?, ?, ?, ?, ?, ?) \qquad (2)$$

where δ_4 denotes modular difference in the 31 most significant bits of the message word X_{16} and δ_1, δ_2, δ_3 denote modular difference in the 8 most significant bits of the message words X_1, X_7, X_{15}. Note that while in the attack on 2 passes we have only differences in the 19 most significant bits, we have now differences in the 8 (respectively 31) most significant bits of the message words.

In order to guarantee that this characteristic holds in the key schedule of Tiger, several conditions have to be fulfilled. In detail, a difference δ_2 in X_7 will lead to a difference in $T_0 = X_0 \boxminus (X_7 \oplus \texttt{A5A5A5A5A5A5A5A5})$ after the first step of the key schedule. By adjusting X_1 accordingly (choosing the difference δ_1 carefully), we can prevent that the difference in T_0 propagates to $T_1 = X_1 \oplus T_0$ and hence, there will be no differences in $T_1, \ldots, T_6$. However, due to the design of the key schedule of Tiger there will be a difference in $T_7 = X_7 \oplus T_6$. In order to prevent the propagation of the differences in T_7 to X_8 we need that $T_6 = \texttt{A5A5A5A5A5A5A5A5}$. Thus, we have that

$$X_8 = T_0 \boxplus T_7$$
$$= X_0 \boxminus (X_7 \oplus \texttt{A5A5A5A5A5A5A5A5}) \boxplus (X_7 \oplus \texttt{A5A5A5A5A5A5A5A5})$$
$$= X_0.$$

We can guarantee that $T_6 = \texttt{A5A5A5A5A5A5A5A5}$ by adjusting X_6 accordingly. Note that by restricting the differences of δ_2 and hence also δ_1 to the 8 most significant there will be only differences in the 8 most significant bits of $T_7 = X_7 \oplus T_6$ and therefore no differences in $X_9 = T_1 \boxminus (X_8 \oplus ((\neg T_7) \lll 19))$ and $X_{10}, \ldots, X_{14}$, only in $X_{15} = T_7 \boxminus (X_{14} \oplus \texttt{0123456789ABCDEF})$ there will be a difference δ_3 in the 8 most significant bits.

However, in the third pass there will be differences in the 31 most significant bits of X_{16} (denoted by δ_4) due to the design of the key schedule of Tiger. It is easy to see that a difference in the 8 most significant bits in X_{15} will result in differences in the 8 most significant bits of $T_0, \ldots, T_5$. Furthermore, since $T_6 = X_{14} \boxminus (T_5 \oplus \neg T_4 \ggg 23)$ we will get differences in the 31 most significant bits of T_6 and hence also in T_7 as well as in $X_{16} = T_0 \boxplus T_7$.

Again, by combining this characteristic for the key schedule of Tiger with a generic meet-in-the-middle approach, we can construct preimages for the compression function of Tiger for more than 2 passes (17 rounds) with a complexity of about 2^{184} compression function evaluations. The attack can be summarized as follows.

1. Suppose we seek a preimage of $h = AA\|BB\|CC$, then we chose $A_{-1} = AA$, $B_{-1} = BB$, and $C_{-1} = CC$. To guarantee that the output after the feed forward is correct, we need that $A_{16} = 0$, $B_{16} = 0$, and $C_{16} = 0$.

2. Choose random values for the message words $X_0, X_1, \ldots, X_7$ such that $T_6 =$ A5A5A5A5A5A5A5A5 after the first step of the key schedule of Tiger. Note that this can be easily done by adjusting X_6 accordingly, i.e. $X_6 = T_6 \boxplus (T_5 \oplus (\neg T_4 \gg 23))$. This is needed to ensure that differences in T_7 will be canceled in the key schedule – leading to the correct value of X_8 after the second step of the key schedule.

3. Next we compute A_6, B_6, C_6 for all 2^{159} choices of A_{-1}, C_{-1} and $B_{-1}[63-33]$ and save the result in a list L. In other words, we get 2^{159} entries in the list L by modifying A_{-1}, C_{-1} and the 31 most significant bits of B_{-1}.

4. For all 2^{159} choices of A_{16}, C_{16} and the 31 most significant bits of B_{16} we compute A_6', B_6', C_6' (by going backward) and check if there is an entry in the list L such that the following conditions are fulfilled:

$$A_6[i] = A_6'[i] \quad \text{for } 0 \leq i \leq 63$$
$$B_6[i] = B_6'[i] \quad \text{for } 0 \leq i \leq 63$$
$$C_6[i] = C_6'[i] \quad \text{for } 0 \leq i \leq 55$$

These conditions will hold with probability of 2^{-184}. Note that we can always adjust the 8 most significant bits of X_7 such that $C_6 = C_6'$ will match. Since there are 2^{159} entries in the list L and we test 2^{159} candidates, we will find $2^{-184} \cdot 2^{318} = 2^{134}$ solutions. In other words, we get 2^{134} solutions with a complexity of about 2^{159} evaluations of the compression function of Tiger and memory requirements of 2^{159}.

5. For each solution, we have to modify the 8 most significant bits of X_1 such that $T_1 = X_1 \oplus T_0$ is correct in the first step of the key schedule for the new value of X_7. Note that by ensuring that T_1 is correct, we will get the same values for $X_8, \ldots, X_{14}$ after applying the key schedule of Tiger, since $T_6 =$ A5A5A5A5A5A5A5A5 due to step 2 of the attack. In order to cancel the differences in the 8 most significant bits of X_1, we have to adjust the 8 most significant bits of A_{-1} accordingly. Furthermore, the 8 most significant bits of X_{15} and the 31 most significant bits of X_{16} will change as well. This results in new values for A_{16}, C_{16} and the 31 most significant bits of B_{16}.

Since, we modify A_{-1}, C_{-1} and the 31 most significant bits of B_{-1} in the attack we get after the feed-forward 2^{134} partial pseudo preimage (partial meaning 33 out of 192 bits) for the compression function of Tiger reduced to 17 rounds.

Hence, we will find 2^{134} partial pseudo preimage (33 out of 192 bits) with a complexity of 2^{159}. By repeating the attack 2^{25} times we will find a preimage for the compression function of Tiger reduced to 17 rounds with a complexity of about $2^{159} \cdot 2^{25} = 2^{184}$ instead of the expected 2^{192} compression function evaluations.

4   Extending the Attacks to the Hash Function

If we want to extend the preimage attack on the compression function of Tiger to the hash function, we encounter two obstacles. In contrast to an attack on the compression function, where the chaining value (or initial value) can be chosen freely, the initial value IV is fixed for the hash function. In other words, for a preimage attack on the hash function we have to find a message m such that $H(IV, m) = h$. Furthermore, we have to ensure that the padding of the message leading to the preimage of h is correct.

First, we choose the message length such that only a single bit of padding will be set in X_6 of the last message block. The last bit of X_6 has to be 1 as specified by the padding rule. Since we use in both attacks characteristics for the key schedule of Tiger where no difference appears in X_6, we can easily guarantee that the last bit of X_6 is 1. However, X_7 of the last message block will contain the message length as a 64-bit integer. While we can choose X_7 free in the attack on 2 passes (16 rounds), this is not the case for the attack on 17 rounds. The 8 most significant bits of X_7 are determined during the attack (*cf.* Section 3.2). However, the remaining bits of X_7 can be chosen freely. Therefore, we can always guarantee that we will have a message length such that the padding of the last block is correct. For the sake of simplicity let us assume for the following discussion that the message (after padding) consists of $\ell + 1$ message blocks.

We show how to construct a preimage for Tiger reduced to 16 rounds consisting of $\ell + 1$ message blocks, *i.e.* $m = M_1 \| M_2 \| \cdots \| M_{\ell+1}$. Note that the attack for Tiger reduced to 17 rounds works similar. It can be summarized as follows.

1. First, we invert the last iteration of the compression function $f(H_\ell, M_{\ell+1}) = h$ to get H_ℓ and $M_{\ell+1}$. Note that this determines the length of our preimage. This step of the attack has a complexity of about 2^{173} compression function evaluations.
2. Once we have fixed the last message block $M_{\ell+1}$ and hence the length of the message m, we have to find a message $m^* = M_1 \| M_2 \| \cdots \| M_\ell$ consisting of ℓ message blocks such that $H(IV, m^* \| M_{\ell+1}) = h$. This can be done once more by using a meet-in-the-middle approach.
   (a) Use the preimage attack on the compression function to generate 2^{10} pairs $(H_{\ell-1}^j, M_\ell^j)$ leading to the chaining value H_ℓ and save them in a list L. This has a complexity of about $2^{10} \cdot 2^{173} = 2^{183}$ compression function evaluations.
   (b) Compute $H_{\ell-1}$ by choosing random values for the message blocks M_i for $1 \le i < \ell$ and check for a match in L. After testing about 2^{182} candidates, we expect to find a match in the list L. Once, we have found a matching entry, we have found a preimage for the hash function Tiger reduced to 16 rounds consisting of $\ell + 1$ message blocks.

Hence, we can construct a preimage for the Tiger hash function reduced to 16 rounds with a complexity of about 2^{183} compression function evaluations. In a similar way we can find a preimage for Tiger reduced to 17 rounds with a complexity of about 2^{188}.

However, due to the special structure of the partial-pseudo-preimages for the compression function of Tiger reduced to 16 and 17 rounds, this complexity can be reduced by using a tree-based approach. This was first used by Mendel and Rijmen in the cryptanalysis of HAS-V [11]. Later variants and extensions of this method were presented in [3,7,8]. With this method, we can construct a preimage for the Tiger hash function reduced to 16 and 17 rounds with a complexity of about 2^{174} and 2^{185} compression function evaluations, respectively. In the following, we will describe this in more detail for Tiger reduced to 16 rounds. Note that the attack for Tiger reduced to 17 rounds works similar.

1. Assume we want to construct a preimage for Tiger reduced to 16 rounds consisting of $\ell + 1$ message blocks.
2. First, compute H_ℓ and $M_{\ell+1}$ by inverting the last iteration of the compression function. Note that this determines the length of our preimage m. This step of the attack has a complexity of about 2^{173} compression function evaluations.
3. Next, we construct a list L containing 2^{39} partial-pseudo-preimages for the compression function of Tiger. Note that all partial-pseudo-preimages will have the following form: $H_i = f(H_{i-1}, M_i)$, where $H_i \wedge \mathtt{mask} = H_{i-1} \wedge \mathtt{mask}$ and $hw(\mathtt{mask}) = 154$, where $hw(x)$ denotes the bit Hamming weight of x. In other words, each preimage for the compression function is also a fixed-point for $192 - 38 = 154$ bits. Note that this is important for the attack to work. Constructing the list L has a complexity of about $2^{39} \cdot 2^{135} = 2^{174}$ compression function evaluations.
4. Next, by using the entries in the list L we build a backward tree starting from H_ℓ. For each node in the tree we expect to get two new nodes on the next level. It is easy to see that since we have 2^{39} entries in the list L, where 154 bits are equal for each entry, we will always have two entries, where H_i is equal. Therefore, we will have about 2^{20} nodes at level 20. In other words, we have about 2^{20} candidates for $H_{\ell-20}$.
5. To find a message consisting of $\ell - 20$ message blocks leading to one of the 2^{20} candidates for $H_{\ell-20}$ we use a meet-in-the-middle approach. First, we choose an arbitrary message (of $\ell-21$ message blocks) leading to some $H_{\ell-21}$. Second, we have to find a message block $M_{\ell-20}$ such that $f(H_{\ell-21}, M_{\ell-20}) = H_{\ell-20}$ for one of the 2^{20} candidates for $H_{\ell-20}$ in the list L. After testing about 2^{172} message blocks $M_{\ell-20}$ we expect to find a matching entry in the tree and hence, a preimage for Tiger reduced to 16 rounds. Thus, this step of the attack has a complexity of about 2^{172} compression function evaluations of Tiger.

Hence, with this method we can find a preimage for the Tiger hash function reduced to 16 rounds with a complexity of about 2^{174} compression function evaluations and memory requirement of 2^{39}. Note that the same method can be used to construct preimages for the Tiger hash function reduced to 17 rounds with a complexity of about 2^{185} compression function evaluations and memory requirements of 2^{160}.

5   Conclusion

In this article, we presented a preimage attack on the compression function of Tiger reduced to 16 and 17 rounds with a complexity of about 2^{173} and 2^{184} compression function evaluations and memory requirements of 2^{38} and 2^{159}, respectively. In the attack, we combined weaknesses in the key schedule of Tiger with a generic meet-in-the-middle approach. Furthermore, we used a tree-based approach to extend the attacks for the compression function to the hash function with a complexity of about 2^{174} and 2^{185} compression function evaluations and memory requirements of 2^{39} and 2^{160}, respectively. Even though the complexities of the presented attacks are only slightly faster than brute force search, they show that the security margins of the Tiger hash function with respect to preimage attacks are not as good as expected.

Acknowledgements

The author wishes to thank Mario Lamberger, Vincent Rijmen, and the anonymous referees for useful comments and discussions. The work in this paper has been supported in part by the European Commission under contract ICT-2007-216646 (ECRYPT II).

References

1. Anderson, R.J., Biham, E.: TIGER: A Fast New Hash Function. In: Gollmann, D. (ed.) FSE 1996. LNCS, vol. 1039, pp. 89–97. Springer, Heidelberg (1996)
2. Aumasson, J.-P., Meier, W., Mendel, F.: Preimage Attacks on 3-Pass HAVAL and Step-Reduced MD5. In: Avanzi, R., Keliher, L., Sica, F. (eds.) SAC 2008. LNCS. Springer, Heidelberg (2008) (to appear)
3. Cannière, C.D., Rechberger, C.: Preimages for Reduced SHA-0 and SHA-1. In: Wagner, D. (ed.) CRYPTO 2008. LNCS, vol. 5157, pp. 179–202. Springer, Heidelberg (2008)
4. Indesteege, S., Preneel, B.: Preimages for Reduced-Round Tiger. In: Lucks, S., Sadeghi, A.-R., Wolf, C. (eds.) WEWoRC 2007. LNCS, vol. 4945, pp. 90–99. Springer, Heidelberg (2008)
5. Isobe, T., Shibutani, K.: Preimage Attacks on Reduced Tiger and SHA-2. In: Dunkelman, O. (ed.) FSE 2009. LNCS, Springer, Heidelberg (2009) (to appear)
6. Kelsey, J., Lucks, S.: Collisions and Near-Collisions for Reduced-Round Tiger. In: Robshaw, M.J.B. (ed.) FSE 2006. LNCS, vol. 4047, pp. 111–125. Springer, Heidelberg (2006)
7. Knudsen, L.R., Mendel, F., Rechberger, C., Thomsen, S.S.: Cryptanalysis of MDC-2. In: Joux, A. (ed.) EUROCRYPT 2009. LNCS, Springer, Heidelberg (2009) (to appear)
8. Leurent, G.: MD4 is Not One-Way. In: Nyberg, K. (ed.) FSE 2008. LNCS, vol. 5086, pp. 412–428. Springer, Heidelberg (2008)

9. Mendel, F., Preneel, B., Rijmen, V., Yoshida, H., Watanabe, D.: Update on Tiger. In: Barua, R., Lange, T. (eds.) INDOCRYPT 2006. LNCS, vol. 4329, pp. 63–79. Springer, Heidelberg (2006)

10. Mendel, F., Rijmen, V.: Cryptanalysis of the Tiger Hash Function. In: Kurosawa, K. (ed.) ASIACRYPT 2007. LNCS, vol. 4833, pp. 536–550. Springer, Heidelberg (2007)

11. Mendel, F., Rijmen, V.: Weaknesses in the HAS-V Compression Function. In: Nam, K.-H., Rhee, G. (eds.) ICISC 2007. LNCS, vol. 4817, pp. 335–345. Springer, Heidelberg (2007)

12. Sasaki, Y., Aoki, K.: Preimage attacks on one-block MD4, 63-step MD5 and more. In: Avanzi, R., Keliher, L., Sica, F. (eds.) SAC 2008. LNCS, vol. 5381. Springer, Heidelberg (to appear)

Generic Attacks on Feistel Networks with Internal Permutations

Joana Treger and Jacques Patarin

University of Versailles, 45 avenue des États-Unis, 78035 Versailles, France
{joana.treger,jacques.patarin}@prism.uvsq.fr

Abstract. In this paper, we describe generic attacks on Feistel networks with internal permutations, instead of Feistel networks with internal functions as designed originally. By generic attacks, we mean that in these attacks the internal permutations are supposed to be random. Despite the fact that some real Feistel ciphers actually use internal permutations like Twofish, Camellia, or DEAL, these ciphers have not been studied much. We will see that they do not always behave like the original Feistel networks with round functions. More precisely, we will see that the attacks (known plaintext attacks or chosen plaintext attacks) are often less efficient, namely on all $3i$ rounds, $i \in \mathbb{N}^*$. For a plaintext of size $2n$ bits, the complexity of the attacks will be strictly less than 2^{2n} when the number of rounds is less than or equal to 5. When the number k of rounds is greater, we also describe some attacks enabling to distinguish a k-round Feistel network generator from a random permutation generator.

1 Introduction

Feistel networks are designed to construct permutations of $\{1, \ldots, 2^{2n}\}$ from functions of $\{1, \ldots, 2^n\}$ (such a function is then called a *round function* or *internal function*). They have been widely used in cryptography to construct pseudorandom permutations, considering a succession of many rounds, with as many internal functions as rounds. The innovating work of Luby and Rackoff [7] on that structure was followed by many others, dealing not only with Feistel networks as described above, but with many derived structures too. For instance, the papers of Jutla [4], Schneier and Kelsey [14], and the one of Patarin and Nachef [11] deal with unbalanced Feistel networks with expanding functions. Also, Patarin Nachef and Berbain [10] focused on attacks on unbalanced Feistel networks with contracting functions.

However, little work has been done on Feistel networks with round permutations, instead of round functions. Knudsen [6] found an attack on a 5-round Feistel network with round permutations, using the fact that some configurations never occur, and more recently Piret [12] interested himself in security proofs for 3 and 4-round Feistel networks with round permutations.

Such a structure is worth being looked at. First, some symmetric ciphers based on Feistel networks using internal permutations have been developed. Twofish

B. Preneel (Ed.): AFRICACRYPT 2009, LNCS 5580, pp. 41–59, 2009.
© Springer-Verlag Berlin Heidelberg 2009

[8], Camellia [2] and DEAL [5] belong to that category. Second, the fact that the internal functions are in fact permutations do have an influence on the study of the security. Rijmen, Preneel and De Win [13] exposed an attack which worked for Feistel networks with internal functions presenting bad surjectivity properties. Biham in [3] gives an example where the bijectiveness of the internal functions leads to an attack. An other instance is given by Knudsen's attack on a 5-round Feistel network with round permutations ([6]).

We take interest in attacks on k-round Feistel networks with internal random permutations of $\{1, \ldots, 2^n\}$. The aim of these attacks, called *generic attacks* because of the supposed randomness of the internal permutations, is to distinguish with high probability such networks from random permutations of $\{1, \ldots, 2^{2n}\}$.

We focus on one type of attacks, namely the *two-point attacks*. These two-point attacks are attacks using correlations between pairs of messages (this includes for example differential cryptanalysis). They are mainly the ones giving the best known generic attacks against Feistel networks with round functions. As we will see, they are efficient against Feistel networks with round permutations too. Indeed, when the number of rounds is small, the bounds of security found by Piret in [12] are the same as the complexities of our best two-point attacks (except for the adaptive chosen plaintext and ciphertext on 3 rounds, where the best attack is a three-point attack). For $k \leq 5$ rounds, we find the same complexities than for classical Feistel networks, except for 3 rounds, where we find a known plaintext attack in $\mathcal{O}(2^n)$ messages instead of $\mathcal{O}(2^{n/2})$ for Feistel ciphers with round functions. Then, a difference appears on all $3i$ rounds, $i \geq 1$. This fact is illustrated in Section 5, Table 2.

The technique used to find these attacks is based on the computation of the so called *H-coefficients*[1]. From these H-coefficients, we deduce attacks using correlations between pairs of messages (two-point attacks). We will find formulae and compute the H-coefficient values for all possible pairs of input/output couples, and as we will know them all, we will deduce the best possible generic two-point attacks.

The attacks presented in this paper are the best known generic attacks. They give an upper bound on the security, but other attacks might be possible because we do not concentrate on proofs of security. However, the complexities of our attacks coincide with the known bounds of security given by Piret in [12].

This article is organized as follows. Section 2 gives some common notations. Section 3 gives detailed generic attacks on the first 5 rounds. Section 4 describes the general method used to get all our attacks and explains how the H-coefficients interfer in finding the best ones. Also, Section 4 gives a table from which we can deduce our attacks. An example of attack obtained by the general method is given there too, illustrating the fact that the attacks given in Section 3 can easily be found again by the general method. Finally, section 5 gives numerical results and section 6 concludes the paper.

---

[1] Here, a H-coefficient, for specific input/output couples, is the number of k-tuples of permutations $(f_1, ..., f_k)$, such that the k-round Feistel network using $f_1, ... f_k$ as internal permutations applied to the inputs gives the corresponding outputs.

2   Notations

We use the following notations.
- $I_n = \{0,1\}^n$ denotes the set of the 2^n binary strings of length n.
For a, $b \in I_n$, $[a,b]$ is the string of length $2n$, concatenation of a and b.
For a, $b \in I_n$, $a \oplus b$ stands for the bitwise exclusive *or* of a and b.
- F_n is the set of all mappings from I_n to I_n. Its cardinal is $|F_n| = 2^{n \cdot 2^n}$.
B_n is the set of all permutations from I_n to I_n. $B_n \subset F_n$ and $|B_n| = (2^n)!$.
- Let $f \in F_n$ and L, R, S, $T \in I_n$. The permutation ψ defined by :

$$\psi(f)([L,R]) = [R, L \oplus f(R)] = [S,T]$$

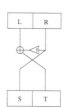

is called a one-round Feistel network. If $f \in B_n$, $\psi(f)$, or simply ψ, is called a one-round Feistel network with internal permutation.
Let $f_1, \ldots, f_k \in F_n$. Then,

$$\psi^k(f_1, \ldots, f_k) := \psi(f_k) \circ \cdots \circ \psi(f_1)$$

Fig. 1. 1-round Feistel network

is called a k-round Feistel network. If $f_1, \ldots, f_k \in B_n$, $\psi^k(f_1, \ldots, f_k)$, or simply ψ^k, is called a k-round Feistel network with internal permutations.

3   Generic Attacks for a Small Numbers of Rounds (≤ 5)

In this Section, the best generic attacks found on a k-round Feistel network with internal permutations, for $k \leq 5$, are directly exposed. Note that the generic attacks on the first two rounds are identical to the ones on Feistel networks with rounds functions of [9]. We skip the presentation and go directly to the analysis of 3 rounds, where the properties of the internal permutations first appear.

3.1   Generic Attacks on 3-Round Feistel Networks with Internal Permutations

The following two-point attacks are different from those described in [9] for Feistel networks with internal functions. Actually, the attacks described there do not work any more once we know that the functions used are in fact permutations.

$CPA - 1$. The attacker chooses some inputs $[L_i, R_i]$ (whose images by p, a permutation, are denoted by $[S_i, T_i]$), such that $L_i = L_j$ and $R_i \neq R_j$ when $i \neq j$. Then he waits until he gets two corresponding outputs $[S_i, T_i]$ and $[S_j, T_j]$ verifying $R_i \oplus R_j = S_i \oplus S_j$. If he computes about $2^{n/2}$ messages and if p is a random permutation, this equality will happen with a probability $\geq 1/2$. If p is a 3-round Feistel network with internal permutations, such a configuration never happens.

To be assured of this last point, let us take a look at Figure 2. Let us assume that there is such an equality, namely $S_i \oplus S_j = R_i \oplus R_j$, $i \neq j$. As f_2 is a permutation, the inner values X_i and X_j are equal. As $L_i = L_j$ and f_1 is a permutation, we get $R_i = R_j$. The assumption leads to a contradiction, therefore this case cannot happen with a 3-round Feistel network with internal permutations.

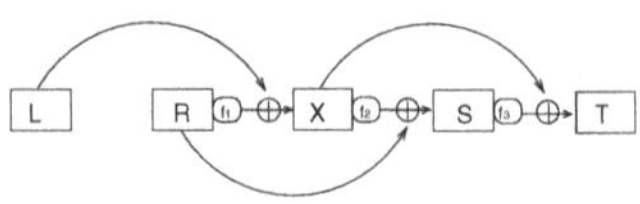

Fig. 2. $\psi^3(f_1, f_2, f_3)$

We then have a non-adaptive chosen plaintext attack $(CPA - 1)$ (thus adaptive chosen plaintext attack $(CPA - 2)$, non-adaptive and adaptive chosen plaintext and ciphertext attack$(CPCA - 1$ and $CPCA - 2)$) working with $\mathcal{O}(2^{n/2})$ messages and $\mathcal{O}(2^{n/2})$ computations.

KPA. For 3 rounds, the best known plaintext attack (KPA) found needs $\mathcal{O}(2^n)$ messages, whereas $\mathcal{O}(2^{n/2})$ messages were enough for the generic attack on a 3-round Feistel network with internal functions.

This attack is an adaptation of the $CPA - 1$. The attacker waits until he gets enough of the desired input pairs (that is to say, he waits until he gets $\mathcal{O}(2^{n/2})$ input pairs verifying: $L_i = L_j$, $R_i \neq R_j$) and then computes the preceding attack. Thus, there is a KPA with $\mathcal{O}(2^n)$ messages.

$CPCA - 2$. The adaptive chosen plaintext and cipher text attack, described in [7], p.385, that needs 3 messages, still works here. It is a three points attack with only three messages. The input/output messages are: $[L_1, R_1]$ and $[S_1, T_1]$, $[L_2, R_1]$ and $[S_2, T_2]$, $[L_3, R_3]$ and $[S_1, T_1 \oplus L_1 \oplus L_2]$. One just can check whether $R_3 = S_2 \oplus S_3 \oplus R_2$ or not. This last equality happens with probability 1 for a 3-rounds Feistel network and probability $1/2^n$ for a random permutation.

Remark: With a number m of messages small compared to $\mathcal{O}(2^{n/2})$, we cannot distinguish random internal permutations from random internal functions. Therefore, the attack here is the same as for a 3-round Feistel network with internal functions.

3.2   Generic Attacks on 4-Round Feistel Networks with Internal Permutations

For 4 rounds, the desired equations on the input/output pairs leading to the best attack is the same as for Feistel networks with internal functions (see [9] or [1]). Still, the attack is not the same as it is based in this case on improbable configurations.

$CPA - 1$. The attacker chooses this time different inputs $[L, R]$, such that $R_i = R_j$. Then he waits until he gets two corresponding outputs $[S_i, T_i]$ and $[S_j, T_j]$ verifying $L_i \oplus L_j = S_i \oplus S_j$. If he computes about $2^{n/2}$ messages and if p is a random permutation, this equality will happen with a probability $\geq 1/2$. If p is a 4-round Feistel network with internal permutations, such a configuration does not happen.

To be assured of this last point, let (f_1, f_2, f_3, f_4) be the internal permutations and X be $f_2(L \oplus f_1(R))$. The equation $X_i = X_j$, when $i \neq j$, never happens because f_1, f_2 are permutations and $L_i \neq L_j$. Thus we also never obtain $R_i \oplus X_i = R_j \oplus X_j$. However, $S_i \oplus S_j = f_3(X_i \oplus R_i) \oplus L_i \oplus f_3(X_j \oplus R_j) \oplus L_j$. As f_3 is a permutation, it is impossible for $S_i \oplus S_j$ to be equal to $L_i \oplus L_j$.

Thus, there is a $CPA - 1$ with $\mathcal{O}(2^{n/2})$ messages.

KPA. The attack is an adaptation of the $CPA - 1$. It consists in waiting to get $\mathcal{O}(2^{n/2})$ input pairs verifying the equality $L_i = L_j$ and then applying the preceding attack. It is an attack with $\mathcal{O}(2^n)$ messages.

3.3   Generic Attacks on 5 Rounds Feistel Networks with Internal Permutations

For 5 rounds, we encounter the same result as Knudsen in [6]. The attack is based on an impossible configuration.

More precisely, if we consider two input/output pairs $([L_1, R_1], [S_1, T_1])$ and $([L_2, R_2], [S_2, T_2])$, the case where $L_1 \neq L_2, R_1 = R_2, S_1 = S_2, T_1 \oplus T_2 = L_1 \oplus L_2$ is impossible to get with a 5-round Feistel network with internal permutation. In other words, there is no $(f_1, \ldots, f_5) \in B_n^5$, such that we get such a pair of input/output couples[2].

$CPA - 1$. The attacker chooses inputs whose left and right blocks verify the above first two conditions and waits for outputs verifying the two last conditions. If he computes about 2^n messages, the equations will hold with a probability $\geq 1/2$ for a random permutation. Thus there is a non-adaptive chosen plaintext attack with $\mathcal{O}(2^n)$ messages.

KPA. As for the previous attacks, this $CPA - 1$ can be turned into a known plaintext attack by waiting for enough wanted pairs of inputs to be generated. This leads to an attack with $\mathcal{O}(2^{3n/2})$ messages.

Remark: Beyond 5 rounds, the attacks are not based on impossible differentials. This can be deduced from the H-coefficient values (see Section 4.1, Definition 2 for a definition, Appendix C for H-values).

4   Generic Attacks for Any Number of Rounds. General Method

In this section, we present a way to systematically analyze all possible two-point attacks, as in [10] for instance. That is, for each relation on the input and output blocks (as in the attacks of Section 3), we compute the complexity of the corresponding attack. We then get the best known generic attacks on Feistel networks with internal permutations.

---

[2] This can also be verified by the H-values (Appendix C).

4.1   Preliminaries

The idea. Let us try to analyze the attacks done in Section 3, in the case of a KPA or CPA. In these attacks, the attacker is confronted to a permutation black box, which is either a random permutation of I_{2n} or a Feistel network with round permutations. His goal is to determine with high probability ($\gg$ $1/2$), which of these two permutations is used. The attacker takes two inputs $[L_1, R_1] \neq [L_2, R_2]$, and gets the corresponding outputs $[S_1, T_1]$ and $[S_2, T_2]$ from the permutation black box. He is waiting for the two input/output couples to verify some specific relations on their input and output blocks (like in the previous Section). These relations happen with some probability, which should differ depending on the nature of the permutation. If the attacker generates enough messages, the difference in the probabilities will result in a difference in the number of couples verifying the specific relations. Therefore, if the number of such couples is significant, he can deduce what permutation is used and achieve his attack.

Thus, we are interested in the probability (for both a random permutation and a Feistel network with internal permutations) for the attacker to have two input/output couples $[L_1, R_1]/[S_1, T_1] \neq [L_2, R_2]/[S_2, T_2]$ verifying some relations on their blocks.

Computation of the probabilities. The probabilities we are interested in depend on the type of attack performed (KPA, CPA, etc). To express them properly, we need the following definition.

Definition 1 (n_e). *Let us consider some equalities between the input and output blocks of two different input/output messages of the attacker $[L_1, R_1]/[S_1, T_1]$ and $[L_2, R_2]/[S_2, T_2]$. (Equalities of the type $L_1 = L_2$, $L_1 \oplus L_2 = T_1 \oplus T_2$, etc). Then, we define the number n_e, which depends on the type of attack we are considering:*

- *in KPA, n_e is the total number of equalities,*
- *in CPA, n_e denotes the total number of equalities minus the number of equalities involving the input blocks only,*
- *in $CPCA$, n_e denotes the total number of equalities minus the number of equalities involving the blocks which have been chosen by the attacker only (unlike those resulting from the computation).*

n_e *will be called the number of* non-imposed *equalities.*

Remark: Among given equalities between input blocks and output blocks that the attacker wants to test on a pair of messages, there are some that can always be verified, due to the type of attack. For example, in CPA, the equality $R_1 = R_2$ happens with probability 1, because the attacker can impose it. This is why the remaining ones are called "non-imposed".

Proposition 1. *For a random permutation, the probability for the attacker to have two input/output couples $[L_1, R_1]/[S_1, T_1] \neq [L_2, R_2]/[S_2, T_2]$ verifying some relations on their blocks can be approximated by:*

$$\frac{2^{(4-n_e)\cdot n}}{2^{2n}(2^{2n} - 1)}.$$

Remark: The exact number of "acceptable" output pairs (in the sense that the wanted relations between the blocks are verified) would be difficult to express, this is why we only give an approximation. However, the approximation is very close to the exact value and sufficient for the use made of it.

Proof. The probability to have a couple of inputs with the right relations on their blocks multiplied by the number of acceptable output couples is approximated by $\frac{1}{2^{n_e \cdot n}} \cdot 2^{4n}$. The probability for the permutation to output one specific couple is $\frac{1}{2^{2n}(2^{2n}-1)}$. The result is obtained by multiplying these two values. □

To get a similar probability for ψ^k, we introduce the following definition :

Definition 2 (H-coefficient). *Let $[L_1, R_1] \neq [L_2, R_2]$ and $[S_1, T_1] \neq [S_2, T_2]$ be four elements of I_{2n}. The H-coefficient for ψ^k (denoted by $H(L_1, R_1, L_2, R_2, S_1, T_1, S_2, T_2)$ or simply H) is the number of k-tuples $(f_1, \ldots, f_k) \in B_n^k$, such that*

$$\begin{cases} \psi^k(f_1, \ldots, f_k)([L_1, R_1]) = [S_1, T_1] \\ \psi^k(f_1, \ldots, f_k)([L_2, R_2]) = [S_2, T_2] \end{cases}.$$

Proposition 2. *For a k-round Feistel network with random internal permutations (ψ^k), the probability for the attacker to have two input/output couples $[L_1, R_1]/[S_1, T_1] \neq [L_2, R_2]/[S_2, T_2]$ verifying some relations on their blocks is approximated by:*

$$\frac{2^{(4-n_e)\cdot n} \cdot H}{|B_n|^k}.$$

Remark: Same remark as for Proposition 1.

Proof. As before, the probability to have a couple of inputs with the right relations on their blocks multiplied by the number of acceptable output pairs is approximated by $2^{(4-n_e)\cdot n}$. The probability for ψ^k to output one specific couple is $\frac{H}{|B_n|^k}$. The result is obtained by multiplying these two values. □

The value of H depends on the following relations:

$$\begin{cases} L_1 = L_2, \text{ or not} \\ R_1 = R_2, \text{ or not} \\ S_1 = S_2, \text{ or not} \\ T_1 = T_2, \text{ or not} \end{cases} \quad \begin{cases} L_1 \oplus L_2 = S_1 \oplus S_2, \text{ or not, when } k \text{ is even} \\ R_1 \oplus R_2 = T_1 \oplus T_2, \text{ or not, when } k \text{ is even} \\ L_1 \oplus L_2 = T_1 \oplus T_2, \text{ or not, when } k \text{ is odd} \\ R_1 \oplus R_2 = S_1 \oplus S_2, \text{ or not, when } k \text{ is odd} \end{cases}$$

This leads to 13 different cases when k is odd and 11 when k is even[3]. The value of H in case j will be denoted by H_j, or simply by H when it is clear that the case j is the one we are working on.

4.2   Explicit Method

Distinguishing a k-round Feistel network from a random permutation.
Let us consider m random plaintext/ciphertext couples.

Let the random variable $X_{\psi^k,j}$ be defined by the number of pairs of these couples belonging to the j-th case of Appendix A ($1 \leq j \leq 13$ when k is odd and $1 \leq j \leq 11$ when k is even), for a k-round Feistel network with internal permutations. Similarly, the random variable $X_{r,j}$ consists in the number of pairs of these couples belonging to the j-th case, for a random permutation.

If the difference $|E(X_{\psi^k,j}) - E(X_{r,j})|$ is larger than about $\sigma(X_{\psi^k,j}) + \sigma(X_{r,j})$, it is possible to distinguish ψ^k (a k-round Feistel network with internal permutations) from a random permutation with $\mathcal{O}(m)$ computations.

For each k, we have to find the cases j, for which $|E(X_{\psi^k,j}) - E(X_{r,j})|$ is larger than $\sigma(X_{\psi^k,j}) + \sigma(X_{r,j})$ for m as small as possible.

The computation of these values is obtained from Propositions 1 and 2. We find that from one case, one can obtain an attack using M pairs, with M such that: $M \geq 2^{n_e \cdot n} \cdot \left(\frac{H \cdot 2^{4n}}{|B_n|^k} - \frac{1}{1 - 1/2^{2n}} \right)^{-2}$.

Distinguishing a k-round Feistel network from a random permutation generator. When m is greater than 2^{2n}, the number of messages needed to attack a single permutation is more than the number of possible inputs to the permutation. However, the method can easily be extended to attack a permutation generator.

We denote by λ the number of permutations. For a given case, the expectation values of the previous paragraph are multiplied by λ and the standard deviation by $\sqrt{\lambda}$. We suppose that we use about the maximum number of pairs of messages possible per permutation ($M = 2^{2n}(2^{2n} - 1)$ in KPA) and solve $|E(X_{\psi^k,j}) - E(X_{r,j})| \geq \sigma(X_{r,j}) + \sigma(X_{\psi^k,j})$ in λ.

The complexity of the attack is then $\mathcal{O}(\lambda \cdot m)$. As the maximum number of pairs of messages per permutation will always be used, m will always be equal to 2^{2n}. Thus the complexity of such an attack is $\mathcal{O}(\lambda \cdot 2^{2n})$.

4.3   Computation of the H-Coefficients

The computation of the formulae for th H-coefficients is rigorously done in Appendix B. Also, values of H are given in the tables of Appendix C. However, let us give an overview of the basic ideas. Let $[L_1, R_1]$, $[L_2, R_2]$ and $[S_1, T_1]$, $[S_2, T_2]$ be four elements of I_{2n} and let us consider the following representation of the network (Fig. 3).

---

[3] There were more in fact, but we could avoid some of them because of the following relation: $\psi^k(f_1, \ldots, f_k)([L, R]) = [S, T] \iff \psi^k(f_k, \ldots, f_1)([T, S]) = [R, L]$.

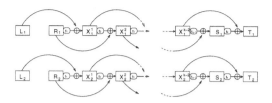

Fig. 3. $\psi^k(f_1, \ldots, f_k)([L_l, R_l]) = [S_l, T_l]$, $l = 1, 2$

We want to determine the number H of $(f_1, \ldots, f_k) \in B_n^k$, such that

$$\begin{cases} \psi^k(f_1, \ldots, f_k)([L_1, R_1]) = [S_1, T_1] \\ \psi^k(f_1, \ldots, f_k)([L_2, R_2]) = [S_2, T_2] \end{cases}$$

For more simplicity, let us also use the notations: $X^{-1} = L$, $X^0 = R$, $X^{k-1} = S$, $X^k = T$. The following results holds:

Theorem 1. *A formula for H is given by*

$$H = \sum_{\text{possible } s} (2^n - 1)!^{e(s)}(2^n - 2)!^{d(s)} \cdot (2^n)^{k-2} \cdot N(d_1) \cdots N(d_{k-2}),$$

where:

- *s denotes a sequence of relations verified by the internal variables, namely $X_1^i = X_2^i$ or $X_1^i \neq X_2^i$, for $i = -1, \ldots, k$,*
- *$e(s)$ is for the number of equalities in s,*
- *$d(s)$ is for the number of differences in s,*
- *$N(d_i)$, for $i = 1, \ldots, k-2$, is the number of possible values for $X_1^i \oplus X_2^i$, for a fixed sequence s.*

Proof. In a first step, we suppose fixed a sequence s as in Theorem 1, and evaluate the number of possibilities $H(s)$ for $(f_1, \ldots, f_k)$. The second step then consists in summing up over all possible sequences s.

Let us detail the first step. In the following, $f_1, \ldots, f_k$ will always denote the permutations we are interested in. A sequence s as in Theorem 1 is fixed.

- For $i = 1 \ldots k - 2$, $N(d_i) \cdot 2^n$ is the number of possibilities for the pair (X_1^i, X_2^i).
- We have $f_i(X^{i-1}) = X^{i-2} \oplus X^i$ for $i = 1, \ldots, k$. Thus, $N(d_1) \cdot 2^n$ is the number of possibilities for the pair $(f_1(X_1^0), f_1(X_2^0))$, and $N(d_i) \cdot 2^n$ for $i = 2, \ldots, k - 2$ is the number of possibilities for the pair $(f_i(X_1^{i-1}), f_i(X_2^{i-1}))$, when $f_1, \ldots, f_{i-1}$ are fixed.

– Let us denote by $F_1(s)$ the number of possibilities for f_1, and $F_i(s)$ for $i = 2, \ldots, k$, the number of possibilities for f_i, when $f_1, \ldots, f_{i-1}$ are fixed. Then we have:

$$\begin{cases} \text{if } X_1^{i-1} \neq X_2^{i-1} : F_i(s) := N(d_i)2^n(2^n - 2)!, \text{for } i = 1, \ldots k - 2, \\ \qquad\qquad F_{k-1}(s) = F_k(s) = (2^n - 2)! \\ \text{if } X_1^{i-1} = X_2^{i-1} : F_i(s) := N(d_i)2^n(2^n - 1)!, \text{for } i = 1, \ldots k - 2, \\ \qquad\qquad F_{k-1}(s) = F_k(s) = (2^n - 1)! \end{cases}$$

This results from the preceding point and from noticing that $(2^n - 2)!$ (respectively $(2^n - 1)!$) is the number of permutations, for which we have already imposed the image of two elements (respectively one element). For $i = k - 1, k$, all values in the equation $f_i(X^{i-1}) = X^{i-2} \oplus X^i$ are fixed, thus the number of possibilities for the pair $(f_i(X_1^{i-1}), f_i(X_2^{i-1}))$ is 1.
– Finally, for a fixed sequence s, the number of possibilities for $(f_1, \ldots, f_k)$ is

$$H(s) = \prod_{i=1}^{k} F_i(s) = (2^n - 1)!^{e(s)}(2^n - 2)!^{d(s)} \cdot (2^n)^{k-2} \cdot N(d_1) \cdots N(d_{k-2}).$$

The final formula for the H-coefficients is then:

$$H = \sum_{\text{possible } s} (2^n - 1)!^{e(s)}(2^n - 2)!^{d(s)} \cdot (2^n)^{k-2} \cdot N(d_1) \cdots N(d_{k-2}),$$

as claimed.   □

Remark: It is also possible to obtain the H-coefficients by induction. This will be done in an extended eprint version of this article.

4.4   Table of Leading Terms of $\frac{H \cdot 2^{4n}}{|B_n|^k} - \frac{1}{1 - 1/2^{2n}}$ and Example of Attack

Table of leading terms of $\frac{H \cdot 2^{4n}}{|B_n|^k} - \frac{1}{1 - 1/2^{2n}}$. At Section 4.2, we saw that the coefficient $\frac{H \cdot 2^{4n}}{|B_n|^k} - \frac{1}{1 - 1/2^{2n}}$ allows us to easily find the best two-point attacks. Table 1 below gives the leading term of $\frac{H \cdot 2^{4n}}{|B_n|^k} - \frac{1}{1 - 1/2^{2n}}$, for each case exposed in Appendix A [4].

Example of attack given by the general method: 3 rounds, KPA. We first examine the number of equalities each case of Appendix A requires. For an odd number of rounds and in the case of a KPA, the case 1 does not require any equations on the inputs and outputs, cases 2 to 5 require 1 equation, cases 6 to 11 require 2 equations, and cases 12 and 13 require 3 equations. Then, from

---

[4] The reader has to be carefull because the case i is not the same depending on the parity of k (see Appendix A).

Table 1. Order of the leading term of $\frac{H \cdot 2^{4n}}{|B_n|^k} - \frac{1}{1-1/2^{2n}}$. From these values, we can easily get the best attacks using correlations between pairs of messages.

number of rounds \ case	1	2	3	4	5	6	7	8	9	10	11	12	13
1	1	1	N^2	1	1	1	1	1	1	1	1	N^3	1
2	$N-1$	1	$N-1$	1	$N-1$	1	1	N	1	1	1		
3	$N-2$	$N-1$	$N-1$	$N-1$	$N-1$	$N-1$	1	1	1	$N-1$	$N-1$	1	1
4	$N-3$	$N-1$	$N-2$	$N-2$	$N-2$	$N-1$	$N-1$	1	$N-1$	$N-2$	1		
5	$N-2$	$N-2$	$N-1$	$N-2$	$N-3$	$N-2$	$N-1$	$N-1$	$N-2$	$N-2$	$N-1$	1	$N-1$
6	$N-3$	$N-3$	$N-3$	$N-2$	$N-3$	$N-2$	$N-2$	$N-2$	$N-3$	$N-2$	$N-1$		
7	$N-3$	$N-3$	$N-2$	$N-3$	$N-3$	$N-3$	$N-3$	$N-2$	$N-2$	$N-3$	$N-2$	$N-1$	$N-3$
8	$N-4$	$N-3$	$N-5$	$N-3$	$N-4$	$N-3$	$N-3$	$N-2$	$N-3$	$N-3$	$N-3$		
9	$N-5$	$N-4$	$N-4$	$N-5$	$N-4$	$N-4$	$N-3$	$N-3$	$N-3$	$N-5$	$N-4$	$N-3$	$N-3$
10	$N-6$	$N-4$	$N-6$	$N-5$	$N-5$	$N-4$	$N-5$	$N-3$	$N-4$	$N-5$	$N-3$		
11	$N-5$	$N-5$	$N-4$	$N-6$	$N-6$	$N-5$	$N-4$	$N-5$	$N-5$	$N-6$	$N-4$	$N-3$	$N-4$
12	$N-6$	$N-6$	$N-6$	$N-5$	$N-6$	$N-5$	$N-6$	$N-5$	$N-6$	$N-5$	$N-4$		

Table 1, we get that the cases leading to the best generic attacks should be cases 7, 8 and 9. For $i = 7, 8, 9$, we have the following values (we use the notations of Section 4): $E_{r,j} \simeq \frac{M}{2^{2n}}$, $|E(X_{\psi^k,j}) - E(X_{r,j})| \simeq \frac{M}{2^{2n}}$, $\sigma(X_{\psi^k,j}) + \sigma(X_{r,j}) \simeq \frac{\sqrt{M}}{2^n}$. Solving the equation $\frac{M}{2^{2n}} \geq \frac{\sqrt{M}}{2^n}$, gives $M \geq 2^{2n}$, thus $m = \mathcal{O}(2^n)$.

Therefore, the best generic attack, exploiting only correlations between pairs of input/output couples works with a complexity $\mathcal{O}(2^n)$, in the case of a KPA.

Remark: It can be noticed that we find this way the attack given in Section 3.1. In fact, the case 9 is the one corresponding to the attack on 3 rounds described in Section 3.1. We also see that the attacks are not unique.

5 Table of Results for Any Number of Rounds

In this section is given a table of results (Table 2), showing the complexities of the attacks we obtained. These results are obtained by computer. All values, except for the special case of Section 3.1, are obtained by computing the formulae for H and applying the reasoning of Section 4 (an example is given in Section 4.4).

Remark: As long as the number of queries to the round functions is small compared to $\mathcal{O}(2^{n/2})$, we cannot distinguish a function of I_n from a permutation of I_n. When the complexities are small compared to $2^{n/2}$, it is then normal to find the same attack complexities for Feistel networks with round permutations than for those with round functions (as for 10 values on the first 3 rounds, in Table 2).

Table 2. Maximum number of computations needed to get an attack on a k-round Feistel network with internal *permutations*. We write $(+)$ when the complexity is worse than for classical Feistel networks.

number k of rounds	KPA	CPA-1	CPA-2	CPCA-1	CPCA-2
1	1	1	1	1	1
2	$2^{n/2}$	2	2	2	2
3	$2^n(+)$	$2^{n/2}$	$2^{n/2}$	$2^{n/2}$	3
4	2^n	$2^{n/2}$	$2^{n/2}$	$2^{n/2}$	$2^{n/2}$
5	$2^{3n/2}$	2^n	2^n	2^n	2^n
6	$2^{3n}(+)$	$2^{3n}(+)$	$2^{3n}(+)$	$2^{3n}(+)$	$2^{3n}(+)$
7	2^{3n}	2^{3n}	2^{3n}	2^{3n}	2^{3n}
8	2^{4n}	2^{4n}	2^{4n}	2^{4n}	2^{4n}
9	$2^{6n}(+)$	$2^{6n}(+)$	$2^{6n}(+)$	$2^{6n}(+)$	$2^{6n}(+)$
10	2^{6n}	2^{6n}	2^{6n}	2^{6n}	2^{6n}
11	2^{7n}	2^{7n}	2^{7n}	2^{7n}	2^{7n}
12	$2^{9n}(+)$	$2^{9n}(+)$	$2^{9n}(+)$	$2^{9n}(+)$	$2^{9n}(+)$
$k\geq 6,\ k=0\ mod\,3$	$2^{(k-3)n}(+)$	$2^{(k-3)n}(+)$	$2^{(k-3)n}(+)$	$2^{(k-3)n}(+)$	$2^{(k-3)n}(+)$
$k\geq 6,\ k=1\ \text{or}\ 2\ mod\,3$	$2^{(k-4)n}$	$2^{(k-4)n}$	$2^{(k-4)n}$	$2^{(k-4)n}$	$2^{(k-4)n}$

6   Conclusion

Whereas Gilles Piret [12] took interest in security proofs for 3 and 4-round Feistel networks with internal permutations, we focused here on generic attacks, and the results obtained are the first ones for these networks. First, the results given here fit with the ones of Piret: to get a chosen plaintext attack on a 3-round Feistel network with round permutation, we found that more than $\mathcal{O}(2^{n/2})$ computations are needed and the same holds to get a chosen plaintext and ciphertext attack on a 4-round Feistel network with round permutations. But here we gave all possible generic attacks on this network, obtained by considering couples of messages. For $k \leq 5$, the final results are similar to the ones on classical Feistel networks, except for the known plaintext attack on 3 rounds. This was not obvious.

We also gave the number of computations needed to distinguish with high probability a k-round Feistel permutation generator (with round permutations), from a random permutation generator. Here, things are a little different than for Feistel networks with round functions. For instance for 6 rounds, the attacks can a priori no longer be done with $\mathcal{O}(2^{2n})$ computations. In fact, more generally for $3i$ rounds $(i \geq 2)$, the attacks seem always harder to perform.

Feistel networks are classical tools in symmetric cryptography for designing block encryption networks. Moreover, many such networks use internal permutations. For these reasons, we believe that it was important to present the best known generic attacks for this construction.

References

1. Aiollo, W., Venkatesan, R.: Foiling Birthday Attacks in Lenght-Doubling Transformations - Benes: A Non-Reversible Alternative to Feistel. In: Maurer, U.M. (ed.) EUROCRYPT 1996. LNCS, vol. 1070, pp. 307–320. Springer, Heidelberg (1996)
2. Aoki, K., Itchikawa, T., Kanda, M., Matsui, M., Nakajima, J., Moriai, S., Tokita, T.: Camellia: A 128-bit Block Cipher Suitable for Multiple Platforms - Design and Analysis. In: Stinson, D.R., Tavares, S. (eds.) SAC 2000. LNCS, vol. 2012, pp. 39–56. Springer, Heidelberg (2001)
3. Biham, E.: Cryptanalysis of Ladder-DES. In: Biham, E. (ed.) FSE 1997. LNCS, vol. 1267, pp. 134–138. Springer, Heidelberg (1997)
4. Jutla, C.S.: Generalised Birthday Attacks on Unbalanced Feistel Networks. In: Krawczyk, H. (ed.) CRYPTO 1998. LNCS, vol. 1462, pp. 186–199. Springer, Heidelberg (1998)
5. Knudsen, L.R.: DEAL - A 128-bit Block Cipher. Technical report number 151, University of Bergen, Norway (1998),
   http://www2.mat.dtu.dk/people/Lars.R.Knudsen/newblock.html
6. Knudsen, L.R.: The Security of Feistel Ciphers with Six Rounds or Less. Journal of Cryptology 15, 207–222 (2002)
7. Luby, M., Rackoff, C.: How to construct pseudorandom permutations from pseudorandom functions. SIAM Journal on Computing 17(2), 373–386 (1988)
8. Nyberg, K.: Linear Approximation of Block Ciphers. In: De Santis, A. (ed.) EUROCRYPT 1994. LNCS, vol. 950, pp. 439–444. Springer, Heidelberg (1995)
9. Patarin, J.: Generic Attacks on Feistel Schemes. In: Boyd, C. (ed.) ASIACRYPT 2001. LNCS, vol. 2248, pp. 222–238. Springer, Heidelberg (2001); An ePrint version of the extended version of this paper is available at,
   http://eprint.iacr.org/2008/036
10. Patarin, J., Nachef, V., Berbain, C.: Generic Attacks on Unbalanced Feistel Schemes with Contracting Functions. In: Lai, X., Chen, K. (eds.) ASIACRYPT 2006. LNCS, vol. 4284, pp. 396–411. Springer, Heidelberg (2006)
11. Patarin, J., Nachef, V., Berbain, C.: Generic Attacks on Unbalanced Feistel Schemes with Expanding Functions, Extended Version. Available from the authors (2006)
12. Piret, G.: Luby-Rackoff revisited: On the Use of Permutations as Inner Functions of a Feistel Scheme. Designs, Codes and Cryptography 39(2), 233–245 (2006)
13. Rijmen, R., Preneel, B., De Win, E.: On Weakness of Non-Surjective Round Functions. Designs, Codes and Criptography 12(3), 253–266 (1997)
14. Schneier, B., Kelsey, J.: Unbalanced Feistel Networks and Block Cipher Design. In: Gollmann, D. (ed.) FSE 1996. LNCS, vol. 1039, pp. 121–144. Springer, Heidelberg (1996)

Appendices

A   Cases Considered

We give here all significant cases. When the number k of rounds is *odd* :

$$L_1{\neq}L_2,R_1{\neq}R_2,S_1{\neq}S_2,T_1{\neq}T_2,L_1{\oplus}L_2{\neq}T_1{\oplus}T_2,R_1{\oplus}R_2{\neq}S_1{\oplus}S_2 \tag{1}$$

$$L_1{\neq}L_2,R_1{\neq}R_2,S_1{\neq}S_2,T_1{\neq}T_2,L_1{\oplus}L_2{=}T_1{\oplus}T_2,R_1{\oplus}R_2{\neq}S_1{\oplus}S_2 \tag{2}$$

$$L_1{\neq}L_2,R_1{\neq}R_2,S_1{\neq}S_2,T_1{\neq}T_2,L_1{\oplus}L_2{\neq}T_1{\oplus}T_2,R_1{\oplus}R_2{=}S_1{\oplus}S_2 \tag{3}$$

$$L_1{\neq}L_2,R_1{\neq}R_2,S_1{=}S_2,T_1{\neq}T_2,L_1{\oplus}L_2{\neq}T_1{\oplus}T_2,R_1{\oplus}R_2{\neq}S_1{\oplus}S_2 \tag{4}$$

$$L_1{=}L_2,R_1{\neq}R_2,S_1{\neq}S_2,T_1{\neq}T_2,L_1{\oplus}L_2{\neq}T_1{\oplus}T_2,R_1{\oplus}R_2{\neq}S_1{\oplus}S_2 \tag{5}$$

$$L_1{=}L_2,R_1{\neq}R_2,S_1{=}S_2,T_1{\neq}T_2,L_1{\oplus}L_2{\neq}T_1{\oplus}T_2,R_1{\oplus}R_2{\neq}S_1{\oplus}S_2 \tag{6}$$

$$L_1{\neq}L_2,R_1{\neq}R_2,S_1{=}S_2,T_1{\neq}T_2,L_1{\oplus}L_2{=}T_1{\oplus}T_2,R_1{\oplus}R_2{\neq}S_1{\oplus}S_2 \tag{7}$$

$$L_1{\neq}L_2,R_1{=}R_2,S_1{=}S_2,T_1{\neq}T_2,L_1{\oplus}L_2{\neq}T_1{\oplus}T_2,R_1{\oplus}R_2{=}S_1{\oplus}S_2 \tag{8}$$

$$L_1{=}L_2,R_1{\neq}R_2,S_1{\neq}S_2,T_1{\neq}T_2,L_1{\oplus}L_2{\neq}T_1{\oplus}T_2,R_1{\oplus}R_2{=}S_1{\oplus}S_2 \tag{9}$$

$$L_1{=}L_2,R_1{\neq}R_2,S_1{\neq}S_2,T_1{=}T_2,L_1{\oplus}L_2{=}T_1{\oplus}T_2,R_1{\oplus}R_2{\neq}S_1{\oplus}S_2 \tag{10}$$

$$L_1{\neq}L_2,R_1{\neq}R_2,S_1{\neq}S_2,T_1{\neq}T_2,L_1{\oplus}L_2{=}T_1{\oplus}T_2,R_1{\oplus}R_2{=}S_1{\oplus}S_2 \tag{11}$$

$$L_1{\neq}L_2,R_1{=}R_2,S_1{=}S_2,T_1{\neq}T_2,L_1{\oplus}L_2{=}T_1{\oplus}T_2,R_1{\oplus}R_2{=}S_1{\oplus}S_2 \tag{12}$$

$$L_1{=}L_2,R_1{\neq}R_2,S_1{\neq}S_2,T_1{=}T_2,L_1{\oplus}L_2{=}T_1{\oplus}T_2,R_1{\oplus}R_2{=}S_1{\oplus}S_2 \tag{13}$$

When the number k of rounds is *even* :

$$L_1{\neq}L_2,R_1{\neq}R_2,S_1{\neq}S_2,T_1{\neq}T_2,L_1{\oplus}L_2{\neq}S_1{\oplus}S_2,R_1{\oplus}R_2{\neq}T_1{\oplus}T_2 \tag{1}$$

$$L_1{\neq}L_2,R_1{=}R_2,S_1{\neq}S_2,T_1{\neq}T_2,L_1{\oplus}L_2{\neq}S_1{\oplus}S_2,R_1{\oplus}R_2{\neq}T_1{\oplus}T_2 \tag{2}$$

$$L_1{=}L_2,R_1{\neq}R_2,S_1{\neq}S_2,T_1{\neq}T_2,L_1{\oplus}L_2{\neq}S_1{\oplus}S_2,R_1{\oplus}R_2{\neq}T_1{\oplus}T_2 \tag{3}$$

$$L_1{\neq}L_2,R_1{\neq}R_2,S_1{\neq}S_2,T_1{\neq}T_2,L_1{\oplus}L_2{=}S_1{\oplus}S_2,R_1{\oplus}R_2{\neq}T_1{\oplus}T_2 \tag{4}$$

$$L_1{=}L_2,R_1{\neq}R_2,S_1{\neq}S_2,T_1{=}T_2,L_1{\oplus}L_2{\neq}S_1{\oplus}S_2,R_1{\oplus}R_2{\neq}T_1{\oplus}T_2 \tag{5}$$

$$L_1{\neq}L_2,R_1{=}R_2,S_1{=}S_2,T_1{\neq}T_2,L_1{\oplus}L_2{\neq}S_1{\oplus}S_2,R_1{\oplus}R_2{\neq}T_1{\oplus}T_2 \tag{6}$$

$$L_1{\neq}L_2,R_1{=}R_2,S_1{\neq}S_2,T_1{=}T_2,L_1{\oplus}L_2{\neq}S_1{\oplus}S_2,R_1{\oplus}R_2{=}T_1{\oplus}T_2 \tag{7}$$

$$L_1{\neq}L_2,R_1{=}R_2,S_1{\neq}S_2,T_1{\neq}T_2,L_1{\oplus}L_2{=}S_1{\oplus}S_2,R_1{\oplus}R_2{\neq}T_1{\oplus}T_2 \tag{8}$$

$$L_1{\neq}L_2,R_1{\neq}R_2,S_1{\neq}S_2,T_1{=}T_2,L_1{\oplus}L_2{=}S_1{\oplus}S_2,R_1{\oplus}R_2{\neq}T_1{\oplus}T_2 \tag{9}$$

$$L_1{\neq}L_2,R_1{\neq}R_2,S_1{\neq}S_2,T_1{\neq}T_2,L_1{\oplus}L_2{=}S_1{\oplus}S_2,R_1{\oplus}R_2{=}T_1{\oplus}T_2 \tag{10}$$

$$L_1{\neq}L_2,R_1{=}R_2,S_1{\neq}S_2,T_1{=}T_2,L_1{\oplus}L_2{=}S_1{\oplus}S_2,R_1{\oplus}R_2{=}T_1{\oplus}T_2 \tag{11}$$

B   How to Find the Exact *H*-Coefficient Formulae

B.1   Preliminaries

Here are some more notations.

- For $-1 \le i \le k$, we denote by d_i the value $X_1^i \oplus X_2^i$.

- Let s be a sequence of relations of length $k+2$, such that s_i is the symbol $=$ if $X_1^i = X_2^i$, s_i is the symbol $\neq$ if $X_1^i \neq X_2^i$. We call $N(s, d_i)$ or simply $N(d_i)$, the number of possible d_i's, for each $-1 \leq i \leq k$.
- Let $s = (s_{-1}, \ldots, s_k)$ be a fixed sequence of $\{=, \neq\}^{k+2}$. We write $e(s_{i_1}, \ldots, s_{i_l})$ ($i_j \in \{-1, \ldots, k\}$, $l \leq k+2$) for the number of $=$'s intervening in the sequence $(s_{i_1}, \ldots s_{i_l})$, and $d(s_{i_1}, \ldots s_{i_l})$ for the number of $\neq$'s intervening in that sequence. For more convenience, $e(s)$ stands for $e(s_{-1}, \ldots, s_k)$ and $d(s)$ stands for $d(s_{-1}, \ldots, s_k)$.

$$\alpha_{odd} = 1 \text{ if } L_1 \oplus L_2 = \begin{cases} S_1 \oplus S_2 \text{ when } k \text{ is even} \\ T_1 \oplus T_2 \text{ when } k \text{ is odd} \end{cases}, \quad 0 \text{ elsewhere}$$

$$\alpha_{ev} = 1 \text{ if } R_1 \oplus R_2 = \begin{cases} T_1 \oplus T_2 \text{ when } k \text{ is even} \\ S_1 \oplus S_2 \text{ when } k \text{ is odd} \end{cases}, \quad 0 \text{ elsewhere}$$

- l_{odd} (respectively l_{ev}) will stand for the number of odd (respectively even) intermediate values (i.e. different from L, R, S, T, in fact, the X^i's of Appendix 4.3, Fig. 3), appearing during the computation of ψ^k.

We already know from Section 4.3, Theorem 1, that the expression

$$F(s) := (2^n - 1)!^{e(s)} \cdot (2^n - 2)!^{d(s)} \cdot (2^n)^{k-2} \cdot N(d_1) \ldots N(d_{k-2})$$

counts the wanted number of k-tuples $(f_1, \ldots, f_k)$, when the particular sequence s is fixed. All possible sequences s and the product $N(d_1) \ldots N(d_{k-2})$ have to be determined for each case.

B.2 Possible Sequences s

By taking a look at Section 4.3, Figure 3 for instance, and taking into account that the f_i's are permutations, we get some conditions on the s_i's and d_i's:

- For $-1 \leq i \leq k$, if s_i is $=$, then s_{i+1}, s_{i+2}, s_{i-1} and s_{i-2} are $\neq$ (when those are well-defined).
- For $-1 \leq i \leq k - 2$, if s_i, s_{i+1} and s_{i+2} are $\neq$, then $d_i \neq d_{i+2}$.
- For $-1 \leq i \leq k$, if $d_i = 0$, then $d_{i-1} = d_{i+1}$ and d_{i+2} can take all values but 0 (when those are well-defined).

These conditions imply that we have different valid sequences s, depending on the initial values s_{-1}, s_0, s_{k-1} and s_k.

B.3 Exact Computation of the Product $N(d_1) \cdots N(d_{k-2})$, for a Fixed Sequence s

Let $s = (s_{-1}, \ldots, s_k)$ be a fixed sequence of $\{=, \neq\}^{k+2}$. Finding the value of the product $\Pi := N(d_1) \cdots N(d_{k-2})$, leads us to considering different situations.

1. When at least one s_i is $=$ amongst the even i's and at least one s_i is $=$ amongst the odd i's:

$$\Pi_1 = (2^n - 1)^{e(s)-2}(2^n - 2)^{k-3e(s)+e(s_0, s_{k-1})+2e(s_{-1}, s_k)}.$$

2. When at least one s_i is $=$ amongst the even i's and none amongst the odd i's:

$$\Pi_2 = \Big(\sum_{j=0}^{M_{odd}} C_{j,odd} \cdot (2^n-2)^{j+\alpha_{odd}} (2^n-3)^{P_{j,odd}} \Big) \cdot (2^n-1)^{e(s)-1} (2^n-2)^{Q_{odd}}.$$

With:

$$M_{odd} = \frac{l_{odd}-\alpha_{odd}-e(s)+e(s_{-1},s_k)}{2}$$
$$C_{j,odd} = \binom{l_{odd}-e(s)+e(s_{-1},s_k)-\alpha_{odd}-j}{j}$$
$$P_{j,odd} = l_{odd} - e(s) + e(s_{-1},s_k) - \alpha_{odd} - 2j$$
$$Q_{odd} = l_{ev} - 2e(s) + e(s_{-1},s_0,s_{k-1},s_k) + 1$$

3. When at least one s_i is $=$ amongst the odd i's and none amongst the even i's:

$$\Pi_3 = \Big(\sum_{j=0}^{M_{ev}} C_{j,ev} \cdot (2^n-2)^{j+\alpha_{ev}} (2^n-3)^{P_{j,ev}} \Big) \cdot (2^n-1)^{e(s)-1} (2^n-2)^{Q_{ev}}.$$

With:

$$M_{ev} = \frac{l_{ev}-\alpha_{ev}-e(s)+e(s_{-1},s_k)}{2}$$
$$C_{j,ev} = \binom{l_{ev}-e(s)+e(s_{-1},s_k)-\alpha_{ev}-j}{j}$$
$$P_{j,ev} = l_{ev} - e(s) + e(s_{-1},s_k) - \alpha_{ev} - 2j$$
$$Q_{ev} = l_{odd} - 2e(s) + e(s_{-1},s_0,s_{k-1},s_k) + 1$$

4. When $e(s) = 0$, we find:

$$\Pi_4 = \Big(\sum_{j=0}^{M_{ev}} C_{j,ev} \cdot (2^n-2)^{j+\alpha_{ev}} \cdot (2^n-3)^{P_{j,ev}} \Big) \cdot \Big(\sum_{j=0}^{M_{odd}} C_{j,odd} \cdot (2^n-2)^{j+\alpha_{odd}} \cdot (2^n-3)^{P_{j,odd}} \Big).$$

In fact, these formulae hold with a special convention sometimes, when the number of rounds is 1, 2 or 3. We will specify it in B.4.

B.4   General Formulae for the H-Coefficients

Here, given two different couples $([L_1, R_1], [S_1, T_1])$ and $([L_2, R_2], [S_2, T_2])$, we state the H-coefficient formulae. The formulae are obtained from Theorem 1 of Section 4.4, where we apply the right expression for $\prod_{i=1}^{k-2} d_i$, depending on the sequence s considered. All possibilities are dispatched in four situations [5].

The formula $F(s)$ for a fixed sequence s only uses the number of $=$'s and $\neq$'s in this sequence (plus the initial equalities on the blocks, see Appendix B.3). Thus, instead of summing up $F(s)$ over all possible sequences s to get H, we sum up over the number A of possible $=$'s in the sequences and multiply the formula $F(s)$ for a fixed s by the number of possible sequences s with A $=$'s.

Remark: Note that $k = 1$ is a very particular case. The H-values for $k = 1$ are easily obtained directly (see Appendix C).

---

[5] The values $\Pi_1, \Pi_2, \Pi_3, \Pi_4$ below are the products of the Appendix B.3, with $e(s) - e(s_{-1},s_0,s_{k-1},s_k) = A$. Also the bound M is $\frac{k-2e(s_0,s_{k-1})-e(s_{-1},s_k)}{3}$

First situation

$$k \text{ even and } L_1 = L_2, \ R_1 \neq R_2, \ S_1 \neq S_2, \ T_1 = T_2$$
$$\text{or } k \text{ even and } L_1 \neq L_2, \ R_1 = R_2, \ S_1 = S_2, \ T_1 \neq T_2$$
$$\text{or } k \text{ odd and } L_1 = L_2, \ R_1 \neq R_2, \ S_1 = S_2, \ T_1 \neq T_2$$
$$\text{or } k \text{ odd and } L_1 \neq L_2, \ R_1 = R_2, \ S_1 \neq S_2, \ T_1 = T_2$$

(cases 5 and 6 for k even, and case 6 for k odd).

$$H = \sum_{A=0}^{M} (2^n - 1)!^{A + e(s_0, s_{k-1})} \cdot (2^n - 2)!^{k - A - e(s_0, s_{k-1})} \cdot 2^{n(k-2)}$$
$$\cdot \binom{k - 2e(s_0, s_{k-1}) - e(s_{-1}, s_k) - 2A}{A} \cdot \Pi_1.$$

This formula is always true for $k > 1$.

Second situation

$$k \text{ even and } L_1 \neq L_2, \ R_1 = R_2, \ S_1 \neq S_2, \ T_1 = T_2$$
$$\text{or } k \text{ even and } L_1 \neq L_2, \ R_1 = R_2, \ S_1 \neq S_2, \ T_1 \neq T_2$$
$$\text{or } k \text{ even and } L_1 \neq L_2, \ R_1 \neq R_2, \ S_1 \neq S_2, \ T_1 = T_2$$
$$\text{or } k \text{ odd and } L_1 \neq L_2, \ R_1 \neq R_2, \ S_1 = S_1, \ T_1 \neq T_2$$
$$\text{or } k \text{ odd and } L_1 \neq L_2, \ R_1 = R_2, \ S_1 \neq S_2, \ T_1 \neq T_2$$
$$\text{or } k \text{ odd and } L_1 \neq L_2, \ R_1 = R_2, \ S_1 = S_2, \ T_1 \neq T_2$$

(cases 2, 7, 8, 9 and 11 for k even, and cases 4, 7, 8 and 12 for k odd).

$$H = \sum_{A=0}^{M} (2^n - 1)!^{A + e(s_0, s_{k-1})} \cdot (2^n - 2)!^{k - A - e(s_0, s_{k-1})} \cdot 2^{n(k-2)} \cdot$$
$$\left[\binom{l_{ev} + 1 - A - e(s_{-1}, s_0, s_{k-1}, s_k)}{A} \cdot \Pi_2 + \left[\binom{k - 2e(s_0, s_{k-1}) - e(s_{-1}, s_k) - 2A}{A} \right. \right. -$$
$$\left. \left. \binom{l_{ev} + 1 - A - e(s_{-1}, s_0, s_{k-1}, s_k)}{A} \right) \right] \cdot \Pi_1 \ \right].$$

This formula is always true for $k > 2$, and also for $k = 2$ if we suppose the sum in Π_2 equal to 1 when $\alpha_{odd} = e(s_0)$.

Third situation

$$k \text{ even and } L_1 \neq L_2, \ R_1 \neq R_2, \ S_1 = S_2, \ T_1 \neq T_2$$
$$\text{or } k \text{ even and } L_1 = L_2, \ R_1 \neq R_2, \ S_1 \neq S_2, \ T_1 \neq T_2$$
$$\text{or } k \text{ even and } L_1 = L_2, \ R_1 \neq R_2, \ S_1 = S_2, \ T_1 \neq T_2$$
$$\text{or } k \text{ odd and } L_1 = L_2, \ R_1 \neq R_2, \ S_1 \neq S_2, \ T_1 \neq T_2$$
$$\text{or } k \text{ odd and } L_1 \neq L_2, \ R_1 \neq R_2, \ S_1 \neq S_2, \ T_1 = T_2$$
$$\text{or } k \text{ odd and } L_1 = L_2, \ R_1 \neq R_2, \ S_1 \neq S_2, \ T_1 = T_2$$

(case 3 for k even, and cases 5, 9, 10 and 13 for k odd).

$$H = \sum_{A=0}^{M} (2^n - 1)!^{A+e(s_0,s_{k-1})} \cdot (2^n - 2)!^{k-A-e(s_{k-1})} \cdot 2^{n(k-2)} \cdot$$

$$\left[\binom{l_{odd}+1-A-e(s_{-1},s_0,s_{k-1},s_k)}{A} \cdot \Pi_3 + \left[\binom{k-2e(s_0,s_{k-1})-e(s_{-1},s_k)-2A}{A} \right.\right.$$

$$\left.\left. - \binom{l_{odd}+1-A-e(s_{-1},s_0,s_{k-1},s_k)}{A} \right) \right] \cdot \Pi_1 \right].$$

This formula is always true for $k > 2$, and also for $k = 2$ if we suppose the sum in Π_3 equal to 1 when $e(s_{-1}, s_{k-1}) = 1$, and $\alpha_{ev} = e(s_{k-1})$.

Fourth situation

$$k \in \mathbb{N}^*, \ L_1 \neq L_2, \ R_1 \neq R_2, \ S_1 \neq S_2, \ T_1 \neq T_2$$

(cases 1, 4 and 10 for k even, and cases 1, 2, 3 and 11 for k odd).

$$H = (2^n - 2)!^k \cdot (2^n)^{k-2} \cdot \Pi_4 + \sum_{A=1}^{k/3} (2^n - 1)!^A \cdot (2^n - 2)!^{k-A} \cdot 2^{n(k-2)} \cdot$$

$$\left[\binom{l_{ev}+1-A}{A} \Pi_2 + \binom{l_{odd}+1-A}{A} \Pi_3 + \left[\binom{k-2A}{A} - \binom{l_{ev}+1-A}{A} - \binom{l_{odd}+1-A}{A} \right] \cdot \Pi_1 \right].$$

This formula is always true for $k > 3$ and also for $k = 2, 3$, if we suppose:

- the sum in Π_4, whose upperbound is M_{ev} is equal to 1, when $k = 2$ and $\alpha_{ev} = 0$,
- the sum in Π_4, whose upperbound is M_{odd} is equal to 1, when $k = 2$ and $\alpha_{ev} = 0$,
- the sum in Π_3 is equal to 1, when $k = 3$ and $\alpha_{ev} = 1 = e(s)$,

C   Exact H-Coefficient Values for $k \leq 5$ Rounds

In Tables 4 and 5 below, are given some exact values of the H-coefficients, computed from the formulae of Appendix B.4.

Remark: For 5 rounds, H in case 12 is 0 (this is linked with Knudsen's attack [6]). However, as soon as $k > 5$, $\forall i \ H_i \neq 0$. The attacks get then more complex.

case	1 round	2 rounds	3 rounds	4 rounds
1	0	$((N-2)!)^2$	$((N-2)!)^3 N(N-3)$	$((N-2)!)^4 N^2(N-3)^2 + 2(N-1)!((N-2)!)^3 N^2$
2	0	0	$((N-2)!)^3 N(N-2)$	$(N-1)!((N-2)!)^3 N^2(N-2)$
3	$(N-2)!$	$((N-2)!)^2$	$(N-1)!((N-2)!)^2 N$	$((N-2)!)^4 N^2(N-2)(N-3)+(N-1)!((N-2)!)^3 N^2$
4	0	0	$(N-1)!((N-2)!)^2 N$	$((N-2)!)^4 N^2(N-2)(N-3)+(N-1)!((N-2)!)^3 N^2$
5	0	$((N-2)!)^2$	$((N-2)!)^3 N(N-2)$	$((N-2)!)^4 N^2(N-2)^2$
6	0	0	$(N-1)!((N-2)!)^2 N$	$((N-1)!)^2((N-2)!)^2 N^2$
7	0	0	0	$(N-1)!((N-2)!)^3 N^2(N-1)$
8	0	$(N-1)!(N-2)!$	0	0
9	0	0	0	$((N-2)!)^4 N^2(N-2)^2+(N-1)!((N-2)!)^3 N^2$
10	0	0	$((N-2)!)^3 N(N-1)$	$((N-2)!)^4 N^2(N-2)^2+(N-1)!((N-2)!)^3 N^2$
11	0	0	$(N-1)!((N-2)!)^2 N$	0
12	$(N-1)!$		0	
13	0		0	

Fig. 4. H-coefficient values in the different cases of A, for 1,2,3 and 4 rounds. Here $N = 2^n$.

case	5 rounds
1	$((N-2)!)^5 N^3\big((N-3)^2+N-2\big)(N-3)+(N-1)!((N-2)!)^4 N^3(3N-7)$
2	$((N-2)!)^5 N^3(N-2)(N-3)^2+(N-1)!((N-2)!)^4 N^3(3N-6)$
3	$((N-2)!)^5 N^3\big((N-3)^2+N-2\big)(N-2)+(N-1)!((N-2)!)^4 N^3(N-3)$
4	$(N-1)!((N-2)!)^4 N^3(N-2)(N-3)+((N-1)!)^2((N-2)!)^3 N^3$
5	$((N-2)!)^5 N^3(N-2)^2(N-3)+(N-1)!((N-2)!)^4 N^3(2N-3)$
6	$(N-1)!((N-2)!)^4 N^3(N-2)^2$
7	$(N-1)!((N-2)!)^4 N^3(N-2)^2+((N-1)!)^2((N-2)!)^3 N^3$
8	$((N-1)!)^2((N-2)!)^3 N^3(N-1)$
9	$((N-2)!)^5 N^3(N-2)^3+(N-1)!((N-2)!)^4 N^3(N-2)$
10	$((N-2)!)^5 N^3(N-3)(N-1)(N-2)+(N-1)!((N-2)!)^4 N^3(N-1)$
11	$((N-2)!)^5 N^3(N-2)^2(N-3)+(N-1)!((N-2)!)^4 N^3(N-2)$
12	0
13	$((N-2)!)^5 N^3(N-2)^2(N-1)+(N-1)!((N-2)!)^4 N^3(N-1)$

Fig. 5. H-coefficient values in the different cases of A, for 5 rounds. Here $N = 2^n$.

Distinguishers for Ciphers and Known Key Attack against Rijndael with Large Blocks

Marine Minier[1], Raphael C.-W. Phan[2,*], and Benjamin Pousse[3]

[1] CITI Laboratory – INSA de Lyon
21 Avenue Jean Capelle, 69621 Villeurbanne Cedex – France
marine.minier@insa-lyon.fr
[2] Electronic and Electrical Engineering, Loughborough University
LE11 3TU Leicestershire – UK
R.Phan@lboro.ac.uk
[3] XLIM (UMR CNRS 6172), Université de Limoges
23 avenue Albert Thomas, F-87060 Limoges Cedex – France
benjamin.pousse@unilim.fr

Abstract. Knudsen and Rijmen introduced the notion of known-key distinguishers in an effort to view block cipher security from an alternative perspective e.g. a block cipher viewed as a primitive underlying some other cryptographic construction such as a hash function; and applied this new concept to construct a 7-round distinguisher for the AES and a 7-round Feistel cipher. In this paper, we give a natural formalization to capture this notion, and present new distinguishers that we then use to construct known-key distinguishers for Rijndael with Large Blocks up to 7 and 8 rounds.

Keywords: Block ciphers, cryptanalysis, known-key distinguishers, Rijndael.

1 Introduction

Rijndael-b is an SPN block cipher designed by Joan Daemen and Vincent Rijmen [4]. It has been chosen as the new advanced encryption standard by the NIST [6] with a 128-bit block size and a variable key length, which can be set to 128, 192 or 256 bits. In its full version, the block lengths b and the key lengths Nk can range from 128 up to 256 bits in steps of 32 bits, as detailed in [4] and in [9]. There are 25 instances of Rijndael. The number of rounds Nr depends on the text size b and on the key size Nk and varies between 10 and 14 (see Table 1 for partial details). For all the versions, the current block at the input of the round r is represented by a $4 \times t$ with $t = (b/32)$ matrix of bytes $A^{(r)}$:

---

* Part of this work done while the author was with the Laboratoire de sécurité et de cryptographie (LASEC), EPFL, Switzerland.

B. Preneel (Ed.): AFRICACRYPT 2009, LNCS 5580, pp. 60–76, 2009.

$$A^{(r)} = \begin{pmatrix} a_{0,0}^{(r)} & a_{0,1}^{(r)} & \cdots & a_{0,t}^{(r)} \\ a_{1,0}^{(r)} & a_{1,1}^{(r)} & \cdots & a_{1,t}^{(r)} \\ a_{2,0}^{(r)} & a_{2,1}^{(r)} & \cdots & a_{2,t}^{(r)} \\ a_{3,0}^{(r)} & a_{3,1}^{(r)} & \cdots & a_{3,t}^{(r)} \end{pmatrix}$$

The round function, repeated $Nr - 1$ times, involves four elementary mappings, all linear except the first one:

- SubBytes: a bytewise transformation that applies on each byte of the current block an 8-bit to 8-bit non linear S-box S.
- ShiftRows: a linear mapping that rotates on the left all the rows of the current matrix. the values of the shifts (given in Table 1) depend on b.
- MixColumns: a linear matrix multiplication; each column of the input matrix is multiplied by the matrix M that provides the corresponding column of the output matrix.
- AddRoundKey: an x-or between the current block and the subkey of the round r K_r.

Those $Nr - 1$ rounds are surrounded at the top by an initial key addition with the subkey K_0 and at the bottom by a final transformation composed by a call to the round function where the MixColumns operation is omitted. The key schedule derives $Nr + 1$ b-bits round keys K_0 to K_{Nr} from the master key K of variable length.

Table 1. Parameters of the Rijndael block cipher where the triplet (i, j, k) for the ShiftRows operation designated the required number of byte shifts for the second row, the third one and the fourth one

	AES	Rijndael-160	Rijndael-192	Rijndael-224	Rijndael-256
ShiftRows	(1,2,3)	(1,2,3)	(1,2,3)	(1,2,4)	(1,3,4)
Nb rounds (Nk=128)	10	11	12	13	14
Nb rounds (Nk=192)	12	12	12	13	14
Nb rounds (Nk=256)	14	14	14	14	14

The idea of exploiting distinguishers for cryptanalyzing block ciphers is well known: a key-recovery attack on block ciphers typically exploits a distinguisher [11]: a structural or statistical property exhibited by a block cipher for a randomly chosen secret key K that is not expected to occur for a randomly chosen permutation. Aside from being used subsequently in key-recovery attacks, so far it seems unclear if there are any other undesirable consequences due to distinguishers, although their existence tends to indicate some certificational weakness in ciphers.

Knudsen and Rijmen [12] recently considered block cipher distinguishers when the cipher key is known to the adversary, and suggested another exploitation of the existence of distinguishers: truncated differential distinguishers lead to near collisions in some hash function compression functions built upon block ciphers, e.g. the Matyas-Meyer-Oseas (MMO) mode [15]. Generalizing, we can similarly say for a compression function constructed from a block cipher in any of the Preneel-Govaerts-Vandewalle (PGV) modes [16], that near collisions in the ciphertext of the underlying cipher translate to near collisions in the compression function's output chaining variable. Knudsen and Rijmen posed as an open problem if a security notion exists that can capture the kind of known-key distinguishers that they proposed, and yet which would rule out non-meaningful and contrived distinguishing attacks.

This paper takes a step to answering this question. We define a security notion to express the existence of known-key distinguishers for block ciphers in Section 2. Rather than settle for a notion that is meaningful solely when the key is known to the adversary, our notion intuitively also gives some indication on the cipher's security in the conventional unknown-key setting.

Many cryptanalyses have been proposed against Rijndael-b, the first one against all the versions of Rijndael-b is due to the algorithm designers themselves and is based upon integral properties ([2], [3], [13]) that allows to efficiently distinguish 3 Rijndael inner rounds from a random permutation. This attack has been improved by Ferguson et al. in [5] allowing to cryptanalyse an 8 rounds version of Rijndael-b with a complexity equal to 2^{204} trial encryptions and $2^{128} - 2^{119}$ plaintexts.

Following the dedicated work of [7], this paper presents new four-round integral properties of Rijndael-b and the resulting 7 and 8 rounds known key distinguishers in Section 3.

2   Notions for Cipher Distinguishers

2.1   Definitions

Consider a family of functions $F : \mathcal{K} \times \mathcal{M} \to \mathcal{R}$ where $\mathcal{K} = \{0,1\}^k$ is the set of keys of F, $\mathcal{M} = \{0,1\}^l$ is the domain of F and $\mathcal{R} = \{0,1\}^L$ is the range of F, where k, l and L are the key, input and output lengths in bits. $F_K(\mathcal{M})$ is shorthand for $F(K, \mathcal{M})$. By $K \xleftarrow{\$} \mathcal{K}$, we denote randomly selecting a string K from $\mathcal{K}$. Similar notations apply for a family of permutations $E : \mathcal{K} \times \mathcal{M} \to \mathcal{M}$ where $\mathcal{K} = \{0,1\}^k$ is the set of keys of E and $\mathcal{M} = \{0,1\}^l$ is the domain and the range of E. Let $\mathrm{Func}(\mathcal{M})$ denotes the set of all functions on $\mathcal{M}$, and $\mathrm{Perm}(\mathcal{M})$ denotes the set of all permutations on $\mathcal{M}$. Let $G \xleftarrow{\$} \mathrm{Perm}(\mathcal{M})$ denotes selecting a random permutation.

The usual security notion one requires from a block cipher is to look like a pseudo-random permutation (PRP), for the keys uniformly drawn. This notion could be formalized as follows: a PRP adversary $\mathcal{A}$ gets access to an oracle, which, on input $P \in \mathcal{M}$, either returns $E_K(P)$ for a random key $k \in \mathcal{K}$ or returns $G(P)$ for a random permutation $G \in \mathrm{Perm}(\mathcal{M})$. The goal of $\mathcal{A}$ is to guess the type of oracle it has - by convention, $\mathcal{A}$ returns 1 if it thinks that the oracle is computing $E_K(\cdot)$. The adversary's advantage is defined by:

$$Adv_E^{PRP}(\mathcal{A}) = |\Pr\left[K \xleftarrow{\$} \mathcal{K} : A^{E_K(\cdot)} = 1\right] - \Pr\left[G \xleftarrow{\$} \mathrm{Perm}(\mathcal{M}) : A^{G(\cdot)} = 1\right]|$$

E is said PRP-secure if for any $\mathcal{A}$ attacking E with resources, the advantage $Adv_E^{PRP}(\mathcal{A})$ is negligible (denoted by ε). The above notion does not take into account the decryption access. Hence, the stronger notion of Super Pseudo-Random Permutation (SPRP): as above, the adversary $\mathcal{A}$ gets access to an oracle, but in this case, the adversary not only accesses the permutations G and E_K but also their inverses G^{-1} and E_K^{-1}:

$$Adv_E^{SPRP}(\mathcal{A}) = |\Pr\left[K \xleftarrow{\$} \mathcal{K} : A^{E_K(\cdot),E_K^{-1}(\cdot)} = 1\right]$$
$$-\Pr\left[G \xleftarrow{\$} \mathrm{Perm}(\mathcal{M}) : A^{G(\cdot),G^{-1}(\cdot)} = 1\right]|$$

As done by Luby and Rackoff in [14], formal results using those notions are stated with concrete bounds. However, in the "real life", we could only say that if a distinguisher exists for a given cipher E_K, it is not a PRP or a SPRP (according the distinguisher used). Note (as done in [17]) that a (adversarial) distinguisher is a (possibly computationally unbounded) Turing machine $\mathcal{A}$ which has access to an oracle $\mathcal{O}$; with the aim to distinguish a cipher E_K from the perfect cipher G by querying the oracle with a limited number n of inputs. The oracle $\mathcal{O}$ implements either E_K (for a key randomly chosen) or G. The attacker must finally answer 0 or 1. We measure the ability to distinguish E_K from G by the advantage $Adv_E(\mathcal{A}) = |p - p^*|$ (that must be true for a large part of the key space) where p (resp. $p*$) is the probability of answering 1 when $\mathcal{O}$ implements E_K (resp. G). Note also that three main classes of distinguishers exist: the non-adaptive distinguishers class (where the n plaintext queries are pre-computed), the adaptive distinguishers class (where the plaintext queries depend on the previous ones) and the super pseudo-random distinguishers one (where the queries are chosen according the previous ones and where the oracle also gets access to inverses of E_K and G).

2.2   Notions for Distinguishers

A generic definition of an n-limited non-adaptive distinguisher is given in [10] and described in Alg. 1. One gives an oracle $\mathcal{O}$ to Algorithm 1, which implements either E_K or G with probability $\frac{1}{2}$ each. The core of the distinguisher is the acceptance region $A^{(n)}$: it defines the set of input values $\mathbf{P} = (P_1, \cdots, P_n)$ which lead to output 0 (i.e. it decides that the oracle implements E_K) or 1 (i.e. it decides that the oracle implements G). The goal of the distinguisher is thus to decide whether $\mathcal{O}$ implements E_K or G. If a particular relation R that defines the acceptance region exists linking together inputs and outputs for a sufficient number of values (this efficiently-computable relation outputs 0 if the link exists and 1 otherwise), the advantage of the non-adaptive distinguisher $Adv_E^{NA}(\mathcal{A})$ will be non-negligible. Note also that this distinguisher must work for a large part of the key space.

Algorithm 1. An n-limited generic non-adaptive distinguisher (NA)

Parameters: a complexity n, an acceptance set $A^{(n)}$
Oracle: an oracle $\mathcal{O}$ implementing a permutation c
  Compute some messages $\mathbf{P} = (P_1, \cdots, P_n)$
  Query $\mathbf{C} = (C_1, \cdots, C_n) = c(P_1, \cdots, P_n)$ to $\mathcal{O}$
  if $\mathbf{C} \in A^{(n)}$ **then**
    Output 1
  else
    Output 0
  end if

This distinguisher is generic and includes the following cases: known plaintexts distinguisher and chosen plaintexts distinguisher. In the second case, the n inputs $\mathbf{P} = (P_1, \cdots, P_n)$ must be pre-defined according a particular filter h_1 (independent from the key). By misuse of language, we use the notation $h_1(\mathbf{P})$ to designate a plaintexts set "correctly" chosen, i.e. that verifies the requirements of h_1. The acceptance region $A^{(n)}$ could be seen as the necessary minimal number of outputs in $\mathbf{C}$ that verify a particular relation $R(h_1(\mathbf{P}), \mathbf{C})$. This relation must be independent from the key or have a high probability to happen for a large class of the keys. In this case, if a such relation R that happens with a certain probability exists between the inputs and outputs sets $h_1(\mathbf{P})$ and $\mathbf{C}$; then, the advantage of the distinguisher could be non-negligible.

To illustrate how this notion applies to the existence of distinguishers for block ciphers, consider E_K as 3-round AES. Let $h_1(\mathbf{P}) = \{P_i\}_{i=0}^{255}$ be a set of 2^8 plaintexts that in one byte each has one of 2^8 possible values, and equal in all other bytes (this defines h_1); and $\mathbf{C} = \{C_i\}_{i=0}^{255}$ denote the corresponding ciphertexts, i.e. $\mathbf{C} = E_K(\mathbf{P})$. Define C_i as a concatenation of 16 bytes i.e. $C_i = C_{i,0}||C_{i,1}||\ldots||C_{i,15}$. Define $R(h_1(\mathbf{P}), \mathbf{C})$ as $\bigoplus_{i=0}^{255} C_{i,j}$ for $j = 0 \ldots 15$ for the particular set $\mathbf{C} = E_K(h_1(\mathbf{P}))$ which outputs 1 (accept) if $\bigoplus_{i=0}^{255} C_{i,j} = 0$ for $j = 0 \ldots 15$ knowing that $\mathbf{P} = \{P_i\}_{i=0}^{255}$ and that $\bigoplus_{i=0}^{255} P_{i,j} = 0$; and outputs 0 otherwise. Thus, for the case of E_K, the probability that $R(h_1(\mathbf{P}), \mathbf{C})$ outputs 1 (accept) is 1, while for the case of a random permutation G, the probability is 2^{-l}. Hence $\mathbf{Adv}_{E_K, G}^{NA\text{-}CPA}(\mathcal{A}) = 1 - 2^{-l} >> \varepsilon$ where $NA\text{-}CPA$ means non-adaptive chosen plaintexts. And so, a distinguisher exists for 3-round AES. In fact, this is the well known 3-round integral distinguisher.

As defined in [17], the super pseudo-random distinguisher (described in Alg. 2) could be defined in a deterministic way because no upper bound on the computational capability of the distinguisher are supposed to be (the only limitation is on the number of queries to the oracle).

In answer to the question posed in [12], we now define a natural extension of the above described n-limited distinguishers, to capture the kind of distinguisher

---

Algorithm 2. An n-limited generic adaptive distinguisher with chosen input plaintexts or output ciphertexts

---

Parameters: functions $g_1, \cdots, g_n$, a set $A^{(n)}$
Oracle: an oracle $\mathcal{O}$ implementing permutations c and c^{-1}

　Select a fixed direction and message $(B_1, Z_1^0) = g_1()$ and get $Z_1^1 = c(Z_1^0)$ if $B_1 = 0$ or $Z_1^1 = c^{-1}(Z_1^0)$ otherwise

　Calculate a direction and a message $(B_2, Z_2^0) = g_2(Z_1^1)$ and get $Z_2^1 = c(Z_2^0)$ if $B_2 = 0$ or $Z_2^1 = c^{-1}(Z_2^0)$ otherwise

　...

　Calculate a direction and a message $(B_n, Z_n^0) = g_n(Z_1^1, \cdots, Z_{n-1}^1)$ and get $Z_n^1 = c(Z_n^0)$ if $B_n = 0$ or $Z_n^1 = c^{-1}(Z_n^0)$ otherwise
　if $(Z_1^1, \cdots, Z_n^1) \in A^{(n)}$ **then**
　　Output 1
　else
　　Output 0
　end if

---

that interests us in this paper and in [12]: the non-adaptive chosen middletexts one. This is shown in Alg. 3. The oracle processes the middletexts supplied by the adversary moving in either/both directions towards plaintext and/or ciphertext ends. This notion also intuitively captures the setting of known-key attacks [12] since the oracle has the same kind of power to that of an adversary having knowledge of the key.

Algorithm 3. An n-limited generic non-adaptive chosen middletexts distinguisher (NA-CMA)

Parameters: a complexity n, an acceptance set $A^{(n)}$
Oracle: an oracle $\mathcal{O}$ implementing internal functions f_1 (resp. f_2) of permutation c that process input middletexts to the plaintext (resp. ciphertext) end

　Compute some middletexts $\mathbf{M} = (M_1, \cdots, M_n)$
　Query $\mathbf{P} = (P_1, \cdots, P_n) = (f_1(M_1), \cdots, f_1(M_n))$ and $\mathbf{C} = (C_1, \cdots, C_n) = (f_2(M_1), \cdots, f_2(M_n))$ to $\mathcal{O}$
　if $(\mathbf{P}, \mathbf{C}) \in A^{(n)}$ **then**
　　Output 1
　else
　　Output 0
　end if

　　To see how this notion properly captures the 7-round known-key distinguisher for AES proposed by Knudsen and Rijmen [12], let E_K be 7-round AES, with no MixColumns in the last round. Let $\mathbf{M} = \{M_i\}_{i=0}^{2^{56}-1}$ denote the set of 2^{56} intermediate texts at the output of round 3 of E_K, that differ in seven bytes for $j = \{0, 1, 2, 3, 5, 10, 15\}$ and which have constant values in the remaining nine bytes. Let $\mathbf{P} = \{P_i\}_{i=0}^{2^{56}-1}$ be a set of 2^{56} plaintexts corresponding to the partial decryption[1] of $\mathbf{M}$ by 3 rounds in reverse thus $\mathbf{P}$ is the plaintext subset input to E_K. Note that $\mathbf{P}$ defined in this way by Knudsen and Rijmen is only computable provided the round keys in rounds 1 to 3 are known to the adversary, or alternatively the key K is known, or one assumes the adversary can start in the middle by choosing the middletexts. This is not a problem since it is allowed by our notion. Let $\mathbf{C} = \{C_i\}_{i=0}^{2^{56}-1}$ denote the corresponding ciphertexts, i.e. $\mathbf{C} = E_K(\mathbf{P})$. Define C_i as a concatenation of 16 bytes i.e. $C_i = C_{i,0}||C_{i,1}||\dots||C_{i,15}$. Define $R(\mathbf{P}, \mathbf{C}) = (\bigoplus_{i=0}^{2^{56}-1} P_{i,j}, \bigoplus_{i=0}^{2^{56}-1} C_{i,j})$ for $j = 0\dots15$ and which outputs 1 if $\bigoplus_{i=0}^{2^{56}-1} C_{i,j} = 0$ for $j = 0\dots15$ knowing that $\mathbf{P} = \{P_i\}_{i=0}^{2^{56}-1}$ and that $\bigoplus_{i=0}^{2^{56}-1} P_{i,j} = 0$; and outputs 0 otherwise. Thus, for the case of E_K, the probability that $R(\mathbf{P}, \mathbf{C})$ outputs 1 (accept) is 1, while for the case of a random permutation G, the probability is 2^{-l}.

Hence $\mathbf{Adv}_{E_K,G}^{NA\text{-}CMA}(\mathcal{A}) = 1 - 2^{-l} >> \varepsilon$. And so, a chosen middletext (a.k.a. known-key [12]) distinguisher exists for 7-round AES.

In the same way, the notion captures the 7-round distinguisher of [12] for a particular kind of Feistel cipher whose round function has the round key exclusive-ORed to the round function input, followed by an arbitrary key-independent transformation.

With this notion, we can also intuitively define security against the existence of distinguishers in the conventional setting where the key is unknown, which can be seen as a special case of $NA\text{-}CMA$.

Note here that it is apparent in the unknown-key setting that f_1 and f_2 are public functions, since it can in no way be dependent on the key, otherwise, the relation $R(\cdot,\cdot)$ becomes impossible to compute and thus verify. We defer the more detailed discussion to subsection 2.3.

2.3   Discussion

Observe that for the conventional setting where the adversary has no knowledge of the key K, it is intuitive that the distinguishing relation R operates on publicly computable functions f_1 and f_2 of the ciphertext set, otherwise if f_1 or f_2 is dependent on K, the relation cannot be verified since K is assumed to be unknown and must be uniformly distributed. Thus, the resultant notion becomes more meaningful and rules out trivial attacks where the adversary obtains a non-negligible advantage but for which is not meaningful.

Consider if the functions f_1 and f_2 in the $NA\text{-}CMA$ notion can be dependent on K, and indeed why not since K is known to A. Then take E_K to be infinitely many rounds of the AES, i.e. the number of rounds $r >> 7$. Then an adversary could still use the 7-round known-key distinguisher described in [12] as follows: peel off any number of rounds since it knows the cipher key and thus round keys to any round, and going backwards in reverse from the ciphertext end it peels off round by round until the output of round 7, and checks that the 7-round distinguisher covering the first 7 rounds is satisfied. Thus his advantage

---

[1] Note that this partial decryption is computable by the adversary since it knows the block cipher key K.

$\mathbf{Adv}_{E_K,G}^{NA\text{-}CMA}(\mathcal{A})$ is trivially non-negligible, although it is clear that we gain no insight into the ciphers security nor insecurity.

Thus, considering known-key distinguishers is a stronger notion than the conventional notion of unknown-key distinguishers, and so the inexistence of known-key distinguishers implies the inexistence of unknown-key distinguishers. Furthermore, it is hoped that this context might tell something about the security margin of a cipher, i.e. if an unknown-key distinguisher exists for the cipher up to s rounds, what is the most number of rounds t of the cipher for which is covered by a distinguisher if the key is known to the adversary. For instance, a distinguisher exists for the AES in the unknown-key setting up to 4 rounds [8], while Knudsen and Rijmen showed that a distinguisher exists for AES in the known-key context up to 7 rounds. This still allows some security margin considering AES has at least 10 rounds. From this perspective, (in)existence of known-key distinguishers can be viewed on the one hand as a certificational strength/weakness of a cipher; and on the other hand one still desires to gain some insight on the (in)security of a cipher from these known-key distinguishers. One obvious aim is what can be learned about unknown-key distinguishers from these known-key distinguishers.

Moreover, the only difference between the $NA\text{-}CPA/NA\text{-}CCA$ and $NA\text{-}CMA$ settings is in the extra control afforded to the latter adversary who effectively can choose his plaintext subset based on knowledge of the key what is really appreciable in the context of a meet in the middle attack. We can hence say that existence of known-key distinguishers exhibits some potentially undesirable structural property of the cipher, but which cannot be exploited in unknown-key distinguishers for key-recovery attacks only in the fact that the adversary is expected to not be able to choose the plaintext subset corresponding to the known-key distinguisher since he does not know the key. For the case of the AES, the adversary cannot turn the 7-round known-key distinguisher of Knudsen and Rijmen into an unknown-key distinguisher because he does not know the rounds keys for rounds 1 to 3.

Interestingly, this in some way relates to the motivation behind the design of cipher key schedules that have known-roundkey security, i.e. knowledge of one or more round keys does not lead to the knowledge of other round keys. In this context, known-key distinguishers can potentially be exploited in actual key-recovery attacks. Taking the 7-round AES as an example but where its key schedule has known-roundkey security: if the adversary already knows the round keys for rounds 1 to 3, and by design he still cannot obtain the other round keys, nevertheless due to the existence of the 7-round known-key distinguisher, he can use his knowledge of round keys 1 to 3 to choose a plaintext subset that corresponds to the distinguisher, and with this distinguisher he can cover all the rounds through to round 7. This can be turned into a key-recovery attack on the round keys for rounds 4 to 7.

Relation to correlation intractability. In motivating the need for a notion for known-key distinguishers, Knudsen and Rijmen discuss the related work of Canetti et al. [1] that considered the notion of *correlation intractability*, that when applied to the context of block ciphers where the key is known to the adversary, can be seen as follows: a correlation intractable cipher is one where there exists no binary relation R between the plaintext and ciphertext that is satisfied with non-negligible probability. Clearly, if the key is known, a relation can always be found and thus a cipher cannot be correlation intractable. Yet, this trivial case is ruled out from our NA-CMA notion because of the restriction we put on R to be independent of the key. Indeed, Canetti et al.'s impossibility example cannot apply in our notion, since they directly input the key to the relation, while this is not allowed for our relation.

3   Known-Key Distinguishers for the Rijndael-b Block Cipher with Large Blocks

We present in this Section known key distinguishers against the Rijndael-b block cipher. Using particular new and old integral properties, the building distinguishers use really few middle-texts and have very low complexity.

In [13], L. Knudsen and D. Wagner analyze integral cryptanalysis as a dual to differential attacks particularly applicable to block ciphers with bijective components. A first-order integral cryptanalysis considers a particular collection of m words in the plaintexts and ciphertexts that differ on a particular component. The aim of this attack is thus to predict the values in the sums (i.e. the integral) of the chosen words after a certain number of rounds of encryption. The same authors also generalize this approach to higher-order integrals: the original set to consider becomes a set of m^d vectors which differ in d components and where the sum of this set is predictable after a certain number of rounds. The sum of this set is called a dth-order integral.

We first introduce and extend the consistent notations proposed in [13] for expressing word-oriented integral attacks. For a first order integral, we have:

- The symbol '$\mathcal{C}$' (for "Constant") in the ith entry, means that the values of all the ith words in the collection of texts are equal.
- The symbol '$\mathcal{A}$' (for "All") means that all words in the collection of texts are different.
- The symbol '?' means that the sum of words can not be predicted.

For a dth order integral cryptanalysis:

- The symbol '$\mathcal{A}^d$' corresponds with the components that participate in a dth-order integral, i.e. if a word can take m different values then $\mathcal{A}^d$ means that in the integral, the particular word takes all values exactly m^{d-1} times.
- The term 'A_i^d' means that in the integral the string concatenation of all words with subscript i take the m^d values exactly once.
- The symbol '$(\mathcal{A}_i^d)^k$' means that in the integral the string concatenation of all words with subscript i take the m^d values exactly k times.

3.1   Known Key Distinguisher for the AES

As mentioned in [12], we could build a 4-th order 4-round AES integral distinguisher considering that the last round does not contain a MixColumns operation. Then, and as shown in Fig. 1.a, all bytes of the ciphertexts are balanced in the 4 rounds integral. Moreover, a backward 3-round property could be built for three complete rounds as shown in Fig. 1.b.

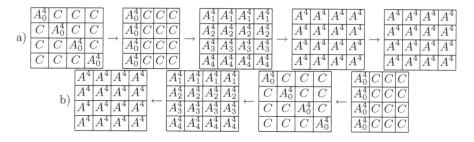

Fig. 1. a) The forward 4-round integral distinguisher with 2^{32} texts. b) A backward integral for three (full) rounds of AES with 2^{32} texts.

By applying the inside-out concatenation technique with the two previous properties, the authors of [12] could build a 7-round known key distinguisher (as shown on Fig. 2) against the AES. One chooses a structure of 2^{56} middle-texts: it has 7 active bytes whereas the other bytes are constant. We thus have 2^{24} sets of 2^{32} middletexts that represent first 2^{24} copies of the 4-round property (of Fig. 1.a) and also 2^{24} copies of the backward 3-round property (of Fig. 1.b). Then, when someone starts in the middle of the cipher, one can compute integral balanced property on both the reverse and forward directions.

$$
\begin{array}{|c|c|c|c|}
\hline A^7 & A^7 & A^7 & A^7 \\
\hline A^7 & A^7 & A^7 & A^7 \\
\hline A^7 & A^7 & A^7 & A^7 \\
\hline A^7 & A^7 & A^7 & A^7 \\
\hline
\end{array}
\xleftarrow{\text{3-round}}
\begin{array}{|c|c|c|c|}
\hline A_0^7 & C & C & C \\
\hline A_0^7 & A_0^7 & C & C \\
\hline A_0^7 & C & A_0^7 & C \\
\hline A_0^7 & C & C & A_0^7 \\
\hline
\end{array}
\xrightarrow{\text{4-round}}
\begin{array}{|c|c|c|c|}
\hline A^7 & A^7 & A^7 & A^7 \\
\hline A^7 & A^7 & A^7 & A^7 \\
\hline A^7 & A^7 & A^7 & A^7 \\
\hline A^7 & A^7 & A^7 & A^7 \\
\hline
\end{array}
$$

Fig. 2. The 7-round AES distinguisher with 2^{56} middle-texts. The 7th round is without MixColumns.

This known-key distinguisher simply records the frequencies in each byte of the plaintexts and ciphertexts, checks whether the values in each byte of the plaintexts and in each byte of the ciphertexts occur equally often. The time complexity is similar to the time it takes to do 2^{56} 7-round AES encryptions and the memory needed is small.

The authors of [12] introduce the k-sum problem (i.e. to find a collection of k texts $x_1, \cdots, x_k$ such as $\sum_{i=1}^{k} f(x_i) = 0$ for a given permutation f) with a running time of $\mathcal{O}(k2^{n/(1+\log_2 k)})$ to conjecture that for a randomly chosen 128-bit permutation such a collection of texts with balanced properties could not be easily (i.e. in a reasonable computational time) found with a large probability. More precisely, the k-sum problem indicates with $n = 128$ and $k = 256$ a running time of 2^{58} operations ignoring small constants and memory. The k-sum problem is the best known approach in this case but does not give the particular balanced properties induced by the distinguisher. Thus, the authors conjecture that they have found a known-key distinguisher for AES reduced to 7 rounds using 2^{56} texts.

3.2 Rijndael-256

Thus, we have studied the same kind of properties for Rijndael-b. For the particular case of Rijndael-256, we have the two following forward and backward integral properties described in Fig. 3 and in Fig. 5. Note that the integral property of Fig. 3 could be extended by one round at the beginning using the method described in Fig. 4. This property is the one described in [7].

First grid:

A_0^3	C	C	C	C	C	C	C
A_0^3	C	C	C	C	C	C	C
A_0^3	C	C	C	C	C	C	C
C	C	C	C	C	C	C	C

$\rightarrow$

A_0^3	C	C	C	C	A_0^3	C	A_0^3
A_1^3	C	C	C	C	A_1^3	C	A_1^3
A_2^3	C	C	C	C	A_2^3	C	A_2^3
A_3^3	C	C	C	C	A_3^3	C	A_3^3

$\rightarrow$

A_0^2	A_0^2	A_0^2	A_4^2	A_4^2	A_0^3	A_0^3	A_0^3
A_1^2	A_1^2	A_1^2	A_5^2	A_5^2	A_1^3	A_1^3	A_1^3
A_2^2	A_2^2	A_2^2	A_6^2	A_6^2	A_2^3	A_2^3	A_2^3
A_3^2	A_3^2	A_3^2	A_7^2	A_7^2	A_3^3	A_3^3	A_3^3

$\rightarrow$

$?$	A^3	A^3	A^3	$?$	A^3	A^3	A^3
$?$	A^3	A^3	A^3	$?$	A^3	A^3	A^3
$?$	A^3	A^3	A^3	$?$	A^3	A^3	A^3
$?$	A^3	A^3	A^3	$?$	A^3	A^3	A^3

$\rightarrow$

$?$	A^3	A^3	A^3	$?$	A^3	A^3	A^3
A^3	A^3	A^3	$?$	A^3	A^3	A^3	$?$
A^3	$?$	A^3	A^3	A^3	$?$	A^3	A^3
$?$	A^3	A^3	A^3	$?$	A^3	A^3	A^3

Fig. 3. 4-round 3th-order forward integral property of Rijndael-256 without the last MixColumns operation

A_0^4	C	C	C	C	C	C	C
C	A_0^4	C	C	C	C	C	C
C	C	A_0^4	C	C	C	C	C
C	C	C	A_0^4	C	C	C	C

$\rightarrow$

A_0^4	C	C	C	C	C	C	C
A_0^4	C	C	C	C	C	C	C
A_0^4	C	C	C	C	C	C	C
A_0^4	C	C	C	C	C	C	C

Fig. 4. Extension of a 4th order integral property by one round at the beginning for Rijndael-256

Using those two properties and the corresponding extension, we could build a 8-round known key distinguisher shown in Fig. 6. The process is exactly the same than the one described in 3.1 and the time complexity is similar to the time it takes to do 2^{40} 8-round Rijndael-256 encryptions and the memory needed is small. If we also use the k-sum problem to estimate the corresponding time to find a k-sum for a 256-bit permutation with $n = 256 - 64$ and $k = 2^{40}$, the

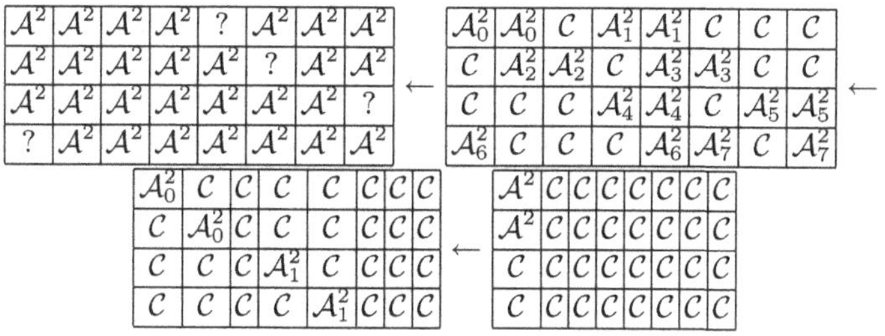

Fig. 5. The 2th-order backward 3-round integral property of Rijndael-256

corresponding complexity is around 2^{44} operations ignoring small constants and memory. Thus, we conjecture that we have found a known-key distinguisher for Rijndael-256 reduced to 8 rounds using 2^{40} middle-texts.

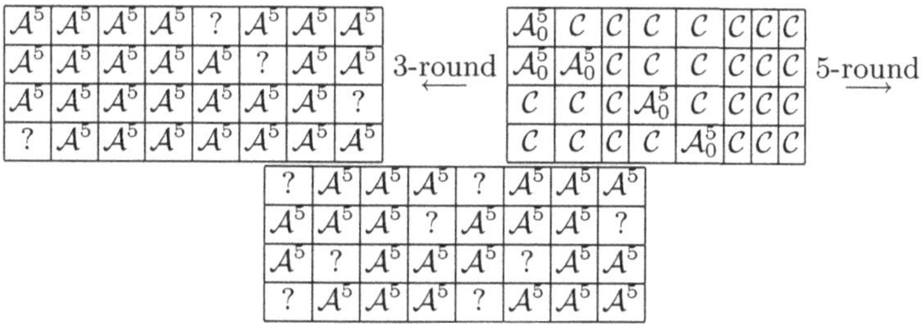

Fig. 6. The 8-round Rijndael-256 known-key distinguisher with 2^{40} middle-texts. The 8th round is without MixColumns.

3.3 Rijndael-224

Similarly, we found a 2th-order 4-round forward integral property for Rijndael-224 as shown in figure 7. We have found 42 2th-order integral properties (essentially the shifted ones). As previously done, this 2th-order four-round property could be extended by one round at the beginning using a 8th-order integral (considering that it represents 2^{48} copies of the 2th-order four-round integral). We also have found the backward 3-round 2-th order integral property for Rijndael-224 shown in Fig. 8.

Using those two properties and the corresponding extension, we could build a 8-round known key distinguisher shown in Fig. 9. The process is exactly the same than the one described in 3.1 and the time complexity is similar to the time it takes to do 2^{72} 8-round Rijndael-224 encryptions and the memory needed is small. If we also use the k-sum problem to estimate the corresponding time to find a k-sum for a 224-bit permutation with $n = 224 - 128$ and $k = 2^{72}$, the corresponding complexity is around $2^{73.8}$ operations ignoring small constants and memory. Thus, we conjecture that we have found a known-key distinguisher for Rijndael-224 reduced to 8 rounds using 2^{72} middle-texts.

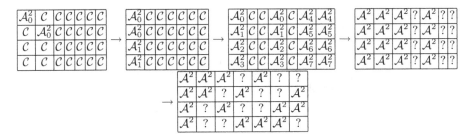

Fig. 7. Four-round 2th-order forward integral property of Rijndael-224. The last Mix-Columns is omitted.

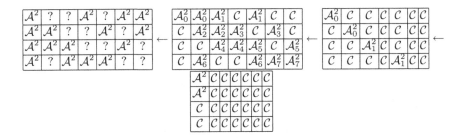

Fig. 8. The 2th-order 3-round backward integral property of Rijndael-224

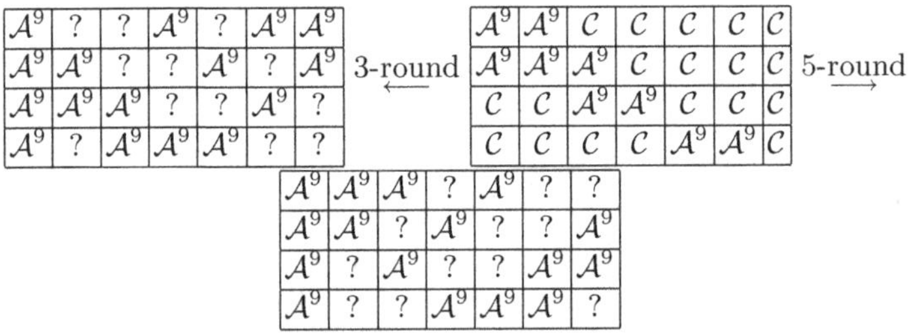

Fig. 9. The 8-round Rijndael-224 known-key distinguisher with 2^{72} middle-texts. The 8th round is without MixColumns.

3.4   Rijndael-192

In the same way, we have found a 3th-order forward 4-round integral property for Rijndael-192 as shown in Fig. 10. We have found 42 3th-order integral properties (essentially the shifted ones). We also have found the backward 2-th order integral property for Rijndael-224 shown in Fig. 11.

Using those two properties and the corresponding extension, we could build a 7-round known key distinguisher shown in Fig. 12. The process is exactly the same than the one described in 3.1 and the time complexity is similar to the time it takes to do 2^{32} 7-round Rijndael-192 encryptions and the memory needed is

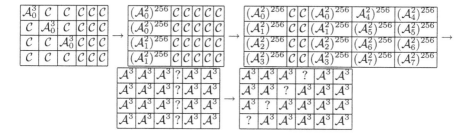

Fig. 10. The 3th-order 4-round forward integral property of Rijndael-192 (the last MixColumns is omitted)

A^2	A^2	?	?	?	A^2
A^2	A^2	A^2	?	?	?
?	A^2	A^2	A^2	?	?
?	?	A^2	A^2	A^2	?

$\leftarrow$

A_0^2	A_0^2	A_1^2	A_1^2	C	C
C	A_2^2	A_2^2	A_3^2	A_3^2	C
C	C	A_4^2	A_4^2	A_5^2	A_5^2
A_6^2	C	C	A_6^2	A_7^2	A_7^2

$\leftarrow$

A_0^2	C	C	C	C	C
C	A_0^2	C	C	C	C
C	C	A_1^2	C	C	C
C	C	C	A_1^2	C	C

$\leftarrow$

A^2	C	C	C	C	C
A^2	C	C	C	C	C
C	C	C	C	C	C
C	C	C	C	C	C

Fig. 11. 2th-order 3-round backward integral property of Rijndael-192

A^4	A^4	?	?	?	A^4
A^4	A^4	A^4	?	?	?
?	A^4	A^4	A^4	?	?
?	?	A^4	A^4	A^4	?

$\xleftarrow{\text{3-round}}$

A^4	C	C	C	C	C
A^4	A^4	C	C	C	C
C	C	A^4	C	C	C
C	C	C	C	C	C

$\xrightarrow{\text{4-round}}$

A^4	A^4	A^4	?	A^4	A^4
A^4	A^4	?	A^4	A^4	A^4
A^4	?	A^4	A^4	A^4	A^4
?	A^4	A^4	A^4	A^4	A^4

Fig. 12. The 7-round Rijndael-192 distinguisher with 2^{32} middle-texts. The 7th round is without MixColumns.

small. If we also use the k-sum problem to estimate the corresponding time to find a k-sum for a 192-bit permutation with $n = 192 - 96$ and $k = 2^{32}$, the corresponding complexity is around $2^{35.9}$ operations ignoring small constants and memory. Thus, we conjecture that we have found a known-key distinguisher for Rijndael-192 reduced to 7 rounds using 2^{32} middle-texts.

3.5 Rijndael-160

We also have found a 3th-order 4-round integral property for Rijndael-160 as shown in Fig. 13. We have found 42 3th-order integral properties (essentially the shifted ones). We also have found the backward 3-th order backward integral property for Rijndael-160 shown in Fig. 14.

Using those two backward and forward properties, we could build a 7-round known key distinguisher shown in Fig. 15. The process is exactly the same than the one described in 3.1 and the time complexity is similar to the time it takes to do 2^{40} 7-round Rijndael-160 encryptions and the memory needed is small. If we also use the k-sum problem to estimate the corresponding time to find a k-sum

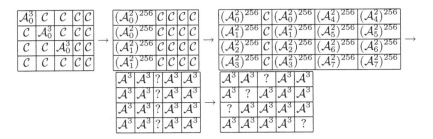

Fig. 13. 3th order 4-round forward integral property of Rijndael-160 (the last round is without MixColumns)

Fig. 14. 3th order integral backward property for Rijndael-160

Fig. 15. The 7-round Rijndael-160 distinguisher with 2^{40} middle-texts. The 7th round is without MixColumns.

Table 2. Summary of known-key distinguishers on Rijndael-b. CM means Chosen Middle-texts.

Cipher	nb rounds	Key sizes	Data	Time Complexity	Memory	Source
AES	7	(all)	2^{56} CM	2^{56}	small	[12]
Rijndael-256	8	(all)	2^{40} CM	2^{40}	small	this paper
Rijndael-224	8	(all)	2^{72} CM	2^{72}	small	this paper
Rijndael-192	7	(all)	2^{32} CM	2^{32}	small	this paper
Rijndael-160	7	(all)	2^{40} CM	2^{40}	small	this paper

for a 160-bit permutation with $n = 160 - 32$ and $k = 2^{40}$, the corresponding complexity is around 2^{43} operations ignoring small constants and memory. Thus, we conjecture that we have found a known-key distinguisher for Rijndael-192 reduced to 7 rounds using 2^{40} middle-texts.

4   Conclusion

In this paper, we have shown how to build known-key distinguisher against the various versions of Rijndael-b using essentially particular new dth-order integral properties in forward and backward sense. Table 2 sums up the results presented in this paper.

Note that we have used as done in [12] the k-sum problem to estimate the corresponding complexity to build such distinguishers for random permutations. This model is less pertinent in our case because we need to estimate such a problem for random functions and no more for random permutations. Thus, we however think that the corresponding complexity is around to be the same even if this stays as an open problem.

References

1. Canetti, R., Goldreich, O., Halevi, S.: On the random oracle methodology, revisited. Journal of the ACM 51(4), 557–594 (2004)
2. Daemen, J., Knudsen, L.R., Rijmen, V.: The block cipher Square. In: Biham, E. (ed.) FSE 1997. LNCS, vol. 1267, pp. 149–165. Springer, Heidelberg (1997)
3. Daemen, J., Rijmen, V.: AES proposal: Rijndael. In: The First Advanced Encryption Standard Candidate Conference. N.I.S.T. (1998)
4. Daemen, J., Rijmen, V.: The Design of Rijndael. Springer, Heidelberg (2002)
5. Ferguson, N., Kelsey, J., Lucks, S., Schneier, B., Stay, M., Wagner, D., Whiting, D.: Improved cryptanalysis of rijndael. In: Schneier, B. (ed.) FSE 2000. LNCS, vol. 1978, pp. 213–230. Springer, Heidelberg (2001)
6. FIPS 197. Advanced Encryption Standard. Federal Information Processing Standards Publication 197, U.S. Department of Commerce/N.I.S.T (2001)
7. Galice, S., Minier, M.: Improving integral attacks against Rijndael-256 up to 9 rounds. In: Vaudenay, S. (ed.) AFRICACRYPT 2008. LNCS, vol. 5023, pp. 1–15. Springer, Heidelberg (2008)
8. Gilbert, H., Minier, M.: A collision attack on 7 rounds of Rijndael. In: AES Candidate Conference, pp. 230–241 (2000)
9. Nakahara Jr., J., de Freitas, D.S., Phan, R.C.-W.: New multiset attacks on Rijndael with large blocks. In: Dawson, E., Vaudenay, S. (eds.) Mycrypt 2005. LNCS, vol. 3715, pp. 277–295. Springer, Heidelberg (2005)
10. Junod, P.: On the optimality of linear, differential, and sequential distinguishers. In: Biham, E. (ed.) EUROCRYPT 2003. LNCS, vol. 2656, pp. 17–32. Springer, Heidelberg (2003)
11. Knudsen, L.R.: Contemporary block ciphers. In: Damgård, I. (ed.) Lectures on Data Security. LNCS, vol. 1561, pp. 105–126. Springer, Heidelberg (1999)
12. Knudsen, L.R., Rijmen, V.: Known-key distinguishers for some block ciphers. In: Kurosawa, K. (ed.) ASIACRYPT 2007. LNCS, vol. 4833, pp. 315–324. Springer, Heidelberg (2007)

13. Knudsen, L.R., Wagner, D.: Integral cryptanalysis. In: Daemen, J., Rijmen, V. (eds.) FSE 2002. LNCS, vol. 2365, pp. 112–127. Springer, Heidelberg (2002)
14. Luby, M., Rackoff, C.: How to construct pseudorandom permutations from pseudorandom functions. SIAM Journal on Computing 17(2), 373–386 (1988)
15. Matyas, S.M., Meyer, C.H., Oseas, J.: Generating strong one-way functions with cryptographic algorithm. IBM Technical Disclosure Buletin 27, 5658–5659 (1985)
16. Preneel, B., Govaerts, R., Vandewalle, J.: Hash functions based on block ciphers: A synthetic approach. In: Stinson, D.R. (ed.) CRYPTO 1993. LNCS, vol. 773, pp. 368–378. Springer, Heidelberg (1994)
17. Vaudenay, S.: Decorrelation: A theory for block cipher security. J. Cryptology 16(4), 249–286 (2003)

Reducing Key Length of the McEliece Cryptosystem

Thierry P. Berger[1], Pierre-Louis Cayrel[2], Philippe Gaborit[1],
and Ayoub Otmani[3]

[1] Université de Limoges, XLIM-DMI,
123, Av. Albert Thomas 87060 Limoges Cedex, France
{thierry.berger,philippe.gaborit}@xlim.fr
[2] Université Paris 8, Département de mathématiques, 2,
rue de la Liberté, 93526 - SAINT-DENIS cedex 02, France
cayrelpierrelouis@gmail.com
[3] GREYC – Ensicaen – Université de Caen,
Boulevard Maréchal Juin, 14050 Caen Cedex, France
Ayoub.Otmani@info.unicaen.fr

Abstract. The McEliece cryptosystem is one of the oldest public-key cryptosystems ever designed. It is also the first public-key cryptosystem based on linear error-correcting codes. Its main advantage is to have very fast encryption and decryption functions. However it suffers from a major drawback. It requires a very large public key which makes it very difficult to use in many practical situations. A possible solution is to advantageously use quasi-cyclic codes because of their compact representation. On the other hand, for a fixed level of security, the use of optimal codes like Maximum Distance Separable ones allows to use smaller codes. The almost only known family of MDS codes with an efficient decoding algorithm is the class of Generalized Reed-Solomon (GRS) codes. However, it is well-known that GRS codes and quasi-cyclic codes do not represent secure solutions. In this paper we propose a new general method to reduce the public key size by constructing quasi-cyclic Alternant codes over a relatively small field like $\mathbb{F}_{2^8}$. We introduce a new method of hiding the structure of a quasi-cyclic GRS code. The idea is to start from a Reed-Solomon code in quasi-cyclic form defined over a large field. We then apply three transformations that preserve the quasi-cyclic feature. First, we randomly block shorten the RS code. Next, we transform it to get a Generalised Reed Solomon, and lastly we take the subfield subcode over a smaller field. We show that all existing structural attacks are infeasible. We also introduce a new NP-complete decision problem called quasi-cyclic syndrome decoding. This result suggests that decoding attack against our variant has little chance to be better than the general one against the classical McEliece cryptosystem. We propose a system with several sizes of parameters from 6,800 to 20,000 bits with a security ranging from 2^{80} to 2^{120}.

Keywords: public-key cryptography, McEliece cryptosystem, Alternant code, quasi-cyclic.

B. Preneel (Ed.): AFRICACRYPT 2009, LNCS 5580, pp. 77–97, 2009.
© Springer-Verlag Berlin Heidelberg 2009

1   Introduction

The McEliece cryptosystem [21] represents one of the oldest public-key cryptosystems ever designed. It is also the first public-key cryptosystem based on linear error-correcting codes. The principle is to select a linear code of length n that is able to efficiently correct t errors. The core idea is to transform it to a random-looking linear code. A description of the original code and the transformations can serve as the private key while a description of the modified code serves as the public key. McEliece's original proposal uses a generator matrix of a binary Goppa code. The encryption function encodes a message according to the public code and adds an error vector of weight t. The decryption function basically decodes the ciphertext by recovering the secret code through the trapdoor that consists of the transformations. Niederreiter [23] also proposed a public-key cryptosystem based on linear codes in which the public key is a parity-check matrix. It is proved in [19] that these two systems are equivalent in terms of security. Any public-key cryptosystem should be made resistant to attacks that seek (from public data) to either recover secret data, or to decrypt an arbitrary ciphertext. Any cryptosystem that fulfils this property is to said to be *one-way secure under a chosen plaintext attack* (OW-CPA). The first category of attacks, which are also called *structural attacks* in code-based cryptography, aims at recovering the secret code or alternatively, constructing an equivalent code that can be efficiently decoded. The other class tries to design decoding algorithms for arbitrary linear codes in order to decrypt a given cipher text. Such an attack is called a *decoding attack*. Currently, the most efficient algorithms used to decode arbitrary linear codes are based on the information set decoding. A first analysis was done by McEliece in [21] then in [17,18,29,6,7,8] and lastly in [3] which is the best refinement up to now. All these algorithms solve the famous search problem of decoding random linear code. It was proved in [2] that decoding an arbitrary linear code is NP-Hard problem. However the security of the McEliece cryptosystem is not equivalent to the general problem of decoding a random linear code due to the following reasons: 1) inverting the McEliece encryption function is a special case of the general problem of decoding where the error weight t is set to a certain value and (2) binary Goppa codes form a subclass of linear codes. Therefore, McEliece cryptosystem is secure as long as there is no efficient algorithm that distinguishes between binary Goppa codes and random binary codes. Nobody has managed to solve this challenging problem for the last thirty years and if ever a solution appears towards that direction, this would toll the knell of the original McEliece cryptosystem. Note that this assumption is not always true for any class of codes that has an efficient decoding algorithm. For instance, Sidel'nikov and Shestakov proved in [28] that the structure of Generalised Reed-Solomon codes of length n can be recovered in $O\left(n^4\right)$. Sendrier proved in [25] that the (permutation) transformation can be extracted for concatenated codes. Minder and Shokrollahi presented in [22] a structural attack that creates a private key against a cryptosystem based on Reed-Muller codes [27].

Despite these attacks on these variants of the McEliece cryptosystem, the original scheme still remains unbroken. The other main advantage of code-based cryptosystems is twofold:

1. high-speed encryption and decryption compared with other public-key cryptosystems which involve for instance modular exponentiations (even faster than the NTRU cryptosystem [15]),
2. resistance to a putative quantum computer.

Unfortunately its major weakness is a huge public key of several hundred thousand bits in general. Currently, the McEliece public key cryptosystem satisfies the OW-CPA criteria for $n \geq 2048$ with appropriate values for t and k such that $W_{2,n,k,t} \geq 2^{100}$ where $W_{q,n,k,t}$ is the work factor of the best algorithm that decodes t errors in a linear code of length n, dimension k over the finite field $\mathbb{F}_q$. For example (See [3]) the work factor is around 2^{128} if we choose $(n, k, t) = (2960, 2288, 56)$. For such parameters, the public key size is about 6.5 Mbits. It is therefore tempting to enhance the McEliece scheme by finding a way to reduce the representation of a linear code as well as the matrices of the transformations.

A possible solution is to take very sparse matrices. This idea has been applied in [5] which examined the implications of using low density parity-check (LDPC) codes. The authors showed that taking sparse matrices for the linear transformations is an unsafe solution. Another recent trend first appeared in [12] tries to use quasi-cyclic codes [12,1,14,13,10]. This particular family of codes offers the advantage of having a very simple and compact description. The first proposal [12] uses subcodes of a primitive BCH cyclic code. The size of the public key for this cryptosystem is only 12Kbits. The other one [1] tries to combine these two positive aspects by requiring quasi-cyclic LDPC codes. The authors propose a public key size that is about 48Kbits. A recent work shows [24] that these two cryptosystems [12,1] can be broken. The main drawbacks of [12] are: (1) the permutation that is supposed to hide the secret generator matrix is very constrained, (2) the use of sub-codes of a *completely known* BCH code. Combining these two flaws lead to a structural attack that recovers the secret permutation by basically solving an over-constrained linear system. The unique solution reveals the secret key.

This present work generalizes and strengthens the point of view developed in [12]. It uses a quasi-cyclic code over a relatively small field (like $\mathbb{F}_{2^8}$ or $\mathbb{F}_{2^{10}}$) to reduce the size of the public keys. In particular it develops new ideas which permit to overcome the weaknesses of [12]. Our proposal is based on a general construction that starts from a family of Maximum Distance Separable (MDS) quasi-cyclic codes defined over a very large field (like $\mathbb{F}_{2^{16}}$ or $\mathbb{F}_{2^{20}}$). An excellent candidate is the family of Generalised Reed-Solomon codes. It represents an important class of MDS cyclic codes equipped with an efficient decoding algorithm. The two main threats that may undermine the security of our variant are the attack of [28], which exploits the structure of generalised Reed-Solomon codes, and the attack of [24] which makes profit of the quasi-cyclic structure. Our first improvement over [12] consists in taking subfield subcodes of Generalised Reed-Solomon codes rather than considering subcodes of a BCH code. Subfield subcodes of Generalised Reed-Solomon codes are also called *Alternant* codes. This approach respects in a sense the McEliece's proposal since binary Goppa codes are a special case of Alternant codes. The first positive effect for using quasi-cyclic Alternant codes is the high number of codes that share the same parameters. The second positive effect is to be immune to the structural attack of Sidelnikov-Shestakov [28] which strictly requires Generalised Reed-Solomon codes to successfully operate. Consequently, the use of Alternant codes permits to break the Reed-Solomon structure and avoids the classical attack of [28]. The second improvement consists in resisting to the attack of [24] by randomly shortening a very long quasi-cyclic Alternant codes. For instance, one constructs codes of length of order 800 from codes of length 2^{16} or 2^{20}. Note that the idea of randomly shortening a code is a new way of hiding the structure of code which exploits the recent result of [31] in which it is proved that deciding whether a code is permutation equivalent to a shortened code is NP-complete. Hence the introduction of the random shortening operation makes harder the recovery of the code structure in two complementary ways. First, it worsens the chance of the Sidelnikov-Shestakov attack to be successful because the original Generalised Reed-Solomon code is even more "degraded". Second, the use a random shortened code permits to avoid the attack of [24] that exploits the quasi-cyclic structure and requires that the public code to be permutation equivalent to a subcode of a *known* code. The fact of considering a random shorter code makes it inapplicable because the shortened code to which the public code is equivalent is unknown.

The main achievement of this paper is to derive a construction which drastically reduces the size of the public key of the McEliece and Niederreiter cryptosystem to about thousands bits. The security of our new variant relies upon two assumptions. First, it is impossible to recover the secret shortened quasi-cyclic Alternant code. We prove that all the existing structural attacks require a prohibitive amount of computation. Assuming that is computationally impossible to structurally recover the secret code, the one-wayness under chosen plaintext attack is guaranteed by the hardness of decoding an *arbitrary quasi-cyclic* linear code. We prove in this paper that the associated decision problem is NP-complete as it is the case for arbitrary linear codes. This important result makes it reasonable to assume that there is no efficient algorithm that decodes any arbitrary quasi-cyclic code. This important assumption establishes the security of our variant.

The paper is organized as follows. In Section 2 we give useful notions of coding theory. We refer the reader to [20] for a detailed treatment of coding theory. In Section 3 we recall basic facts about code-based cryptography (See [11] for more details). Section 4 deals with a description of our new variant. In Sections 5 we provide different parameters for our scheme. In Section 6 we analyze the security of our scheme. Section 7 covers the performance of our proposal.

2   Coding Theory Background

2.1   Generalised Reed-Solomon Codes

Definition 1. *Let m be a positive integer and let q be a power of a prime number. Let α be a primitive element of the finite field $\mathbb{F}_{q^m}$ and assume that $n = q^m - 1$. Let d and k be integers such that where $d = n - k + 1$.*

A code $\mathscr{R}_{n,k}$ of length n and dimension k over $\mathbb{F}_{q^m}$ is a Reed-Solomon code if, up to a column permutation, it is defined by the parity-check matrix

$$H = \begin{bmatrix} 1 & \alpha & \alpha^2 & \cdots & \alpha^{n-1} \\ 1 & \alpha^2 & (\alpha^2)^2 & \cdots & (\alpha^2)^{n-1} \\ \vdots & \vdots & \vdots & & \vdots \\ 1 & \alpha^{d-1} & (\alpha^{d-1})^2 & & (\alpha^{d-1})^{n-1} \end{bmatrix} \tag{1}$$

Reed-Solomon codes represent an important class of Maximum Distance Separable (MDS) cyclic codes. It is well-known that d is actually its minimum distance. Moreover, Reed-Solomon codes admit a t-bounded decoding algorithm as long as $2t \leq d - 1$. They can also be seen as a sub-family of Generalised Reed-Solomon codes.

Definition 2. *Let $\boldsymbol{\lambda}$ be an n-tuple of nonzero elements in $\mathbb{F}_{q^m}$ where $n = q^m - 1$. Let $\mathscr{R}_{n,k}$ be the Reed-Solomon code of length n and dimension k. The Generalised Reed-Solomon $\mathscr{G}_{n,k}(\boldsymbol{\lambda})$ code over $\mathbb{F}_{q^m}$ of length n and dimension k is the set:*

$$\mathscr{G}_{n,k}(\boldsymbol{\lambda}) = \left\{ (\lambda_1 v_1, \dots, \lambda_n v_n) \ / \ (v_1, \dots, v_n) \in \mathscr{R}_{n,k} \right\}. \tag{2}$$

Remark 1. Generalised Reed-Solomon codes are decoded in the same way as Reed-Solomon codes by means of the classical Berlekamp-Massey algorithm.

2.2   Subfield Subcode Construction

We describe an operation that takes as input a code over $\mathbb{F}_{q^m}$ and outputs a linear code over $\mathbb{F}_q$. In our applications $q = 2^r$ for a positive integer $r \geq 1$.

Definition 3. *Let $\mathscr{C}$ be a code over $\mathbb{F}_{q^m}$. The subfield subcode $\tilde{\mathscr{C}}$ of $\mathscr{C}$ over $\mathbb{F}_q$ is the vector space $\mathscr{C} \cap \mathbb{F}_q^n$.*

The construction of a subfield subcode from a code $\mathscr{C}$ is easy by using the dual code. Indeed, it is well-known that the dual of $\tilde{\mathscr{C}}$ is the trace of the dual of $\mathscr{C}$ [20]. Concretely this means that from a basis of $\mathbb{F}_{q^m}$ over $\mathbb{F}_q$, at each element $a \in \mathbb{F}_{q^m}$, we can associate the column vector composed by its coordinates and denoted by $[a]$. From an $(n - k) \times n$ generator matrix $\boldsymbol{H} = (h_{i,j})$ of the dual of $\mathscr{C}$ over $\mathbb{F}_{q^m}$, we can construct an $m(n - k) \times n$ matrix $\tilde{\boldsymbol{H}}$ over $\mathbb{F}_q$. The rows of $\tilde{\boldsymbol{H}}$ form a system of generators for the dual of $\tilde{\mathscr{C}}$. Note that $\tilde{\boldsymbol{H}}$ is not necessary of full rank and must be reduced by a Gaussian elimination.

Example 1. Let us consider $q = 2$, $m = 2$ and $(1, \alpha)$ a basis of $\mathbb{F}_4$ over $\mathbb{F}_2$ with $\alpha^2 = \alpha + 1$. Set $\boldsymbol{H} = \begin{pmatrix} 1 & \alpha & \alpha^2 & 1 \\ 0 & \alpha^2 & 1 & \alpha \end{pmatrix}$. Since $[0] = \begin{pmatrix} 0 \\ 0 \end{pmatrix}$ $[1] = \begin{pmatrix} 1 \\ 0 \end{pmatrix}$ $[\alpha] = \begin{pmatrix} 0 \\ 1 \end{pmatrix}$ and

$[\alpha^2] = \begin{pmatrix} 1 \\ 1 \end{pmatrix}$, we obtain $\tilde{\boldsymbol{H}} = \begin{pmatrix} 1 & 0 & 1 & 1 \\ 0 & 1 & 1 & 0 \\ 0 & 1 & 1 & 0 \\ 0 & 1 & 0 & 1 \end{pmatrix}$. After a row elimination, a generator

matrix of $\mathscr{C}'^{\perp}$ is $\boldsymbol{H}' = \begin{pmatrix} 1 & 0 & 1 & 1 \\ 0 & 1 & 1 & 0 \\ 0 & 0 & 1 & 1 \end{pmatrix}$.

The subfield subcode construction is a very important tool to obtain a new class of codes. Another advantage of restricting ourselves to such subcodes is the possibility to obtain codes with a better minimum distance which hence enables to correct more errors. Note that Goppa codes are subfield subcodes of generalised Reed-Solomon codes. They are also called *Alternant* codes.

Definition 4 (Alternant code). *Let m be a positive and let $\mathbb{F}_q$ be the field with q elements. Let $\mathscr{G}_{n,k}(\boldsymbol{\lambda})$ be a Generalised Reed-Solomon code of dimension k over $\mathbb{F}_{q^m}$. The Alternant code $\mathscr{A}_n(\boldsymbol{\lambda})$ over $\mathbb{F}_q$ is the subfield subcode of $\mathscr{G}_{n,k}(\boldsymbol{\lambda})$ over $\mathbb{F}_q$.*

Remark 2. Recall that the dimension of $\mathscr{A}_n(\boldsymbol{\lambda})$ is $\geq n - m(n - k)$.

2.3 Quasi-cyclic Codes

Definition 5. *Let $N = \ell N_0$. The quasi-cyclic permutation σ_ℓ on $\{0, ..., N - 1\}$ of order ℓ (and index N_0) is the permutation defined by the orbits $\{(0, ..., \ell - 1), (\ell, ... 2\ell - 1), ..., ((N_0 - 1)\ell, ..., N - 1)\}$.*

Definition 6. *A linear code $\mathscr{C}$ of length $N = \ell N_0$ is a quasi-cyclic code of order ℓ (and index N_0) if it is globally invariant under the action of σ_ℓ.*

Note that, up to a change of order, this definition is equivalent to the more classical invariance under σ^ℓ, where σ is the cyclic shift permutation defined by $\sigma(\boldsymbol{v}) = (v_{N-1}, v_0, ..., v_{N-2})$ for any vector $\boldsymbol{v} \in \mathbb{F}^N$ where $\mathbb{F}$ is a finite field. However this definition corresponds to the description of quasi-cyclic codes with block circulant matrices which leads to a compact description (See [12] for more details).

Example 2. Let $\boldsymbol{M}$ be the following 6×15 binary matrix:

$$\boldsymbol{M} = \begin{pmatrix} 1\,0\,0 & 0\,0\,0 & 1\,0\,1 & 1\,1\,0 & 0\,1\,1 \\ 0\,1\,0 & 0\,0\,0 & 1\,1\,0 & 0\,1\,1 & 1\,0\,1 \\ 0\,0\,1 & 0\,0\,0 & 0\,1\,1 & 1\,0\,1 & 1\,1\,0 \\ 0\,0\,0 & 1\,0\,0 & 1\,1\,1 & 1\,0\,1 & 0\,0\,0 \\ 0\,0\,0 & 0\,1\,0 & 1\,1\,1 & 1\,1\,0 & 0\,0\,0 \\ 0\,0\,0 & 0\,0\,1 & 1\,1\,1 & 0\,1\,1 & 0\,0\,0 \end{pmatrix}.$$

It is a generator matrix of a $[15, 6]$ quasi-cyclic code of order 3 (as it can be easily checked by its block circulant matrices form). It is completely described by the first row of each block:

$$\begin{bmatrix} [1\,0\,0] & [0\,0\,0] & [1\,0\,1] & [1\,1\,0] & [0\,1\,1] \\ [0\,0\,0] & [1\,0\,0] & [1\,1\,1] & [1\,0\,1] & [0\,0\,0] \end{bmatrix}.$$

Moreover, using the systematic form of the generator matrix, this description can be reduced to its redundancy part:

$$\begin{bmatrix} [1\,0\,1] & [1\,1\,0] & [0\,1\,1] \\ [1\,1\,1] & [1\,0\,1] & [0\,0\,0] \end{bmatrix}.$$

Remark 3. Since the dual of a quasi-cyclic code is also a quasi-cyclic code, it is possible to define a quasi-cyclic code through block circulant parity-check matrices.

Remark 4. A cyclic code is obviously quasi-cyclic of any order that is a divisor of N.

2.4   Reed-Solomon Codes in Quasi-cyclic Form

We have seen that Reed-Solomon codes are cyclic codes. We now propose to reorder the support in order to obtain quasi-cyclic Reed-Solomon codes. Let $\mathbb{F}_q$ the finite field where q is power of a prime number p (actually $p = 2$ and $q = 2^r$). Let m be an positive integer. Let us define α as a primitive element of $\mathbb{F}_{q^m}$ and set $N = q^m - 1$. Assume that $N = N_0 \ell$ and define $\beta = \alpha^{N_0}$. Note that β is of order ℓ. We denote by $\boldsymbol{\beta}_\ell$ the ℓ-tuple $(1, \beta, \ldots, \beta^{\ell-1})$. Let t be a positive integer and let $\boldsymbol{U}_{2t} = (\boldsymbol{A}_0 \mid \cdots \mid \boldsymbol{A}_{N_0-1})$ be the following block parity-check matrix where for $0 \leq j \leq N_0 - 1$:

$$\boldsymbol{A}_j = \begin{pmatrix} 1 & 1 & \cdots & 1 \\ \alpha^j & \alpha^j \beta & \cdots & \alpha^j \beta^{\ell-1} \\ \vdots & \vdots & & \vdots \\ (\alpha^j)^{2t-1} & (\alpha^j \beta)^{2t-1} & \cdots & (\alpha^j \beta^{\ell-1})^{2t-1} \end{pmatrix}. \tag{3}$$

Proposition 1. *The code defined by $\boldsymbol{U}_{2t}$ is a Reed-Solomon code $\mathscr{R}_{N,K}$ with $N - K = 2t + 1$ that is quasi-cyclic of order ℓ.*

3   Code-Based Cryptography

3.1   Public-Key Cryptography

A public-key cryptosystem uses a *trapdoor one-way function* E that will serve as the *encryption function*. The calculation of the inverse E^{-1} also called the *decryption function* is only possible thanks to a secret (the trapdoor) K. This concept of trapdoor one-way function forms the basis of the public-key cryptography in which the *private key* is K and the public key is E. More precisely a public-key cryptosystem should provide three algorithms namely KeyGen, Encrypt and Decrypt algorithms. KeyGen is a probabilistic polynomial-time algorithm which given an input 1^κ, where $\kappa \geq 0$ is a security parameter, outputs a pair (pk, sk) of public/private key. The KeyGen also specifies a finite *message space* M_{pk}. The Encrypt is a probabilistic polynomial-time algorithm that on inputs 1^κ, pk and a word x in M_{pk} outputs a word c. The decryption is a deterministic polynomial-time algorithm that on inputs 1^κ, sk and a word c outputs a word x. The cryptosystem should satisfy the *correctness* property which means that the decryption must undo the encryption.

3.2   McEliece Cryptosystem

The McEliece cryptosystem [21] utilizes error-correcting codes that have an efficient decoding algorithm in order to build trapdoor one-way functions. McEliece proposed binary Goppa codes as the underlying family of codes. The parameters of a binary Goppa code are $[2^m, 2^m - mt, \geq 2t + 1]$ where m and t are non-negative integers. Additionally, a binary Goppa code can be decoded by an efficient t-bounded decoding algorithm [20]. The principle of the McEliece cryptosystem is to randomly pick a code $\mathscr{C}$ among the family of binary Goppa codes. The private key is the Goppa polynomial of $\mathscr{C}$. The public key will be a generator matrix that is obtained from the private key and by two random linear transformations: the *scrambling* transformation S, which sends the secret matrix G to another generator matrix, and a permutation transformation which reorders the columns of the secret matrix. In the rest of the paper we denote by $W_{q,n,k,t}$ the work factor of the best attack against the McEliece cryptosystem when the code is of length n, dimension k over $\mathbb{F}_q$ and the number of added errors is t.

Algorithm 1. McEliece.KeyGen(1^κ)

---

1 Choose n, k and t such that $W_{2,n,k,t} \geq 2^\kappa$
2 Randomly pick a generator matrix $\boldsymbol{G_0}$ of an $[n, k, 2t + 1]$ binary Goppa code $\mathscr{C}$
3 Randomly pick a $n \times n$ permutation matrix $\boldsymbol{P}$
4 Randomly pick a $k \times k$ invertible matrix $\boldsymbol{S}$
5 Calculate $\boldsymbol{G} = \boldsymbol{S} \times \boldsymbol{G_0} \times \boldsymbol{P}$
6 Output pk $= (\boldsymbol{G}, t)$ and sk $= (\boldsymbol{S}, \boldsymbol{G_0}, \boldsymbol{P}, \gamma)$ where γ is a t-bounded decoding algorithm of $\mathscr{C}$

3.3   Niederreiter Cryptosystem

A dual encryption scheme is the Niederreiter cryptosystem [23] which is equivalent in terms of security [19] to the McEliece cryptosystem. The main difference between McEliece and Niederreiter cryptosystems lies in the description of the codes. The Niederreiter encryption scheme describes codes through parity-check matrices. But both schemes has to hide any structure through a scrambling transformation and a permutation transformation. The encryption algorithm takes as input words of weight t where t is the number of errors that can be decoded. We denote by $\mathcal{W}_{q,n,t}$ the words of $\mathbb{F}_q^n$ of weight t.

Algorithm 2. McEliece.Encrypt(pk,$m \in \mathbb{F}_2^k$)

1 Randomly pick e in $\mathbb{F}_2$ of weight t
2 Calculate $c = m \times G + e$
3 Output c

Algorithm 3. McEliece.Decrypt(sk,$c \in \mathbb{F}_2^n$)

1 Calculate $z = c \times P^{-1}$
2 Calculate $y = \gamma(z)$
3 Output $m = y \times S^{-1}$

3.4   Security of the McEliece Cryptosystem

Any public-key cryptosystem primarily requires to be resistant against an adversary that manages either to extract the private data given only public data, or to invert the trapdoor encryption function given the ciphertexts of his choice (and public data). This security notion is also named OW-CPA for *One-Wayness under Chosen Plaintext Attack*.

Algorithm 4. Niederreiter.KeyGen(1^κ)

1 Choose n, k and t such that $W_{2,n,k,t} \geq 2^\kappa$
2 Randomly pick a $(n-k) \times n$ parity-check matrix H_0 of an $[n, k, 2t+1]$ binary Goppa code $\mathscr{C}$
3 Randomly pick a $n \times n$ permutation matrix P
4 Randomly pick a $(n-k) \times (n-k)$ invertible matrix S
5 Calculate $H = S \times H_0 \times P$
6 Output pk $= (H, t)$ and sk $= (S, H_0, P, \gamma)$ where γ is a t-bounded decoding algorithm of $\mathscr{C}$

Structural attacks. If we consider irreducible binary Goppa codes then there is no efficient algorithm that extracts the secret key from the public key in the McEliece or the Niederreiter cryptosystem provided that weak keys are avoided.

Decoding attack. plus The *one-wayness* property is the weakest security notion that any public-key cryptosystem must satisfy. It essentially states that inverting the encryption function is computationally impossible without a secret called the *trapdoor*. In the case of McEliece (or Niederreiter) cryptosystem, it consists in decoding a given word into a codeword of the public code. Another equivalent way of stating this problem is the *syndrome decoding problem* which constitutes an important algorithmic problem in coding theory. It was proved NP-Complete in [2].

Algorithm 5. Niederreiter.Encrypt(pk,$m \in \mathcal{W}_{2,n,t}$)

1 Calculate $c = H \times m^T$
2 Output c

Algorithm 6. Niederreiter.Decrypt(sk,$c \in \mathbb{F}_2^{n-k}$)

1 Calculate $z = S^{-1} \times c$
2 Calculate $y = \gamma(z)$
3 Output $m = y \times P$

Definition 7 (Syndrome decoding). *Given an $r \times n$ matrix H over $\mathbb{F}_q$, a positive integer $w \leq n$ and an r-tuple z in $\mathbb{F}_q^r$, does there exist e in $\mathbb{F}_q^n$ such that $\mathsf{wt}(e) \leq w$ and $H \times e^T = z$?*

This important result means that decoding an arbitrary linear code is difficult (in the worst case). The issue of decoding can also be translated into the problem of finding a low-weight codeword in an appropriate code. We assume that we encrypt by means of the McEliece cryptosystem. We have a public generator G and a given ciphertext z. We know that there exist a codeword c in $\mathscr{C}$ and a vector $e \in \mathbb{F}_q^n$ of weight t such that $z = c + e$. By hypothesis, the minimum distance of $\mathscr{C}$ is $2t + 1$. Therefore the linear code $\tilde{\mathscr{C}}$ defined by the generator matrix $\tilde{G} = \begin{bmatrix} G \\ c \end{bmatrix}$ contains a codeword of weight t namely e. Thus inverting the encryption amounts to find codewords of low weight in a given code. We know that this problem is NP-Hard for an arbitrary linear code.

The best practical algorithms for searching low weight codewords of any linear code are all derived from the *Information Set Decoding*. An information set for a generator matrix G of a code of dimension k is a set J of k positions such that the restriction G_J of G over J is invertible. The redundancy part is then the complementary set. The algorithm looks for a codeword of weight w. The principle is to randomly pick a permutation matrix P and a matrix S such that $G' = SGP = (I_k | B)$ is a row reduced echelon form. It then computes all the sums of g rows or less (where g is small) of G' and stops as soon as a codeword of weight w is obtained. Otherwise the algorithm picks another permutation matrix P and continues again. Lee and Brickell [17] were the first to use to McEliece cryptosystem. Leon [18] proposed to looks only for codewords that are zero over a window of size σ in the redundancy part. Stern [29] improved the algorithm by dividing the information set into parts and by looking for codewords that are identical over a window of size σ in the redundancy part. Another improvement proposed by van Tilburg in [30] consists in not choosing a completely new information set J at each new iteration but rather permuting only a small number c of columns of G' and keeping intact the $(k - c)$ other columns of the information set. This permits to decrease the amount of computation of the row-reduction operation. Canteaut and Chabaud gathered all these refinements with $c = 1$ (also suggested in [6] and [7]) and studied the complexity of the algorithm in [8]. This last attempt represented the most effective attack against the McEliece cryptosystem (in the case of *binary linear codes* [9]). Recently, Bernstein *et al.* in [3] further enhanced the attack to a level that makes it practically possible to break the original parameters of the McEliece cryptosystem.

It is worthwhile remarking that the improvements of [8] and [3] are essentially applied to binary codes. In particular both analyze the complexity of the algorithms specifically in the binary case. It would be interesting to study the behavior of [3] when non-binary codes are considered.

4   A New Variant of the McEliece Cryptosystem

The new variant of the McEliece (or Niederreiter) cryptosystem we propose is not based on classical binary Goppa codes but rather on shortened quasi-cyclic Alternant codes. For a cryptographic purpose, we need a large family of codes with the following requirement: a description as compact as possible, a secret efficient decoding algorithm (i.e. a hidden structure) and a resistance to known (and unknowns...) attacks. As seen previously, quasi-cyclic codes offer the advantage of having a very compact representation which can lead to shorter keys. In an other hand, for a fixed level of security, the use of optimal codes (in particular, large minimum distance for fixed length and dimension) allows to use smaller codes. The so-called Maximum Distance Separable (MDS) codes are good candidates. The almost only known family of MDS codes with an efficient decoding algorithm is the class of Generalized Reed-Solomon (GRS) codes. However, it is well-known (see eg [28]) that GRS codes cannot be used directly in cryptography.

The aim is to obtain a family of quasi-cyclic Alternant code defined over relatively small field $\mathbb{F}_q$ (like $\mathbb{F}_{2^{10}}$ or $\mathbb{F}_{2^5}$). The idea is to start from a Reed-Solomon code in quasi-cyclic form defined over a large alphabet $\mathbb{F}_{q^m}$ (for instance $\mathbb{F}_{2^{20}}$) in quasi-cyclic form. Then we randomly delete the majority of circulant blocks to counter known attacks against the recovery of the quasi-cyclic structure. We transform this code into a quasi-cyclic shortened Generalized Reed Solomon code. Finally, we construct the subfield subcode over an intermediate field $\mathbb{F}_q$ to mask the GRS structure. Recall that subfield subcodes of GRS codes constitute the family of Alternant codes. Note that the strategy of subfield subcodes in order to mask the structure of GRS code is not new since the class of Goppa codes used in the original McEliece cryptosystem is a subclass of Alternant codes. In our case, we use a large subfield in order to increase the performance of our codes.

4.1   Construction of a Family of Quasi-cyclic Alternant Codes

We keep the notation of Section 2.4 throughout this section. Let t be a positive integer and let $\mathscr{R}_{N,K}$ be a Reed-Solomon in quasi-cyclic form over $\mathbb{F}_{q^m}$ of length

$N = \ell N_0$ and such that $N - K = 2t+1$. The idea is to start from a Reed-Solomon code $\mathscr{R}_{N,K}$ in quasi-cyclic form defined over $\mathbb{F}_{q^m}$ and successively applying three operations that preserve the quasi-cyclic feature, namely (1) randomly block shortening the code in order to obtain a code of length $n = n_0\ell$ with $n_0 < N_0$, (2) transforming it to get a Generalised Reed Solomon code of length n and (3) taking the subfield subcode over $\mathbb{F}_q$.

In term of security the first operation (1) adds a combinatorial complexity by hiding our code into a larger code. Thus recovering the correct blocks requires to search for a relatively small number n_0 (say n_0 is greater than 6) of blocks, among a relatively large number N_0 of blocks (say 1000 to 100.000 blocks) which makes it very difficult in practice. This operation permits to resist to a recent cryptanalysis on quasi-cyclic codes by the combinatorial masking of the dual code. Operation (2) is a classical algebraic choice of blocks multiplication or cyclic permutations which increases the number of possible generalised Reed-Solomon. At last operation (3) scrambles the structure of the code. In particular it permits to resist to the Sidelnikov-Shestakov attack which can be used on generalised Reed-Solomon codes. But it is inefficient on subfield subcodes even if they are derived from generalised Reed-Solomon codes. Notice than we could also have considered another hiding procedure by considering a subcode as in [12]. But this operation would increase the size of the key by a factor 2 we omitted it here. We present now the details of our strategy.

(1) Random Block Shortening. This step consists in randomly choosing n_0 (with $n_0 < N_0$) circulant blocks of the parity-check matrix $\boldsymbol{U}_{2t}$. More precisely, let $\boldsymbol{j} = (j_1, \ldots, j_{n_0})$ be an n_0-tuple of different non-negative integers less than or equal to N_0. Note that we do not have necessarily that $j_1 \leq j_2 \leq \cdots \leq j_{n_0}$. We then consider the code $\mathscr{R}_{n,K}(\boldsymbol{j})$ of length $n = \ell n_0$ over $\mathbb{F}_{q^m}$ defined by the following parity-check matrix $\boldsymbol{U}_{2t}(\boldsymbol{j}) = (\,\boldsymbol{A}_{j_1} \mid \boldsymbol{A}_{j_2} \mid \cdots \mid \boldsymbol{A}_{j_{n_0}}\,)$.

Proposition 2. $\mathscr{R}_{n,K}(\boldsymbol{j})$ is a quasi-cyclic code of order ℓ.

Remark 5. One can easily check that random block shortening consists in randomly block permuting and puncturing the parity-check matrix $\boldsymbol{U}_{2t}$.

(2) GRS Transformation. The random block shortened Reed-Solomon $\mathscr{R}_{n,K}(\boldsymbol{j})$ code is transformed to obtain a quasi-cyclic Generalised Reed-Solomon code. Let $\boldsymbol{D}_{\boldsymbol{\beta}_\ell} = (d_{i,j})$ be the $\ell \times \ell$ diagonal matrix such $d_{i,i} = \beta^{i-1}$. For any integer s, we have that:

$$\boldsymbol{D}_{\boldsymbol{\beta}_\ell}^s = \begin{pmatrix} 1 & & & \\ & \beta^s & & \\ & & \ddots & \\ & & & (\beta^s)^{\ell-1} \end{pmatrix}. \tag{4}$$

We consider an n_0-tuple $\boldsymbol{a} = (a_1, \ldots, a_{n_0})$ of nonzero elements of $\mathbb{F}_{q^m}$ and an integer s with $1 \leq s \leq \ell - 1$. Let $\mathscr{R}_{n,K}(\boldsymbol{j}, \boldsymbol{a}, s)$ be the code defined by the block parity-check matrix $U_{2t}(j, a, s) = (B_1 \mid \cdots \mid B_{n_0})$ where $B_i = a_i \, A_{j_i} \times D_{\beta_\ell}^s$ for $1 \leq i \leq n_0$, or equivalently:

$$
B_i = \begin{pmatrix}
a_i & a_i \beta^s & \cdots & a_i (\beta^s)^{\ell-1} \\
a_i \alpha^{j_i} & a_i \beta^s \alpha^{j_i} \beta & \cdots & a_i (\beta^s)^{\ell-1} \alpha^{j_i} \beta^{\ell-1} \\
\vdots & \vdots & & \vdots \\
a_i (\alpha^{j_i})^{2t-1} & a_i \beta^s (\alpha^{j_i} \beta)^{2t-1} & \cdots & a_i (\beta^s)^{\ell-1} (\alpha^{j_i} \beta^{\ell-1})^{2t-1}
\end{pmatrix}. \tag{5}
$$

Proposition 3. $\mathscr{R}_{n,K}(j, a, s)$ is a quasi-cyclic code of order ℓ.

Proof. It is easy to see that if we apply the quasi-cyclic permutation σ_ℓ to the i-th row u_i of $U_{2t}(j, a, s)$ then we have $\sigma_\ell(u_i) = (\beta^{\ell-1})^{i+s} u_i$, which proves that $\mathscr{R}_{n,K}(j, a, s)$ is globally invariant under the action of σ_ℓ.

(3) Subfield Subcode Operation. We then consider the subfield subcode over $\mathbb{F}_q$ of $\mathscr{R}_{n,K}(j, a, s)$. We denote it by $\mathscr{A}_n(j, a, s)$ and is defined by a block $2tm \times n$ parity-check matrix $\tilde{H}$ that is the trace matrix of $U_{2t}(j, a, s)$ (See Section 2.2 for more details).

4.2   Description of the New Variant

We have seen in the previous section how to get a family of quasi-cyclic Alternant codes over $\mathbb{F}_q$ from a unique reed-Solomon code $\mathscr{R}_{N,K}$ over $\mathbb{F}_{q^m}$. It is easy to see that the number of possible codes is equal to $n_0! \binom{N_0}{n_0} \times (q^m - 1)^{n_0} \times (\ell - 1)$. We now describe the KeyGen algorithm of our variant for a security parameter κ. We assume that q and m are given as input to the algorithm. In particular N_0 and ℓ are also known.

In Step 7, we further hide the structure of our codes by applying column permutations as it is done in the original McEliece cryptosystem. However we need to take a special kind of permutations that preserve the quasi-cyclic feature. This can be achieved by permuting columns inside each circulant block by means of a power σ^i of the cyclic shift σ of $\{0, \ldots, \ell - 1\}$. Therefore we randomly pick an n_0-tuple $i = (i_1, \ldots, i_{n_0})$ of non-negative integer smaller that ℓ. Note that M^π is the result of a column permutation π over a matrix M.

The goal of Step 7 is to choose a compact representation. This can be done by transforming H' into a $\delta\ell \times n$ block circulant matrix in systematic form H where δ is the smallest integer such that $\delta\ell \geq 2mt$. This means that there exists then $\delta\ell \times \delta\ell$ matrix S such that $H = S \times H'$. H is then completely described by $(n_0 - \delta)\delta$ vectors of $\mathbb{F}_q^\ell$, that is to say $(n_0 - \delta)\delta\ell r$ bits (recall that $q = 2^r$). We refer to [12] for a more detailed algorithm.

Size of key If one starts from a masked $[n, k, d]$ code over F_{2^r} with cyclicity index l, such that $n = n_0 l$ and $k = l\delta$ the size of the public key obtained from the code in systematic form is $(n_0 l - \delta l)\delta r$, which can be seen as $\frac{(n-k)kr}{l}$. The latter formula can be seen roughly (up to a factor $k/(n - k)$) as the cost of one row times the repetition factor δ which may remain smaller than 5.

Algorithm 7. KeyGen(1^κ)

---

1 Choose t and n_0 such that $W_{2,n,k,t} \geq 2^\kappa$ where $n = \ell n_0$ and $k = n - 2mt$
2 Randomly pick an n_0-tuple $\boldsymbol{j} = (j_1, \ldots, j_{n_0})$ of distinct non-negative integers
   $\leq N_0 - 1$.
3 Randomly pick an n_0-tuple $\boldsymbol{a}$ of non zero elements of $\mathbb{F}_{q^m}$.
4 Let α be a primitive element of $\mathbb{F}_{q^m}$ and set $\beta = \alpha^{N_0}$. Randomly pick an integer
   $1 \leq s \leq \ell - 1$.
5 Let $\tilde{\boldsymbol{H}} = (\tilde{\boldsymbol{B}}_1 \mid \cdots \mid \tilde{\boldsymbol{B}}_{n_0})$ be the trace matrix of $U_{2t}(\boldsymbol{j}, \boldsymbol{a}, s)$.
6 Randomly pick an n_0-tuple $\boldsymbol{i} = (i_1, \ldots, i_{n_0})$ of non-negative integer smaller that
   ℓ. Compute $\boldsymbol{H}' = (\tilde{\boldsymbol{B}}_1^{\sigma^{i_1}} \mid \cdots \mid \tilde{\boldsymbol{B}}_{n_0}^{\sigma^{i_{n_0}}})$.
7 Transform $\boldsymbol{H}'$ into a block circulant matrix in systematic form $\boldsymbol{H}$.
8 Output pk $= (\boldsymbol{H}, t)$ and sk $= (\boldsymbol{j}, \boldsymbol{a}, s, \boldsymbol{i})$

---

5   Suggested Parameters

We illustrate our construction with different sets of parameters in Table 1. Notice that the parameters n and k corresponds to the generator matrix G, the size of the public is computed from a matrix G or dual matrix H in systematic form if a semantic conversion is also used (See [16,4,26]). The table also sums up the security entry that is computed from the Stern's algorithm applied over *non-binary* field. Notice that in Table 1, we focused on codes on rather large fields like $\mathbb{F}_{2^8}$ or $\mathbb{F}_{2^{10}}$, because it gives the best size of key. Meanwhile our construction is general and may also work for smaller fieds like $\mathbb{F}_2$ or $\mathbb{F}_4$, the limitation of smaller fields is that general alternant codes are less interesting when the subfield is smaller, leading to a size of key, about two times larger than what is obtained for larger fields. We give parameters with security from 2^{80} to 2^{120} with for instance public key sizes ranging from $6,750$ bits for a 2^{80} security to $20,000$ bits for a 2^{120} security. These parameters show the adaptability and the scalability of our system with a key size that moderately increases according to the security level. The security is taken to be the best attack between structural attack and decoding attack, but in all cases we considered, the structural attacks have a greater complexity than decoding attack.

Example 3. Consider the set of parameters A_{16} where $2^{16} - 1 = 51 \times 1285$. We take $l = 51$ and $N_0 = (2^{16} - 1)/51 = 1285$. We consider the subfield $\mathbb{F}_{2^8}$ of $\mathbb{F}_{2^{16}}$, so $r = 2$. We choose a GRS cyclic code C_0 with $t = 50$, which gives $N = 2^{16} - 1$ and $K = 2^{16} - 1 - 2 \times 50$. Therefore, we get $r(N - K) = 100 \sim 2.\ell$ with $n_r = 2$ (the number of row eventually needed). Then we take $n_0 = 9$ and keep only 9 blocks of size ℓ among the 1285 possible blocks. This operation corresponds to step (1) of Section 4. We obtain a quasi-cyclic GRS $[9.51, 9.51 - 100, t = 50]_{2^{16}}$ code of order 51. We then apply linear transformations (2) and (3) of Section 4 to get a code C_1 over $\mathbb{F}_{2^{16}}$ from which we take the subfield subcode over $\mathbb{F}_{2^8}$. We hence obtain a $[9.51, 9.51 - 200, t = 25]_{2^8} = [459, 255, t = 50]_{2^8}$ code. We can check from the Canteaut-Chabaud and the Bernstein et al. algorithms that the

Table 1. Suggested parameters for different security levels

$q^m - 1 = N = N_0\ell$				Public code $\mathscr{C}[n, k]$ over $\mathbb{F}_q$							
q^m	ℓ	N_0	t	Code	n	k	q	n_0	δ	Security	Public key size (bits)
2^{16}	51	1,285	50	A_{16}	459	255	2^8	9	4	80	8,160
	51	1,285	50	B_{16}	510	306	2^8	10	4	90	9,792
	51	1,285	50	C_{16}	612	408	2^8	12	4	100	13,056
	51	1285	50	D_{16}	765	510	2^8	15	5	120	20,400
2^{20}	75	13,981	56	A_{20}	450	225	2^{10}	6	3	80	6,750
	93	11,275	63	B_{20}	558	279	2^{10}	6	3	90	8,370
	93	11,275	54	C_{20}	744	372	2^{10}	8	4	110	14,880

security (taking account of the quasi-cyclic order) is 2^{80}. The size of the public key is $8,160$ bits because we consider the parity check matrix put in systematic form.

6   Security Analysis

6.1   Structural Attacks

We review now all the existing attacks that aim at recovering the private key from public data. These attacks are listed below.

Brute force key search attack. Recall that the private key is sk $= (j, a, s, i)$ where j is an n_0-tuple of distinct non-negative integers $\leq N_0 - 1$, $a = (a_1, \ldots, a_{n_0})$ is an n_0-tuple of non zero elements of $\mathbb{F}_{q^m}$, s is a non-negative integer $\leq \ell - 1$, and i is an n_0-tuple of non negative integer $\leq \ell - 1$. Note that a_1 can be chosen to be equal to 1. Thus the private key space contains $n_0! \binom{N_0}{n_0} \times (q^m - 1)^{n_0-1} \times (\ell - 1)^{n_0}$ elements. This implies that the private key is out of reach by an exhaustive search.

Recall also that the public parity-check matrix H is obtained by putting $H' = (B'_1 \mid \cdots \mid B'_{n_0})$ in systematic form. There exists therefore a $\delta\ell \times \delta\ell$ matrix S such that $H = S \times H'$. A possible strategy is to guess the δ first blocks $\Delta = (B'_1 \mid \cdots \mid B'_\delta)$. This matrix Δ is reduced in row echelon form. This achieved by left multiplying Δ with a matrix Γ. If one guesses the correct blocks then $\Gamma = S$. This can be checked if there exists at least a column c of H_{2t} and an nonzero element a in $\mathbb{F}_{q^m}$ such that the column vector $a\Gamma \times c$ appears in the public matrix. If it is not the case, one choose another guess for Δ. The number of candidates for Δ is $\delta! \binom{N_0}{\delta}(q^m - 1)^{\delta-1}\ell^{\delta+1}$. The cost to transform Δ in row-reduced echelon form is $O\left(mr(\delta\ell)^3\right)$. Since H_{2t} has N columns, the overall complexity of the attack is then $O\left(\delta! \binom{N_0}{\delta}(q^m - 1)^{\delta-1}\ell^{\delta+1}\left(mr(\delta\ell)^3 + N(q^m - 1)mr(\delta\ell)^2\right)\right)$.

Attack exploiting the generalised Reed-Solomon structure. Sidelnikov and Shestakov proved in [28] that it is possible to completely recover the structure of Generalised Reed-Solomon codes with time complexity in $O(n^4)$ where n is the length of the public code. This attack can be used against any type of generalized Reed-Solomon code but uses the fact that the underlying matrix of the Reed-Solomon code is completely known. In our case we do not directly use a GRS code but a (random shortened) subfield subcode, which hence makes this attack unfeasible. It is worthwhile remarking that if such an efficient (or a generalized attack) was to exist for Alternant codes, then it could be potentially used to break the McEliece cryptosystem since Goppa codes are a special case of binary Alternant codes.

Attack exploiting the quasi-cyclic structure. Recently a new structural attack appeared in [24] that extracts the private key of the variant presented in [12]. This cryptosystem takes a binary quasi-cyclic subcode of a BCH code of length n as the secret code. The structure is hidden by a strongly constrained permutation in order to produce a quasi-cyclic public code. This implies that the permutation transformation is completely described with n_0^2 binary entries where n_0 is the quasi-cyclic index rather than n^2 entries. The attack consists in taking advantage of the fact that the secret is a subcode of completely known BCH code. One generates linear equations by exploiting the public generator matrix and a known parity-check matrix of the BCH code so that one gets an over-constrained linear system satisfied by the unknown permutation matrix.

We show how to adapt this attack to our variant. We start from the parity-check matrix U_{2t} of the Reed-Solomon code $\mathscr{R}_{K,N}$ in quasi-cyclic form as defined in Section 2.4. We consider the matrix $G = (G_p \mid 0)$ where G_p is a generator matrix of the public code of length n and 0 is a zero matrix with $N - n$ columns. Clearly, there exists an $N \times N$ matrix X such that:

$$U_{2t} \times X \times G^T = 0. \tag{6}$$

where G^T is the transpose of G. The matrix X has a block structure where each block is either the $\ell \times \ell$ zero matrix or a matrix of the form $aD_{\beta_\ell}^s$ where a is a nonzero element of $\mathbb{F}_{q^m}$ and s is an integer smaller than ℓ. The matrix X actually gathers all the secret operations that have been made to obtain the public quasi-cyclic Alternant code. Thus the integer s is the *same* for any nonzero block whereas a may vary. Therefore solving the linear system given by Equation (6) reveals the secret key. Note that X is thus totally determined by N_0^2 matrices of size $\ell \times \ell$. Additionally, as s is very small, one may assume that $D_{\beta_\ell}^s$ is known. Therefore the number of unknowns to totally describe X is *exactly* N_0^2. On the other hand, Equation 6 provides $(n - 2mt) \times 2t$ equations (each row of G gives $2t$ check equations from U_{2t}). But if we consider the trace matrix U'_{2t} instead of U_{2t}, then m times more parity equations for each row of G. We can get a total of $2mt(n - 2mt)$ linear equations for N_0^2 unknowns.

An attacker has therefore to reduce the number of unknowns in order to guess X. Another strategy is to set $N_0 - \tilde{N}$ random block columns of X to zero where

$\tilde{N}^2 = 2mt(n - 2mt)$. This method would give the right solution if only if the n_0 columns that occur in the construction of the shortened Alternant code are not set to zero. The success probability of this method is therefore $\frac{\binom{N_0-n_0}{\tilde{N}-n_0}}{\binom{N_0}{\tilde{N}}}$. Additionally, for each random choice, one has to solve a linear system with a time complexity $O((rm)^2\tilde{N}^3)$ with coefficients in $\mathbb{F}_{2^{rm}}$. In practice this attack does not give better results than direct decoding attack.

6.2 Decoding Attack

We prove in this section that the primitive of our variant is also OW-CPA under the assumption that it is computationally impossible to recover the secret Alternant quasi-cyclic code. We show that decoding an arbitrary quasi-cyclic code is also an NP-Hard problem. For doing so, we propose another decision problem called *quasi-cyclic syndrome decoding*. We prove that this new problem is also NP-Complete.

Definition 8 (Quasi-cyclic syndrome decoding). *Given $\ell > 1$ (we avoid the case $l = 1$ which corresponds to a degenerate case) matrices $A_1, \ldots, A_\ell$ of size $r^* \times n^*$ over $\mathbb{F}_q$, an integer $w < \ell n^*$ and a word z in $\mathbb{F}_q^{\ell r^*}$. Let A be the $\ell r^* \times \ell n^*$ matrix:*

$$A = \begin{bmatrix} A_1 & \cdots & \cdots & A_\ell \\ A_\ell & A_1 & \cdots & A_{\ell-1} \\ \vdots & \ddots & \ddots & \vdots \\ A_2 & \cdots & A_\ell & A_1 \end{bmatrix}$$

Does there exist e in $\mathbb{F}_q^{\ell n^*}$ of weight $\mathsf{wt}(e) \le w$ such that $A \times e^T = z$?

Proposition 4. *The quasi-cyclic syndrome decoding problem is NP-Complete.*

Proof. We consider an instance H, w and z of the syndrome decoding problem. We define $w^* = 2w$, the $2r$-tuple $z^* = (z, z)$ and the following $2r \times 2n$ matrix A:

$$A = \begin{bmatrix} H & 0 \\ 0 & H \end{bmatrix}.$$

Clearly A, z^* and w^* are constructed in polynomial time. Assume now that there exist e in $\mathbb{F}_q^n$ of weight $\mathsf{wt}(e) \le w$ such that $H \times e^T = z$. Then $\mathsf{wt}(e^*) \le w^*$ and $A \times e^{*T} = z^*$.

Conversely assume that there exists e^* in $\mathbb{F}_q^{2n}$ of weight $\mathsf{wt}(e^*) \le w^*$ such that $A \times e^{*T} = z^*$. If we denote $e^* = (e_1, e_2)$ where e_1 is formed by the first n symbols and e_2 by the last n symbols. Obviously we have either e_1 or e_2 of weight $\le w^*/2$ and for both of them $H \times e_j^T = z$.

This result ensures that (in the worse case) decoding an arbitrary quasi-cyclic code is a difficult task. One may think that classical algorithms can be modified by taking into account the quasi-cyclic structure. This fact can be favorably used to indeed improve performances of generic algorithms, and one can reasonably expect to decrease by a factor the order of quasi-cyclicity ℓ the work factor of the general decoding algorithms. Note that the proposed parameters in Table 5 take into account this fact. However our proof shows that if ever an algorithm appears that efficiently solves the quasi-cyclic syndrome decoding problem, then it will be give another efficient one for the more general problem of syndrome decoding. This result would represent a major breakthrough in the field of coding theory. This result also suggests that decoding attacks against our variant have little chance to be better than the general ones against the classical McEliece cryptosystem.

6.3   Discussion on the Practical Difficulty of Decoding Quasi-cyclic Codes

In fact up to now, although decoding a general random QC codes may seem easier than decoding a random code without any visible structure, the best known algorithm remains the general algorithms for random codes (up to the order of quasi-cyclicity which is a small factor). The situation is in some sense very similar to the case of lattice based cryptography and the LLL algorithm. For instance the NTRU lattice has a similar quasi-cyclic structure but no attack related to the LLL algorithm did really took advantage of this structure except by a very small improvement of a constant, although the NTRU cryptosystem has been known for more than 10 years and was really scrutinized. For coding theory the same situation seems to occur and general random QC codes seem to be as hard as general random codes (up to the size of the order of quasi-cyclicity).

7   Performance

We implemented the encryption with the system A_{20} on $\mathbb{F}_{2^8}$ on 2.4 Ghz computer with a 64-bit architecture. The multiplication over $\mathbb{F}_{2^8}$ was tabulated in cache memory and all operations were done byte by byte as a matrix vector product so that eventually the 64 bits structure was not really enhanced. Overall the encryption speed was 15Mo/s, which corresponds to about 128 cycles per byte.

This speed compares well with the implementation of [4] which speed was inferior with a factor 2 to what we obtain. Moreover our implementation can still be improved (probably by a factor 2 or 3) by taking account of the cyclic structure. Indeed rather than tabulating the multiplication over the field ($\mathbb{F}_{2^8}$ in this case) it is possible to put in cache memory the multiplication of the first row of the code by all elements of the base field. Since all the rows of the generator matrix are obtained as cyclic shift from the first row it then possible to deduce all multiplied rows from shifts of the multiplied first rows in cache. So that it is possible to profit by the 64 bits structure by summation (but in 64 bits) of the multiplied shifted first rows and even enhance our performance. For decryption we obtain similar speed of a few hundred cycle per byte as in [4] although we did not yet optimized our implementation.

Our implementation speed compares of course very well with RSA-1024 and Elliptic Curve cryptosystems. An interesting question is the comparison with

NTRU. Although exact performance results are hard to find, in term of encryption speed our system seems better with a factor 10 by comparison to known performance of NTRU [4]. These good results comes from the fact that besides the matrix-vector product in NTRU, one also needs to encrypt vectors with given weight which becomes in fact the main cost (notice that the same problem arises for Niederreiter version of the scheme, but in our case we only considered the McEliece scheme).

8 Conclusion

In this paper we presented a new way to reduce the size of the public key for code-based cryptosystems like McEliece or Niederreiter schemes. We use quasi-cyclic Alternant codes over a non-binary small field. We introduced new methods to hide the structure of quasi-cyclic Alternant codes. We showed that all structural attacks cannot cope with our parameters. We prove that decoding quasi-cyclic codes is a new NP-Hard problem. This result makes decoding attack inappropriate. Our scheme permits to reach a public key size as low as 6,500 bits for a security of 2^{80} and 20,000 bits for 2^{120}. Lastly, an implementation of our scheme A_{16} ran at 120 Mb/s (on Intel Core2, 32 bits)for encryption speed which makes it far better than RSA or NTRU cryptosystems (see [4] for comparisons) and 13 Mb/s for decryption.. Such low parameters together with the high speed of the system open the doors to new potential applications for code-based cryptography like smart cards, key exchange or authentication.

Acknowledgements

The authors would like to warmly thank J.P. Tillich for many valuable comments that improved the preliminary versions of this paper.

References

1. Baldi, M., Chiaraluce, G.F.: Cryptanalysis of a new instance of McEliece cryptosystem based on QC-LDPC codes. In: IEEE International Symposium on Information Theory, Nice, France, March 2007, pp. 2591–2595 (2007)
2. Berlekamp, E., McEliece, R., van Tilborg, H.: On the inherent intractability of certain coding problems. IEEE Transactions on Information Theory 24(3), 384–386 (1978)
3. Bernstein, D.J., Lange, T., Peters, C.: Attacking and defending the mceliece cryptosystem. In: Buchmann, J., Ding, J. (eds.) PQCrypto 2008. LNCS, vol. 5299, pp. 31–46. Springer, Heidelberg (2008)
4. Biswas, B., Sendrier, N.: Mceliece cryptosystem implementation: theory and practice. In: Buchmann, J., Ding, J. (eds.) PQCrypto 2008. LNCS, vol. 5299, pp. 47–62. Springer, Heidelberg (2008)
5. Shokrollahi, A., Monico, C., Rosenthal, J.: Using low density parity check codes in the McEliece cryptosystem. In: IEEE International Symposium on Information Theory (ISIT 2000), Sorrento, Italy, p. 215 (2000)

6. Canteaut, A., Chabanne, H.: A further improvement of the work factor in an attempt at breaking McEliece's cryptosystem. In: EUROCODE 1994, pp. 169–173. INRIA (1994)

7. Canteaut, A., Chabaud, F.: Improvements of the attacks on cryptosystems based on error-correcting codes. Technical Report 95–21, INRIA (1995)

8. Canteaut, A., Chabaud, F.: A new algorithm for finding minimum-weight words in a linear code: Application to McEliece's cryptosystem and to narrow-sense BCH codes of length 511. IEEE Transactions on Information Theory 44(1), 367–378 (1998)

9. Canteaut, A., Sendrier, N.: Cryptanalysis of the original McEliece cryptosystem. In: Ohta, K., Pei, D. (eds.) ASIACRYPT 1998. LNCS, vol. 1514, pp. 187–199. Springer, Heidelberg (1998)

10. Cayrel, P.L., Otmani, A., Vergnaud, D.: On Kabatianskii-Krouk-Smeets Signatures. In: Carlet, C., Sunar, B. (eds.) WAIFI 2007. LNCS, vol. 4547, pp. 237–251. Springer, Heidelberg (2007)

11. Engelbert, D., Overbeck, R., Schmidt, A.: A summary of McEliece-type cryptosystems and their security. Journal of Mathematical Cryptology 1, 151–199 (2007)

12. Gaborit, P.: Shorter keys for code based cryptography. In: Proceedings of the 2005 International Workshop on Coding and Cryptography (WCC 2005), Bergen, Norway, pp. 81–91 (March 2005)

13. Gaborit, P., Girault, M.: Lightweight code-based authentication and signature. In: IEEE International Symposium on Information Theory (ISIT 2007), Nice, France, March 2007, pp. 191–195 (2007)

14. Gaborit, P., Lauradoux, C., Sendrier, N.: Synd: a fast code-based stream cipher with a security reduction. In: IEEE International Symposium on Information Theory (ISIT 2007), Nice, France, March 2007, pp. 186–190 (2007)

15. Hoffstein, J., Pipher, J., Silverman, J.H.: NTRU: A ring-based public key cryptosystem. In: Buhler, J. (ed.) ANTS 1998. LNCS, vol. 1423, pp. 267–288. Springer, Heidelberg (1998)

16. Kobara, K., Imai, H.: Semantically secure mceliece public-key cryptosystems-conversions for mceliece pkc. In: Kim, K. (ed.) PKC 2001. LNCS, vol. 1992, pp. 19–35. Springer, Heidelberg (2001)

17. Lee, P.J., Brickell, E.F.: An observation on the security of mcEliece's public-key cryptosystem. In: Günther, C.G. (ed.) EUROCRYPT 1988. LNCS, vol. 330, pp. 275–280. Springer, Heidelberg (1988)

18. Leon, J.S.: A probabilistic algorithm for computing minimum weights of large error-correcting codes. IEEE Transactions on Information Theory 34(5), 1354–1359 (1988)

19. Li, Y.X., Deng, R.H., Wang, X.-M.: On the equivalence of McEliece's and Niederreiter's public-key cryptosystems. IEEE Transactions on Information Theory 40(1), 271–273 (1994)

20. MacWilliams, F.J., Sloane, N.J.A.: The Theory of Error-Correcting Codes, 5th edn. North-Holland, Amsterdam (1986)

21. McEliece, R.J.: A Public-Key System Based on Algebraic Coding Theory, pp. 114–116. Jet Propulsion Lab. (1978); DSN Progress Report 44

22. Minder, L., Shokrollahi, A.: Cryptanalysis of the Sidelnikov cryptosystem. In: Naor, M. (ed.) EUROCRYPT 2007. LNCS, vol. 4515, pp. 347–360. Springer, Heidelberg (2007)

23. Niederreiter, H.: Knapsack-type cryptosystems and algebraic coding theory. Problems Control Inform. Theory 15(2), 159–166 (1986)

24. Otmani, A., Tillich, J.P., Dallot, L.: Cryptanalysis of two McEliece cryptosystems based on quasi-cyclic codes (2008) (preprint)
25. Sendrier, N.: On the concatenated structure of a linear code. Appl. Algebra Eng. Commun. Comput. (AAECC) 9(3), 221–242 (1998)
26. Sendrier, N.: Cryptosystèmes à clé publique basés sur les codes correcteurs d'erreurs. Ph.D thesis, Université Paris 6, France (2002)
27. Sidelnikov, V.M.: A public-key cryptosystem based on binary Reed-Muller codes. Discrete Mathematics and Applications 4(3) (1994)
28. Sidelnikov, V.M., Shestakov, S.O.: On the insecurity of cryptosystems based on generalized Reed-Solomon codes. Discrete Mathematics and Applications 1(4), 439–444 (1992)
29. Stern, J.: A method for finding codewords of small weight. In: Cohen, G.D., Wolfmann, J. (eds.) Coding Theory 1988. LNCS, vol. 388, pp. 106–113. Springer, Heidelberg (1989)
30. van Tilburg, J.: On the mceliece public-key cryptosystem. In: Goldwasser, S. (ed.) CRYPTO 1988. LNCS, vol. 403, pp. 119–131. Springer, Heidelberg (1990)
31. Wieschebrink, C.: Two NP-complete problems in coding theory with an application in code based cryptography. In: IEEE International Symposium on Information Theory, July 2006, pp. 1733–1737 (2006)

Cryptanalysis of RSA Using the Ratio of the Primes

Abderrahmane Nitaj

Laboratoire de Mathématiques Nicolas Oresme
Université de Caen, France
nitaj@math.unicaen.fr
http://www.math.unicaen.fr/~nitaj

Abstract. Let $N = pq$ be an RSA modulus, i.e. the product of two large unknown primes of equal bit-size. In the $X9.31$-1997 standard for public key cryptography, Section 4.1.2, there are a number of recommendations for the generation of the primes of an RSA modulus. Among them, the ratio of the primes shall not be close to the ratio of small integers. In this paper, we show that if the public exponent e satisfies an equation $eX - (N - (ap + bq))Y = Z$ with suitably small integers X, Y, Z, where $\frac{a}{b}$ is an unknown convergent of the continued fraction expansion of $\frac{q}{p}$, then N can be factored efficiently. In addition, we show that the number of such exponents is at least $N^{\frac{3}{4}-\varepsilon}$ where ε is arbitrarily small for large N.

Keywords: RSA, Cryptanalysis, Factorization, Coppersmith's Method, Continued Fraction.

1 Introduction

The RSA public-key cryptosystem [15] was invented by Rivest, Shamir, and Adleman in 1978. Since then, the RSA system has been the most widely accepted public key cryptosystem. In the RSA cryptosystem, the modulus $N = pq$ is a product of two primes of equal bit-size. Let e be an integer coprime with $\phi(N) = (p-1)(q-1)$, the Euler function of N. Let d be the integer solution of the equation $ed \equiv 1 \pmod{\phi(N)}$ with $d < \phi(N)$. We call e the public exponent and d the private exponent. The pair (N, e) is called the public key and the pair (N, d) is the corresponding private key.

RSA is computationally expensive as it requires exponentiations modulo the large RSA modulus N. For efficient modular exponentiation in the decryption/signing phase, one may be tempted to choose a small d. Unfortunately, Wiener [17] showed in 1990 that using continued fractions, one can efficiently recover the secret exponent d from the public key (N, e) as long as $d < \frac{1}{3}N^{\frac{1}{4}}$. Wiener's attack is based on solving the equation $ex - \phi(N)y = 1$ where $x < \frac{1}{3}N^{\frac{1}{4}}$. Since then, attacking RSA using information encoded in the public key (N, e) has been a stimulating area of research.

B. Preneel (Ed.): AFRICACRYPT 2009, LNCS 5580, pp. 98–115, 2009.

Based on the lattice basis reduction, Boneh and Durfee [2] proposed in 1999 a new attack on the use of short secret exponent d, namely, they improved the bound to $d < N^{0.292}$.

In 2004, Blömer and May [1] showed that N can be factored in polynomial time for every public key (N, e) satisfying an equation $ex - (N+1-(p+q))k = y$, with $x < \frac{1}{3}N^{\frac{1}{4}}$ and $|y| < N^{-\frac{3}{4}}ex$.

Another attack using information encoded in (N, e) was recently proposed by Nitaj in [13]. The idea of [13] is based on solving the equation satisfied by the public exponent e. Suppose e satisfies an equation $eX - (p - u)(q - v)Y = 1$ with $1 \leq Y \leq X < 2^{-\frac{1}{4}}N^{\frac{1}{4}}$, $|u| < N^{\frac{1}{4}}$ and $v = \left[-\frac{qu}{p-u}\right]$. If the prime factors of $p - u$ or $q - v$ are less than 10^{50}, then N can be factored efficiently.

In this paper, we propose new attacks on RSA. Let $N = pq$ be an RSA modulus with $q < p < 2q$. Let $\frac{a}{b}$ be an unknown convergent of the continued fraction expansion of $\frac{q}{p}$. Define α such that $ap + bq = N^{\frac{1}{2}+\alpha}$ with $0 < \alpha < \frac{1}{2}$. We focus on the class of the public exponents satisfying an equation

$$eX - (N - (ap + bq))Y = Z,$$

with small parameters X, Y, Z satisfying

$$1 \leq Y \leq X < \frac{1}{2}N^{\frac{1}{4}-\frac{\alpha}{2}}, \ \gcd(X,Y) = 1,$$

and Z depends on the size of $|ap - bq|$. We present three attacks according to the size of the difference $|ap - bq|$. The first attack concerns small difference, i.e. $|ap - bq| < (abN)^{\frac{1}{4}}$, the second attack will work for medium difference, i.e. $(abN)^{\frac{1}{4}} < |ap - bq| < aN^{\frac{1}{4}}$, and the third attack concerns large difference, i.e. $|ap - bq| > aN^{\frac{1}{4}}$. The first attack always lead to the factorization of N. The second and the third attacks work if, in addition, $b \leq 10^{52}$. This corresponds to the current limit of the Elliptic Curve Method [8] to find large factors of integers.

The attacks combine techniques from the theory of continued fractions, Coppersmith's method [5] for finding small roots of bivariate polynomial equations and the Elliptic Curve Method [8] for Integer Factorization. We also show that the set of exponents e for which our approach works is at least $N^{\frac{3}{4}-\varepsilon}$ where ε is a small positive constant depending only on N.

Our approach is more efficient if $\frac{q}{p}$ is close to $\frac{a}{b}$ with small integers a and b. This is a step in the direction of the recommendations of the X9.31-1997 standard for public key cryptography (Section 4.1.2) which requires that the ratio of the primes shall not be close to the ratio of small integers. It is important to notice that, since $q < p < 2q$, then $\frac{0}{1}$ and $\frac{1}{1}$ are among the convergents of the continued fraction expansion of $\frac{q}{p}$ (see Section 2). For $a = 0$, $b = 1$, the equation $eX - (N - (ap + bq))Y = Z$ becomes

$$eX - q(p - 1)Y = Z.$$

and was studied by Nitaj [12] with suitably small parameters X, Y, Z. Consequently, in this paper, we focus on the convergents $\frac{a}{b}$ with $a \geq 1$. For $a = b = 1$, our third attack applies and matches the attack of Blömer and May [1].

The rest of the paper is organized as follows. In Section 2 we give a brief introduction to continued fractions, Coppersmith's lattice-based method for finding small roots of polynomials [5] and the Elliptic Curve Method of Factorization. In Section 3 we study the properties of the convergents of the continued fraction expansion of the ratio of the primes of $N = pq$. In Section 4 we present the new attacks. In Section 5, we give an estimate for the size of the set of the public exponents for which our attacks work. Section 6 concludes the paper.

2   Preliminaries on Continued Fractions, Coppersmith's Method and the Elliptic Curve Method (ECM)

We first introduce some notation. The integer closest to x is denoted $[x]$ and the largest integer less than or equal to x is denoted $\lfloor x \rfloor$.

2.1   Continued Fractions and the Euclidean Algorithm

We briefly recall some basic definitions and facts that we use about continued fractions and the Euclidean algorithm, which can be found in [6].

The process of finding the continued fraction expansion of a rational number $\frac{q}{p}$ involves the same series of long divisions that are used in the application of the Euclidean algorithm to the pair of integers (q, p). Starting with $r_{-2} = q$ and $r_{-1} = p$, define the recursions

$$a_i = \left\lfloor \frac{r_{i-2}}{r_{i-1}} \right\rfloor, \quad r_i = r_{i-2} - a_i r_{i-1}, \ i \geq 0, \tag{1}$$

where a_i is the integer quotient $\lfloor r_{i-2}/r_{i-1} \rfloor$ and r_i is the integer remainder that satisfies $0 \leq r_i < r_{i-1}$. The Euclidean algorithm terminates with a series of remainders satisfying

$$0 = r_m < r_{m-1} < \cdots < r_2 < r_1 < r_0 < r_{-1} = p.$$

The continued fraction expansion of $\frac{q}{p}$ is then

$$\frac{q}{p} = a_0 + \cfrac{1}{a_1 + \cfrac{1}{a_2 + \cfrac{1}{\cdots + \cfrac{1}{a_m}}}},$$

or alternatively, $\frac{q}{p} = [a_0, a_1, \cdots, a_m]$. The rational number $[a_0, a_1, \cdots, a_i]$ with $0 \leq i \leq m$ is called the i-th convergent of $\frac{q}{p}$ and satisfies

$$[a_0, a_1, \cdots, a_i] = \frac{p_i}{q_i},$$

where the integers p_i and q_i are coprime positive integers. Note that the integers p_i and q_i are also defined by the double recursions

$$p_{-2} = 0, \quad p_{-1} = 1, \quad p_i = a_i p_{i-1} + p_{i-2}, \ i \geq 0, \tag{2}$$

$$q_{-2} = 1, \quad q_{-1} = 0, \quad q_i = a_i q_{i-1} + q_{i-2}, \ i \geq 0. \tag{3}$$

Since $q < p < 2q$, we have $\frac{q}{p} < 1$ and taking $i = 0$ in (1), (2) and (3), we get

$$a_0 = \left\lfloor \frac{r_{-2}}{r_{-1}} \right\rfloor = \left\lfloor \frac{q}{p} \right\rfloor = 0, \quad r_0 = q, \quad p_0 = 0, \quad q_0 = 1.$$

Similarly, we have $1 < \frac{p}{q} < 2$ and taking $i = 1$ in (1), (2) and (3), we get

$$a_1 = \left\lfloor \frac{r_{-1}}{r_0} \right\rfloor = \left\lfloor \frac{p}{q} \right\rfloor = 1, \quad p_1 = 1, \quad q_1 = 1.$$

From this we deduce that the first convergents of the continued fraction expansion of $\frac{q}{p}$ are $\frac{0}{1}$ and $\frac{1}{1}$.

Proposition 1. *Let $\frac{q}{p} = [a_0, a_1, \cdots, a_m]$ be a continued fraction. For $0 \leq i < m$, we have*

$$\left| \frac{q}{p} - \frac{p_i}{q_i} \right| < \frac{1}{q_i^2}.$$

We terminate with a famous result on good rational approximations.

Theorem 1. *Let $\frac{q}{p} = [a_0, a_1, \cdots, a_m]$. If a and b are coprime positive integers such that $b < p$ and*

$$\left| \frac{q}{p} - \frac{a}{b} \right| < \frac{1}{2b^2},$$

then $a = p_i$ and $b = q_i$ for some i with $0 \leq i \leq m$.

2.2   Coppersmith's Method

At Eurocrypt'96, Coppersmith [5] introduced two lattice reduction based techniques to find small roots of polynomial diophantine equations. The first technique works for modular univariate polynomials, the second for bivariate integer polynomial equations. Since then, Coppersmith's techniques have been used in a huge variety of cryptanalytic applications. Coppersmith illustrated his technique for solving bivariate integer polynomial equations with the problem of finding the factors of $n = xy$ if we are given the high order $\frac{1}{4} \log_2 n$ bits of y.

Theorem 2. *Let $n = xy$ be the product of two unknown integers such that $x < y < 2x$. Given an approximation of y with additive error at most $n^{\frac{1}{4}}$, then x and y can be found in polynomial time.*

2.3   The Elliptic Curve Method of Factorization

The difficulty of factoring a large number is an element of the security of the RSA system. In the recent years, the limits of the best factorization algorithms have been extended greatly. There are two classes of algorithms for finding a nontrivial factor p of a composite integer n. The algorithms in which the run time depends on the size of n: Lehmans algorithm [7], the Continued Fraction algorithm [11], the Multiple Polynomial Quadratic Sieve algorithm [16], the Number Field Sieve [9]. And the algorithms in which the run time depends on the size of p: Trial Division, Pollard's "rho" algorithm [14], Lenstra's Elliptic Curve Method [8].

The Elliptic Curve Method (ECM for short) was originally proposed by H.W. Lenstra [8] and subsequently extended by Brent [3], [4], and Montgomery [10]. The original part of the algorithm proposed by Lenstra is typically referred to as Phase 1, and the extension by Brent and Montgomery is called Phase 2. ECM is suited to find small factors p of large numbers n and has complexity

$$\mathcal{O}\left(\exp\left\{c\sqrt{\log p \log \log p}\right\} M(n)\right),$$

where $c > 0$ and $M(n)$ denotes the cost of multiplication (mod n). R. Brent [4] extrapolated that the Elliptic Curve Method record will be a D-digit factor in year $Y(D) = 9.3\sqrt{D} + 1932.3$. According to this formula, $Y(50) \approx 1998$ and $Y(67) \approx 2008$. A table of the largest factors found using the ECM is maintained by Zimmermann [18]. The largest prime factor found using the ECM had 67 decimal digits and was found by B. Dodson on August 24, 2006.

3   Useful Lemmas and Properties

First we recall a very useful lemma (see [13]).

Lemma 1. *Let $N = pq$ be an RSA modulus with $q < p < 2q$. Then*

$$2^{-\frac{1}{2}}N^{\frac{1}{2}} < q < N^{\frac{1}{2}} < p < 2^{\frac{1}{2}}N^{\frac{1}{2}}.$$

The following lemma shows that a and b are of the same bit-size.

Lemma 2. *Let $N = pq$ be an RSA modulus with $q < p < 2q$. If $\frac{a}{b}$ is a convergent of $\frac{q}{p}$ with $a \geq 1$, then $a \leq b \leq 2a$.*

Proof. If $b = 1$, then $a = 1$ and the inequalities $a \leq b \leq 2a$ are satisfied. Next, suppose $b \geq 2$. Observe that if $\frac{a}{b}$ is a convergent of $\frac{q}{p}$ then by Proposition 1 we have $|ap - bq| \leq \frac{p}{b} \leq \frac{p}{2}$. Isolating bq and dividing by q, we get

$$a\frac{p}{q} - \frac{p}{2q} \leq b \leq a\frac{p}{q} + \frac{p}{2q}.$$

Combining this with $1 < \frac{p}{q} < 2$, we get

$$a - \frac{p}{2q} < a\frac{p}{q} - \frac{p}{2q} \leq b \leq a\frac{p}{q} + \frac{p}{2q} < 2a + \frac{p}{2q}.$$

Since $p < 2q$, then $0 < \frac{p}{2q} < 1$. Hence $a \leq b \leq 2a$ which completes the proof.   □

The following lemma plays an important role in this paper. Recall that the integer closest to x is denoted $[x]$.

Lemma 3. *Let $N = pq$ be an RSA modulus with $q < p < 2q$ and $\frac{a}{b}$ a convergent of the continued fraction expansion of $\frac{q}{p}$ with $a \geq 1$. Let $ap + bq = N^{\frac{1}{2}+\alpha}$ with $\alpha < \frac{1}{2}$. If $|ap + bq - M| < \frac{1}{2}N^{\frac{1}{2}-\alpha}$, then*

$$ab = \left[\frac{M^2}{4N}\right].$$

Proof. Set $M = ap + bq + x$. Using $(ap - bq)^2 = (ap + bq)^2 - 4abN$, we get, after rearrangement,

$$M^2 - 4abN = (ap + bq + x)^2 - 4abN = (ap - bq)^2 + 2(ap + bq)x + x^2. \quad (4)$$

Consider the term $(ap - bq)^2$ on the right side of (4). If $b = 1$, then by Lemma 2, $a = 1$. Hence, since $q < p < 2q$, we have $|ap - bq| = |p - q| = p - q < \frac{p}{2}$. If $b \geq 2$, then by Proposition 1, we have $|ap - bq| < \frac{p}{b} \leq \frac{p}{2}$. Combining with Lemma 1, we get in both cases

$$(ap - bq)^2 < \left(\frac{p}{2}\right)^2 < \left(\frac{2^{\frac{1}{2}}N^{\frac{1}{2}}}{2}\right)^2 = \frac{N}{2}.$$

Hence, using $|x| < \frac{1}{2}N^{\frac{1}{2}-\alpha}$, the right side of (4) becomes

$$\left|(ap - bq)^2 + 2(ap + bq)x + x^2\right| \leq (ap - bq)^2 + 2(ap + bq)|x| + x^2$$
$$< \frac{N}{2} + 2N^{\frac{1}{2}+\alpha} \cdot \frac{1}{2}N^{\frac{1}{2}-\alpha} + \frac{1}{4}N^{1-2\alpha}$$
$$= \left(\frac{1}{2} + 1 + \frac{1}{4}N^{-2\alpha}\right)N$$
$$< 2N,$$

where we used $\alpha > 0$. Plugging this in (4) and dividing by $4N$, we get

$$\left|\frac{M^2}{4N} - ab\right| = \frac{|M^2 - 4abN|}{4N} = \frac{|(ap - bq)^2 + 2(ap + bq)x + x^2|}{4N} < \frac{2N}{4N} = \frac{1}{2}.$$

It follows that $ab = \left[\frac{M^2}{4N}\right]$ which terminates the proof.   □

The following lemma indicates that ap and bq are of the same bit-size.

Lemma 4. *Let $N = pq$ be an RSA modulus with $q < p < 2q$ and $\frac{a}{b}$ a convergent of the continued fraction expansion of $\frac{q}{p}$ with $a \geq 1$. Then*

$$ap < bq < 2ap \quad or \quad bq < ap < 2bq$$

Proof. First, assume $ap < bq$. By Lemma 2, we have $b \leq 2a$. Combining this with $q < p$, we get $bq < 2ap$, and consequently $ap < bq < 2ap$.

Next, assume $bq < ap$. By Lemma 2, we have $a \leq b$. Combining this with $p < 2q$, we get $ap < 2bq$ and finally $bq < ap < 2bq$. This terminates the proof.   □

4   The New Attacks on RSA

In this section, we show how to factor the RSA modulus N if (N, e) is a public key satisfying an equation $eX - (N - (ap + bq))Y = Z$ with small parameters X, Y and Z where $\frac{a}{b}$ is an unknown convergent of $\frac{q}{p}$ with $a \geq 1$. We shall consider separately the cases when the difference $|ap - bq|$ is small, i.e. $|ap - bq| < (abN)^{\frac{1}{4}}$, medium, i.e. $(abN)^{\frac{1}{4}} < |ap - bq| < aN^{\frac{1}{4}}$, and large, i.e. $|ap - bq| > aN^{\frac{1}{4}}$. This corresponds approximately to $b > 2^{\frac{1}{2}} N^{\frac{1}{6}}$, $2^{\frac{1}{2}} N^{\frac{1}{6}} > b > 2^{\frac{1}{4}} N^{\frac{1}{8}}$ and $b < 2^{\frac{1}{4}} N^{\frac{1}{8}}$ respectively.

First we present a result based on continued fractions.

Lemma 5. *Let $N = pq$ be an RSA modulus with $q < p < 2q$. Let a, b be coprime positive integers such that $ap + bq = N^{\frac{1}{2}+\alpha}$ with $\alpha < \frac{1}{2}$. Let e be a public exponent satisfying the equation $eX - (N - (ap + bq))Y = Z$ with $\gcd(X, Y) = 1$. If $|Z| < N^{\frac{1}{2}+\alpha}X$ and $1 \leq Y \leq X < \frac{1}{2}N^{\frac{1}{4}-\frac{\alpha}{2}}$, then $\frac{Y}{X}$ is a convergent of $\frac{e}{N}$.*

Proof. Set $ap + bq = N^{\frac{1}{2}+\alpha}$ with $\alpha < \frac{1}{2}$. Rewrite $eX - (N - ap - bq)Y = Z$ as $eX - NY = Z - (ap + bq)Y$. Now suppose $|Z| < N^{\frac{1}{2}+\alpha}X$, $1 \leq Y \leq X$ and $\gcd(X, Y) = 1$. Then

$$
\begin{aligned}
\left| \frac{e}{N} - \frac{Y}{X} \right| &= \frac{|eX - NY|}{NX} \\
&= \frac{|Z - (ap + bq)Y|}{NX} \\
&\leq \frac{|Z|}{NX} + \frac{(ap + bq)Y}{NX} \\
&< \frac{N^{\frac{1}{2}+\alpha}}{N} + \frac{N^{\frac{1}{2}+\alpha}}{N} \\
&= 2N^{-\frac{1}{2}+\alpha}.
\end{aligned}
$$

Since $X < \frac{1}{2}N^{\frac{1}{4}-\frac{\alpha}{2}}$, then $2N^{-\frac{1}{2}+\alpha} < \frac{1}{2X^2}$. Hence, by Theorem 1, $\frac{Y}{X}$ is one of the convergents of the continued fraction expansion of $\frac{e}{N}$. □

4.1   An Attack for Small Difference $|ap - bq|$

We now present the first attack.

Theorem 3. *Let $N = pq$ be an RSA modulus with unknown factors p, q such that $q < p < 2q$. Let $\frac{a}{b}$ be an unknown convergent of the continued fraction expansion of $\frac{q}{p}$ with $a \geq 1$ and $|ap - bq| < (abN)^{\frac{1}{4}}$. Let e be a public exponent satisfying an equation $eX - (N - ap - bq)Y = Z$ with $\gcd(X, Y) = 1$. Set $ap + bq = N^{\frac{1}{2}+\alpha}$ with $0 < \alpha < \frac{1}{2}$. If $1 \leq Y \leq X < \frac{1}{2}N^{\frac{1}{4}-\frac{\alpha}{2}}$ and $|Z| < \inf\left((abN)^{\frac{1}{4}}, \frac{1}{2}N^{\frac{1}{2}-\alpha}\right)Y$, then N can be factored in polynomial time.*

Proof. Assume $|Z| < \inf\left((abN)^{\frac{1}{4}}, \frac{1}{2}N^{\frac{1}{2}-\alpha}\right)Y$, $1 \leq Y \leq X$ with $\gcd(X,Y) = 1$. Then

$$|Z| < \inf\left((abN)^{\frac{1}{4}}, \frac{1}{2}N^{\frac{1}{2}-\alpha}\right)X \leq \frac{1}{2}N^{\frac{1}{2}-\alpha}X < N^{\frac{1}{2}+\alpha}X.$$

Hence by Lemma 5, $\frac{Y}{X}$ is one of the convergents of $\frac{e}{N}$. Set $M = N - \frac{eX}{Y}$. Starting with the equation $eX - (N - (ap + bq))Y = Z$, we get

$$|ap + bq - M| = \frac{|Z|}{Y} < \inf\left((abN)^{\frac{1}{4}}, \frac{1}{2}N^{\frac{1}{2}-\alpha}\right) < \frac{1}{2}N^{\frac{1}{2}-\alpha}.$$

Hence, by Lemma 3, we find $ab = \left[\frac{M^2}{4N}\right]$. On the other hand, we have

$$|ap + bq - M| < \inf\left((abN)^{\frac{1}{4}}, \frac{1}{2}N^{\frac{1}{2}-\alpha}\right) < (abN)^{\frac{1}{4}}.$$

Moreover, if $|ap - bq| < (abN)^{\frac{1}{4}}$, then

$$\left|ap - \frac{M}{2}\right| \leq \frac{1}{2}|ap + bq - M| + \frac{1}{2}|ap - bq| < \frac{1}{2}(abN)^{\frac{1}{4}} + \frac{1}{2}(abN)^{\frac{1}{4}} = (abN)^{\frac{1}{4}}.$$

It follows that the term $\frac{M}{2}$ is an approximation of the factor ap of $n = abN$ with additive error at most $n^{\frac{1}{4}}$. In addition, by Lemma 4, the factors ap and bq of n are of the same bit-size. Hence, using Theorem 2 with n and $\frac{M}{2}$, we find ap, and since $a < q$, we get $p = \gcd(N, ap)$ which terminates the proof. □

Let us summarize the first factorization algorithm.

Algorithm 1. Small $|ap - bq|$

Input: a public key (N, e) satisfying $N = pq$, $q < p < 2q$ and $eX - (N-(ap+bq))Y = Z$ for small parameters X, Y, Z where $\frac{a}{b}$ is an unknown convergent of $\frac{q}{p}$ with $a \geq 1$.
Output: the prime factors p and q.
1: Compute the continued fraction expansion of $\frac{e}{N}$.
2: For every convergent $\frac{Y}{X}$ of $\frac{e}{N}$ with $X < \frac{1}{2}N^{\frac{1}{4}}$:
3: Compute $M = N - \frac{eX}{Y}$ and $N_0 = \left[\frac{M^2}{4N}\right]$.
4: Apply Coppersmith's algorithm (Theorem 2) with $n = N_0N$ and $\frac{M}{2}$ as an approximation of y.
5: Compute $g = \gcd(y, N)$. If $1 < g < N$, then stop.

4.2   An Attack for Medium Difference $|ap - bq|$

Here we present the second attack. It is based on the Elliptic Curve Method (ECM) which can find factors of about 52-digits. Assuming the efficiency of ECM, every step in this attack can be done in polynomial time and the number of convergents is bounded by $\mathcal{O}(\log N)$. To express this fact, the term *efficient* is used.

Theorem 4. *Let $N = pq$ be an RSA modulus with unknown factors p, q such that $q < p < 2q$. Let $\frac{a}{b}$ be an unknown convergent of the continued fraction expansion of $\frac{q}{p}$ such that $a \geq 1$, $(abN)^{\frac{1}{4}} < |ap - bq| < aN^{\frac{1}{4}}$ and $b \leq 10^{52}$. Let e be a public exponent satisfying an equation $eX - (N - ap - bq)Y = Z$ with $\gcd(X,Y) = 1$. Set $M = N - \frac{eX}{Y}$ and $ap + bq = N^{\frac{1}{2}+\alpha}$ with $0 < \alpha < \frac{1}{2}$. If $1 \leq Y \leq X < \frac{1}{2}N^{\frac{1}{4}-\frac{\alpha}{2}}$ and $|Z| < \min\left(aN^{\frac{1}{4}}, \frac{1}{2}N^{\frac{1}{2}-\alpha}\right)Y$, then, under ECM, N can be factored efficiently.*

Proof. Assume $|Z| < \min\left(aN^{\frac{1}{4}}, \frac{1}{2}N^{\frac{1}{2}-\alpha}\right)Y$, $1 \leq Y \leq X$ and $\gcd(X,Y) = 1$. Then

$$|Z| < \min\left(aN^{\frac{1}{4}}, \frac{1}{2}N^{\frac{1}{2}-\alpha}\right)X \leq \frac{1}{2}N^{\frac{1}{2}-\alpha}X < N^{\frac{1}{2}+\alpha}X.$$

It follows, by Lemma 5, that $\frac{Y}{X}$ is among the convergents of $\frac{e}{N}$.
Next, set $M = N - \frac{eX}{Y}$. Using the equation $eX - (N - (ap + bq))Y = Z$, we get

$$|ap + bq - M| = \frac{|Z|}{Y} < \min\left(aN^{\frac{1}{4}}, \frac{1}{2}N^{\frac{1}{2}-\alpha}\right) \leq \frac{1}{2}N^{\frac{1}{2}-\alpha}.$$

Hence, by Lemma 3, we find $ab = \left[\frac{M^2}{4N}\right]$ and by Lemma 2, we know that a and b are of equal bit-size. Hence, applying the Elliptic Curve Method with $\left[\frac{M^2}{4N}\right]$, we can efficiently find a and b assuming $b \leq 10^{52}$.
From $|ap + bq - M| < aN^{\frac{1}{4}}$, we get

$$\left|p + \frac{bq}{a} - \frac{M}{a}\right| < \frac{aN^{\frac{1}{4}}}{a} = N^{\frac{1}{4}}. \tag{5}$$

On the other hand, by assumption, $|ap - bq| < aN^{\frac{1}{4}}$. Then $\left|p - \frac{bq}{a}\right| < N^{\frac{1}{4}}$, and combining with (5), we get

$$\begin{aligned}
\left|p - \frac{M}{2a}\right| &= \left|\frac{1}{2}\left(p + \frac{bq}{a} - \frac{M}{a}\right) + \frac{1}{2}\left(p - \frac{bq}{a}\right)\right| \\
&\leq \frac{1}{2}\left|p + \frac{bq}{a} - \frac{M}{a}\right| + \frac{1}{2}\left|p - \frac{bq}{a}\right| \\
&< \frac{1}{2}N^{\frac{1}{4}} + \frac{1}{2}N^{\frac{1}{4}} \\
&= N^{\frac{1}{4}}.
\end{aligned}$$

This implies that $\frac{M}{2a}$ is an approximation of p with additive error at most $N^{\frac{1}{4}}$. Then, using Theorem 2, this gives p which terminates the proof.   □

Here we summarize the second factorization algorithm.

Algorithm 2. Medium $|ap - bq|$

Input: a public key (N, e) satisfying $N = pq$, $q < p < 2q$ and $eX - (N - (ap + bq))Y = Z$
for small parameters X, Y, Z where $\frac{a}{b}$ is an unknown convergent of $\frac{q}{p}$ with $a \geq 1$.
Output: the prime factors p and q.
1: Compute the continued fraction expansion of $\frac{e}{N}$.
2: For every convergent $\frac{Y}{X}$ of $\frac{e}{N}$ with $X < \frac{1}{2}N^{\frac{1}{4}}$:
3: Compute $M = N - \frac{eX}{Y}$ and $N_0 = \left[\frac{M^2}{4N}\right]$.
4: **if** $N_0 < 10^{104}$ **then**
5:     Apply ECM to find a and b such that $N_0 = ab$ and $a \leq b \leq 2a$.
6:     Apply Coppersmith's algorithm (Theorem 2) with $n = N$ and $\frac{M}{2a}$ as an approx-
       imation of y. If Coppersmith's algorithm outputs the factors p and q of N, then
       stop.
7: **end if**

4.3   An Attack for Large Difference $|ap - bq|$

Here we present the last attack. We suppose $|ap - bq| > aN^{\frac{1}{4}}$ so that the Small
and the Medium difference attacks should not succeed. This attack depends on
the efficiency of the Elliptic Curve Method (ECM) to find factors up to 10^{52}.
Assuming the efficiency of ECM, the term *efficient* is also used to express the
fact that every step in this attack can be done in polynomial time.

Theorem 5. *Let $N = pq$ be an RSA modulus with unknown factors p, q such
that $q < p < 2q$. Let $\frac{a}{b}$ be an unknown convergent of the continued fraction
expansion of $\frac{q}{p}$ such that $a \geq 1$ and $b \leq 10^{52}$. Let e be a public exponent satisfying
an equation $eX - (N - (ap + bq))Y = Z$ with $\gcd(X, Y) = 1$. Let $M = N - \frac{eX}{Y}$.
Set $D = \sqrt{|M^2 - 4abN|}$ and $ap + bq = N^{\frac{1}{2} + \alpha}$ with $0 < \alpha < \frac{1}{2}$. If $1 \leq Y \leq X <
\frac{1}{2}N^{\frac{1}{4} - \frac{\alpha}{2}}$ and $|Z| < \frac{1}{3}a|ap - bq|N^{-\frac{1}{4} - \alpha}Y$ then, under ECM, N can be factored
efficiently.*

Proof. Combining Proposition 1 and Lemma 1, we have

$$|ap - bq| < \frac{p}{b} < \frac{2^{\frac{1}{2}}N^{\frac{1}{2}}}{b}.$$

Hence, since $a \leq b$, this gives

$$\frac{1}{3}a|ap - bq|N^{-\frac{1}{4} - \alpha} < \frac{1}{3}a \cdot \frac{2^{\frac{1}{2}}N^{\frac{1}{2}}}{b} \cdot N^{-\frac{1}{4} - \alpha} \leq \frac{2^{\frac{1}{2}}}{3}N^{\frac{1}{4} - \alpha}. \tag{6}$$

Now, suppose $|Z| < \frac{1}{3}a|ap - bq|N^{-\frac{1}{4} - \alpha}Y$, $1 \leq Y \leq X$ and $\gcd(X, Y) = 1$. Then
using (6), we get

$$|Z| < \frac{2^{\frac{1}{2}}}{3}N^{\frac{1}{4} - \alpha}X < N^{\frac{1}{2} + \alpha}X.$$

Consequently, by Lemma 5, $\frac{Y}{X}$ is a convergent of $\frac{e}{N}$. Next, set $M = N - \frac{eX}{Y}$. Using the equation $eX - (N - ap - bq)Y = Z$, we get

$$|ap + bq - M| = \frac{|Z|}{Y} < \frac{1}{3}a|ap - bq|N^{-\frac{1}{4}-\alpha}. \tag{7}$$

Then using (6), we get

$$|ap + bq - M| < \frac{2^{\frac{1}{2}}}{3}N^{\frac{1}{4}-\alpha} < \frac{1}{2}N^{\frac{1}{2}-\alpha}.$$

Hence, by Lemma 3, $ab = \left[\frac{M^2}{4N}\right]$ and by Lemma 2, we know that a and b are of the same bit-size. Hence, if $b \leq 10^{52}$, then applying the Elliptic Curve Method with $\left[\frac{M^2}{4N}\right]$, we can find a and b.

Next, using $|ap - bq| < 2N^{\frac{1}{2}}$, we can rewrite (7) as

$$|ap + bq - M| < \frac{1}{3}a \cdot 2N^{\frac{1}{2}} \cdot N^{-\frac{1}{4}-\alpha} = \frac{2}{3}aN^{\frac{1}{4}-\alpha} < aN^{\frac{1}{4}}. \tag{8}$$

Now, let $D = \sqrt{|M^2 - 4abN|}$. Then

$$\begin{aligned}
\left||ap - bq|^2 - D^2\right| &= \left||ap - bq|^2 - |M^2 - 4abN|\right| \\
&\leq \left|(ap - bq)^2 - M^2 + 4abN\right| \\
&= \left|(ap + bq)^2 - M^2\right|.
\end{aligned}$$

From this we deduce

$$\left||ap - bq| - D\right| \leq \frac{|ap + bq - M||ap + bq + M|}{|ap - bq| + D}.$$

Next, by (8), we have $|ap + bq - M| < aN^{\frac{1}{4}}$. Then $M < ap + bq + aN^{\frac{1}{4}}$ and

$$ap + bq + M < 2(ap + bq) + aN^{\frac{1}{4}} < 3(ap + bq) = 3N^{\frac{1}{2}+\alpha}.$$

Combining with (7), this leads to

$$\left||ap - bq| - D\right| < \frac{3 \cdot \frac{1}{3}a|ap - bq|N^{-\frac{1}{4}-\alpha}N^{\frac{1}{2}+\alpha}}{|ap - bq|} = aN^{\frac{1}{4}}.$$

If $ap - bq > 0$, then combining with (8), we get

$$\begin{aligned}
|2ap - M - D| &= \left|ap + bq - M + |ap - bq| - D\right| \\
&\leq |ap + bq - M| + \left||ap - bq| - D\right| \\
&< 2aN^{\frac{1}{4}}.
\end{aligned}$$

Dividing by $2a$, we find that $\frac{M+D}{2a}$ is an approximation of p with additive error at most $N^{\frac{1}{4}}$.

If $ap - bq < 0$, then combining with (8), we get

$$|2ap - M + D| = |ap + bq - M - (bq - ap - D)|$$
$$< |ap + bq - M| + ||ap - bq| - D|$$
$$< 2aN^{\frac{1}{4}}.$$

Dividing again by $2a$, we find that $\frac{M-D}{2a}$ is an approximation of p with additive error at most $N^{\frac{1}{4}}$. We can then apply Theorem 2 to the values $\frac{M \pm D}{2a}$. The correct term will lead to the factorization of N. □

Now we summarize the third factorization algorithm.

Algorithm 3. Large $|ap - bq|$

Input: a public key (N, e) satisfying $N = pq$, $q < p < 2q$ and $eX - (N - (ap+bq))Y = Z$ for small parameters X, Y, Z where $\frac{a}{b}$ is an unknown convergent of $\frac{q}{p}$ with $a \geq 1$.
Output: the prime factors p and q.
1: Compute the continued fraction expansion of $\frac{e}{N}$.
2: For every convergent $\frac{Y}{X}$ of $\frac{e}{N}$ with $X < \frac{1}{2}N^{\frac{1}{4}}$:
3: Compute $M = N - \frac{eX}{Y}$ and $N_0 = \left[\frac{M^2}{4N}\right]$.
4: **if** $N_0 < 10^{104}$ **then**
5:    Apply ECM to find a and b such that $N_0 = ab$ and $a \leq b \leq 2a$.
6:    Compute $D = \sqrt{|M^2 - 4N_0 N|}$.
7:    Compute $m_1 = \frac{M+D}{2a}$ and $m_2 = \frac{M-D}{2a}$.
8:    Apply Coppersmith's algorithm (Theorem 2) with $n = N$ and m_1 and m_2 as approximations of y. If Coppersmith's algorithm outputs the factors p and q, then stop.
9: **end if**

---

5   Estimation of the Public Exponents for which the Attacks Apply

In this Section, we will study the size of the class of the public keys for which our attacks can be applied. Let $\frac{a}{b}$ be a convergent of $\frac{q}{p}$ with $a \geq 1$. Define α by $ap + bq = N^{\frac{1}{2}+\alpha}$ with $0 < \alpha < \frac{1}{2}$ and let

$$\mathcal{P}(a, b) = \left\{ (X, Y, z) \mid 1 \leq Y \leq X < \frac{1}{2}N^{\frac{1}{4} - \frac{\alpha}{2}}, \ \gcd(X, Y) = 1, \ |z| < N^{\frac{1}{4} - \frac{\alpha}{2}} \right\},$$

be the set of the parameters and

$$\mathcal{E}(a, b) = \left\{ e \mid e = \left[(N - (ap + bq))\frac{Y}{X} \right] + z, \ (X, Y, z) \in \mathcal{P}(a, b) \right\},$$

the set of the exponents. We will show that much of these exponents are vulnerable to our attacks. To find a lower bound for the size of the sets $\mathcal{E}(a, b)$, we show that different convergents $\frac{a}{b}$ of $\frac{q}{p}$ and different parameters in the set $\mathcal{P}(a,b)$ define different exponents in the sets $\mathcal{E}(a,b)$.

First, we show that our attacks will work for the exponents in $\mathcal{E}(a,b)$: given an exponent in $\mathcal{E}(a,b)$, it is possible to find the factorization of N according to Theorem 3, Theorem 4 or Theorem 5. First, we start with a result for small difference $|ap - bq|$.

Corollary 1. *Let $N = pq$ be an RSA modulus with $q < p < 2q$. Let $\frac{a}{b}$ be an unknown convergent of $\frac{q}{p}$ with $a \geq 1$ and $|ap - bq| < (abN)^{\frac{1}{4}}$. Let X, Y be unknown coprime positive integers with $1 \leq Y \leq X < \frac{1}{2}N^{\frac{1}{4}-\frac{\alpha}{2}}$ where $ap + bq = N^{\frac{1}{2}+\alpha}$ and $0 < \alpha < \frac{1}{2}$. If $e = \lfloor (N - (ap + bq))\frac{Y}{X} \rfloor + z$ is a public exponent with*

$$|z| < \inf\left((abN)^{\frac{1}{4}}\frac{Y}{X}, N^{\frac{1}{4}-\frac{\alpha}{2}} \right),$$

then N can be factored in polynomial time.

Proof. Set $e_0 = \lfloor (N - (ap + bq))\frac{Y}{X} \rfloor$, $e = e_0 + z$, $Z = eX - (N - (ap + bq))Y$. We want to show that the conditions of Theorem 3 are satisfied. Assume that $|z| < \inf\left((abN)^{\frac{1}{4}}\frac{Y}{X}, N^{\frac{1}{4}-\frac{\alpha}{2}} \right)$. Then, since

$$\left| (N - (ap + bq))\frac{Y}{X} - e_0 \right| < 1,$$

we get

$$\begin{aligned}
|Z| = |eX - (N - (ap + bq))Y| &= |(e_0 + z)X - (N - (ap + bq))Y| \\
&\leq |e_0 X - (N - (ap + bq))Y| + |z|X \\
&< (1 + |z|)X.
\end{aligned}$$

Observe that $(1 + |z|)X < (abN)^{\frac{1}{4}}Y$ and, assuming $X < \frac{1}{2}N^{\frac{1}{4}-\frac{\alpha}{2}}$, we find

$$(1 + |z|)X < N^{\frac{1}{4}-\frac{\alpha}{2}} \cdot \frac{1}{2}N^{\frac{1}{4}-\frac{\alpha}{2}} \leq \frac{1}{2}N^{\frac{1}{2}-\alpha}Y.$$

From this, we deduce $|Z| < \inf\left((abN)^{\frac{1}{4}}, \frac{1}{2}N^{\frac{1}{2}-\alpha} \right)Y$. It follows that the conditions of Theorem 3 are fulfilled which leads to the factorization of N. $\square$

Next, we give a result for medium difference $|ap - bq|$.

Corollary 2. *Let $N = pq$ be an RSA modulus with $q < p < 2q$. Let $\frac{a}{b}$ be an unknown convergent of $\frac{q}{p}$ with $a \geq 1$, $b \leq 10^{52}$ and $(abN)^{\frac{1}{4}} < |ap - bq| < aN^{\frac{1}{4}}$. Let X, Y be unknown coprime positive integers with $1 \leq Y \leq X < \frac{1}{2}N^{\frac{1}{4}-\frac{\alpha}{2}}$ where $ap + bq = N^{\frac{1}{2}+\alpha}$ and $0 < \alpha < \frac{1}{2}$. If $e = \lfloor (N - (ap + bq))\frac{Y}{X} \rfloor + z$ is a public exponent with*

$$|z| < \inf\left(aN^{\frac{1}{4}}\frac{Y}{X}, N^{\frac{1}{4}-\frac{\alpha}{2}} \right),$$

then, under ECM, N can be factored efficiently.

Proof. The proof is similar to that of Corollary 1 and the parameters satisfy the condition of Theorem 4. □

Finally, we give a result which concerns large difference $|ap - bq|$.

Corollary 3. *Let* $N = pq$ *be an RSA modulus with* $q < p < 2q$. *Let* $\frac{a}{b}$ *be an unknown convergent of* $\frac{q}{p}$ *with* $a \geq 1$, $b \leq 10^{52}$ *and* $|ap - bq| > aN^{\frac{1}{4}}$. *Let* X, Y *be unknown coprime positive integers with* $1 \leq Y \leq X < \frac{1}{2}N^{\frac{1}{4}-\frac{\alpha}{2}}$ *where* $ap + bq = N^{\frac{1}{2}+\alpha}$ *and* $0 < \alpha < \frac{1}{2}$. *If* $e = \lfloor(N - (ap + bq))\frac{Y}{X}\rfloor + z$ *is a public exponent with*

$$|z| < \min\left(\frac{1}{3}a|ap - bq|N^{-\frac{1}{4}-\alpha}\frac{Y}{X}, N^{\frac{1}{4}-\frac{\alpha}{2}}\right),$$

then, under ECM, N *can be factored efficiently.*

Proof. Let $Z = eX - (N - (ap + bq))Y$, $e = \lfloor(N - (ap + bq))\frac{Y}{X}\rfloor + z$ with $|z| < \min\left(\frac{1}{3}a|ap - bq|N^{-\frac{1}{4}-\alpha}\frac{Y}{X}, N^{\frac{1}{4}-\frac{\alpha}{2}}\right)$ and $1 \leq Y \leq X < \frac{1}{2}N^{\frac{1}{4}-\frac{\alpha}{2}}$. Using the same arguments as in the proof of Corollary 1, we get

$$|Z| < (1 + |z|)X < \frac{1}{3}a|ap - bq|N^{-\frac{1}{4}-\alpha}Y.$$

It follows that all the conditions of Theorem 5 are fulfilled which leads to the factorization of N. □

The following result shows that distinct parameters from $\mathcal{P}(a, b)$ define different exponents in $\mathcal{E}(a, b)$.

Lemma 6. *Let* $N = pq$ *be an RSA modulus with* $q < p < 2q$. *Let* $\frac{a}{b}$ *be a convergent of* $\frac{q}{p}$ *with* $a \geq 1$ *and* $ap + bq = N^{\frac{1}{2}+\alpha}$. *Let* $(X, Y, z), (X', Y', z') \in \mathcal{P}(a, b)$. *Let*

$$e = \left\lfloor(N - (ap + bq))\frac{Y}{X}\right\rfloor + z, \quad e' = \left\lfloor(N - (ap + bq))\frac{Y'}{X'}\right\rfloor + z'.$$

If $e = e'$ *then* $X = X'$, $Y = Y'$ *and* $z = z'$.

Proof. Let $e_0 = \lfloor(N - (ap + bq))\frac{Y}{X}\rfloor$, $e_0' = \lfloor(N - (ap + bq))\frac{Y'}{X'}\rfloor$. If $e = e_0 + z$ and $e' = e_0' + z'$ then

$$\left|(N - (ap + bq))\left(\frac{Y'}{X'} - \frac{Y}{X}\right) - e' + e\right|$$
$$\leq \left|(N - (ap + bq))\frac{Y'}{X'} - e_0' - z'\right| + \left|(N - (ap + bq))\frac{Y}{X} - e_0 - z\right|$$
$$\leq \left|(N - (ap + bq))\frac{Y'}{X'} - e_0'\right| + |z'| + \left|(N - (ap + bq))\frac{Y}{X} - e_0\right| + |z|$$
$$< 2 + |z| + |z'|.$$

Suppose $e = e'$. Then, multiplying by XX', we get

$$(N - (ap + bq)) |Y'X - YX'| < (2 + |z| + |z'|)XX'. \tag{9}$$

We want to compare the sides of (9). Assume that $X, X' < \frac{1}{2}N^{\frac{1}{4} - \frac{\alpha}{2}}$ and $|z|, |z'| < N^{\frac{1}{4} - \frac{\alpha}{2}}$. Then

$$(2 + |z| + |z'|)XX' < 2N^{\frac{1}{4} - \frac{\alpha}{2}} \cdot \frac{1}{4} N^{\frac{1}{2} - \alpha} = \frac{1}{2} N^{\frac{3}{4} - \frac{3\alpha}{2}}.$$

On the other hand, we have

$$N - (ap + bq) = N - N^{\frac{1}{2} + \alpha} = N^{\frac{3}{4} - \frac{3\alpha}{2}} \left(N^{\frac{1}{4} + \frac{3\alpha}{2}} - N^{-\frac{1}{4} + \frac{5\alpha}{2}} \right).$$

Since $0 < \alpha < \frac{1}{2}$, then $\frac{1}{4} + \frac{3\alpha}{2} > -\frac{1}{4} + \frac{5\alpha}{2}$ and $N^{\frac{1}{4} + \frac{3\alpha}{2}} > N^{-\frac{1}{4} + \frac{5\alpha}{2}} + 1$. Hence $N - (ap + bq) > N^{\frac{3}{4} - \frac{3\alpha}{2}}$. From our comparison of the sides of (9), we conclude that $Y'X - YX' = 0$. Since $\gcd(X, Y) = 1$ and $\gcd(X', Y') = 1$, we find $X = X'$ and $Y = Y'$ and consequently $z = z'$. This terminates the proof. $\qquad\square$

Finally, the following result shows that different convergents of $\frac{q}{p}$ lead to different exponents in $\mathcal{E}(a, b)$.

Lemma 7. *Let $N = pq$ be an RSA modulus with $q < p < 2q$. Let $\frac{a}{b}$ and $\frac{a'}{b'}$ be convergents of $\frac{q}{p}$ with $a \geq 1$, $a' \geq 1$, $ap + bq = N^{\frac{1}{2} + \alpha}$ and $a'p + b'q = N^{\frac{1}{2} + \alpha'}$. Let $(X, Y, z) \in \mathcal{P}(a, b)$ and $(X', Y', z') \in \mathcal{P}(a', b')$. Let*

$$e = \left\lfloor (N - (ap + bq)) \frac{Y}{X} \right\rfloor + z, \quad e' = \left\lfloor (N - (a'p + b'q)) \frac{Y'}{X'} \right\rfloor + z'.$$

If $e = e'$ then $X = X'$, $Y = Y'$, $a = a'$, $b = b'$ and $z = z'$.

Proof. Assume for contradiction that $a \neq a'$, $a < a'$ say. Then $b < b'$. Hence $ap + bq < a'p + b'q$ and $\alpha < \alpha'$. Combining with Lemma 1, we get

$$N - (ap + bq) - (N - (a'p + b'q)) = (a' - a)p + (b' - b)q > p + q > p > N^{\frac{1}{2}},$$

which leads to

$$N - (ap + bq) > N - (a'p + b'q) + N^{\frac{1}{2}} \tag{10}$$

Now, set $e = \left\lfloor (N - (ap + bq)) \frac{Y}{X} \right\rfloor + z$, $e' = \left\lfloor (N - (a'p + b'q)) \frac{Y'}{X'} \right\rfloor + z'$ and assume $e = e'$. Then, since $|z| < N^{\frac{1}{4} - \frac{\alpha}{2}}$ and $|z'| < N^{\frac{1}{4} - \frac{\alpha'}{2}} < N^{\frac{1}{4} - \frac{\alpha}{2}}$, we get

$$\left| (N - (a'p + b'q)) \frac{Y'}{X'} - (N - (ap + bq)) \frac{Y}{X} \right| < 2 + |z| + |z'| < 2N^{\frac{1}{4} - \frac{\alpha}{2}}. \tag{11}$$

On the other hand, we know that $\frac{Y}{X}$ and $\frac{Y'}{X'}$ are convergents of the continued fraction expansion of $\frac{e}{N}$. Hence $\frac{Y}{X} \approx \frac{Y'}{X'}$ and, combining (10) with $X < \frac{1}{2}N^{\frac{1}{4} - \frac{\alpha}{2}}$, we get

$$(N - (ap + bq))\frac{Y}{X} > (N - (a'p + b'q))\frac{Y}{X} + N^{\frac{1}{2}}\frac{Y}{X}$$

$$> (N - (a'p + b'q))\frac{Y}{X} + N^{\frac{1}{2}} \cdot \frac{1}{\frac{1}{2}N^{\frac{1}{4} - \frac{\alpha}{2}}}$$

$$\approx (N - (a'p + b'q))\frac{Y'}{X'} + 2N^{\frac{1}{4} + \frac{\alpha}{2}}$$

It follows that

$$\left| (N - (a'p + b'q))\frac{Y'}{X'} - (N - (ap + bq))\frac{Y}{X} \right| > 2N^{\frac{1}{4} + \frac{\alpha}{2}}.$$

Comparing with (11), we get a contradiction. Hence $a = a'$ and $b = b'$. Now, we have $\lfloor (N - (ap + bq))\frac{Y}{X} \rfloor + z = \lfloor (N - (ap + bq))\frac{Y'}{X'} \rfloor + z'$. By Lemma 6, we conclude that $X = X'$, $Y = Y'$ and $z = z'$. This terminates the proof. □

Let us now prove a lower bound for the size of the number of the exponents e that are vulnerable to our approach. Note that we do not require $\gcd(e, \phi(N)) = 1$ as usual.

Theorem 6. *Let $N = pq$ be an RSA modulus with $q < p < 2q$. Then the number of the exponents $e \in \mathcal{E}(a, b)$ that are vulnerable to the attacks for some convergent $\frac{a}{b} \neq \frac{0}{1}$ of $\frac{q}{p}$ is at least $N^{\frac{3}{4} - \varepsilon}$ where ε is arbitrarily small for suitably large N.*

Proof. We focus on $\mathcal{E}(1, 1)$ since the total number of exponents is much higher. Let α_0 such that $p + q = N^{\frac{1}{2} + \alpha_0}$. Since $q < p$, then $2q < p + q < 2p$ and by Lemma 1, we get $2^{\frac{1}{2}}N^{\frac{1}{2}} < N^{\frac{1}{2} + \alpha_0} < 2^{\frac{3}{2}}N^{\frac{1}{2}}$. From this we deduce $\alpha_0 \approx 0$. On the other hand, by Corollary 3, we need

$$|z| < \min\left(\frac{1}{3}|p - q|N^{-\frac{1}{4} - \alpha_0}\frac{Y}{X}, N^{\frac{1}{4} - \frac{\alpha_0}{2}} \right),$$

where $1 \leq Y \leq X < \frac{1}{2}N^{\frac{1}{4} - \frac{\alpha_0}{2}}$ and $\gcd(X, Y) = 1$. Observe that for the normal RSA, we have $p - q > cN^{\frac{1}{2}}$ with a constant $c > 0$. So let

$$|z| < \min\left(\frac{c}{3}N^{\frac{1}{4} - \alpha_0}\frac{Y}{X}, N^{\frac{1}{4} - \frac{\alpha_0}{2}} \right),$$

and put

$$X_0 = \left\lfloor \frac{1}{2}N^{\frac{1}{4} - \frac{\alpha_0}{2}} \right\rfloor.$$

We want to estimate

$$\#\mathcal{E}(1, 1) = \sum_{X=1}^{X_0} \sum_{\substack{Y=1 \\ \gcd(X,Y)=1}}^{X-1} |z|.$$

Taking $|z| < \frac{c}{3} N^{\frac{1}{4} - \alpha_0} \frac{Y}{X}$, we get

$$
\#\mathcal{E}(1,1) = \frac{c}{3} N^{\frac{1}{4} - \alpha_0} \sum_{X=1}^{X_0} \sum_{\substack{Y=1 \\ \gcd(X,Y)=1}}^{X-1} \frac{Y}{X} = \frac{c}{6} N^{\frac{1}{4} - \alpha_0} \sum_{X=1}^{X_0} \phi(X), \qquad (12)
$$

where we used the well known identity

$$
\sum_{\substack{Y=1 \\ \gcd(X,Y)=1}}^{X-1} Y = \frac{1}{2} X \phi(X).
$$

Similarly, taking $|z| < N^{\frac{1}{4} - \frac{\alpha_0}{2}}$, we get

$$
\#\mathcal{E}(1,1) = N^{\frac{1}{4} - \frac{\alpha_0}{2}} \sum_{X=1}^{X_0} \sum_{\substack{Y=1 \\ \gcd(X,Y)=1}}^{X-1} 1 = N^{\frac{1}{4} - \frac{\alpha_0}{2}} \sum_{X=1}^{X_0} \phi(X). \qquad (13)
$$

We can rewrite (12) and (13) in a single expression

$$
\#\mathcal{E}(1,1) = N^{\frac{1}{4} - \varepsilon_0} \sum_{X=1}^{X_0} \phi(X),
$$

for a suitable $\varepsilon_0 > 0$. It is well known (see Theorem 328 of [6]), that

$$
\phi(X) > \frac{CX}{\log \log X},
$$

where C is a positive constant. Since $X < N$, then $\phi(X) > X N^{-\varepsilon_1}$ for a small positive constant ε_1. From this, we deduce

$$
\#\mathcal{E}(1,1) > N^{\frac{1}{4} - \varepsilon_0 - \varepsilon_1} \sum_{X=1}^{X_0} X > N^{\frac{1}{4} - \varepsilon_0 - \varepsilon_1} \frac{X_0^2}{2} > \frac{1}{8} N^{\frac{3}{4} - \alpha_0 - \varepsilon_0 - \varepsilon_1},
$$

where we used $X_0 \approx \frac{1}{2} N^{\frac{1}{4} - \frac{\alpha_0}{2}}$. We get finally $\#\mathcal{E}(1,1) > N^{\frac{3}{4} - \varepsilon}$, with a constant $\varepsilon \approx \alpha_0 + \varepsilon_0 + \varepsilon_1$ depending only on N. This terminates the proof. $\qquad \square$

6   Conclusion

In this paper, we showed how to perform three attacks on RSA using the ratio of the primes. The attacks apply when the public key (N, e) satisfies an equation $eX - (N - (ap + bq))Y = Z$ with suitably small parameters X, Y and Z where $\frac{a}{b}$ is an unknown convergent of $\frac{q}{p}$ with $a \geq 1$. The attacks combine a variety of techniques, including continued fractions, Coppersmith's lattice based method and H.W. Lenstra's Elliptic Curve Method for Factoring (ECM). Our results illustrate once again the fact that we should be very cautious when using RSA with specific exponents. Moreover, we showed that the number of such exponents is at least $N^{\frac{3}{4} - \varepsilon}$. Using the notion of weak keys, as defined by Blömer and May [1], the results of this paper show that this set of RSA public keys is a class of weak keys.

References

1. Blömer, J., May, A.: A generalized Wiener attack on RSA. In: Bao, F., Deng, R., Zhou, J. (eds.) PKC 2004. LNCS, vol. 2947, pp. 1–13. Springer, Heidelberg (2004)
2. Boneh, D., Durfee, G.: Cryptanalysis of RSA with private key d less than $N^{0.292}$. In: Stern, J. (ed.) EUROCRYPT 1999. LNCS, vol. 1592, pp. 1–11. Springer, Heidelberg (1999)
3. Brent, R.P.: Some integer factorization algorithms using elliptic curves. Australian Computer Science Communications 8, 149–163 (1986)
4. Brent, R.P.: Recent progress and prospects for integer factorisation algorithms. In: Du, D.-Z., Eades, P., Sharma, A.K., Lin, X., Estivill-Castro, V. (eds.) COCOON 2000. LNCS, vol. 1858, pp. 3–22. Springer, Heidelberg (2000)
5. Coppersmith, D.: Small solutions to polynomial equations, and low exponent RSA vulnerabilities. Journal of Cryptology 10(4), 233–260 (1997)
6. Hardy, G.H., Wright, E.M.: An Introduction to the Theory of Numbers. Oxford University Press, London (1965)
7. Lehman, R.S.: Factoring large integers. Mathematics of Computation 28, 637–646 (1974)
8. Lenstra, H.W.: Factoring integers with elliptic curves. Annals of Mathematics 126, 649–673 (1987)
9. Lenstra, A.K., Lenstra, H.W., Manasse, M.S., Pollard, J.M.: The number field sieve. In: Proc. 22nd Annual ACM Conference on Theory of Computing, Baltimore, Maryland, pp. 564–572 (1990)
10. Montgomery, P.L.: Speeding the Pollard and elliptic curve methods of factorization. Mathematics of Computation 48, 243–264 (1987)
11. Morrison, M.A., Brillhart, J.: A method of factoring and the factorization of F7. Math. of Comput., 29, 183–205 (1975)
12. Nitaj, A.: Cryptanalysis of RSA with constrained keys. International Journal of Number Theory (to appear)
13. Nitaj, A.: Another generalization of Wiener's attack on RSA. In: Vaudenay, S. (ed.) AFRICACRYPT 2008. LNCS, vol. 5023, pp. 174–190. Springer, Heidelberg (2008)
14. Pollard, J.M.: A Monte Carlo method for factorization. BIT 15, 331–334 (1975)
15. Rivest, R., Shamir, A., Adleman, L.: A method for obtaining digital signatures and public-key cryptosystems. Communications of the ACM 21(2), 120–126 (1978)
16. Silverman, R.D.: The multiple polynomial quadratic sieve. Mathematics of Computation 48, 329–339 (1987)
17. Wiener, M.: Cryptanalysis of short RSA secret exponents. IEEE Transactions on Information Theory 36, 553–558 (1990)
18. Zimmermann, P.: 50 largest factors found by ECM, http://www.loria.fr/~zimmerma/records/top50.html

New RSA-Based (Selectively) Convertible Undeniable Signature Schemes

Le Trieu Phong[1], Kaoru Kurosawa[2], and Wakaha Ogata[1]

[1] Tokyo Institute of Technology, Japan
[2] Ibaraki University, Japan

Abstract. In this paper, we design and analyze some new and practical (selectively) convertible undeniable signature (SCUS) schemes in both random oracle and standard model, which enjoy several merits over existing schemes in the literature. In particular, we design *the first* practical RSA-based SCUS schemes secure in the standard model. On the path, we also introduce two moduli RSA assumptions, including the strong twin RSA assumption, which is the RSA symmetry of the strong twin Diffie-Hellman assumption (Eurocrypt '08).

Keywords: RSA, undeniable signature, selective/all conversion, standard model.

1 Introduction

1.1 Background

In 1989, Chaum and van Antwerpen [5] introduced the concept of undeniable signature (US) scheme, where a signature is not publicly verifiable, which is in contrast to ordinary signature schemes. The verification of an undeniable signature requires the cooperation of the signer through the confirmation protocol (for validity of signature) and disavowal protocol (for invalidity of signature). A mandatory property of a US scheme thus is invisibility, namely without interacting with the signer, it is hard to decide whether a signature is valid or not.

US schemes are useful in licensing software [5], electronic cash [8], confidential business agreement [11]. There have been a wide range of works proposing various US schemes with different properties and degrees of security [3,6,11,14,15,17,19, 22], to list just a few. Most of them are discrete-logarithm based schemes, and there are some RSA-based schemes [14,15,17,19]. The RSA-based US schemes in [15,17] do not satisfy invisibility, as shown in [14]. Most of RSA-based schemes so far were analysed in the random oracle model (ROM), and the first effort to achieve security in the standard model with efficient constructions was made by Kurosawa and Takagi [19]. In particular, the paper [19] suggests two efficient schemes, called KT_0 and KT_1 in this paper, and the former's security is in ROM, while the latter's is in the standard model. The scheme KT_1 is the only known

B. Preneel (Ed.): AFRICACRYPT 2009, LNCS 5580, pp. 116–134, 2009.

RSA-based practical scheme whose security is considered in the standard model so far. However, in Section 6.2, we revisit the scheme KT_1 and show that it *does not* satisfy invisibility as claimed.

The US schemes KT_0 and KT_1 are also *selectively-convertible*, a notion invented by Boya et al [3], where the signer can release a piece of information (a converter, for short) in order to convert an undeniable signature into a regular signature at later time. The converter length in KT_0 and KT_1 is short, which is $k = 1024$ bits, compared to previous RSA-based schemes which require Fiat-Shamir heuristic. However, KT_0 and KT_1 do not practically support *all conversion*, also defined by Boya et al [3], where the signer can convert all undeniable signature into regular ones.

1.2 Our Contribution

We propose practical RSA-based US schemes supporting *both* selective and all conversion, called $SCUS_0$ in ROM, and $SCUS_1$, $SCUS_2$ in the standard model. A comparison of some aspects between our proposals with some other schemes is in Table 1. Below are explanations.

Table 1. Comparisons between our US schemes and those of [14, 19]

Schemes	Signature size (bits)	Assumption	Conversion		Model
			Selective, $|cvt|$ (bits)	All	
GM [14]	$2k + \lfloor k/3 \rfloor$	Factoring + CDDH	yes, $160 + 2k$	yes	rom
KT_0 [19]	k	CNR + DNR	yes, k	no	rom
Our $SCUS_0$	k	RSA + dtm-RSA	yes, k	yes	rom
KT_1 [19]	$160 + 3k$	broken (Sect.6.2)	yes, k	no	std
Our $SCUS_1$	$80 + 2k$	sRSA + DNR	yes, k	yes	std
Our $SCUS_2$	$2k$	sRSA + DIV + DNR	yes, k	yes	std

In the table, $|cvt|$ means the converter size, and $k = 1024$, typically. For assumption, CNR (resp, sRSA, DIV, CDDH, dtm-RSA, DNR) = computational N-th residuosity (resp, strong RSA, division intractability, composite decision Diffie-Hellman, decisional two moduli RSA, decisional N-th residuosity). And rom = random oracle model, std = standard model.

In the random oracle model, $SCUS_0$ gives very short signature and converter (compared to that of [14]), while maintaining the RSA assumption for unforgeability.

In the standard model, $SCUS_1$ and $SCUS_2$ offer short signature, and short converter (even if compared with the ROM schemes), and we do not need the strong computational N-th residuosity (sCNR) assumption, first formalized in [19]. (Although KT_1 has a problem in invisibility, we keep it in the table just for comparison.)

In addition to invisibility, our schemes also have anonymity, considered in Section 6, ensuring that no adversary can decide *who* is the author of a given message-signature pair.

The main demerit of our proposals is in the confirmation and disavowal protocol costs for computation and communication, which are respectively at most $O(k^2)$ in term of modular multiplications and bits. Namely, we needs a number of exponentiations to compute and some kilobytes to transfer in the protocols, which is a bit less efficient than those of (the best) KT_0, which is $O(k)$ asymptotically.

Let us now look at the methods for achieving the above results.

Strong Twin RSA Assumption. Consider a pair of RSA modulus and exponent (N_1, e_1) and (N_2, e_2) with $N_1 < N_2$. We call $(y_1, y_2) \in Z^*_{N_1} \times Z^*_{N_2}$ a twin RSA tuple if for some $0 < x < N_1$: $y_1 = x^{e_1} \bmod N_1$ and $y_2 = x^{e_2} \bmod N_2$. The twin RSA problem is to find y_2 from $(N_1, N_2, e_1, e_2, y_1)$ such that (y_1, y_2) is a twin RSA tuple. The twin RSA assumption then states that the problem is hard.

We are now ready to describe the *strong* twin RSA assumption. It states that the twin RSA assumption holds *even if* we have access to a help oracle who on input arbitrary (z_1, z_2) returns the root x as above if (z_1, z_2) is a twin RSA tuple; and returns $\perp$ otherwise. Interestingly, as a warm-up result, we prove (see Theorem 1) that the strong twin RSA assumption is equivalent to the ordinary RSA assumption. The equivalence makes us choose the name *strong twin RSA* for the assumption, since it is similar to a situation in discrete-logarithm based setting [7].

A SCUS Scheme in ROM. Our SCUS_0 in ROM is as follows: the public key is (N_1, N_2, e_1, e_2) and a hash function $H : \{0, 1\}^* \to Z^*_{N_1}$. The secret key is the factorizations of (N_1, N_2). The undeniable signature on a message $m \in \{0, 1\}^*$ is the value σ such that $(H(m), \sigma)$ is a twin RSA tuple with respect to (N_1, N_2, e_1, e_2). This is done by first solving $H(m) = x^{e_1} \bmod N_1$ for $0 < x < N_1$ and then setting $\sigma = x^{e_2} \bmod N_2$. Intuitively, we can see that no one (except the signer) can compute σ from $(N_1, N_2, e_1, e_2, H(m))$ thanks to the strong twin RSA assumption. Furthermore, the help oracle makes the loss factor in security reduction better. The unforgeability of SCUS_0 is equivalent to the RSA assumption, and the invisibility is equivalent to the newly-formed decisional two moduli RSA (dtm-RSA) assumption. The full description and security in ROM of the scheme are in Section 3.2.

SCUS Schemes in the Standard Model. Another viewpoint on SCUS_0 scheme is as follows: we first sign on m by the RSA signature scheme, namely RSA-FDH [9,4], to get the signature x and then encrypt the signature x by plain RSA encryption. The RSA-FDH signature scheme causes the use of the random oracle model. We thus replace the RSA-FDH scheme by the recent GRSA signature scheme [18], which is unforgeable in the standard model, and has a signature of the form $(e, y = H(m)^{1/e} \bmod N_1)$ for random prime e and a standard model hash function H. Now that plain RSA encryption seems too weak to provide invisibility in the standard model, we use the Paillier encryption scheme [21], which has semantic security under the DNR assumption, to encrypt y: $\tau = r^{N_2}(1 + yN_2) \bmod N_2^2$ for randomness r. The undeniable signature will be $\sigma = (e, \tau)$, and r will be used to decrypt τ to release y and thus convert σ into an ordinary signature. Intuitively, the scheme satisfies unforgeability because GRSA is unforgeable, and the signature σ is invisible since e is random and τ hides all information about y.

However, the above naive replacements do not work, and we cannot prove the invisibility of the above scheme. Interestingly, going back into the literature, we find that Damgard and Peterson [11] have used a similar approach for discrete-log based schemes, where a part of the ElGamal signature is encrypted by Rabin or ElGamal encryption scheme, but they could not formally prove the invisibility of their SCUS schemes either.

One of the main difficulties when proving invisibility of our scheme is as follows: if $\sigma = (e, \tau)$ is valid then $\sigma' = (e, s^{N_2}\tau)$ is also valid for randomly chosen $s \in Z_{N_2}^*$, and we cannot convert σ' for unknown s. We thus overcome this problem by also signing on the randomness r, namely now $y = H(m \parallel r)^{1/e} \bmod N_1$, so that σ' becomes invalid and we just return $\perp$ when being asked for converting it. But another problem arises: we do not know how to prove in zero-knowledge the equation $y^e = H(m \parallel r) \bmod N_1$ with two secrets y and r in the confirmation protocol. We again overcome the problem by authenticating $r^{N_2} \bmod N_2$ $(= \tau \bmod N_2)$, namely now $y = H\big(m \parallel (r^{N_2} \bmod N_2)\big)^{1/e} \bmod N_1$, which is equivalent to authenticating r since $r \mapsto r^{N_2} \bmod N_2$ is a permutation on $Z_{N_2}^*$. Now in the confirmation protocol, we can prove $y^e = H\big(m \parallel (\tau \bmod N_2)\big) \bmod N_1$ with one secret y. The discussion thus far is the idea behind the construction of our SCUS$_1$ scheme, which is given in Section 4.

The same ideas still work if we replace the GRSA signature scheme by the GHR signature scheme [13], yielding the second scheme SCUS$_2$ in Section 5 which is also secure in the standard model.

2   Syntax and Security Definitions

Definition 1 (SCUS scheme). *A selectively–convertible undeniable scheme* SCUS = (KeyGen, Usign, Convert, Verify, Confirm, Disavowal) *consists of four algorithms and two protocols whose descriptions are as follows.*

– KeyGen(1^k) → (pk, sk): This algorithm generates the pubic key pk and the secret key (signing key) sk for user.

– USign(sk, m) → σ: Using the secret key sk, this algorithm produces a signature σ on a message m.

– Convert(sk, m, σ) → cvt/ $\perp$: Using sk, this algorithm releases a converter cvt of a valid pair message-signature (m, σ), and release $\perp$ if the pair is invalid.

– Verify(pk, m, σ, cvt) → $0/1$: Using the converter cvt, everyone can check the validity of (m, σ) by this algorithm.

– Confirm: This is a protocol between the signer and a verifier, on common input (pk, m, σ), the signer with sk proves that (m, σ) is a valid message-signature pair in zero-knowledge.

– Disavowal: This is a protocol between the signer and a verifier, on common input (pk, m, σ), the signer with sk proves that (m, σ) is an invalid message-signature pair in zero-knowledge.

The following definitions describe securities that a SCUS scheme should meet.

Definition 2 (Unforgeability and strong unforgeability of SCUS). *A selectively convertible undeniable scheme* SCUS *is said to be existential unforgeable under adaptive chosen message attack if no probabilistic polynomial time (p.p.t) forger* $\mathcal{F}$ *has a non-negligible advantage in the following game: First,* $\mathcal{F}$ *is given the public key pk. Then,* $\mathcal{F}$ *is permitted to issue a series of queries shown below:*

– Signing queries: $\mathcal{F}$ *submits a message* m *to the signing oracle and receives a signature* σ *on* m. *These queries are adaptive, namely the next query can depend on previous ones.*

– Convert queries: $\mathcal{F}$ *submits a message-signature pair* (m, σ) *to the convert oracle, and receives a converter cvt or* $\perp$. *These queries are also adaptive.*

– Confirmation/disavowal queries: $\mathcal{F}$ *submits a message-signature pair* (m, σ) *to the confirmation/disavowal oracle. We will consider active attack, where the oracle first checks the validity of* (m, σ). *If it is a valid pair, the oracle returns 1 and executes the confirmation protocol with* $\mathcal{F}$ *(acting as a cheating verifier). Otherwise, the oracle returns 0 and executes the disavowal protocol with* $\mathcal{F}$.

At the end of the game, $\mathcal{F}$ *outputs a pair* (m^*, σ^*). *In the definition of unforgeability, the forger* $\mathcal{F}$ *wins the game if the pair* (m^*, σ^*) *is a valid message-signature pair, and* m^* *has never been queried to the signing oracle. The advantage of* $\mathcal{F}$ *is defined to be* $\mathbf{Adv}_{SCUS}^{forge}(\mathcal{F}) = \Pr[\mathcal{F} \text{ wins}]$.

In the definition of strong unforgeability, the only different point is that (m^*, σ^*) *does not coincide with any* (m, σ) *at signing queries. We denote* $\mathcal{F}$'s *advantage in this case by* $\mathbf{Adv}_{SCUS}^{sforge}(\mathcal{F}) = \Pr[\mathcal{F} \text{ wins}]$.

Definition 3 (Invisibility [14]). *A selectively-convertible undeniable scheme* SCUS *satisfies invisibility under adaptive chosen message attack if no p.p.t distinguisher* $\mathcal{D}$ *has a non-negligible advantage in the following game: (1)* $\mathcal{D}$ *is given the public key pk. (2)* $\mathcal{D}$ *is permitted to issue a series of queries: signing queries, convert queries, confirmation/disavowal queries, as in Definition 2. (3) At some point,* $\mathcal{D}$ *outputs a message* m^* *which has never been queried to the signing oracle, and requests a challenge signature* σ^* *on* m^*. *The challenge signature* σ^* *is generated based on a hidden bit b. If* $b = 0$, *then* σ^* *is generated as usual using the signing algorithm; otherwise* σ^* *is chosen randomly from the signature space of the scheme. (4) The distinguisher* $\mathcal{D}$ *performs some signing queries, convert queries, confirmation/disavowal queries again with the restriction that no signing query* m^* *is allowed, no confirmation/disavowal query and convert query* (m^*, σ^*) *are allowed. (5) At the end,* $\mathcal{D}$ *outputs a bit* b' *as the guess for b.*

The distinguisher wins the game if $b' = b$ *and its advantage in this game is defined as* $\mathbf{Adv}_{SCUS}^{iv}(\mathcal{D}) = |\Pr[b' = b] - 1/2|$.

We now briefly recall the syntax and security notions of an ordinary signature scheme.

Definition 4 (Standard signature scheme and its security). *A signature scheme* S = (Kg, Sign, Vrf) *is as follows. On input* 1^k, *the key generation algorithm* Kg *produces the public key pk and the secret signing key sk. On input sk* *and a message m, the signing algorithm* Sign *produces a signature σ, which is publicly-verifiable using the verification algorithm* Vrf *on input pk and σ.*

The unforgeability under chosen message attack (uf-cma security) of a signature scheme S is defined essentially the same as that of SCUS in Definition 2, except that the forger $\mathcal{F}$ against S only issues signing queries. We denote the advantage of $\mathcal{F}$ by $\mathbf{Adv}_S^{uf-cma}(\mathcal{F}) = \Pr[\mathcal{F}\ wins]$. The strong unforgeability (suf-cma security) is defined in a similar manner and we have the advantage $\mathbf{Adv}_S^{suf-cma}(\mathcal{F}) = \Pr[\mathcal{F}\ wins]$.

3   Two Moduli RSA Assumptions and SCUS$_0$

3.1   Two Moduli RSA Assumptions

RSA Key Generator. A p.p.t algorithm $\mathcal{K}_{rsa}$ is said to be a RSA key generator if on input 1^k, its output is a triple (N, e, d) of positive integers, where N is a product of two distinct primes, N is k bits long, $(e, \phi(N)) = 1$ and $ed = 1$ mod $\phi(N)(= (p-1)(q-1))$.

RSA and Strong RSA Assumption. We define the advantage of an adversary $\mathcal{A}_{rsa}$ against $\mathcal{K}_{rsa}$ as $\mathbf{Adv}_{\mathcal{K}_{rsa}}^{rsa}(\mathcal{A}_{rsa}) = \Pr[(N, e, d) \xleftarrow{\$} \mathcal{K}_{rsa}(1^k); y \xleftarrow{\$} Z_N^*; x \xleftarrow{\$} \mathcal{A}_{rsa}(N, e, y) : x^e = y \bmod N]$. The RSA assumption asserts that the advantage is negligible for all p.p.t adversary $\mathcal{A}_{rsa}$. The strong RSA assumption is defined in the same manner, except that the value $e > 1$ is chosen by the adversary.

Two Moduli RSA Key Generator. This generator, which is denoted as $\mathcal{K}_{tm-rsa}(1^k)$, first runs $\mathcal{K}_{rsa}(1^k)$ twice to obtain (N_1, e_1, d_1) and (N_2, e_2, d_2) with $N_1 < N_2$. The output of $\mathcal{K}_{tm-rsa}(1^k)$ is $(N_1, N_2, e_1, e_2, d_1, d_2)$.

STRONG TWIN RSA ASSUMPTION. We first define the advantage of an adversary $\mathcal{A}_{st-rsa}$ solving the strong twin RSA problem as

$$\mathbf{Adv}_{\mathcal{K}_{tm-rsa}}^{st-rsa}(\mathcal{A}_{st-rsa})$$
$$= \Pr\Big[(N_1, N_2, e_1, e_2, d_1, d_2) \xleftarrow{\$} \mathcal{K}_{tm-rsa}(1^k); y_1 \xleftarrow{\$} Z_{N_1}^*;$$
$$y_2 \xleftarrow{\$} \mathcal{A}_{st-rsa}^{\mathcal{O}(\cdot,\cdot)}(N_1, N_2, e_1, e_2, y_1) : \exists 0 < x < N_1,$$
$$y_1 = x^{e_1} \bmod N_1 \wedge y_2 = x^{e_2} \bmod N_2\Big],$$

where $\mathcal{O}(\cdot, \cdot)$ is a help oracle, on input (z_1, z_2), returning u if $0 < z_1 < N_1$, $0 < z_2 < N_2$, and $z_1 = u^{e_1} \bmod N_1$ and $z_2 = u^{e_2} \bmod N_2$ for some $0 < u < N_1$; returning $\perp$ otherwise. The strong twin RSA assumption then asserts that the above advantage is negligible for all p.p.t adversary $\mathcal{A}_{st-rsa}$.

The following is the equivalence between the strong twin RSA assumption and the standard RSA assumption.

Theorem 1 (RSA = Strong Twin RSA). *The strong twin RSA assumption is equivalent to the (standard) RSA assumption. Moreover, if there is a strong twin RSA adversary $\mathcal{A}_{st-rsa}$, then there also exists a RSA adversary $\mathcal{A}_{rsa}$ whose running time is polynomial in that of $\mathcal{A}_{st-rsa}$ such that*

$$\mathbf{Adv}^{st-rsa}_{\mathcal{K}_{tm-rsa}}(\mathcal{A}_{st-rsa}) \leq \mathbf{Adv}^{rsa}_{\mathcal{K}_{rsa}}(\mathcal{A}_{rsa}).$$

Proof. If we can solve the RSA problem, then it is clear that we can solve the strong twin RSA problem. We now consider the converse. Suppose there exists a strong twin RSA adversary $\mathcal{A}_{st-rsa}$, we construct a RSA adversary $\mathcal{A}_{rsa}$, on input (N, e, y), running $\mathcal{A}_{st-rsa}$ as a subroutine. The adversary $\mathcal{A}_{rsa}$ proceeds as follows: (1) It sets $N_1 \leftarrow N, e_1 \leftarrow e, y_1 \leftarrow y$. (2) It chooses random primes p_2, q_2 such that $N_2 = p_2 q_2 \geq N_1$, and e_2, d_2 such that $\gcd(e_2, \phi(N_2)) = 1$ and $e_2 d_2 = 1 \mod \phi(N_2)$. (3) It runs $\mathcal{A}_{st-rsa}$ on input $(N_1, N_2, e_1, e_2, y_1)$, simulating the decisional oracle $\mathcal{O}(z_1, z_2)$ as follows: $\mathcal{A}_{rsa}$ first tests whether $0 < z_1 < N_1, 0 < z_2 < N_2$, and returns $\perp$ if that is not the case. Otherwise, $\mathcal{A}_{rsa}$ sets $u = z_2^{d_2} \mod N_2$. If $0 < u < N_1$ and $z_1 = u^{e_1} \mod N_1$ then return u, else return $\perp$. (4) Finally, $\mathcal{A}_{st-rsa}$ outputs y_2. $\mathcal{A}_{rsa}$ in turn sets $x = y_2^{d_2} \mod N_2$ and then outputs $x \mod N_1$.

We proceed to prove that $(x \mod N_1)^e = y \mod N$, or equivalently, $x^{e_1} = y_1 \mod N_1$. If $\mathcal{A}_{st-rsa}$ succeeds, then there exists x' such that: $0 < x' < N_1, y_1 = (x')^{e_1} \mod N_1 \wedge y_2 = (x')^{e_2} \mod N_2$. Thus $(x')^{e_2} \mod N_2 = y_2 = x^{e_2} \mod N_2$, and hence $x' = x \mod N_2$. Since $0 < x' < N_1 < N_2$ and $0 < x < N_2$, we have $x' = x$ (without modulus). Thus $y_1 = (x')^{e_1} \mod N_1 = x^{e_1} \mod N_1$, as desired. The advantage of $\mathcal{A}_{rsa}$ is the same as that of $\mathcal{A}_{st-rsa}$, and the running time of $\mathcal{A}_{rsa}$ is polynomial in that of $\mathcal{A}_{st-rsa}$.

Decisional Two Moduli RSA Assumption. We define the advantage of an adversary $\mathcal{A}_{dtm-rsa}$ solving the decisional two moduli RSA problem as

$$\mathbf{Adv}^{dtm-rsa}_{\mathcal{K}_{tm-rsa}}(\mathcal{A}_{dtm-rsa})$$
$$= \left| \Pr \left[(N_1, N_2, e_1, e_2, d_1, d_2) \xleftarrow{\$} \mathcal{K}_{tm-rsa}(1^k); x, x' \xleftarrow{\$} Z^*_{N_1}; \right. \right.$$
$$Y_0 = (x^{e_1} \mod N_1, x^{e_2} \mod N_2);$$
$$Y_1 = (x^{e_1} \mod N_1, (x')^{e_2} \mod N_2); b \xleftarrow{\$} \{0,1\};$$
$$\left. \left. b' \xleftarrow{\$} \mathcal{A}_{dtm-rsa}(N_1, N_2, e_1, e_2, Y_b) : b' = b \right] - \frac{1}{2} \right|.$$

The decisional two moduli RSA assumption then asserts that the advantage is negligible for all p.p.t adversary $\mathcal{A}_{dtm-rsa}$.

Note that simple Jacobi symbol attack cannot break the decisional two moduli RSA assumption. In fact, given $Y_b = (Y_b[1], Y_b[2])$, the adversary $\mathcal{A}_{dtm-rsa}$ can certainly compute the Yacobi symbols $\left(\frac{Y_b[1]}{N_1}\right)$ and $\left(\frac{Y_b[2]}{N_2}\right)$, which are either $\left(\frac{x}{N_1}\right)$ and $\left(\frac{x}{N_2}\right)$ if $b = 0$, or $\left(\frac{x}{N_1}\right)$ and $\left(\frac{x'}{N_2}\right)$ if $b = 1$. However, these pieces of information seem useless, since $N_1 \neq N_2$.

3.2   The Scheme SCUS_0

Our proposed selectively-convertible undeniable scheme $\mathsf{SCUS}_0 = (\mathsf{KeyGen}, \mathsf{Usign}, \mathsf{Convert}, \mathsf{Verify}, \mathsf{Confirm}, \mathsf{Disavowal})$ is described as follows.

– $\mathsf{KeyGen}(1^k) \to (pk, sk)$: Run $\mathcal{K}_{tm-rsa}(1^k)$ to obtain $(N_1, N_2, e_1, e_2, d_1, d_2)$. Choose a hash function $H : \{0,1\}^* \to Z^*_{N_1}$. Output the public key $pk = (N_1, N_2, e_1, e_2, H)$ and the secret key $sk = (d_1, d_2)$.

– $\mathsf{USign}(sk, m) \to \sigma$: Using d_1 in sk, solve the equation $H(m) = x^{e_1} \bmod N_1$ for $0 < x < N_1$ (namely, let $x \leftarrow H(m)^{d_1} \bmod N_1$). Output $\sigma = x^{e_2} \bmod N_2$.

– $\mathsf{Convert}(sk, m, \sigma) \to cvt/\perp$: Using d_2 in sk, solve the equation $\sigma = x^{e_2} \bmod N_2$ for $0 < x < N_2$ (namely, let $x \leftarrow \sigma^{d_2} \bmod N_2$). If $0 < x < N_1$ and $H(m) = x^{e_1} \bmod N_1$ then output $cvt = x$, else output $\perp$.

For all conversion, the signer just release the secret d_2 to convert SCUS_0 into the standard RSA signature scheme, namely RSA-FDH, since the algorithm Convert only needs that secret to work for all pairs (m, σ).

– $\mathsf{Verify}(pk, m, \sigma, cvt = x) \to 0/1$: Return 1 if and only if $0 < x < N_1$ and $H(m) = x^{e_1} \bmod N_1$ and $\sigma = x^{e_2} \bmod N_2$.

– Confirm: On common input $pk = (N_1, N_2, e_1, e_2, H)$, and (m, σ), the signer and the verifier execute $\mathrm{PoK}\big\{(x) : H(m) = x^{e_1} \bmod N_1 \wedge \sigma = x^{e_2} \bmod N_2 \wedge x \in [1, N_1 - 1]\big\}$.

Namely, the signer (with sk) proves that he knows the secret value $x \in [1, N_1 - 1]$ satisfying the equations. The value x can be computed using sk. The disavowal protocol below is translated similarly, and the implementation of the protocols will be given in the next section. (Note that the soundness of the PoK is implicitly used to ensure whether a signature is valid or not.)

– Disavowal: On common input $pk = (N_1, N_2, e_1, e_2, H)$, and (m, σ), the signer and the verifier execute

$$\mathrm{PoK}\Big\{(x_1, x_2, x_3) \quad : \quad H(m) = x_1^{e_1} \bmod N_1 \wedge x_1 \in [1, N_1 - 1]$$
$$\wedge \quad \sigma = x_2^{e_2} \bmod N_2 \quad \wedge x_2 \in [1, N_2 - 1]$$
$$\wedge \quad x_3 = x_1 - x_2 \bmod N_2 \wedge x_3 \in [1, N_2 - 1]\Big\}.$$

Remark 1 (Inequality proof of roots of modular polynomials). *Note that $x_3 \in [1, N_2 - 1]$ implies $x_1 \neq x_2$. This is a trick we develop to use the results of [1] and [12] (see Section 3.3 below for details). Generally, using the idea, one can prove the inequality for roots of modular polynomials, say $\mathrm{PoK}\{(x_1, x_2) : f(x_1) = 0 \bmod N_1 \wedge x_1 \in [0, N_1 - 1] \wedge g(x_2) = 0 \bmod N_2 \wedge x_2 \in [0, N_2 - 1] \wedge x_1 \neq x_2\}.$*

3.3   Implementation of Confirmation/Disavowal Protocol

In this section part, we show how to efficiently implement the zero-knowledge proofs of knowledge of our confirmation and disavowal protocol. For "efficiently", we mean the costs for computation (in term of modular multiplications) and communication (in term of bits) in the protocol are *at most* $O(k^2)$ (and linear in k for small e_1, e_2 such as 3), where the security parameter k is typically 1024.

BUILDING BLOCKS. We first describe some building blocks, which are the results of Boudot [1], and Fujisaki, Okamoto [12] (with some refinements [10]).

- A basic commitment scheme [12,10]: To commit $x \in [0, N-1]$ for $|N| = k$, a committer computes and sends to a verifier the value $\mathsf{Com}(x, r) = b_0^x b_1^r \bmod N$, where $b_0, b_1 \in Z_N^*$, and $r \xleftarrow{\$} [0, 2^k N)$. The values (N, b_0, b_1) are set up by either the verifier or a trusted party so that N's factoring is unknown to the committer. We refer the readers to the papers [12,10] for the precise set-up of (N, b_0, b_1). This commitment scheme is statistically hiding and computationally binding under the factoring assumption. Below, we will omit the randomness r to write $\mathsf{Com}(x)$ for clarity.
- Proof of modular polynomial relation [12]: Succinctly, Fujisaki and Okamoto showed how to efficiently implement the following

$$\mathrm{PoK}\big\{(x_1, \ldots, x_t) : E_1 = \mathsf{Com}(x_1) \wedge \cdots \wedge E_t = \mathsf{Com}(x_t),$$
$$f(x_1, \ldots, x_t) = 0 \bmod n\big\}.$$

  In other words, the prover first commits its secret values $x_1, \ldots, x_t \in [0, N-1]$ by sending $E_1, \ldots, E_t$ to the verifier. It then proves to the verifier that the committed values satisfy the modular polynomial equation $f(x_1, \ldots, x_t) = 0 \bmod n$ for publicly-known polynomial f and modulus n, where $|n| = O(k)$. Note that n may differ from N.

  The proof is statistical zero-knowledge and its soundness relies on the strong RSA assumption. The computation cost is $O(k \cdot \log_2(degf))$ modular multiplications over modulus N for small t (we will use $t \leq 3$). The communication cost is $O(k \cdot \log_2(degf))$ bits, also for small t.
- Proof of interval [1]: Succinctly, the result of Boudot is the following $\mathrm{PoK}\big\{(x) : E = \mathsf{Com}(x) \wedge x \in [a, b]\big\}$. The computation and communication costs are respectively less than 20 modular exponentiations modulus N and less than 2 KBytes, as estimated in [1]. We can also count them as $O(k)$ modular multiplications and $O(k)$ bits. This proof also relies on the strong RSA assumption.

We are now ready to describe our implementation of the confirmation and disavowal protocol of the scheme.

Implementation of the Confirmation Protocol. The signer first commits x by $E = \mathsf{Com}(x)$ and sequentially executes the following proofs with the signature verifier on common input $pk = (N_1, N_2, e_1, e_2, H)$, and a message-signature pair (m, σ): (1) $\mathrm{PoK}\big\{(x) : E = \mathsf{Com}(x) \wedge x \in [1, N_1 - 1]\big\}$, (2) $\mathrm{PoK}\big\{(x) : E = \mathsf{Com}(x) \wedge H(m) = x^{e_1} \bmod N_1\big\}$, (3) $\mathrm{PoK}\big\{(x) : E = \mathsf{Com}(x) \wedge \sigma = x^{e_2} \bmod N_2\big\}$.

The first PoK utilises the result of Boudot, where we need the condition $N \geq N_1$, which is easy to be fulfilled. The second and third PoK use the result of Fujisaki and Okamoto with polynomials $f_1(x) = x^{e_1} - H(m) \bmod N_1$ and $f_2(x) = x^{e_2} - \sigma \bmod N_2$.

The whole computation cost for the implementation is $O\big(k(|e_1|+|e_2|)\big)$ modular multiplications over modulus N. For small e_1 and e_2, the cost becomes linear in k. The whole communication cost is $O\big(k(|e_1|+|e_2|)\big)$ bits, which is also linear in k for small e_1 and e_2.

Implementation of the disavowal Protocol. The signer first commits x_1, x_2, x_3 by $E_1 = \mathsf{Com}(x_1)$, $E_2 = \mathsf{Com}(x_2)$, $E_3 = \mathsf{Com}(x_3)$ and sequentially executes the following proofs with the signature verifier on common input $pk = (N_1, N_2, e_1, e_2, H)$, and a message-signature pair (m, σ): (1) $\mathrm{PoK}\big\{(x_1) : E_1 = \mathsf{Com}(x_1) \wedge x_1 \in [1, N_1 - 1]\big\}$, (2) $\mathrm{PoK}\big\{(x_2) : E_2 = \mathsf{Com}(x_2) \wedge x_2 \in [1, N_2 - 1]\big\}$, (3) $\mathrm{PoK}\big\{(x_3) : E_3 = \mathsf{Com}(x_3) \wedge x_3 \in [1, N_2 - 1]\big\}$, (4) $\mathrm{PoK}\big\{(x_1) : E_1 = \mathsf{Com}(x_1) \wedge H(m) = x_1^{e_1} \bmod N_1\big\}$, (5) $\mathrm{PoK}\big\{(x_2) : E_2 = \mathsf{Com}(x_2) \wedge \sigma = x_2^{e_2} \bmod N_2\big\}$, (6) $\mathrm{PoK}\big\{(x_1, x_2, x_3) : E_i = \mathsf{Com}(x_i) \forall i = 1, 2, 3 \wedge x_3 = x_1 - x_2 \bmod N_2\big\}$.

The first three PoKs apply Boudot's result, thus we need to set $N \geq N_2 (> N_1)$ and hence $|N| = k$. The other PoK uses the result of Fujisaki and Okamoto with the polynomials $f_1(x_1) = x_1^{e_1} - H(m) \bmod N_1$, $f_2(x_2) = x_2^{e_2} - \sigma \bmod N_2$ and $f_3(x_1, x_2, x_3) = x_3 - (x_1 - x_2) \bmod N_2$.

The disavowal protocol cost is a bit higher than that of the confirmation one, but still in the same order. The protocol needs $O\big(k(|e_1| + |e_2|)\big)$ modular multiplications over modulus N and $O\big(k(|e_1| + |e_2|)\big)$ bits of communication, so that the costs become linear in k for small e_1 and e_2.

3.4   Securities of SCUS$_0$

We first consider unforgeability of SCUS$_0$. It is possible to directly prove the security of the scheme under the strong twin RSA assumption (and hence the RSA assumption), but for modularity, we will reduce the unforgeability to the uf-cma security of the well-known RSA-FDH [4, 9] signature scheme. For the purpose, let us briefly recall the RSA-FDH scheme and its uf-cma security.

RSA-FDH And Its Security. The public key is (N_1, e_1, H), and the secret signing key is d_1, where $(N_1, e_1, d_1) \xleftarrow{\$} \mathcal{K}_{rsa}(1^k)$ and $H : \{0, 1\}^* \to Z_{N_1}^*$. On a message $m \in \{0, 1\}^*$ and the secret key d_1, the signing algorithm returns the signature $x = H(m)^{d_1} \bmod N_1$. On a message-signature pair (m, x) and the public key (N_1, e_1, H), everyone can check the validity using the equation $x^{e_1} = H(m) \bmod N_1$. It is proved in [9] that RSA-FDH is uf-cma-secure under the RSA assumption.

Theorem 2 (Unforgeability of SCUS$_0$). *The unforgeability of SCUS$_0$ is equivalent to the RSA assumption in the random oracle model. Moreover, given a forger $\mathcal{F}$ against SCUS$_0$, there exists another forger $\mathcal{G}$, whose running time is essentially the same of that of $\mathcal{F}$, against the RSA-FDH signature scheme such that*

$$\mathbf{Adv}_{SCUS_0}^{forge}(\mathcal{F}) \leq \mathbf{Adv}_{RSA\text{-}FDH}^{uf-cma}(\mathcal{G})$$

Proof. If the RSA assumption fails, then it is easy to forge the selectively-convertible undeniable scheme $\mathsf{SCUS_0}$. In fact, just take any message m, solve $H(m) = x^{e_1} \bmod N_1$ for x and let $\sigma = x^{e_2} \bmod N_2$ be the signature on m.

We proceed to prove the converse. Suppose there is a forger $\mathcal{F}$ against the unforgeability of $\mathsf{SCUS_0}$. We build the forger $\mathcal{G}$ against RSA-FDH. The forger $\mathcal{G}$ receives input (N_1, e_1, H). It then chooses $N_2 = p_2 q_2 > N_1$, and e_2, d_2 as in $\mathcal{K}_{rsa}$ and simulates $\mathcal{F}$'s oracle queries as follows.

– Signing query m: $\mathcal{G}$ calls its own signing oracle on m to obtain $x = H(m)^{d_1} \bmod N_1$, and then returns $\sigma = x^{e_2} \bmod N_2$ to $\mathcal{F}$.
– Convert query (m, σ): $\mathcal{G}$ computes $x = \sigma^{d_2} \bmod N_2$. If $0 < x < N_1$ and $x^{e_1} = H(m) \bmod N_1$ then return x, else return $\bot$.
– Confirmation/disavowal query (m, σ): $\mathcal{G}$ computes $x = \sigma^{d_2} \bmod N_2$. If $0 < x < N_1$ and $x^{e_1} = H(m)$ then return 1, and use x to execute the confirmation protocol with $\mathcal{F}$ (acting as a cheating verifier). Otherwise, return 0 and execute the disavowal protocol with $\mathcal{F}$. The disavowal protocol is simulatable using rewinding technique [16], since it is zero-knowledge.

At the end, $\mathcal{F}$ outputs (m^*, σ^*). If $\mathcal{F}$ wins, then there exists $0 < x^* < N_1$ such that: $H(m^*) = (x^*)^{e_1} \bmod N_1$ and $\sigma^* = (x^*)^{e_2} \bmod N_2$. The forger $\mathcal{G}$ of RSA-FDH now sets $x^* = (\sigma^*)^{d_2} \bmod N_2$ and returns (m^*, x^*) as the forgery pair of RSA-FDH. Thus $\mathcal{G}$ succeeds whenever $\mathcal{F}$ succeeds, completing the proof.

We now consider the invisibility of $\mathsf{SCUS_0}$, which is ensured by the following theorem, whose proof will be given in the full version of this paper.

Theorem 3 (Invisibility of $\mathsf{SCUS_0}$). *The invisibility of the proposed scheme $SCUS_0$ is equivalent to the decisional two moduli RSA assumption in the random oracle model. Moreover, if there is a distinguisher $\mathcal{D}$ against the invisibility of the scheme $SCUS_0$, then there also exists a decisional two moduli RSA adversary $\mathcal{A}_{dtm-rsa}$ whose running time is polynomial in that of $\mathcal{D}$ such that*

$$\mathbf{Adv}_{SCUS_0}^{iv}(\mathcal{D}) \lesssim (q_{hash} + 1)\mathbf{Adv}_{\mathcal{K}_{tm-rsa}}^{dtm-rsa}(\mathcal{A}_{dtm-rsa}),$$

where q_{hash} is the number of hash queries $\mathcal{D}$ made.

4   The Scheme $\mathsf{SCUS_1}$

We first need the following ingredients.

Generic RSA Signature Scheme GRSA [18]. The signature scheme GRSA = (GRSA.Kg, GRSA.Sign, GRSA.Vrf) is briefly recalled with some minor modifications as follows.

GRSA.Kg(1^k) $\rightarrow$ (pk_1, sk_1): Generate $N_1 = p_1 q_1$ for safe distinct primes p_1, q_1, where the length $|N_1| = k$, and a function $H : \{0,1\}^* \rightarrow QR_{N_1}$. The secret signing key $sk_1 = (p_1, q_1)$ and the public key $pk_1 = (N_1, H, \eta)$, where $\eta = 80$.

GRSA.Sign(sk_1, m) $\rightarrow$ (e, y): The signature (e, y) on a message $m \in \{0,1\}^*$ is generated by choosing a prime e of η bits and set $y = H(m)^{1/e} \bmod N_1$, which is computed using p_1, q_1 in sk_1.

GRSA.Vrf$(pk_1, m, (e, y)) \rightarrow 0/1$: Given (e, y), check that $(e, y) \in \{0, 1\}^\eta \times Z_{N_1}$, e is odd, and $y^e = H(m) \bmod N_1$. Return 1 if all checks pass, else return 0.

It is known that the GRSA signature scheme is suf-cma-secure under the strong RSA assumption. To be complete, the proof given in [18] is for the uf-cma case, but works well even in the case of suf-cma if we strictly check the range $(e, y) \in \{0, 1\}^\eta \times Z_{N_1}$.

The DNR Assumption and Paillier Encryption. On input 1^k, the DNR key generator $\mathcal{K}_{dnr}$ outputs (N_2, p_2, q_2), where $|N_2| = k$ and p_2, q_2 are distinct primes. The advantage of an adversary $\mathcal{A}_{dnr}$ against $\mathcal{K}_{dnr}$ is defined as:

$$\mathbf{Adv}_{\mathcal{K}_{dnr}}^{dnr} (\mathcal{A}_{dnr})$$

$$= \left| \Pr \left[(N_2, p_2, q_2) \xleftarrow{\$} \mathcal{K}_{dnr}(1^k); r^* \xleftarrow{\$} Z_{N_2}^*; Y_0 \leftarrow (r^*)^{N_2} \bmod N_2^2; \right. \right.$$

$$\left. \left. Y_1 \xleftarrow{\$} Z_{N_2^2}^*; b \xleftarrow{\$} \{0, 1\}; b' \xleftarrow{\$} \mathcal{A}_{dnr}(1^k, N_2, Y_b) : b' = b \right] - \frac{1}{2} \right|.$$

The DNR assumption asserts that the above advantage is negligible in k for every p.p.t adversary $\mathcal{A}_{dnr}$.

We now describe the Paillier encryption scheme (K, E, D). The key generator K(1^k) runs $\mathcal{K}_{dnr}(1^k)$ and sets the public key $pk_2 = N_2$ and the secret key $sk_2 = (p_2, q_2)$. The encryption of a message $m \in Z_{N_2}$ under public key $pk_2 = N_2$ is E$(m; r) = r^{N_2}(1 + mN_2) \bmod N_2^2$, for randomness $r \in Z_{N_2}^*$. Given the secret key sk_2 and a ciphertext C, the decryption D(sk_2, C) works as follows: first solve $C = r^{N_2} \bmod N_2$ for $r \in Z_{N_2}^*$, then compute m as $\left((C(r^{-1})^{N_2} \bmod N_2^2) - 1 \right)/N_2$ over the integers.

We are now ready to describe our scheme SCUS$_1$, which is secure in the standard model.

Our Proposal SCUS$_1$. The scheme is described as follows.

- KeyGen$(1^k) \rightarrow (pk, sk)$: Let $\left(pk_1 = (N_1, H, \eta), sk_1 = (p_1, q_1) \right) \xleftarrow{\$}$ GRSA.Kg(1^k), and $\left(pk_2 = N_2, sk_2 = (p_2, q_2) \right) \xleftarrow{\$}$ K(1^k) such that $N_1 \leq N_2$. Set the secret key $sk = (sk_1, sk_2)$, and the public key $pk = (pk_1, pk_2)$. The message space is $\mathcal{M} = \{0, 1\}^*$ and the signature space is $\mathcal{S} = \mathcal{P} \times Z_{N_2^2}$, where $\mathcal{P}$ is the set of all primes of η bits.

- USign$(sk, m) \rightarrow \sigma$: First, choose $r \xleftarrow{\$} Z_{N_2}^*$ and set $\overline{m} \leftarrow m \| (r^{N_2} \bmod N_2)$. Next, take GRSA signature on $\overline{m}$: $(e, y = H(\overline{m})^{1/e} \bmod N_1) \xleftarrow{\$}$ GRSA.Sign$(sk, \overline{m})$. Then, encrypt y: $\tau \xleftarrow{\$}$ E$(y; r)$. Finally, return $\sigma = (e, \tau)$ as the undeniable signature.

- Convert$(sk, m, \sigma = (e, \tau)) \rightarrow cvt/ \bot$: Let $(r, y) \leftarrow$ D(sk_2, τ). If GRSA.Vrf$(pk_1, m \| (\tau \bmod N_2), (e, y)) = 1$, then return $cvt = r$; otherwise return $\bot$.

- Verify$(pk, m, \sigma, r) \rightarrow 0/1$: Parse σ as $(e, \tau) \in \{0, 1\}^\eta \times Z_{N_2^2}$. From $r \in Z_{N_2}^*$ and $\tau =$E$(y; r)$, compute $y \in Z_{N_2}$ as in the decryption algorithm of Paillier scheme, namely $y = \left((\tau(r^{-1})^{N_2} \bmod N_2^2) - 1 \right)/N_2$. Return GRSA.Vrf$(pk_1, m \| (\tau \bmod N_2), (e, y))$.

- Confirm: On common input $(m, \sigma = (e, \tau)) \in \mathcal{M} \times \mathcal{S}$ and pk, execute

$$\text{PoK}\{(r, y) \ : \ r \in [1, N_2 - 1] \wedge y \in [1, N_1 - 1]$$
$$\wedge \ y^e = H(m \parallel (\tau \bmod N_2)) \bmod N_1 \wedge \tau = \mathsf{E}(y; r)\}.$$

- Disavowal: On common input $(m, \sigma = (e, \tau)) \in \mathcal{M} \times \mathcal{S}$ and pk, execute

$$\text{PoK}\{(r, y_1, y_2, \delta) :$$
$$r \in [1, N_2 - 1] \wedge y_1 \in [1, N_1 - 1] \wedge y_2 \in [1, N_2 - 1]$$
$$\wedge \ y_1^e = H(m \parallel (\tau \bmod N_2)) \bmod N_1$$
$$\wedge \ \tau = \mathsf{E}(y_2; r) \wedge \delta = y_2 - y_1 \bmod N_2 \wedge \delta \in [1, N_2 - 1]\}.$$

The implementations of the confirmation and disavowal protocol are similar to those of SCUS_0 given in Section 3.3. The costs for computation and communication are now $O(k^2)$ in term of modular multiplications and bits respectively, since Paillier encryption is viewed as a polynomial having degree N_2 of k bits. All conversion can be done by releasing the secret $sk_2 = (p_2, q_2)$, which suffices for the algorithm Convert to work.

Theorem 4 (Strong unforgeability of SCUS_1). *If GRSA is suf-cma secure then SCUS_1 is strongly unforgeable. Furthermore, given a forger $\mathcal{F}$ against SCUS_1, there exists another forger $\mathcal{F}'$ against GRSA whose running time is essentially the same as that of $\mathcal{F}$ and*

$$\mathbf{Adv}_{\mathsf{SCUS}_1}^{sforge}(\mathcal{F}) \leq \mathbf{Adv}_{\mathsf{GRSA}}^{suf-cma}(\mathcal{F}').$$

Proof. Given the forger $\mathcal{F}$, we will build the forger $\mathcal{F}'$, who receives input $pk_1 = (N_1, H, \eta)$ and simulates the environment for $\mathcal{F}$. Firstly, $\mathcal{F}'$ generates $N_2 = p_2 q_2$ $(\geq N_1)$ with primes p_2, q_2, sets $pk_2 = N_2$ and $sk_2 = (p_2, q_2)$. It then gives $\mathcal{F}$ the public key $pk = (pk_1, pk_2)$ and simulates $\mathcal{F}$'s oracle queries as follows.

- Signing query m: $\mathcal{F}'$ first chooses $r \xleftarrow{\$} Z_{N_2}^*$ and calls its own signing oracle GRSA.Sign$(sk_1, \cdot)$ on $m \parallel (r^{N_2} \bmod N_2)$ to obtain (e, y). $\mathcal{F}'$ then computes $\tau = \mathsf{E}(y; r)$ and returns $\sigma = (e, \tau)$ to $\mathcal{F}$. Let $\mathcal{Q}$ be the set of all pairs (m, σ) of $\mathcal{F}$.
- Convert query (m, σ): Check that $\sigma = (e, \tau) \in \{0, 1\}^\eta \times Z_{N_2^2}$, then decrypt τ using sk_2, namely $(y, r) \leftarrow \mathsf{D}(sk_2, \tau)$. Check that GRSA.Vrf$(pk_1, m \parallel (\tau \bmod N_2), (e, y)) = 1$. If all checks pass then return $cvt = r$, else return $\perp$.
- Confirmation/disavowal query (m, σ): Proceed as in the simulation of convert query. The different point is that if all checks pass, then return 1 and use (r, y) to simulate the confirmation protocol, otherwise return 0 and simulate the disavowal protocol using the rewinding technique [16].

At the end, $\mathcal{F}$ outputs a forged pair $(m^*, \sigma^* = (e^*, \tau^*)) \notin \mathcal{Q}$. Note that if $\mathcal{F}$ succeeds, then $(m^*, \sigma^* = (e^*, \tau^*))$ is valid, so that there exist (r^*, y^*) satisfying: $r^* \in [1, N_2 - 1], y^* \in [1, N_1 - 1], (y^*)^{e^*} = H(m^* \parallel (\tau^* \bmod N_2)) \bmod N_1$, and $\tau^* = \mathsf{E}(y^*; r^*)$.

Using sk_2, $\mathcal{F}'$ decrypts τ^* to get y^* and then outputs $(m^* \| (\tau^* \bmod N_2),$ $(e^*, y^*))$ as a forged message-signature pair of GRSA.

We now proceed to prove that $(m^* \| (\tau^* \bmod N_2), (e^*, y^*))$ is a valid message-signature pair of GRSA in the suf-cma sense. It is clear that the pair is a valid, so we just need to prove that it is different from all query-answer pairs at the signing oracle GRSA.Sign$(sk_1, \cdot)$. By contrary, suppose that $(m^* \| (\tau^* \bmod N_2), (e^*, y^*)) = (m \| (r^{N_2} \bmod N_2), (e, y))$, a previously appeared query-answer at the oracle. Since $\tau^* = \mathsf{E}(y^*; r^*)$, and $\tau^* = r^{N_2} \bmod N_2$, we have $(r^*)^{N_2} = r^{N_2} \bmod N_2$ and hence $r^* = r$ (without modulus). Thus $\tau^* = \mathsf{E}(y^*; r^*) = \mathsf{E}(y; r) = \tau$, so that $\mathcal{F}$'s forgery $(m^*, \sigma^* = (e^*, \tau^*)) = (m, \sigma = (e, \tau)) \in \mathcal{Q}$, which is a contradiction to the success of $\mathcal{F}$. This completes the proof.

Theorem 5 (Invisibility of SCUS$_1$). *The scheme SCUS$_1$ satisfies invisibility, if the DNR assumption holds and SCUS$_1$ is strongly unforgeable. Furthermore, given a distinguisher $\mathcal{D}$ against SCUS$_1$, there exists an adversary $\mathcal{A}_{dnr}$ whose running time is essentially the same as that of $\mathcal{D}$ such that*

$$\mathbf{Adv}^{iv}_{SCUS_1}(\mathcal{D}) \lesssim \mathbf{Adv}^{dnr}_{\mathcal{K}_{dnr}}(\mathcal{A}_{dnr}).$$

Proof. Given $\mathcal{D}$, we will build $\mathcal{A}_{dnr}$ against the DNR assumption. The input of $\mathcal{A}_{dnr}$ is $(1^k, N_2, Y_b)$ and it has to simulate the environment for $\mathcal{D}$. The adversary $\mathcal{A}_{dnr}$ first runs GRSA.Kg(1^k) to obtain $(pk_1 = (N_1, H, \eta), sk_1 = (p_1, q_1))$. It sets $pk = (pk_1, pk_2 = N_2)$ as the public key given to $\mathcal{D}$, and begins simulating the environment for $\mathcal{D}$ as follows.

– Signing query m: Choose $r \xleftarrow{\$} Z^*_{N_2}$, and prime $e \xleftarrow{\$} \{0, 1\}^\eta$. Let $\overline{m} = m \|$ $(r^{N_2} \bmod N_2)$ and set $y \leftarrow H(\overline{m})^{1/e} \bmod N_1$ using sk_1. Compute $\tau = \mathsf{E}(y; r)$, and return the undeniable signature $\sigma = (e, \tau)$ to $\mathcal{D}$, and keep the values r, y for later use. Denote $\mathcal{Q}$ (initially empty) as the set of all pairs (m, σ) appeared so far.

– Convert query (m, σ): If $(m, \sigma) \in \mathcal{Q}$ then return the recorded value r to $\mathcal{D}$. Otherwise, we have $(m, \sigma) \notin \mathcal{Q}$, then return $\bot$ to $\mathcal{D}$. The reasoning behind this simulation is that, if $(m, \sigma) \notin \mathcal{Q}$ then the pair must be invalid thanks to the strong unforgeability of SCUS$_1$.

– Confirmation/disavowal query (m, σ): If $(m, \sigma) \in \mathcal{Q}$ then return 1 and run the confirmation protocol with $\mathcal{D}$ using the knowledge of recorded values r, y. Otherwise, return 0 and run the disavowal protocol with $\mathcal{D}$ using the rewinding technique [16].

– Challenge query m^*: Choose prime $e^* \xleftarrow{\$} \{0, 1\}^\eta$, set $y^* \leftarrow H(m \| (Y_b \bmod N_2))^{1/e^*} \bmod N_1$, and $\tau^* = Y_b(1 + y^* N_2) \bmod N_2^2$. Return $\sigma^* = (e^*, \tau^*)$ to $\mathcal{D}$. Note that if $b = 0$, then $Y_0 = (r^*)^{N_2} \bmod N_2^2$ so that σ^* is a valid undeniable signature on m^*. If $b = 1$ then Y_1 is random over $Z^*_{N_2^2}$, so that τ^* is also random over $Z^*_{N_2^2}$, which is indistinguishable from $Z_{N_2^2}$, and hence σ^* is random over $\mathcal{S} = \mathcal{P} \times Z_{N_2^2}$, where $\mathcal{P}$ is the set of all η-bit primes.

Finally, $\mathcal{D}$ outputs a bit b'. The adversary $\mathcal{A}_{dnr}$ also returns b' as its guess of b. It is clear that if $\mathcal{D}$ succeeds then $\mathcal{A}_{dnr}$ also succeeds with advantage

$\mathbf{Adv}^{dnr}_{\mathcal{K}_{dnr}}(\mathcal{A}_{dnr}) \gtrsim \mathbf{Adv}^{iv}_{\mathsf{SCUS}_1}(\mathcal{D})$. We use the symbol $\gtrsim$ to capture the use of strong unforgeability of SCUS_1 in the simulation.

5   The Scheme SCUS$_2$

We will use the following result.

The GHR Signature Scheme [13]. The ordinary signature scheme GHR = (GHR.Kg, GHR.Sign, GHR.Vrf) is briefly recalled with some minor modifications as follows.

> GHR.Kg$(1^k) \to (pk_1, sk_1)$: Generate $N_1 = p_1 q_1$ for safe distinct primes p_1, q_1, where the length $|N_1| = k$, $s \in Z^*_{N_1}$, and a function $H : \{0,1\}^* \to Z^*_{N_1}$. The secret signing key $sk_1 = (p_1, q_1)$ and the public key $pk_1 = (s, N_1, H)$.
>
> GHR.Sign$(sk_1, m) \to y$: The signature y on a message $m \in \{0,1\}^*$ is $y = s^{\frac{1}{H(m)}} \bmod N_1$, which is computed using p_1, q_1 in sk_1.
>
> GHR.Vrf$(pk_1, m, y) \to 0/1$: Given y, check that $y \in Z_{N_1}$, and $y^{H(m)} = s \bmod N_1$. Return 1 if all checks pass, else return 0.

It is known that the GHR signature scheme is suf-cma under the strong RSA assumption and the DIV assumption, where the latter intuitively asserts that, given the hash function H, it is hard to find distinct values $X_1, \ldots, X_n$, and X such that $H(X)$ divides the product $H(X_1) \cdots H(X_n)$.

Our Proposal SCUS$_2$. The scheme is described as follows.

– KeyGen$(1^k) \to (pk, sk)$: Let $\big(pk_1 = (s, N_1, H), sk_1 = (p_1, q_1)\big) \xleftarrow{\$}$ GHR.Kg(1^k), and $\big(pk_2 = N_2, sk_2 = (p_2, q_2)\big) \xleftarrow{\$} \mathsf{K}(1^k)$ such that $N_1 \leq N_2$. Set the secret key $sk = (sk_1, sk_2)$, and the public key $pk = (pk_1, pk_2)$. The message space is $\mathcal{M} = \{0,1\}^*$ and the signature space is $\mathcal{S} = Z_{N_2^2}$.

– USign$(sk, m) \to \sigma$: First, choose $r \xleftarrow{\$} Z^*_{N_2}$ and set $\overline{m} \leftarrow m \parallel (r^{N_2} \bmod N_2)$. Next, take GHR signature on $\overline{m}$: $y = s^{1/H(\overline{m})} \bmod N_1 \xleftarrow{\$}$ GHR.Sign$(sk, \overline{m})$. Then, encrypt y: $\tau \xleftarrow{\$} \mathsf{E}(y; r)$. Finally, return $\sigma = \tau$ as the undeniable signature.

– Convert$(sk, m, \sigma = \tau) \to cvt/\perp$: Let $(r, y) \leftarrow \mathsf{D}(sk_2, \tau)$. If having GHR.Vrf$\big(pk_1, m \parallel (\tau \bmod N_2), y\big) = 1$, then return $cvt = r$; otherwise return $\perp$.

– Verify$(pk, m, \sigma, r) \to 0/1$: From $r \in Z^*_{N_2}$ and $\sigma = \mathsf{E}(y; r)$, compute $y \in Z_{N_2}$ as in the decryption algorithm of Paillier scheme, namely $y = \big((\sigma(r^{-1})^{N_2} \bmod N_2^2) - 1\big)/N_2$. Return GHR.Vrf$(pk_1, m \parallel (\sigma \bmod N_2), y)$.

– Confirm: On common input $(m, \sigma = \tau) \in \mathcal{M} \times \mathcal{S}$ and pk, execute $\mathrm{PoK}\big\{(r, y) : r \in [1, N_2 - 1] \wedge y \in [1, N_1 - 1] \wedge y^{H(m \parallel (\tau \bmod N_2))} = s \bmod N_1 \wedge \tau = \mathsf{E}(y; r)\big\}$.

– Disavowal: On common input $(m, \sigma = \tau) \in \mathcal{M} \times \mathcal{S}$ and pk, execute
PoK$\big\{(r, y_1, y_2, \delta) : r \in [1, N_2 - 1] \wedge y_1 \in [1, N_1 - 1] \wedge y_2 \in [1, N_2 - 1] \wedge$
$y_1^{H\big(m\|(\tau \bmod N_2)\big)} = s \bmod N_1 \wedge \tau = \mathsf{E}(y_2; r) \wedge \delta = y_2 - y_1 \bmod N_2 \wedge \delta \in$
$[1, N_2 - 1]\big\}$.

All conversion is done by publish sk_2 and the protocol costs are similar to those of SCUS$_1$. Security theorems for SCUS$_2$ are similar to those of SCUS$_1$, and will be given in the full version of this paper.

6 Anonymity of Our Schemes and KT$_1$

The notion of anonymity for undeniable signature schemes was presented by Galbraith and Mao [14], intuitively ensuring that no-one can tell *who* is the author of a message-signature pair (m, σ) in a multi-signer setting. In particular, given two signers A and B, it is hard to decide whether A or B produces (m, σ). Invisibility and anonymity are closely related, and it is known that invisibility implies anonymity, provided that all signers share the same signature space (see [14], Section 8). In short, we have: invisibility + common-signature-space $\longrightarrow$ anonymity.

Unfortunately, signers using RSA-based schemes usually do not have the same signature space. (For example, in our SCUS schemes, the signature space varies with the modulus N_2.) One needs to overcome this problem to ensure anonymity. Two solutions are known in the literature: (1) stretching the signature to a fixed length relying only on the system parameter k via padding (e.g., [14] and references therein); or (2) shortening the signature to a fixed length via repeated computation [2]. The first approach suffers from longer signatures and the second suffers from increased cost for computation. Below we suggest another simple way to provide anonymity for our schemes without increasing signature size or computation cost.

6.1 Anonymity for Our Schemes

Let us consider SCUS$_1$, whose signature spaces for signers A and B are respectively $\mathcal{S}_A = \mathcal{P} \times Z_{N_{2,A}^2}$ and $\mathcal{S}_B = \mathcal{P} \times Z_{N_{2,B}^2}$, where $\mathcal{P}$ is the set of all primes of $\eta = 80$ bits. Without loss of generality, we assume $N_{2,A} \leq N_{2,B}$. Our solution for anonymity is simple: we choose "close enough" $N_{2,A}$ and $N_{2,B}$ so that given a random signature $\sigma = (e, \tau)$ on a message m of either A or B, we have $\sigma \in \mathcal{S}_A \cap \mathcal{S}_B = \mathcal{P} \times Z_{N_{2,A}^2}$ with overwhelming probability. The common signature space is thus $\mathcal{S}_A \cap \mathcal{S}_B$ with overwhelming probability. Namely, we require the probability $\Pr[N_{2,A}^2 \leq \tau < N_{2,B}^2]$ is negligible for random τ. For 80-bit security, we need

$$\Pr[N_{2,A}^2 \leq \tau < N_{2,B}^2] = \frac{N_{2,B}^2 - N_{2,A}^2}{N_{2,B}^2} \leq 2^{-80}.$$

Since $|N_{2,A}| = |N_{2,B}| = k$, we have $2^{2(k-1)} \leq N_{2,B}^2$ and $N_{2,A} + N_{2,B} < 2^{k+1}$. The above inequality will be satisfied if we choose $N_{2,A}, N_{2,B}$ such that

$0 \leq N_{2,B} - N_{2,A} \leq 2^{k-83}$. This means 83 leading bits of $N_{2,A}$ and $N_{2,B}$ are equal. Fortunately, the method of choosing RSA moduli with (about $k/2$) common-leading-bits is known in the literature, e.g., [20] (with application like making RSA public key short). The above discussion, applied as well to the schemes SCUS_0 and SCUS_2, is a new application of that method in the context of undeniable signature.

6.2 Anonymity and Invisibility of KT_1, Revisited

We revisit the SCUS scheme in the standard model of [19], called KT_1, and show an evidence that it does not have invisibility (as claimed in Theorem 10 of [19]) in the sense of [14]. To do so, we point out that the scheme does not satisfy anonymity, *even if* signers share a common signature space.

We first recall the undeniable signature on a message m of KT_1, which is of the form (e, y, x', ω), where e, x', ω are random-like values. The point here is y is not a random-like value, and it must satisfy the following equation

$$y^e = x \cdot h_2^{H(x')} \bmod N_2,$$

for public values x, h_2, and hash function H of the current signer. In fact (e, y) is an *ordinary* signature on x' (this signature scheme is similar to the GRSA signature scheme and is weakly-secure), so that everyone can check the validity of (e, y) on x' using the above equation.

Now it is easy to decide whether the signer A or B produces a pair $\big(m, (e, y, x', \omega)\big)$, just by checking whether (e, y) is the ordinary signature on x' with respect to public key of A and B. If (e, y) is valid on x' with respect to public key of A, we bet that A produces $\big(m, (e, y, x', \omega)\big)$; otherwise we bet B does.

The above observation works even if the signers A and B have the same or "close enough" signature spaces so that (e, y, x', ω) lies in both. What we really need is that the signers have independent public keys, which is a trivial fact.

7 Conclusion

We construct several RSA-based undeniable signature schemes in random oracle and standard model, directly offering selective/all conversion at the same time. The standard model schemes are the first RSA-based ones achieving invisibility and anonymity in the sense of [14]. Furthermore, the signature and converter of our schemes are very short among RSA-based schemes. The costs for confirmation and disavowal protocol in our schemes is a bit higher, but still practical. We also point out that the standard model scheme in [19] does not satisfy invisibility and anonymity in the sense of [14].

Acknowledgement. We thank Dennis Hofheinz for information on suf-cma security of GRSA. We also thank the anonymous reviewers for comprehensive comments.

References

1. Boudot, F.: Efficient Proofs that a Committed Number Lies in an Interval. In: Preneel, B. (ed.) EUROCRYPT 2000. LNCS, vol. 1807, pp. 431–444. Springer, Heidelberg (2000)
2. Bellare, M., Boldyreva, A., Desai, A., Pointcheval, D.: Key-Privacy in Public-Key Encryption. In: Boyd, C. (ed.) ASIACRYPT 2001. LNCS, vol. 2248, pp. 566–582. Springer, Heidelberg (2001)
3. Boyar, J., Chaum, D., Damgard, I., Pedersen, T.: Convertible undeniable signatures. In: Menezes, A., Vanstone, S.A. (eds.) CRYPTO 1990. LNCS, vol. 537, pp. 189–205. Springer, Heidelberg (1991)
4. Bellare, M., Rogaway, P.: The Exact Security of Digital Signatures - How to Sign with RSA and Rabin. In: Maurer, U.M. (ed.) EUROCRYPT 1996. LNCS, vol. 1070, pp. 399–416. Springer, Heidelberg (1996)
5. Chaum, D., van Antwerpen, H.: Undeniable signatures. In: Brassard, G. (ed.) CRYPTO 1989. LNCS, vol. 435, pp. 212–216. Springer, Heidelberg (1990)
6. Chaum, D., van Heijst, E., Pfitzmann, B.: Cryptographically strong undeniable signatures, unconditionally secure for the signer. In: Feigenbaum, J. (ed.) CRYPTO 1991. LNCS, vol. 576, pp. 470–484. Springer, Heidelberg (1992)
7. Cash, D., Kiltz, E., Shoup, V.: The Twin Diffie-Hellman Problem and Applications. In: Smart, N.P. (ed.) EUROCRYPT 2008. LNCS, vol. 4965, pp. 127–145. Springer, Heidelberg (2008)
8. Chaum, T., Pedersen, T.P.: Wallet databases with observers. In: Brickell, E.F. (ed.) CRYPTO 1992. LNCS, vol. 740, pp. 89–105. Springer, Heidelberg (1993)
9. Coron, J.: On the Exact Security of Full Domain Hash. In: Bellare, M. (ed.) CRYPTO 2000. LNCS, vol. 1880, pp. 229–235. Springer, Heidelberg (2000)
10. Damgard, I., Fujisaki, E.: A Statistically-Hiding Integer Commitment Scheme Based on Groups with Hidden Order. In: Zheng, Y. (ed.) ASIACRYPT 2002. LNCS, vol. 2501, pp. 125–142. Springer, Heidelberg (2002)
11. Damgard, I., Pedersen, T.: New convertible undeniable signature schemes. In: Maurer, U.M. (ed.) EUROCRYPT 1996. LNCS, vol. 1070, pp. 372–386. Springer, Heidelberg (1996)
12. Fujisaki, E., Okamoto, T.: Statistical Zero Knowledge Protocols to Prove Modular Polynomial Relations. In: Kaliski Jr., B.S. (ed.) CRYPTO 1997. LNCS, vol. 1294, pp. 16–30. Springer, Heidelberg (1997)
13. Gennaro, R., Halevi, S., Rabin, T.: Secure Hash-and-Sign Signatures Without the Random Oracle. In: Stern, J. (ed.) EUROCRYPT 1999. LNCS, vol. 1592, pp. 123–139. Springer, Heidelberg (1999)
14. Galbraith, S., Mao, W.: Invisibility and anonymity of undeniable and confirmer signatures. In: Joye, M. (ed.) CT-RSA 2003. LNCS, vol. 2612, pp. 80–97. Springer, Heidelberg (2003)
15. Galbraith, S., Mao, W., Paterson, K.G.: RSA-based undeniable signatures for general moduli. In: Preneel, B. (ed.) CT-RSA 2002. LNCS, vol. 2271, pp. 200–217. Springer, Heidelberg (2002)
16. Goldreich, O., Oren, Y.: Definitions and properties of zero-knowledge proof systems. Journal of Cryptology 7(1), 1–32 (1994)
17. Gennaro, R., Rabin, T., Krawczyk, H.: RSA-based undeniable signatures. Journal of Cryptology 13(4), 397–416 (2000)
18. Hofheinz, D., Kiltz, E.: Programmable Hash Functions and Their Applications. In: Wagner, D. (ed.) CRYPTO 2008. LNCS, vol. 5157, pp. 21–38. Springer, Heidelberg (2008)

19. Kurosawa, K., Takagi, T.: New Approach for Selectively Convertible Undeniable Signature Schemes. In: Lai, X., Chen, K. (eds.) ASIACRYPT 2006. LNCS, vol. 4284, pp. 428–443. Springer, Heidelberg (2006)
20. Lenstra, A.K.: Generating RSA Moduli with a Predetermined Portion. In: Ohta, K., Pei, D. (eds.) ASIACRYPT 1998. LNCS, vol. 1514, pp. 1–10. Springer, Heidelberg (1998)
21. Paillier, P.: Public-Key Cryptosystems Based on Composite Degree Residuosity Classes. In: Stern, J. (ed.) EUROCRYPT 1999. LNCS, vol. 1592, pp. 223–238. Springer, Heidelberg (1999)
22. Ogata, W., Kurosawa, K., Heng, S.: The security of the FDH variant of Chaum's undeniable signature scheme. IEEE Transactions on Information Theory 52(5), 2006–2017 (2006)

A Schnorr-Like Lightweight Identity-Based Signature Scheme

David Galindo[1,*] and Flavio D. Garcia[2,**]

[1] University of Luxembourg
david.galindo@uni.lu
[2] Institute for Computing and Information Sciences,
Radboud University Nijmegen, The Netherlands
flaviog@cs.ru.nl

Abstract. The use of concatenated Schnorr signatures [Sch91] for the hierarchical delegation of public keys is a well-known technique. In this paper we carry out a thorough analysis of the identity-based signature scheme that this technique yields. The resulting scheme is of interest since it is intuitive, simple and does not require pairings. We prove that the scheme is secure against existential forgery on adaptive chosen message and adaptive identity attacks using a variant of the Forking Lemma [PS00]. The security is proven in the Random Oracle Model under the discrete logarithm assumption. Next, we provide an estimation of its performance, including a comparison with the state of the art on identity-based signatures. We draw the conclusion that the Schnorr-like identity-based signature scheme is arguably the most efficient such scheme known to date.

Keywords: identity-based signature, lightweight cryptography, provable security, Schnorr, random oracle model.

1 Introduction

Digital signatures are fundamental primitives in public key cryptography. They ensure the authenticity of the originator of a digital document and the integrity of that document, while they also prevent that the originator can repudiate that very same document later on. A signature on a given bit-string is valid if it passes the associated verification test, which takes as inputs a verification key, a purported signature, and a document. In traditional signature schemes the verification key is a mathematical object taken at random from some set. An

---

* Work done while at University of Malaga. Partially funded by the Spanish Ministry of Science and Education through the projects ARES (CSD2007-00004) and CRISIS (TIN2006-09242).

** partially supported by the research program Sentinels (www.sentinels.nl), project PEARL (7639). Sentinels is being financed by Technology Foundation STW, the for Scientific Research (NWO), and the Dutch Ministry of Economic Affairs.

B. Preneel (Ed.): AFRICACRYPT 2009, LNCS 5580, pp. 135–148, 2009.

external binding between the verification key and the signing entity is therefore needed. This binding takes the form of a certificate, which is created by a Certification Authority.

The concept of identity-based signature (IBS) was introduced by Shamir [Sha85]. The idea is that the identity of the signer is used as the verification key. This dispenses with the need of an external binding. The identity can be any string that singles out the entity, for instance, a social security number or an IP address. Identity based systems have a drawback, namely, the entity cannot build the signing key by itself. Instead, a Trusted Third Party (TTP) is in charge of assigning and delivering, via a confidential channel, the secret key to each user. This in sharp contrast to the traditional setting, where the secret key is generated by the user itself and kept secret. Still scenarios where a TTP creates a signing/verification key pair on behalf of the user are quite often found in practice. This is the case of the Spanish Electronic Identity Card eDNI [oIA08, EG08] and the digital signature token services provided by DigiNotar [Dig08]. There, the key pair is stored together with a certificate in a smart card to be owned by the relevant user.

Several IBS schemes based on factoring or RSA were proposed in the late eighties and early nineties, for instance [Sha85, FS87, GQ90, Oka93] to name just a few. After the revival of research on identity-based cryptography due to the use of pairings (cf. [BSS05] Chapter 5 for an introduction to pairings), many other IBS schemes have been proposed, for instance [SOK00, Hes03, CC02]. As a consequence of this revival, Bellare, Namprempre and Neven [BNN04] provided a framework for deriving security proofs for identity-based signature and identification schemes, and compiled most previous work on IBS. The most efficient scheme of those studied in [BNN04] turns out to be an IBS scheme obtained from a conventional signature scheme by Beth [Bet88]. Unfortunately, Beth's scheme lacks a security proof.

Our contribution. This work studies the identity-based signature scheme resulting from sequentially delegating Schnorr signatures [Sch91]. Such a technique is certainly well-known and it has been in use for many years and in different contexts, e.g. [Gir91, PH97, AO99, CJT04, BSNS05, BFPW07]. We prove that the scheme is secure against existential forgery on adaptive chosen message and adaptive identity attacks using a variant of the Forking Lemma [PS00] by Boldyreva, Palacio and Warinschi [BPW03]. The security is proven in the Random Oracle Model under the discrete logarithm assumption. We show that the resulting scheme is among the most efficient provably-secure IBS schemes known to date, be it based on factoring, discrete logarithm or pairings. In particular it has the same performance that the aforementioned Beth IBS scheme. This makes it attractive for application in resource-constrained environments where saving in computation, communication and implementation code area are a premium.

Organization of this paper. Section 2 introduces standard definitions from the literature. Section 3 describes our new identity-based signature scheme. In

Section 4 we prove the security of the scheme. Section 5 compares the computational and length complexity of our scheme to other schemes from the literature. Finally Section 6 concludes the paper.

2 Preliminaries

This section introduces the syntax and security definitions of identity-based signatures and the discrete logarithm assumption. Most of it is standard, we refer the reader to [BNN04] for a thorough explanation. We introduce some basic notation. If S is a set then $s_1,\ldots,s_n \overset{\$}{\leftarrow} S$ denotes the operation of picking n elements s_i of S independently and uniformly at random. We write $\mathcal{A}(x,y,\ldots)$ to indicate that $\mathcal{A}$ is an algorithm with inputs $x,y,\ldots$ and by $z \leftarrow \mathcal{A}(x,y,\ldots)$ we denote the operation of running $\mathcal{A}$ with inputs $(x,y,\ldots)$ and letting z be the output.

2.1 Identity-Based Signatures

An *identity-based signature scheme* (IBS) is a quadruple $(\mathcal{G},\mathcal{E},\mathcal{S},\mathcal{V})$ of probabilistic polynomial-time algorithms, where

- $\mathcal{G}$ is the *parameter-generation algorithm*. It takes as input the security parameter η (a positive integer and outputs the system public parameters mpk and the master secret key msk.

- $\mathcal{E}$ is the *key-extraction algorithm* that takes as input parameters mpk, a master key msk and an identity id and outputs a private key $\mathsf{sk_{id}}$ corresponding to the user with this identity.

- $\mathcal{S}$ is a *signing algorithm* that takes as input parameters mpk, a private key $\mathsf{sk_{id}}$ and a message m and outputs a signature σ.

- $\mathcal{V}$ is a deterministic *signature verification algorithm* that takes as input parameters mpk, a signature σ, a message m and an identity id and outputs whether or not σ is a valid signature of m relative to $(\mathsf{mpk},\mathsf{id})$.

Definition 1 (EUF-IBS-CMA). An identity-based signature scheme $\Sigma = (\mathcal{G}, \mathcal{E},\mathcal{S},\mathcal{V})$ is said to be secure against *existential forgery on adaptively chosen message and identity attacks* if for all probabilistic polynomial-time adversaries $\mathcal{A}$, the probability of the experiment **EUF-IBS-CMA**$_\Sigma(\mathcal{A}) = 1$ defined below is a negligible function of η. During this experiment $\mathcal{A}$ has access to two oracles: a key-extraction oracle $\mathcal{O}_\mathcal{E}$ that takes as input an identity id and outputs $\mathcal{E}(\mathsf{mpk},\mathsf{msk},\mathsf{id})$; and a signature oracle $\mathcal{O}_\mathcal{S}$ that takes as input an identity id and a message m and returns a signature $\mathcal{S}(\mathsf{mpk},\mathsf{sk_{id}},m)$.

$$
\begin{array}{|l|}
\hline
\textbf{EUF-IBS-CMA}_\Sigma(\mathcal{A})\text{:} \\
(\mathsf{mpk},\mathsf{msk}) \leftarrow \mathcal{G}(\eta) \\
(\mathsf{id}^\star,m^\star,\sigma^\star) \leftarrow \mathcal{A}^{\mathcal{O}_\mathcal{E}(\cdot),\mathcal{O}_\mathcal{S}(\cdot,\cdot)}(\mathsf{mpk}) \\
\textbf{return }\ \mathcal{V}(\mathsf{mpk},\sigma^\star,m^\star,\mathsf{id}^\star) \\
\hline
\end{array}
$$

Some restrictions apply. In particular, it is required that $\mathsf{id}^\star$ and $(\mathsf{id}^\star, m^\star)$ are not equal to any query made to the oracles $\mathcal{O}_\mathcal{E}(\cdot)$ and $\mathcal{O}_\mathcal{S}(\cdot, \cdot)$ respectively, and that the same id cannot be queried to $\mathcal{O}_\mathcal{E}(\cdot)$ twice (see [BNN04] for details).

Definition 2 (Discrete Logarithm Assumption). We say that a group generation function $(\mathcal{G}, g, q) \leftarrow \mathsf{Gen}(\eta)$ generates DL-secure groups if for all probabilistic polynomial-time algorithms $\mathcal{A}$, the probability

$$\mathbb{P}[(\mathcal{G}, g, q) \leftarrow \mathsf{Gen}(\eta); a \xleftarrow{\$} \mathbb{Z}_q : a \leftarrow \mathcal{A}(\mathcal{G}, q, g, g^a)]$$

is a negligible function of η.

3   The Construction

The idea behind the construction is to use two concatenated Schnorr signatures [Sch91]. This technique is certainly not new and has been used elsewhere [Gir91, PH97, AO99, CJT04, BSNS05, BFPW07]. This said, the benefits of this technique in terms of efficiency and simplicity for the design of identity-based signatures seem to have gone unnoticed in the relevant literature as far as we are aware of (see Section 5).

Roughly speaking, the scheme works as follows. Firstly, the TTP produces a Schnorr signature on the identity of the user by using the master secret key. This signature implicitly defines a unique Schnorr-like public key for which only the user knows the corresponding private key. Next, the user builds a second Schnorr-like signature, this time on the message, by using its private key. The verification of a signature on message m under identity id implicitly checks whether the two concatenated Schnorr signatures are correct. Details are given below.

- The global parameters generation algorithm $\mathcal{G}$ on input η outputs public parameters mpk and a master secret key msk where: $(\mathcal{G}, g, q)$ is generated by calling the group generation algorithm Gen on input η. $\mathcal{G}$ is the description of a group of prime order $q = q(\eta)$ with $2^\eta \leq q < 2^{\eta+1}$ and g is a generator of $\mathcal{G}$. $G: \{0,1\}^* \to \mathbb{Z}_q, H: \{0,1\}^* \to \mathbb{Z}_q$ are descriptions of hash functions. Let z be a random element from $\mathbb{Z}_q$ and set $(\mathsf{mpk}, \mathsf{msk}) = ((\mathcal{G}, g, q, g^z, G, H), z)$.

- The key-extraction algorithm $\mathcal{E}$ on input global parameters mpk, a master secret key $\mathsf{msk} = z$ and an identity id, picks $r \xleftarrow{\$} \mathbb{Z}_q$ and sets $y = r + z \cdot H(g^r, \mathsf{id}) \mod q$. Then it outputs the secret key $\mathsf{sk}_\mathsf{id} = (y, g^r)$. Note that g^r is actually public information even though it is part of the secret key.

- The signing algorithm $\mathcal{S}$ on input parameters mpk, user private key $\mathsf{sk}_\mathsf{id} = (y, g^r)$ and a message m proceeds as follows. It selects $a \xleftarrow{\$} \mathbb{Z}_q$ and computes $b = a + y \cdot G(\mathsf{id}, g^a, m)$. Then it outputs the signature $\sigma = (g^a, b, g^r)$.

- The verification algorithm $\mathcal{V}$ on input parameters $\mathsf{mpk} = (\mathcal{G}, g, q, G, H, g^z)$, a signature $\sigma = (g^a, b, g^r)$, a message m and an identity id proceeds as follows. It outputs whether or not the equation $g^b = g^a(g^r g^{zc})^d$ holds, where $c = H(g^r, \mathsf{id})$ and $d = G(\mathsf{id}, g^a, m)$.

4   Security of the Construction

This section analyzes the security of the proposed scheme in the random oracle model.

Theorem 3. *The construction above is secure with respect to Definition 1, in the random oracle model, if the group generation function* Gen *generates discrete logarithm secure groups. More precisely, there exist adversaries* $\mathcal{B}_1, \mathcal{B}_2$ *satisfying either*

$$\mathsf{Adv}_{\mathcal{B}_1}^{DL} \geq \frac{\mathsf{Adv}_{\mathcal{A}}^{\text{EUF-IBS-CMA}}(\eta)}{Q_G} \left(\frac{\mathsf{Adv}_{\mathcal{A}}^{\text{EUF-IBS-CMA}}(\eta)}{Q_G^2} - \frac{1}{2^\eta} \right)$$

or

$$\mathsf{Adv}_{\mathcal{B}_2}^{DL}(\eta) \geq \mathsf{Adv}_{\mathcal{A}}^{\text{EUF-IBS-CMA}}(\eta) \left(\frac{(\mathsf{Adv}_{\mathcal{A}}^{\text{EUF-IBS-CMA}}(\eta))^3}{(Q_G Q_H)^6} - \frac{3}{2^\eta} \right).$$

Proof. Assume there is an adversary $\mathcal{A}$ that wins the game EUF-IBS-CMA with non-negligible probability. Eventually, $\mathcal{A}$ outputs an attempted forgery of the form $\sigma = (A, B, R)$. Let E be the event that σ is a valid signature and R was contained in an answer of the signing oracle $\mathcal{O}_S$. Let NE be the event that σ is a valid signature and R was never part of an answer of $\mathcal{O}_S$. Clearly, $\mathsf{Adv}_{\mathcal{A}}^{\text{EUF-IBS-CMA}}(\eta) = \mathbb{P}[E] + \mathbb{P}[NE]$.

Next we build the following adversaries $\mathcal{B}_1, \mathcal{B}_2$ against the discrete logarithm assumption. Intuitively, $\mathcal{B}_1$ breaks the DL-assumption when the event E happens and $\mathcal{B}_2$ in case of NE.

$\mathcal{B}_1$ takes as argument the description of a group $(\mathcal{G}, q, g)$ and a challenge g^r with $r \xleftarrow{\$} \mathbb{Z}_q$ and tries to extract the discrete logarithm r. To do so it will run the adversary $\mathcal{A}$, for which simulates its environment as follows:

- $\mathcal{B}_1$ picks a random $i \leftarrow [Q_G]$, where Q_G is the maximal number of queries that the adversary $\mathcal{A}$ performs to the random oracle G. Let id* (the target identity) be the identity in the i-th query to the random oracle G. Next, $\mathcal{B}_1$ chooses $z \xleftarrow{\$} \mathbb{Z}_q$ and sets public parameters mpk $= (\mathcal{G}, q, g, G, H, \mathsf{mpk} = g^z)$ where G, H are descriptions of hash functions modeled as random oracles. As usual, $\mathcal{B}_1$ simulates these oracles by keeping two lists L_G, L_H containing the queried values together with the answers given to $\mathcal{A}$.

- Every time $\mathcal{A}$ queries the key extraction oracle $\mathcal{O}_{\mathcal{E}}$, for user id, $\mathcal{B}_1$ chooses $c, y \xleftarrow{\$} \mathbb{Z}_q$, sets $R = g^{-zc}g^y$ and adds $((R, \text{id}), c)$ to the list L_H. Then it returns the key (y, R) to $\mathcal{A}$.

- When $\mathcal{A}$ makes a call to the signature oracle (id, m) with id $\neq$ id*, $\mathcal{B}_1$ simply computes id's private key as described in the previous bullet. Then it runs the signing algorithm $\mathcal{S}$ and returns the produced signature to $\mathcal{A}$.

- When $\mathcal{A}$ makes a call to the signature oracle (id, m) with id $=$ id*, $\mathcal{B}_1$ chooses $t \xleftarrow{\$} \mathbb{Z}_q, B \xleftarrow{\$} \mathcal{G}$, sets $R = g^{-zc}(g^r)^t$, $c = H(\text{id}, g^r)$ and $A = B(g^r g^{zc})^{-d}$. Then it returns the signature (A, B, R) to the adversary $\mathcal{A}$.

- $\mathcal{B}_1$ runs the algorithm $\mathsf{MF}_{Y,1}(\mathsf{mpk})$ as described in Lemma 1. Here algorithm Y is simply a wrapper that takes as explicit input the answers from the random oracles. Then it calls $\mathcal{A}$ and returns its output together with two integers I, J. This integers are the indexes of $\mathcal{A}$'s calls to the random oracles G, H with the target identity $\mathsf{id}^\star$.

---

algorithm $\mathsf{MF}_{Y,1}(\mathsf{mpk})$:

Pick random coins ρ for Y

$s_1 \ldots s_{Q_G} \xleftarrow{\$} \mathbb{Z}_q$

$(I, J, \sigma_0) \leftarrow Y(\mathsf{mpk}, s_1 \ldots s_{Q_G}; \rho)$

if $(I = 0 \vee J = 0)$ **then return** $\perp$

$s_I^1 \ldots s_{Q_G}^1 \xleftarrow{\$} \mathbb{Z}_q$

$(I_1, J_1, \sigma_1) \leftarrow Y(\mathsf{mpk}, s_1, \ldots s_{I-1}, s_I^1 \ldots s_{Q_G}^1; \rho)$

if $((I, J) \neq (I_1, J_1) \vee s_I = s_I^1)$ **then return** $\perp$

else return σ_0, σ_1

In this way we get two forgeries of the form $\sigma_0 = (\mathsf{id}, m, (A, B_1, R))$ and $\sigma_1 = (\mathsf{id}, m, (A, B_2, R))$. Let d_1 be the answer from the random oracle G given to $\mathcal{A}$ in the first execution, i.e., s_I in $\mathsf{MF}_{Y,1}(\mathsf{mpk})$, and let d_2 be the second answer s_I^1. If the identity id is not equal to the target identity $\mathsf{id}^\star$ then $\mathcal{B}_1$ aborts. Otherwise it terminates and outputs the attempted discrete logarithm $(B_1 - B_2)(td_1 - td_2)^{-1}$.

Next we show that $\mathcal{B}_1$'s output is indeed the discrete logarithm r of the challenge G^r. Since both signatures are valid we get that

$$g^{B_1} = A(Rg^{zc})^{d_1} \qquad \text{and} \qquad g^{B_2} = A(Rg^{zc})^{d_2}$$

where $c = H(g^r, \mathsf{id})$ and $R = g^{-zc}g^{rt}$. Hence, $B_1 = \log A + rtd_1$ and $B_2 = \log A + rtd_2$ and therefore $r = (B_1 - B_2)(td_1 - td_2)^{-1}$.

Next we need to lower bound success probability of $\mathcal{B}_1$ against the discrete logarithm assumption. To this aim we use the Multiple-Forking Lemma of Boldyreva, Palacio and Warinschi [BPW03], which we include for the sake of self-containment. The lemma is a further generalization of the General Forking Lemma proposed by Bellare and Neven [BN06] to multiple random oracles and signatures. The General Forking Lemma itself is a generalization of the Forking Lemma, originally proposed in [PS00]. Intuitively, this lemma states that, in the random oracle model, if there is an algorithm (a forger) that can produce a forgery, then it is possible to get a different forgery on the same message (by changing the answer of the random oracle).

Lemma 1 (Multiple-Forking Lemma [BPW03]). *Fix $\alpha \in \mathbb{Z}^+$ and a set S such that $|S| \geq 2$. Let Y be a randomized algorithm that on input x and a sequence of elements $s_1 \ldots s_\alpha \in S$, returns a triple (I, J, σ) consisting of two*

```
algorithm  MF_{Y,n}(x):
  Pick random coins ρ for Y
  s_1 ... s_α  ←$ S
  (I, J, σ_0) ← Y(x, s_1 ... s_α; ρ)
  if (I = 0 ∨ J = 0) then return ⊥
  s_I^1 ... s_α^1  ←$ S
  (I_1, J_1, σ_1) ← Y(x, s_1, ... s_{I-1}, s_I^1 ... s_α^1; ρ)
  if ((I, J) ≠ (I_1, J_1) ∨ s_I = s_I^1) then return ⊥
  for  i = 2 ; i < n ; i = i + 2  do
    s_1^i ... s_α^i  ←$ S
    (I_i, J_i, σ_i) ← Y(x, s_1 ... s_{J-1}, s_J^i, ..., s_α^i; ρ)
    if ((I_i, J_i) ≠ (I, J) ∨ s_J^i = s_J^{i-1}) then return ⊥
    s_I^{i+1} ... s_α^{i+1}  ←$ S
    (I_{i+1}, J_{i+1}, σ_{i+1}) ← Y(x, s_1 ... s_{J-1}, s_J^i, ..., s_{I-1}^i, s_I^{i+1}, ..., s_α^{i+1}; ρ)
    if ((I_{i+1}, J_{i+1}) ≠ (I, J) ∨ s_J^{i+1} = s_J^i) then return ⊥
  endfor
  return σ_0 ... σ_n
```

Fig. 1. Multiple-forking algorithm

integers associated to Y and a bitstring σ. Let $n \geq 1$ be an odd integer and x a bitstring. The multiple-forking algorithm $\mathsf{MF}_{Y,n}$ associated to Y and n is defined as in Figure 1. Let

$$\mathsf{acc} = \mathbb{P}[x \xleftarrow{\$} PG; s_1 \ldots s_\alpha \xleftarrow{\$} S; (I, J, \sigma) \leftarrow Y(x, s_1 \ldots s_\alpha) : I \geq 1 \wedge J \geq 1]$$

$$\mathsf{frk} = \mathbb{P}[x \xleftarrow{\$} PG : \mathsf{MF}_{Y,n}(x) \neq \bot].$$

Then

$$\mathsf{frk} \geq \mathsf{acc} \left(\frac{\mathsf{acc}^n}{\alpha^{2n}} - \frac{n}{|S|} \right)$$

and therefore

$$\mathsf{acc} \leq \sqrt[n+1]{\alpha^{2n} \, \mathsf{frk}} + \sqrt[n+1]{\frac{n \alpha^{2n}}{|S|}}$$

To bound the success probability of $\mathcal{B}_1$ against de discrete logarithm assumption first notice that with probability $1/Q_G$ the target identity $\mathsf{id}^\star$ equals the identity id output by the adversary. Then, it follows from Lemma 1 that

$$\mathsf{Adv}_{\mathcal{B}_1}^{DL} \geq \frac{\mathsf{frk}}{Q_G} \geq \frac{\mathsf{Adv}_{\mathcal{A}}^{\text{EUF-IBS-CMA}}(\eta)}{Q_G} \left(\frac{\mathsf{Adv}_{\mathcal{A}}^{\text{EUF-IBS-CMA}}(\eta)}{Q_G^2} - \frac{1}{2^\eta} \right)$$

It lacks to bound the success probability of NE. This case is slightly more involved as it uses nested rewindings of the adversary. In this case $\mathcal{B}_2$ attacks the public key of the trusted authority g^z. It takes as argument the description of a group $(\mathcal{G}, q, g)$ and a challenge g^z with $z \xleftarrow{\$} \mathbb{Z}_q$ and outputs its discrete logarithm z. To do so it will run the adversary $\mathcal{A}$ simulating its environment as follows:

- At the beginning of the experiment, $\mathcal{B}_2$ sets public parameters $\mathsf{mpk} = (\mathcal{G}, q, g, G, H, \mathsf{mpk} = g^z)$ where G, H are description of hash functions modeled as random oracles and g^z is the challenge. As usual, $\mathcal{B}_2$ simulates these oracles by keeping two lists L_G, L_H containing the queried values together with the answers given to $\mathcal{A}$.
- Every time $\mathcal{A}$ queries the key extraction oracle $\mathcal{O}_\mathcal{E}$, for user id, $\mathcal{B}_2$ chooses $c, y \xleftarrow{\$} \mathbb{Z}_q$, sets $R = g^{-zc}g^y$ and adds $((R, \mathsf{id}), c)$ to the list L_H. Then it returns the key (y, R) to $\mathcal{A}$.
- When $\mathcal{A}$ makes a call to the signature oracle (id, m), $\mathcal{B}_2$ simply computes id's secret key as described in the previous step. Then computes a signature by calling $\mathcal{S}$, adding the respective call to the oracle G, $((\mathsf{id}, g^a, m), d)$ to the list L_G and gives the resulting signature to the adversary.
- $\mathcal{B}_2$ runs the algorithm $\mathsf{MF}_{\mathcal{A},3}(\mathsf{mpk})$ as described in Lemma 1. In this way, either $\mathcal{B}_2$ aborts prematurely or we get, for some identity id, some message m and some R, four forgeries $(\mathsf{id}, m, (A_k, B_k, R))$, $k = 1 \ldots 4$ with $A_1 = A_2$ and $A_3 = A_4$. As all these signatures are valid, the following equations hold

$$B_1 = \log A_1 + (\log R + c_1 z)d_1 \qquad B_2 = \log A_2 + (\log R + c_1 z)d_2$$
$$B_3 = \log A_3 + (\log R + c_2 z)d_3 \qquad B_4 = \log A_4 + (\log R + c_2 z)d_4$$

with $c_1 \neq c_2$, $d_1 \neq d_2$ and $d_3 \neq d_4$. Since we know $c_1, c_2, d_1, \ldots, d_4$, a simple computation yields

$$z = \frac{B_3 + B_2 - B_1 - B_4}{c_2(d_3 - d_4) - c_1(d_1 - d_2)}.$$

It follows that the success probability of $\mathcal{B}_2$ is bounded by

$$\mathsf{Adv}_{\mathcal{B}_2}^{DL}(\eta) \geq \mathsf{frk} \geq \mathsf{Adv}_{\mathcal{A}}^{\text{EUF-IBS-CMA}}(\eta) \left(\frac{(\mathsf{Adv}_{\mathcal{A}}^{\text{EUF-IBS-CMA}}(\eta))^3}{(Q_G Q_H)^6} - \frac{3}{2^\eta} \right) \qquad \square$$

5   Comparison to Previous Schemes

This section compares the efficiency of our scheme with previous provably secure IBS schemes in terms of computational complexity and signature size. As it is common for cryptographic schemes with security reductions in the random oracle model, we ignore the tightness of the corresponding security reductions when computing the length of the parameters of the schemes. Given the considerable number of IBS schemes in the literature, we choose not to list all the existing schemes one by one. Instead we classify previous schemes in three categories, depending on whether those schemes are based on factoring, elliptic curve discrete logarithm (ECDL) or pairing. Then we show that our scheme can be considered as the benchmark in efficiency in each of these categories. The reader interested in the state of the art of IBS schemes is referred to [BNN04].

We need to introduce some terminology. In what follows, an exponentiation refers to computing g^r for randomly taken $g \xleftarrow{\$} G$ and $r \xleftarrow{\$} \mathbb{Z}_t$, where G denotes a finite group, t is its order and r is an integer. A multi-exponentiation $\text{mexp}(l)$ refers to computing $g_1^{r_1} \cdots g_l^{r_l}$, where $g_1, \ldots, g_l \xleftarrow{\$} G$ and $r_1, \ldots, r_l \xleftarrow{\$} \mathbb{Z}_t$. This operation can be computed more efficiently than just computing l single exponentiations due to an algorithm by Strauss [Str64], which is sometimes referred as Shamir's trick in the literature. Finally, we write $|G|$ to denote the number of bits needed to represent an element in G.

To start with the comparison, we need to state the efficiency properties of our scheme. For reasons of efficiency we consider that our scheme should be built upon a group of points $\mathcal{G}$ of prime order q of a suitable elliptic curve. A signature on our scheme consists of two elements in $\mathcal{G}$ and one element in $\mathbb{Z}_q$. To sign, one needs to compute one exponentiation (a small number of multiplications in $\mathbb{Z}_q$ can be ignored). To verify, one needs to compute one multi-exponentiation $\text{mexp}(3)$. For the sake of concreteness, we fix $|\mathcal{G}| \approx |\mathbb{Z}_q| \approx 256$ bits for a security level equivalent to a 128-bit symmetric key for AES (cf. [ECR]). According to Brumley [Bru06], a multi-exponentiation $\text{mexp}(3)$ has a cost of about 1.5 times that of a single exponentiation.

In the first category of our classification we find *factoring-based schemes*. As it is well-known, key sizes for factoring-based schemes are much larger than key sizes for ECDL-based schemes for an equivalent security level. This implies in particular that performing a group exponentiation in a factoring-based scheme, where $G = \mathbb{Z}_n^*$ for an RSA-modulus n, is more expensive than an exponentiation in an ECDL-based scheme. For instance, for the 128-bit security level, $|\mathbb{Z}_n| \approx 3072$ bits. This already forces us to restrict our attention to the schemes in this category that present the shortest signature length, since otherwise they would be prohibitive in certain resource-constrained devices where communication is expensive, like Wireless Sensor Networks [GST07]. The shortest signature size for existing factoring-based IBS schemes is equivalent to representing two elements in $|\mathbb{Z}_n|$ (cf. [BNN04]), and thus signature size is still considerably bigger than in our scheme. They do present a higher computational cost in signing and verifying, since the most efficient algorithms require at least two exponentiations in $\mathbb{Z}_n$ for signing, and one $\text{mexp}(2)$ in verifying. Computing one $\text{mexp}(2)$ in $\mathbb{Z}_n$ is more costly than computing one $\text{mexp}(3)$ in $\mathcal{G}$ due to the larger key size in factoring-based schemes.

In the second place, let us consider previous provably secure *ECDL-based schemes*. A comparison to previous schemes yields that our scheme is the most efficient in all three features, namely, in signature size, signing cost and verifying cost. Indeed, our scheme enjoys the same efficiency as the scheme named as Beth IBS scheme in [BNN04], which to our knowledge was the most efficient ECDL-based IBS scheme to date. In sharp contrast to our case, the security of the Beth IBS scheme is still unproven (cf. [BNN04]).

It remains to compare our scheme with existing *pairing-based schemes*. To this end, we need to recall some facts about pairings (see [BSS05] for a comprehensive introduction). Let $\mathcal{G}_1, \mathcal{G}_2$ and $\mathcal{G}_T$ be finite Abelian groups in which the discrete logarithm is believed to be hard. A *pairing* is a bilinear function $e : \mathcal{G}_1 \times \mathcal{G}_2 \rightarrow \mathcal{G}_T$. Let $\mathcal{G}_1 = E(\mathbb{F}_p)$ denote an elliptic curve over the finite field $\mathbb{F}_p$. Let the order of $\mathcal{G}_1$ be divisible by a primer r such that r also divides $p^\alpha - 1$, where α is the order of p in $\mathbb{Z}_r^*$ and is called the MOV embedding degree. The modified Tate pairing $e(\cdot, \cdot)$, which is the bilinear map usually recommended, takes values in the subgroup $\mathcal{G}_T$ of $\mathbb{F}_{p^\alpha}^*$ of order r and is defined in two ways, depending on whether E is a supersingular or ordinary curve.

In the supersingular case $\mathcal{G}_1 = \mathcal{G}_2 = E(\mathbb{F}_p)$. For supersingular curves the best parameter choices are in characteristic 3, and in this case $\alpha = 6$ and $|\mathbb{F}_{p^\alpha}^*| \approx 3072$. As a consequence, $|\mathcal{G}_1| \geq 3072/6 = 512$, since in groups equipped with a pairing an algorithm solving the (finite-field) discrete logarithm in $\mathcal{G}_T$ can be used to compute the ECDL in $\mathcal{G}_1$.

Ordinary curves can also be used to implement pairings, but in this case $\mathcal{G}_2$ is set to be a subgroup of $E(\mathbb{F}_{p^\alpha})$. Several techniques have been presented [BKLS02, SB06, GPS06, DSD07] in order to improve efficiency and bandwidth. For instance, by using an appropriate map, certain points in $E(\mathbb{F}_{p^{12}})$ can be compressed to points in a sextic twist $E(\mathbb{F}_{p^2})$. Therefore, typical parameters would be $|\mathcal{G}_1| \geq 256$, $\alpha = 12$ and $|\mathcal{G}_2| \geq 2 \cdot 256 = 512$.

In the pairing-based IBS category, the shortest-length signature schemes for pairings over supersingular curves consist of two elements in $\mathcal{G}_1$. In the ordinary curves category, the shortest signatures consist of two elements in $\mathcal{G}_1$, as in [Her06], or alternatively one element in $\mathcal{G}_1$ and one element in $\mathbb{Z}_q$, where $q = |\mathcal{G}_1|$, as in the scheme [BLMQ05] by Barreto et al. In the supersingular case, a signature has length of at least 1024 bits. This is in favor of our scheme, since our signatures are 768 bits long.

A more detailed comparison is needed in the case of Herranz and Barreto *et al.* schemes when pairings are implemented using ordinary curves. Then in both schemes the signature size is smaller than ours in about 256 bits, which is in our disadvantage. However, our scheme significantly outperforms the above mentioned schemes in computational time. We discuss that below.

5.1   The Case of Herranz and Barreto *et al.* Schemes

In the scheme proposed by Herranz, signing requires 1 exponentiation in $\mathcal{G}_1$ plus 1 hash-to-group evaluation, which generally has a non-negligible computational cost [BF03]. In contrast, our scheme only needs 1 exponentiation in $\mathcal{G}_1$, and therefore our signing algorithm is marginally more efficient than Herranz's. As for verifying, Herranz's scheme needs to compute 1 exponentiation in $\mathcal{G}_1$, 1 hash-to-group operation and 2 pairings. The latter operation can be replaced by the product of two pairings, which by a trick by Granger and Smart [GS06], has a cost of about 1.46 times that of a single pairing.

Thus we need a figure on the computational cost of pairings. The most efficient implementation we are aware of is due to Devegili, Scott and Dahab [DSD07]. They report to compute a pairing for the 128-bit security level using Barreto-Naehrig curves in about $9.04 \cdot 10^7$ clock cycles with the Philips HiPerSmart™ smartcard. The latter is an instantiation of the MIPS32®-based SmartMIPS® architecture with various instruction set enhancements to facilitate the implementation of popular cryptographic algorithms.

In order to compare the cost of computing a pairing versus the cost of exponentiating in $\mathcal{G}_1$, we use a result by Großschädl, Szekely and Tillich [GST07]. They report that computing an exponentiation in the NIST curve P-256 (that allows for efficient elliptic curve computations) requires $4.25 \cdot 10^6$ clock cycles in the 133 MHz StrongARM processor, the latter being used in resource-constrained devices like wireless sensor networks. Therefore, we can estimate that currently computing a pairing can be as expensive as 21 exponentiations in $\mathcal{G}_1$ at the 128-bit security level. We conclude that Herranz's scheme verification algorithm requires a computational effort equivalent to 31 exponentiations in the NIST curve P-256.

Barreto *et al.* scheme needs to compute one exponentiation in $\mathcal{G}_1$ and one exponentiation in $\mathcal{G}_T$ for signing. For verification, it computes 1 exponentiation in $\mathcal{G}_2$, 1 exponentiation in $\mathcal{G}_T$ plus 1 pairing. Exponentiating in $\mathcal{G}_2$ and $\mathcal{G}_T$ requires a higher computational cost than computing an exponentiation in $\mathcal{G}_1$, the exact cost depending on how the arithmetic on those groups is implemented. We were not able to find explicit estimates for these particular operations, so we leave them un-quantified. In any case, while our scheme requires 1.5 exponentiations in $\mathcal{G}_1$ for signing, Barreto *et al.* requires a computational effort equivalent to 21 exponentiations in $\mathcal{G}_1$ (for the pairing) plus one exponentiation in $\mathcal{G}_2$ plus one exponentiation in $\mathcal{G}_T$.

Figure 2 summarizes the performance comparison between the Schnorr-like scheme and [BLMQ05, Her06]. $\exp_{\mathcal{G}_*}$ indicates the cost of computing an exponentiation in $\mathcal{G}_*$, and assumes that a pairing costs about 21 exponentiations in $\mathcal{G}_1$. As a result, our scheme outperforms previous schemes in the pairing-based category, except for signature size, where the benchmark is still [BLMQ05, Her06]. The new scheme drastically outperforms IBS pairing-based schemes in verification time.

Scheme	Signature size	Sign	Verification
Schnorr-like	768 bits	$1 \exp_{\mathcal{G}_1}$	$1.5 \exp_{\mathcal{G}_1}$
Barreto *et al.*	512 bits	$1 \exp_{\mathcal{G}_2} + 1 \exp_{\mathcal{G}_T}$	$21 \exp_{\mathcal{G}_1} + 1 \exp_{\mathcal{G}_2} + 1 \exp_{\mathcal{G}_T}$
Herranz	512 bits	$> 1 \exp_{\mathcal{G}_1}$	$31 \exp_{\mathcal{G}_1}$

Fig. 2. Efficiency comparison among [BLMQ05, Her06] and Schnorr-like schemes for the 128-bit security level. $\exp_{\mathcal{G}_*}$ indicates the cost of computing an exponentiation in $\mathcal{G}_*$. For the sake of comparison, $\exp_{\mathcal{G}_1} \leq \exp_{\mathcal{G}_2}$ and $\exp_{\mathcal{G}_1} \leq \exp_{\mathcal{G}_T}$

6   Conclusion

In this work we have studied in detail the identity-based signature scheme yield by the concatenation of two Schnorr signatures. We have proven it secure under the discrete logarithm assumption using the random oracle methodology. The Schnorr-like IBS scheme outperforms in computational cost and underlying security assumption every single previously proposed provably secure identity-based signature scheme. Moreover, our signatures have also smaller bit complexity than any other provably secure scheme in the literature, with the sole exception of [BLMQ05, Her06]. Last but not least, the new scheme avoids the heavy code machinery need by pairing-based schemes. The properties enjoyed by our new scheme make it specially suited for deployment in resource-constrained devices where savings in computation and communication are a premium, e.g. wireless sensor networks.

Acknowledgements

The authors wish to thank the anonymous reviewers for helpful comments.

References

[AO99]   Abe, M., Okamoto, T.: Delegation chains secure up to constant length. In: Varadharajan, V., Mu, Y. (eds.) ICICS 1999. LNCS, vol. 1726, pp. 144–156. Springer, Heidelberg (1999)

[Bet88]   Beth, T.: Efficient zero-knowledge identification scheme for smart cards. In: Günther, C.G. (ed.) EUROCRYPT 1988. LNCS, vol. 330, pp. 77–84. Springer, Heidelberg (1988)

[BF03]   Boneh, D., Franklin, M.K.: Identity-Based encryption from the Weil pairing. SIAM Journal of Computing 32(3), 586–615 (2003); This is the full version of an extended abstract of the same title presented in: Kilian, J. (ed.) CRYPTO 2001. LNCS, vol. 2139, pp. 213–615. Springer, Heidelberg (2001)

[BFPW07] Boldyreva, A., Fischlin, M., Palacio, A., Warinschi, B.: A closer look at PKI: Security and efficiency. In: Okamoto, T., Wang, X. (eds.) PKC 2007. LNCS, vol. 4450, pp. 458–475. Springer, Heidelberg (2007)

[BKLS02] Barreto, P.S.L.M., Kim, H.Y., Lynn, B., Scott, M.: Efficient algorithms for pairing-based cryptosystems. In: Yung, M. (ed.) CRYPTO 2002. LNCS, vol. 2442, pp. 354–368. Springer, Heidelberg (2002)

[BLMQ05] Barreto, P.S.L.M., Libert, B., McCullagh, N., Quisquater, J.-J.: Efficient and provably-secure identity-based signatures and signcryption from bilinear maps. In: Roy, B. (ed.) ASIACRYPT 2005. LNCS, vol. 3788, pp. 515–532. Springer, Heidelberg (2005)

[BN06]   Bellare, M., Neven, G.: Multi-signatures in the plain public-key model and a general forking lemma. In: Proceedings of the 13th ACM conference on Computer and communications security (CCS 2006), pp. 390–399. ACM, New York (2006)

[BNN04]   Bellare, M., Namprempre, C., Neven, G.: Security proofs for identity-based identification and signature schemes. In: Cachin, C., Camenisch, J.L. (eds.) EUROCRYPT 2004. LNCS, vol. 3027, pp. 268–286. Springer, Heidelberg (2004); The full version appears in Cryptology ePrint Archive: Report 2004/252

[BPW03]   Boldyreva, A., Palacio, A., Warinschi, B.: Secure proxy signature schemes for delegation of signing rights. Cryptology ePrint Archive, Report 2003/096 (2003), http://eprint.iacr.org/

[Bru06]   Brumley, B.B.: Efficient three-term simultaneous elliptic scalar multiplication with applications. In: Fåk, V. (ed.) Proceedings of the 11th Nordic Workshop on Secure IT Systems—NordSec 2006, Linköping, Sweden, October 2006, pp. 105–116 (2006)

[BSNS05]   Baek, J., Safavi-Naini, R., Susilo, W.: Certificateless public key encryption without pairing. In: Zhou, J., López, J., Deng, R.H., Bao, F. (eds.) ISC 2005. LNCS, vol. 3650, pp. 134–148. Springer, Heidelberg (2005)

[BSS05]   Blake, I.F., Seroussi, G., Smart, N.: Advances in Elliptic Curve Cryptography. London Mathematical Society Lecture Note Series, vol. 317. Cambridge University Press, Cambridge (2005)

[CC02]   Cha, J.C., Cheon, J.H.: An identity-based signature from gap Diffie-Hellman groups. In: Desmedt, Y.G. (ed.) PKC 2003. LNCS, vol. 2567, pp. 18–30. Springer, Heidelberg (2002)

[CJT04]   Castelluccia, C., Jarecki, S., Tsudik, G.: Secret handshakes from CA-oblivious encryption. In: Lee, P.J. (ed.) ASIACRYPT 2004. LNCS, vol. 3329, pp. 293–307. Springer, Heidelberg (2004)

[Dig08]   DigiNotar. Diginotar internet trust services (2008), http://www.diginotar.com

[DSD07]   Devegili, A.J., Scott, M., Dahab, R.: Implementing cryptographic pairings over barreto-naehrig curves. In: Takagi, T., Okamoto, T., Okamoto, E., Okamoto, T. (eds.) Pairing 2007. LNCS, vol. 4575, pp. 197–207. Springer, Heidelberg (2007)

[ECR]   ECRYPT. Ecrypt yearly report on algorithms and key lengths (2006), http://www.ecrypt.eu.org/documents/D.SPA.21-1.1.pdf revision 1.1 (January 29, 2007)

[EG08]   Espinosa-Garcia, J.: The new Spanish electronic identity card: DNI-e. In: Conference on Cryptology and Digital Content Security (2008), http://www.crm.cat/Cryptology/Slides/Espinosa.pdf

[FS87]   Fiat, A., Shamir, A.: How to prove yourself: Practical solutions to identification and signature problems. In: Odlyzko, A.M. (ed.) CRYPTO 1986. LNCS, vol. 263, pp. 186–194. Springer, Heidelberg (1987)

[Gir91]   Girault, M.: Self-certified public keys. In: Davies, D.W. (ed.) EUROCRYPT 1991. LNCS, vol. 547, pp. 490–497. Springer, Heidelberg (1991)

[GPS06]   Granger, R., Page, D., Smart, N.P.: High security pairing-based cryptography revisited. In: Hess, F., Pauli, S., Pohst, M. (eds.) ANTS 2006. LNCS, vol. 4076, pp. 480–494. Springer, Heidelberg (2006)

[GQ90]   Guillou, L.C., Quisquater, J.-J.: A "paradoxical" identity-based signature scheme resulting from zero-knowledge. In: Goldwasser, S. (ed.) CRYPTO 1988. LNCS, vol. 403, pp. 216–231. Springer, Heidelberg (1990)

[GS06]   Granger, R., Smart, N.: On computing products of pairings. Cryptology ePrint Archive, Report 2006/172 (2006), http://eprint.iacr.org/

[GST07]   Großschädl, J., Szekely, A., Tillich, S.: The energy cost of crypto-
          graphic key establishment in wireless sensor networks. In: ASIACCS 2007,
          pp. 380–382. ACM, New York (2007)
[Her06]   Herranz, J.: Deterministic identity-based signatures for partial aggrega-
          tion. Comput. J. 49(3), 322–330 (2006)
[Hes03]   Hess, F.: Efficient identity based signature schemes based on pairings.
          In: Nyberg, K., Heys, H.M. (eds.) SAC 2002. LNCS, vol. 2595, pp. 310–324.
          Springer, Heidelberg (2003)
[oIA08]   Spanish Ministry of Internal Affairs. Electronic identity card (2008)
          (in Spanish), http://www.dnielectronico.es/
[Oka93]   Okamoto, T.: Provably secure and practical identification schemes and
          corresponding signature schemes. In: Brickell, E.F. (ed.) CRYPTO 1992.
          LNCS, vol. 740, pp. 31–53. Springer, Heidelberg (1993)
[PH97]   Petersen, H., Horster, P.: Self-certified keys – concepts and applications.
          In: Communications and Multimedia Security 1997, pp. 102–116 (1997)
[PS00]   Pointcheval, D., Stern, J.: Security arguments for digital signatures and
          blind signatures. Journal of Cryptology 13(3), 361–396 (2000)
[SB06]   Scott, M., Barreto, P.S.L.M.: Generating more mnt elliptic curves. Des.
          Codes Cryptography 38(2), 209–217 (2006)
[Sch91]   Schnorr, C.-P.: Efficient signature generation by smart cards. Journal of
          Cryptology 4(3), 161–174 (1991)
[Sha85]   Shamir, A.: Identity-based cryptosystems and signature schemes.
          In: Blakely, G.R., Chaum, D. (eds.) CRYPTO 1984. LNCS, vol. 196,
          pp. 47–53. Springer, Heidelberg (1985)
[SOK00]   Sakai, R., Ohgishi, K., Kasahara, M.: Cryptosystems based on pairing.
          In: The 2000 Symposium on Cryptography and Information Security, Oiso,
          Japan (2000)
[Str64]   Strauss: Addition chains of vectors. American Mathematical
          Monthly 71(7), 806–808 (1964)

On the Theoretical Gap between Group Signatures with and without Unlinkability

Go Ohtake[1], Arisa Fujii[1], Goichiro Hanaoka[2], and Kazuto Ogawa[1]

[1] Japan Broadcasting Corporation
1-10-11 Kinuta, Setagaya-ku, Tokyo 157-8510, Japan
{ohtake.g-fw,fujii.a-fw,ogawa.k-cm}@nhk.or.jp
[2] National Institute of Advanced Industrial Science and Technology
1102 Akihabara Daibiru, 1-18-13 Sotokanda, Chiyoda-ku, Tokyo 101-0021, Japan
hanaoka-goichiro@aist.go.jp

Abstract. We investigate a theoretical gap between unlinkability of group signature schemes and their other requirements, and show that this gap is significantly large. Specifically, we clarify that *if unlinkability is achieved from any other property of group signature schemes, then it becomes possible to construct a chosen-ciphertext secure cryptosystem from any one-way function.* This result implies that it would be possible to drastically improve efficiency of group signature schemes if unlinkability is not taken into account. We also demonstrate to construct a significantly more efficient scheme (without unlinkability) than the best known full-fledged scheme.

Keywords: group signature, anonymity, unlinkability.

1 Introduction

Background. In various applications, it is often required that privileged users should prove their membership while still preserving anonymity. *Group signature* (*GS*) schemes [10] are considered powerful tools for satisfying these requirements. Roughly speaking, a *GS* scheme provides two functionalities, which are as follows: (1) any one can verify whether a given signature is generated by one of privileged users, and (2) no one (except for the group manager) can detect who is its signer. The latter functionality is called *anonymity*. For a strong (but conventional) definition of anonymity, we further take into account *unlinkability*, such that (3) no one can guess whether two given signatures are generated by the same signer or not.

There are in fact many applications which require unlinkability. However, there are also many applications which do not always require unlinkability, e.g. (certain types of) electronic auctions, on-line payments, bulletin boards and etc.[1] If the theoretical gap between a full-fledged *GS* (FA-*GS*) scheme and a *GS* scheme without unlinkability (QA-*GS*) is significantly large, and if the QA-*GS* is

---

[1] In these applications, unlinkability would be even harmful for guaranteeing consistency of data sources.

B. Preneel (Ed.): AFRICACRYPT 2009, LNCS 5580, pp. 149–166, 2009.

efficient, then we should use weakened *GS* schemes for such applications. *Our main motivation is to investigate such a theoretical gap.*

Our Contribution. We show that the theoretical gap between *GS* schemes with full-functionalities and that without unlinkability is significantly large. More precisely, this gap is equivalent to that between one-way functions and chosen-ciphertext secure public key encryption schemes. This implies that it would be possible to construct a significantly more efficient cryptographic tool which provides all functionalities of *GS* schemes except for unlinkability. We further demonstrate to construct an example of such cryptographic tools. Actually, our scheme (without unlinkability) is considerably more efficient than the best known full-fledged scheme, and consequently, our scheme is more suitable for various applications which do not require unlinkability (but other functionalities of *GS* schemes).

Concretely, our results are as follows. First, we give the security definition of *GS* schemes which addresses all security notions except for unlinkability. Our definition is based on the definition of full-fledged *GS* schemes due to Bellare, Micciancio and Warinschi [3], and we carefully modify Bellare et al.'s definition without loosing any security property except for unlinkability. We call a *GS* scheme which is secure in the sense of our definition *quasi group signature* scheme.

Next, we show that a quasi *GS* scheme can be generically constructed from any (weak) identity-based signature scheme. Consequently, we immediately see that a *GS* scheme which provides all functionalities of *GS* schemes except for unlinkability can be constructed from any one-way function generically.

Then, we also show that a (publicly verifiable) chosen-ciphertext secure public key encryption scheme can be obtained from any full-fledged *GS* scheme generically. By combining this and the above results, we further have the following (informal) theorem:

Theorem 1 (Informal). *If it is possible to construct a* FA-*GS scheme from any* QA-*GS scheme generically, then it is also possible to construct a chosen-ciphertext secure public key encryption scheme from any one-way function generically.*

Impagliazzo and Rudich's result [17] indirectly says that constructing a chosen-plaintext secure public key encryption scheme from an arbitrary one-way function is as hard as solving the P vs. NP problem and that it is impossible if $P = NP$. Moreover, Reingold, Travisan and Vadhan [31] offered that the construction would definitely have to rely on non-black-box reduction techniques regardless of an assumption $P = NP$. Furthermore, Gertner, Malkin and Myers' result [15] says that for a large class of public key encryption schemes it is impossible to construct a chosen-ciphertext secure public key encryption scheme from chosen-plaintext secure one in a black-box manner. Hence, *for achieving unlinkability from any other functionality of GS schemes, it is necessary to simultaneously overcome these two extremely hard barriers.* Hence, we see that the theoretical gap between unlinkability and other functionalities is significantly large.

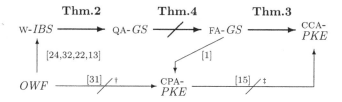

Fig. 1. Relationships among cryptographic primitives. $A \rightarrow B$ denotes that construct-ing B from A generically is possible. $A \nrightarrow B$ denotes that constructing B from A generically is almost impossible. OWF, w-IBS, QA-GS, FA-GS, CCA-PKE and CPA-PKE denote one-way function, (weak) identity-based signature scheme, quasi group signature scheme, full-fledged group signature scheme, chosen-ciphertext secure public key encryption scheme, and chosen-plaintext secure public key encryption scheme, re-spectively. Separation result with † means that its non-black-box construction has not been found and that its black-box construction is impossible. Separation result with ‡ means that its non-black-box construction has not been found and that its black-box construction is impossible unless the encryption algorithm of CPA-PKE scheme is used in the decryption algorithm of CCA-PKE scheme.

Fig. 1 illustrates implications and separations among (full-fledged and quasi) GS schemes, (chosen-plaintext and chosen-ciphertext secure) public key encryp-tion schemes, one-way functions, and some other notions.

Finally, we demonstrate to construct an efficient quasi GS scheme, and address its efficiency. This scheme is significantly more efficient than the best known full-fledged GS scheme.

We conclude that theoretical and practical gaps between GS schemes with and without unlinkability are both significant, and therefore, it is important to choose appropriate cryptographic tools for individual applications according to their necessity of unlinkability.

Related Works. The notion of GS schemes was firstly advocated by Chaum and van Heyst [10]. Following this seminal work, there have been many attempts for designing secure and efficient GS schemes, e.g. [10,2,25,12]. Camenisch and Stadler [8] proposed the first GS scheme whose efficiency is independent of the number of users, and Ateniese, Camenisch, Joye and Tsudik [2] demonstrated to construct another scheme with such efficiency as well as provable security.

However, in these earlier works including [2], security definitions were given in ad-hoc manners, and therefore, it was not easy to confirm whether these schemes have proper security or not. Then, Bellare, Micciancio and Warinschi [3] proposed a simple security definition which is stronger than any previous one. Therefore, the definition in [3] is currently considered as the basic one. Bellare, Shi and Zhang [4] further proposed a variant of the definition of [3] in which dynamic join of users is allowed.

It is also important to investigate minimum assumptions for constructing GS schemes. In, for example, [9,27,29], some generic constructions of GS schemes from weak primitives are demonstrated. However, these constructions are based on ad-hoc and weak definitions. Under the definition of [3], Abdalla and Warinschi showed that a chosen-plaintext secure public key encryption scheme can be derived from any *GS* scheme [1]. Our result is stronger than theirs since we show that *chosen-ciphertext secure* public key encryption scheme is crucial for constructing *GS* schemes, and furthermore, clarify that this is required for only unlinkability.

As a notion of anonymity which ignores unlinkability, *pseudonymity* is often discussed in the literatures. Actually, one of its possible definitions is equivalent to our security definition of QA-*GS* schemes, and a widely-known technique to achieve pseudonymity from standard digital signatures is considered as a special case of our generic construction in Sect. 3.2 (see a footnote in Sect. 3.1). However, there have never been a formal discussion which clarifies whether pseudonymity captures all requirements of a *GS* scheme except for unlinkability (and our work answers to this question in affirmative).

In [21] and [20], some *GS* schemes with linkability are discussed. However, these schemes are designed under specific definitions and therefore, it is not clear what theoretical gap is caused by unlinkability.

2   Standard Definitions

We review several conventional definitions of group signature (*GS*) scheme, digital signature scheme, identity-based signature (*IBS*) scheme, public key encryption (*PKE*) scheme, one-way function (*OWF*) and their security. In addition, we review the computational Diffie-Hellman (CDH) assumption.

Group Signature. We introduce the model and security definition of *GS* schemes due to Bellare et al. [3]. In the model, a group manager (GM) initially generates a group master secret key $gmsk$ and a verification key gpk, and signing keys $gsk_1, \cdots, gsk_n$, and distributes gsk_i to a group member i for $1 \leq i \leq n$. The member signs a message by using gsk_i, and this signature can be verified by using gpk. When specification of a signed member is necessary, GM specifies him/her by using $gmsk$. The formal definition is as follows:

A *GS* scheme $\mathcal{GS}$ consists of the following four polynomial time algorithms $\mathcal{GS} =(\mathsf{GKg},\mathsf{GSig},\mathsf{GVf},\mathsf{Open})$:

GKg: A probabilistic key generation algorithm. This algorithm takes as inputs a security parameter $k \in \mathbb{N}$ and the number of members of the group $n \in \mathbb{N}$. It returns a group public key gpk, a group master secret key $gmsk$ and secret signing keys $gsk_1, \cdots, gsk_n$.

GSig: A probabilistic group signing algorithm. This algorithm takes as inputs gsk_i ($i \in \{1, \cdots, n\}$) and a message m. It returns a signature $\sigma_{\mathcal{GS}}$ of m under gsk_i.

Experiment $Exp_{\mathcal{GS},\mathcal{A}}^{f\_anon-b}(1^k,n)$

$\quad (gpk, gmsk, gsk_1, \cdots, gsk_n) \leftarrow \mathsf{GKg}(1^k, n);$

$\quad (St, i_0, i_1, m^*) \leftarrow \mathcal{A}^{Open(\cdot)}(\textbf{choose}, gpk, gsk_{i..n}); \sigma_{\mathcal{GS}}^* \leftarrow \mathsf{GSig}(gsk_{i_b}, m^*);$

$\quad d \leftarrow \mathcal{A}^{Open(\cdot)}(\textbf{guess}, St, \sigma_{\mathcal{GS}}^*);$

$\quad$ If $\mathcal{A}$ did not query $Open(\cdot)$ with $\sigma_{\mathcal{GS}}^*$ in neither the **choose** and nor the **guess** stages then return d EndIf;

$\quad$ Return 0.

Fig. 2. Experiment used to define full-anonymity of GS scheme: $\mathcal{GS} = (\mathsf{GKg}, \mathsf{GSig}, \mathsf{GVf}, \mathsf{Open})$. Here $\mathcal{A}$ is an adversary, $gsk_{i..n}$ denotes $gsk_1, \cdots, gsk_n$, $b \in_R \{0, 1\}$, and St denotes state information passed by $\mathcal{A}$ between **choose** and **guess** phases.

GVf: A deterministic group signature verification algorithm. This algorithm takes as inputs gpk, m and $\sigma_{\mathcal{GS}}$. It returns either 1 or 0.

Open: A deterministic opening algorithm. This algorithm takes as inputs $gmsk$ and $\sigma_{\mathcal{GS}}$. It returns an identity i or a special symbol $\bot$ to indicate failure.

For simplicity we are assigning the members consecutive integer identities $1, \cdots$, n. We say that $\sigma_{\mathcal{GS}}$ is a *true* signature of m if there exists $i \in \{1, \cdots, n\}$ such that $\sigma_{\mathcal{GS}} \in [\mathsf{GSig}(gsk_i, m)]$ and that $\sigma_{\mathcal{GS}}$ is a *valid* signature of m with respect to gpk if $\mathsf{GVf}(gpk, m, \sigma_{\mathcal{GS}}) = 1$, and the following conditions are required for all $k, n \in \mathbb{N}$, all $(gpk, gmsk, gsk_1, \cdots, gsk_n) \in [\mathsf{GKg}(1^k, n)]$, all $i \in \{1, \cdots, n\}$ and all $m \in \{0, 1\}^*$:

$$\mathsf{GVf}(gpk, m, \mathsf{GSig}(gsk_i, m)) = 1 \text{ and } \mathsf{Open}(gmsk, \mathsf{GSig}(gsk_i, m)) = i .$$

We then review security definitions of GS schemes with *full-anonymity*. Informally, full-anonymity requires that an adversary $\mathcal{A}$ not in possession of $gmsk$ finds it hard to recover the signer's identity from a signature even if $\mathcal{A}$ can get signer's identity of any other signature.

To formally define full-anonymity, we associate experiments given in Fig. 2. In this experiment, we give $\mathcal{A}$ access to an *opening oracle* $Open(\cdot)$, which answers with $\mathsf{Open}(gmsk, \sigma_{\mathcal{GS}})$ when queried with a signature $\sigma_{\mathcal{GS}}$.

Full-anonymity is defined by using this experiment as follows:

Definition 1. Let $\mathcal{GS} = (\mathsf{GKg}, \mathsf{GSig}, \mathsf{GVf}, \mathsf{Open})$ be a GS scheme. Let $\mathcal{A}$ be an adversary. Define the advantage of $\mathcal{A}$ as follows:

$$Adv_{\mathcal{GS},\mathcal{A}}^{f\_anon}(1^k, n) \stackrel{\text{def}}{=} \left| Pr[Exp_{\mathcal{GS},\mathcal{A}}^{f\_anon-1}(1^k, n) = 1] - Pr[Exp_{\mathcal{GS},\mathcal{A}}^{f\_anon-0}(1^k, n) = 1] \right| .$$

Then $\mathcal{GS}$ is (τ, q, ϵ)-*fully-anonymous* if $Adv_{\mathcal{GS},\mathcal{A}}^{f\_anon}(\cdot, \cdot) < \epsilon$ holds for any τ-time adversary $\mathcal{A}$ who makes q queries to $Open(\cdot)$, and $\mathcal{GS}$ is *fully-anonymous* if ϵ is a negligible function in k for all polynomial-time adversaries $\mathcal{A}$.

We then address full-traceability. Informally, full-traceability means that no colluding set S of group members can create signatures that cannot be opened, or signatures that cannot be traced back to some member of the coalition. To formally define it, we give $\mathcal{A}$ access to a *group signing oracle* $GSig(\cdot, \cdot)$ and a *group signing key exposure oracle* $GExp(\cdot)$. $GSig(\cdot, \cdot)$ takes as inputs a member's identity i and a message m and returns a signature σ_{GS}. $GExp(\cdot)$ takes as input a member's identity i and returns its signing key gsk_i. $\mathcal{A}$ outputs a forged group signature σ_{GS}^* of i^* for a message m^*. Then, full-traceability is defined as follows:

Definition 2. Let GS=(GKg,GSig,GVf,Open) be a GS scheme. Let $\mathcal{A}$ be an adversary. Define the advantage of $\mathcal{A}$ as follows:

$$
\begin{aligned}
Adv_{GS,\mathcal{A}}^{trace}(1^k, n) = Pr[&(gpk, gmsk, gsk_1, \cdots, gsk_n) \leftarrow \mathsf{GKg}(1^k, n); \\
&(m^*, \sigma_{GS}^*) \leftarrow \mathcal{A}^{GSig(\cdot,\cdot), GExp(\cdot)}(gpk): \\
&\mathsf{GVf}(gpk, m^*, \sigma_{GS}^*) = 1 \wedge \\
&\quad (((\mathsf{Open}(gmsk, \sigma_{GS}^*) = i^*) \wedge (i^* \in \{1, \cdots, n\} \backslash \{id_1, \cdots, id_p\})) \\
&\quad \vee \mathsf{Open}(gmsk, \sigma_{GS}^*) = \bot)],
\end{aligned}
$$

where id_j ($1 \le j \le p \le n$) is an identity $\mathcal{A}$ submitted to $GExp(\cdot)$, and $\mathcal{A}$ is not allowed to query i^* and (i^*, m^*) to $GExp(\cdot)$ and $GSig(\cdot, \cdot)$, respectively. Then GS is $(\tau, q_s, q_e, \epsilon)$-*fully-traceable* if $Adv_{GS,\mathcal{A}}^{trace}(\cdot, \cdot) < \epsilon$ holds for any τ-time adversary $\mathcal{A}$ who makes q_s and q_e queries to $GSig(\cdot, \cdot)$ and $GExp(\cdot)$, respectively, and we say that GS is *fully-traceable* if ϵ is a negligible function in k for all polynomial-time adversaries $\mathcal{A}$.

We then define a GS scheme which has full-functionalities: full-anonymity and full-traceability. We call the GS scheme a FA-GS scheme.

Definition 3. We say that a GS scheme is a FA-GS scheme if it is fully-anonymous and fully-traceable.

Digital Signature. A digital signature scheme $\mathcal{DS}$ consists of the following three polynomial time algorithms $\mathcal{DS}$=(Kg,Sig,Vf).

Kg: The probabilistic key generation algorithm. This algorithm takes as an input a security parameter $k \in \mathbb{N}$. It returns a verification key pk and a signing key sk.

Sig: The probabilistic signing algorithm. This algorithm takes as inputs sk and a message m. It returns a signature σ of m under sk.

Vf: The deterministic signature verification algorithm. This algorithm takes as inputs pk, m and σ. It returns either 1 or 0.

We then define the security of digital signature schemes. Precisely, we define existentially unforgeability against chosen message attack (*CMA* security in short). To formally define it, we give an adversary $\mathcal{A}$ access to a signing oracle $Sig(\cdot)$, which takes as inputs a message m and returns a signature σ. CMA security is defined as follows:

Definition 4. Let $\mathcal{DS} = (\mathsf{Kg}, \mathsf{Sig}, \mathsf{Vf})$ be a digital signature scheme. Let $\mathcal{A}$ be an adversary. Define the advantage of forgery of the signature σ as follows:

$$Adv_{\mathcal{A},\mathcal{DS}}(1^k) \stackrel{\text{def}}{=} Pr[(pk, sk) \leftarrow \mathsf{Kg}(1^k); (m^*, \sigma^*) \leftarrow \mathcal{A}^{Sig(sk,\cdot)}(pk) :$$
$$\mathsf{Vf}(pk, m^*, \sigma^*) = 1] \, ,$$

where $\mathcal{A}$ is not allowed to query the target message m^* to $Sig(\cdot)$. Then $\mathcal{DS}$ is (τ, q, ϵ)-CMA secure if $Adv_{\mathcal{A},\mathcal{DS}}(k) < \epsilon$ holds for any τ-time adversary $\mathcal{A}$ who makes q signing queries to $Sig(\cdot)$, and we say that $\mathcal{DS}$ is CMA secure if ϵ is a negligible function in k for all polynomial-time adversaries $\mathcal{A}$.

weak Identity-Based Digital Signature. An *IBS* scheme $\mathcal{IBS}$ consists of the following four polynomial time algorithms $\mathcal{IBS} = (\mathsf{Setup}, \mathsf{IExt}, \mathsf{ISig}, \mathsf{IVf})$.

Setup: A probabilistic parameter setup algorithm. This algorithm takes as input a security parameter $k \in \mathbb{N}$. It returns a verification key $vkey$ and a master secret key $mkey$.

IExt: A probabilistic key extraction algorithm. This algorithm takes as inputs $vkey$, $mkey$ and a public identity ID. It returns the corresponding private key sk_{ID}.

ISig: A probabilistic signing algorithm. This algorithm takes as inputs sk_{ID} and a message m. It returns a signature σ_{ID} of m.

IVf: A deterministic signature verification algorithm. This algorithm takes as inputs $vkey$, m and σ_{ID}. It returns either 1 or 0.

We then address security of *IBS* schemes, existentially unforgeability against selective identity chosen message attack (we refer it to *sID-CMA* security in short). In this attack, an adversary $\mathcal{A}$ has to select target identity ahead of time to publish the verification key.

To formally define it, we give $\mathcal{A}$ access to an *IBS* signing oracle $ISig(\cdot, \cdot)$ and to an *IBS* key extraction oracle $IExt(\cdot)$. $ISig(\cdot, \cdot)$ answers with $\mathsf{ISig}(sk_{ID}, m)$ when queried with an identity ID and a message m and $IExt(\cdot)$ answers with $\mathsf{IExt}(vkey, mkey, ID)$ when queried with ID.

We describe the formal definition of the sID-CMA security as follows:

Definition 5. Let $\mathcal{IBS} = (\mathsf{Setup}, \mathsf{IExt}, \mathsf{ISig}, \mathsf{IVf})$ be an *IBS* scheme. Let $\mathcal{A}$ be an adversary. Define the advantage of $\mathcal{A}$ as follows:

$$Adv_{\mathcal{A},\mathcal{IBS}}(1^k) \stackrel{\text{def}}{=} Pr[ID^* \leftarrow \mathcal{A}(1^k); (vkey, mkey) \leftarrow \mathsf{Setup}(1^k);$$
$$(m^*, \sigma^*_{\mathcal{IBS},ID^*}) \leftarrow \mathcal{A}^{ISig(\cdot,\cdot),IExt(\cdot)}(vkey) : \mathsf{IVf}(vkey, m^*, \sigma^*_{\mathcal{IBS},ID^*}) = 1] \, ,$$

where $\mathcal{A}$ is not allowed to query the target m^* and ID^* to $ISig(\cdot, \cdot)$ and is not allowed to query ID^* to $IExt(\cdot)$. Then $\mathcal{IBS}$ is $(\tau, q_s, q_e, \epsilon)$-sID-CMA secure if $Adv_{\mathcal{A},\mathcal{IBS}}(1^k) < \epsilon$ holds for any τ-time adversary $\mathcal{A}$ with q_s IBS signing and q_e IBS key extraction queries, and $\mathcal{IBS}$ is sID-CMA secure if ϵ is a negligible function in k for all polynomial-time adversaries $\mathcal{A}$.

In our discussion, we need only more weakened *IBS* schemes whose identity space is polynomially small, and these are stated as follows:

Definition 6. Let $\mathcal{IBS}$ be an *IBS* scheme and let $\mathcal{ID}$ be identity space used for $\mathcal{IBS}$. We say that $\mathcal{IBS}$ is *wID-CMA secure* if $\mathcal{IBS}$ is sID-CMA secure and if $\mathcal{ID} = \{1, \cdots, n\}$ such that n is a polynomial in k.

In what follows, we call a wID-CMA secure *IBS* scheme a w-*IBS* scheme.

Public Key Encryption. A *PKE* scheme $\mathcal{PKE}$ consists of the following three polynomial time algorithms $\mathcal{PKE} =$ (Gen, Enc, Dec).

Gen: A probabilistic key generation algorithm. This algorithm takes as an input a security parameter $k \in \mathbb{N}$. It returns an encryption key *ekey* and a decryption key *dkey*.

Enc: A probabilistic encryption algorithm. This algorithm takes as inputs *ekey* and a plaintext m. It returns the corresponding ciphertext C.

Dec: A deterministic decryption algorithm. This algorithm takes as inputs *ekey*, *dkey* and C. It returns a plaintext m or a special symbol $\perp$.

We then define security of *PKE* schemes, namely indistinguishability against chosen ciphertext attack (*CCA* security) [30,11] and semantic security (*CPA* security) [16].

Definition 7. Let $\mathcal{PKE} = $ (Gen, Enc, Dec) be a *PKE* scheme. Let $\mathcal{A}$ be an adversary. Define the advantage of $\mathcal{A}$ as follows:

$$Adv_{\mathcal{A},\mathcal{PKE}}(1^k) \stackrel{\text{def}}{=} \Big| Pr[(dkey, ekey) \leftarrow \mathsf{Gen}(1^k);$$

$$(M_0, M_1, St) \leftarrow \mathcal{A}^{Dec(\cdot)}(\mathsf{choose}, ekey); b \in_R \{0,1\};$$

$$C^* \leftarrow \mathsf{Enc}(ekey, M_b); b' \leftarrow \mathcal{A}^{Dec(\cdot)}(\mathsf{guess}, ekey, St, C^*) : b' = b] - \frac{1}{2} \Big|,$$

where St is side information obtained in **choose** phase, and $\mathcal{A}$ is allowed to request to $Dec(\cdot)$ except for the target ciphertext C^*. Then $\mathcal{PKE}$ is *(τ, q, ϵ)-CCA secure* if $Adv_{\mathcal{A},\mathcal{PKE}}(1^k) < \epsilon$ holds for any τ-time adversary $\mathcal{A}$ with q decryption queries, and $\mathcal{PKE}$ is *CCA secure* if ϵ is a negligible function in k for all polynomial-time adversaries $\mathcal{A}$. In addition, $\mathcal{PKE}$ is *CPA secure* if $\mathcal{PKE}$ is CCA secure with $q = 0$.

In what follows, we refer a CCA secure *PKE* scheme and a CPA secure *PKE* scheme to a CCA-*PKE* scheme and a CPA-*PKE* scheme, respectively.

One-Way Function. We define security of *OWF* as follows.

Definition 8. Let $\mathcal{OWF} = f(x)$ be a *OWF*. Let $\mathcal{A}$ be an adversary. Define the advantage to invert $f(\cdot)$ as follows:

$$Adv_{\mathcal{A},\mathcal{OWF}}(1^k) \stackrel{\text{def}}{=} Pr[f(z) = f(x) | x \in \{0,1\}^k, y \leftarrow f(x), z \leftarrow \mathcal{A}(y, 1^k)] .$$

Then $\mathcal{OWF}$ is *(τ, ϵ)-OWF* if $Adv_{\mathcal{A},\mathcal{OWF}} < \epsilon$ holds for any τ-time adversary $\mathcal{A}$. Namely, any probabilistic $\mathcal{A}$ succeeds within a polynomial time τ with only a negligible probability ϵ in finding any valid preimage z of $y = f(x)$.

Computational Diffie-Hellman Assumption. We also review the definition of the computational Diffie-Hellman (CDH) assumption.

Definition 9. Given a group $\mathbb{G}$ of prime order p with generator g and elements $g^a, g^b \in \mathbb{G}$ where a, b are selected uniformly at random from $\mathbb{Z}_p^*$, the CDH problem in $\mathbb{G}$ is to compute g^{ab}. We say that *the (ϵ, τ)-CDH assumption holds in a* group $\mathbb{G}$ if no algorithm running in time at most τ can solve the CDH problem in $\mathbb{G}$ with probability at least ϵ.

3   Quasi Group Signature: Definitions and Construction

For discussing a necessary condition for providing all functionalities of a full-fledged *GS* scheme except for unlinkability, we carefully address a slight modification of *GS* schemes in which all security notions except for unlinkability are captured. More specifically, we introduce the notion of *quasi-anonymity* which meets the condition, and call a *GS* scheme *quasi group signature (QA-GS) scheme* if it has quasi-anonymity and full-traceability.

In order to define quasi-anonymity, we take into consideration Bellare, Micciancio and Warinschi's explanation on (not full-) "anonymity" which is as follows: (Paragraph ANONYMITY, Sect. 3, p.624 in [3])

> *Anonymity is just a weaker form of full-anonymity, in which the adversary does not have access to the opening oracle, and also, has no information of member's secret keys.*

It is natural to think that this is the requirement for quasi-anonymity since in [3] only "anonymity" (in the above sense) and unlinkability are described as the existing notions on user's privacy in the context of *GS* schemes (and full-anonymity is a sufficient condition for both notions). Therefore, based on this notion, we carefully address our quasi-anonymity. More precisely, we do not use the above "anonymity" as it is, but make it as strong as possible. We note that for obtaining a stronger separation result it is important to address a stronger definition for quasi-anonymity. Actually, in our definition of quasi-anonymity, an adversary is allowed to access the opening oracle (with a limitation) and possess all members' secret signing keys. Consequently, the difference between full-anonymity and quasi-anonymity becomes very slight, and specifically it is as follows:

– In the notion of quasi-anonymity, the same attack environment as that for full-anonymity is given to the adversary except that all members' secret signing keys are given to the adversary as a set without any information on relationship between each key and its owner's identity. Furthermore, it does not win if one of answers from the opening oracle is identical to i_0 or i_1.

See Figs. 2 and 3 for the difference in more detail.

Experiment $Exp_{\mathcal{GS},\mathcal{A}}^{q\text{-}anon\text{-}b}(1^k,n)$

   $(gpk, gmsk, gsk_1, \cdots, gsk_n) \leftarrow \mathsf{GKg}(1^k, n);$

   $(St, i_0, i_1, m^*) \leftarrow \mathcal{A}^{Open(\cdot)}(\textbf{choose}, gpk, \{gsk_i\});$

   $\sigma_{\mathcal{GS}}^* \leftarrow \mathsf{GSig}(gsk_{i_b}, m^*); \; d \leftarrow \mathcal{A}^{Open(\cdot)}(\textbf{guess}, St, \sigma_{\mathcal{GS}}^*);$

   <u>If $\mathcal{A}$ did not query $Open(\cdot)$ with $\sigma_{\mathcal{GS}}$ signed by gsk_{i_0} nor gsk_{i_1} in both **choose** and</u>
<u>**guess** phases</u> nor with $\sigma_{\mathcal{GS}}^*$ in **guess** phase, then return d EndIf;

   Return 0.

Fig. 3. Experiment used to define quasi-anonymity of QA-GS scheme: $\mathcal{GS} = ($GKg,GSig, GVf,Open$)$. The parts, which are underlined, are different from Fig. 2. Here $\mathcal{A}$ is an adversary, $\{gsk_i\}$ denotes a set of $\{gsk_1, \cdots, gsk_n\}$, $b \in_R \{0, 1\}$, and St denotes state information passed by the adversary between **choose** and **guess** phases.

3.1 Security Definition

We then define *quasi-anonymity*. Informally, quasi-anonymity requires that an adversary $\mathcal{A}$ not in possession of $gmsk$ finds it hard to recover the signer's identity from a signature under the condition that $\mathcal{A}$ has not got the signer's identity of the signatures signed by the same signer. [2] It remarks that for full-anonymity $\mathcal{A}$ can get any signer's identity of any signature except for the target signature, and only this condition is the difference between quasi-anonymity and full-anonymity.

To formally define the quasi-anonymity, we associate an experiment given in Fig. 3. In the experiment, we give $\mathcal{A}$ access to an *opening oracle* $Open(\cdot)$. Using this experiment, we define the quasi-anonymity as follows:

Definition 10. Let $\mathcal{GS}=($GKg,GSig,GVf,Open$)$ be a GS scheme. Let $\mathcal{A}$ be an adversary. Define the advantage of $\mathcal{A}$ as follows:

$$Adv_{\mathcal{GS},\mathcal{A}}^{q\text{-}anon}(1^k, n) \overset{\text{def}}{=} \left| \Pr[Exp_{\mathcal{GS},\mathcal{A}}^{q\text{-}anon\text{-}1}(1^k, n) = 1] - \Pr[Exp_{\mathcal{GS},\mathcal{A}}^{q\text{-}anon\text{-}0}(1^k, n) = 1] \right|.$$

Then $\mathcal{GS}$ is *(τ, q, ϵ)-quasi-anonymous* if $Adv_{\mathcal{GS},\mathcal{A}}^{q\text{-}anon}(\cdot, \cdot) < \epsilon$ holds for any τ-time adversary $\mathcal{A}$ with q opening queries, and $\mathcal{GS}$ is *quasi-anonymous* if ϵ is a negligible function in k for all polynomial-time adversaries $\mathcal{A}$.

We then define GS schemes, which do not have a property of unlinkability but which have quasi-anonymity and full-traceability. We call such GS schemes QA-GS schemes.

---

[2] This notion is similar to what is called *pseudonymity* in the literatures. As far as we know, there is no rigorous definition of pseudonymity, but it is considered that this security notion is fulfilled by the following method: A group member i holds his verification key vk_i, which is signed by the GM, and corresponding signing key sk_i. Let σ_{GM} be the GM's signature for vk_i. Only the GM has the list of members' identities and their verification keys. For signing a message m, the member i signs on a message m by using sk_i as $\sigma = \mathsf{Sig}(m, sk_i)$, and sends $(m, \sigma, vk_i, \sigma_{GM})$ to a verifier. The verifier accepts this message if and only if both σ and σ_{GM} are valid signatures under vk_i and the GM's verification key, respectively.

We note that this scheme also satisfies security requirements for a QA-GS scheme.

Definition 11. We say that a *GS* scheme is a QA-*GS* scheme if it is quasi-anonymous and fully-traceable.

3.2 Construction of Quasi Group Signature from Identity-Based Signature

We show that there exists a black-box construction of a QA-*GS* scheme from any W-*IBS* scheme. The generic construction method of a QA-*GS* scheme $\mathcal{GS}^- =$ (GKg, GSig, GVf, Open) from a W-*IBS* scheme $\mathcal{IBS}$ =(Setup, IExt, ISig, IVf) is as follows:

GKg($1^k, n$): Given a security parameter k and a number of users n, $\pi \in \mathcal{P}_n$ is selected, where $\mathcal{P}_n$ denotes the set of all permutations of $\{1, \cdots, n\}$. Then it runs $(vkey, mkey) \leftarrow$ Setup(1^k), and $gpk \leftarrow vkey$ and $gmsk \leftarrow \pi^{-1}$ are performed. In addition, the following processes are performed for all i ($1 \leq i \leq n$): $ID_i \leftarrow \pi(i)$; $sk_{ID_i} \leftarrow$ IExt($vkey, mkey, ID_i$); $gsk_i \leftarrow (ID_i, sk_{ID_i})$, where gsk_i is a signing key of user i.

GSig(gsk_i, m): Given i's signing key gsk_i and a message m, gsk_i is parsed into (ID_i, sk_{ID_i}). It then runs $\sigma_{ID_i} \leftarrow$ ISig(sk_{ID_i}, m) and $\sigma_{\mathcal{GS}} \leftarrow (ID_i, \sigma_{ID_i})$ is performed.

GVf($gpk, m, \sigma_{\mathcal{GS}}$): Given gpk, m and $\sigma_{\mathcal{GS}}$, $\sigma_{\mathcal{GS}}$ is parsed into (ID_i, σ_{ID_i}). If $ID_i \notin \{ID_1, \cdots, ID_n\}$, then it outputs 0, otherwise it outputs $b =$ IVf(gpk, m, σ_{ID_i}).

Open($gmsk, \sigma_{\mathcal{GS}}$): Given $gmsk$ and $\sigma_{\mathcal{GS}}$, $\sigma_{\mathcal{GS}}$ is parsed into (ID_i, σ_{ID_i}). It outputs $i = \pi^{-1}(ID_i)$.

We then address security of this scheme, i.e. quasi-anonymity and full-traceability. More precisely, we prove the following theorem:

Theorem 2. *There exists a black-box construction of a* $(\infty, \infty, 0)$-*quasi-anonymous and* $(\tau - (q_s + q_e) \cdot O(p(k)), q_s, q_e, \frac{n^2}{n - q_e}\epsilon)$-*full-traceable* QA-*GS scheme from any* $(\tau, q_s, q_e, \epsilon)$-W-*IBS scheme, where* $p(k)$ *is a polynomial of* k.

To prove Theorem 2, it is sufficient to prove that the scheme has quasi-anonymity and full-traceability defined in Definitions 10 and 2, respectively. Formally, we prove the following Lemmas 1 and 2.

Lemma 1. $\mathcal{GS}^-$ *has* $(\infty, \infty, 0)$-*quasi-anonymity. We note that quasi-anonymity of the above scheme is guaranteed for even computationally unbounded adversaries.*

Lemma 2. $\mathcal{GS}^-$ *has* $(\tau - (q_s + q_e) \cdot O(p(k)), q_s, q_e, \frac{n^2}{n - q_e}\epsilon)$-*full-traceability, where* $p(k)$ *is a polynomial of* k.

We will show the proofs of Lemmas 1 and 2 in the full version of this paper.

4 Construction of Public Key Encryption from Group Signature

We show that there exists a black-box construction of a CCA-*PKE* scheme from any FA-*GS* scheme. This result is integrated with previously known results and the result in Sect. 3.2, and derives the implication and separation results described in Sect. 5.

The generic construction of a one-bit message CCA-*PKE* scheme $\mathcal{PKE}$ = (Gen, Enc, Dec) from a FA-*GS* scheme $\mathcal{GS}$ =(GKg,GSig,GVf,Open) is as follows:

Gen(1^k): Given a security parameter k, it runs $(gpk, gsk_1, gsk_2, gmsk) \leftarrow$ Setup($1^k, n = 2$) and sets an encryption key $ekey := (gpk, gsk_1, gsk_2, M)$ and a decryption key $dkey := gmsk$, where M is a fixed message. It outputs $ekey$ and $dkey$.

Enc($ekey, m$): Given $ekey = (gpk, gsk_1, gsk_2, M)$ and a plaintext $m \in \{0,1\}$, it runs $\sigma \leftarrow$ GSig(gsk_{m+1}, M). Then it outputs a ciphertext $C := \sigma$.

Dec($ekey, dkey, C$): Given $ekey = (gpk, gsk_1, gsk_2, M)$, $dkey = gmsk$ and $C = \sigma$, it runs $b \leftarrow$ GVf(gpk, M, σ). If $b = 0$, it returns a special symbol $\perp$. Otherwise, it outputs m'=Open($gmsk, \sigma$)-1.

The above *PKE* scheme is CCA secure. Formally, the following theorem holds.

Theorem 3. *Let $\mathcal{GS}$ be a GS scheme with (τ, q, ϵ)-full-anonymity. Let $\mathcal{PKE}$ be the associated PKE scheme. Then $\mathcal{PKE}$ has $(\tau - q \cdot O(p(k)), q, \epsilon)$-CCA security, where $p(k)$ is a polynomial of k.*

We will provide its proof in the full version of this paper.

It remarks that the *PKE* scheme, which is constructed with the above method, has public verifiability. Namely, anyone can verify validity of a ciphertext by using gpk. In addition, it is well-known that a threshold encryption scheme can be constructed from a publicly verifiable *PKE* scheme. It results that a threshold encryption scheme might be generically constructed from any FA-*GS* scheme by using it as a black-box.

5 Implications and Separations

We address the theoretical gap between FA-*GS* and QA-*GS* schemes by observing relationship among our implication results in Sects. 3 and 4, and known separation and implication results. Roughly speaking, we show that *it is hard to construct a FA-GS scheme from any QA-GS scheme generically*. More precisely, if a FA-*GS* scheme can be constructed from any QA-*GS* scheme generically, then a CCA-*PKE* scheme can be obtained from any *OWF* generically. However, the generic construction of a CCA-*PKE* scheme from any *OWF* is a long-standing open problem. That is, the generic construction of a FA-*GS* scheme from any QA-*GS* scheme is another open problem.

Furthermore, we show that *it is unconditionally impossible to obtain a FA-GS scheme from any QA-GS scheme in a black-box manner*. More precisely, if a FA-*GS* scheme can be constructed from any QA-*GS* scheme in a black-box manner, then a CCA-*PKE* scheme can be obtained from any *OWF* in the same manner. However, a black-box construction of a CCA-*PKE* scheme from any *OWF* is unconditionally impossible [31] and only possibility for non-black-box construction remains. It results that a black-box construction of a FA-*GS* scheme from any QA-*GS* scheme is unconditionally impossible.

Our implication and separation results can be interpreted that there is a huge theoretical gap between *GS* schemes with and without unlinkability, and thus we should be careful for selecting appropriate cryptographic tools for providing anonymity in individual applications since there would be significantly much more efficient tools than full-fledged *GS* schemes if unlinkability is not necessary.

In the rest of this section, we first explain known implication results, and next show our implication results which are derived from Sects. 3 and 4, and finally give a separation result between FA-*GS* and QA-*GS* schemes.

5.1   Known Implication Results

We review two previous implication results among *OWF*, digital signature scheme and *IBS* scheme briefly.

(1) There exists a black-box construction of a CMA secure digital signature scheme from any *OWF*.
(2) There exists a black-box construction of a sID-CMA secure *IBS* scheme from any CMA secure digital signature scheme.

The first one is derived from previous results [18,24,32]. More precisely, Lamport showed a conversion from a *OWF* to a one-time digital signature scheme in [18], Naor and Yung showed that, if there exist a universal one-way hash function [24] and a one-time digital signature scheme, a CMA secure digital signature scheme can be constructed in [24] and Rompel showed a conversion from a *OWF* into a universal one-way hash function in [32]. Consequently, we obtain the first implication result.

The second implication result is a well-known folklore [5,13,14,22,23]. We will describe the concrete construction method of a sID-CMA secure *IBS* scheme from any CMA secure digital signature scheme and its security proof in the full version of this paper.

5.2   New Implications from Sections 3 and 4

We then discuss new implication results derived from Sects. 3 and 4. These are as follows:

(3) There exists a black-box construction of a QA-*GS* scheme from any *OWF*.
(4) There exists a black-box construction of a CCA-*PKE* scheme from any FA-*GS* scheme.

The above implication result (3) is derived from Theorem 2 and the known implication results (1) and (2) in the previous section. The implication result (4) can be obtained directly from Theorem 3 of the result of Sect. 4.

5.3   Separation Result

We show that constructing a FA-GS scheme from any QA-GS scheme in a black-box manner is unconditionally impossible and that it seems difficult to generically construct a FA-GS scheme from any QA-GS scheme. More precisely, we obtain the following theorem:

Theorem 4. *Constructing a* FA-GS *scheme from any* QA-GS *scheme generically is at least as hard as constructing a* CCA-PKE *scheme from any* OWF *generically.*

We will give the proof of this theorem in the full version of this paper.

In addition, we discuss this separation result in more detail. We then introduce CPA-PKE schemes as intermediate schemes between QA- and FA-GS schemes and discuss possibility for construction of a CPA-PKE scheme from any QA-GS scheme and that for construction of a FA-GS scheme from any CPA-PKE scheme. Subsequently, we show the height of the hurdle between QA- and FA-GS schemes is extremely high.

From the above implication (3) and (4), we obtain the following theorems.

Theorem 5. *Constructing a* CPA-PKE *scheme from any* QA-GS *scheme in a black-box manner is unconditionally impossible.*

Theorem 6. *Constructing a* FA-GS *scheme from any* CPA-PKE *scheme generically is at least as hard as constructing a* CCA-PKE *scheme from any* CPA-PKE *scheme generically.*

We will give the proofs of these theorems in the full version of this paper.

From these theorems, we can say that there are two high hurdles to construct a FA-GS scheme from any QA-GS scheme. That is, one is the hurdle to construct a CPA-PKE scheme from any QA-GS scheme generically, and the next is the hurdle to construct a FA-GS scheme from any CPA-PKE scheme generically. Consequently, these two hurdles are high and it results that the gap between QA- and FA-GS schemes are extremely large.

More precisely, from Theorem 5, we can say that there only remains possibility of a non-black-box construction of a CPA-PKE scheme from any QA-GS scheme. However, it is at least as hard as the generic non-black-box construction of a CPA-PKE scheme from any OWF and the latter construction is not known. As a result, the hurdle to cross in order to construct a CPA-PKE scheme from any QA-GS scheme generically is high. In addition, from Theorem 6, we have that the hurdle to cross in order to construct a FA-GS scheme from any CPA-PKE scheme generically is also high.

In other words, there are two unsurmountable difficulties in order to construct a CCA-PKE scheme from any OWF and Theorem 4 implies that the constructing a FA-GS scheme from any QA-GS scheme is at least as difficult as this. As a result, we see that the theoretical gap between unlinkability and other functionalities is significantly large. See Fig. 1, which shows all of implication and separation results.

6   Our Construction of Efficient Quasi Group Signature

In Sect. 5, we showed that the theoretical gap between hardness for constructing
a QA-GS scheme and that for constructing a FA-GS scheme is significantly large.
Therefore, there is possibility to construct a drastically efficient QA-GS scheme.
We then demonstrate that a significantly more efficient QA-GS scheme than the
best known full-fledged GS scheme can actually be constructed. As shown in
Sect. 3.2, a QA-GS scheme can be immediately constructed from any W-IBS
scheme, and therefore, we concentrate on designing an efficient W-IBS scheme.

6.1   Our Efficient wID-CMA Secure ID-Based Signature

We construct an efficient W-IBS scheme by simplifying Paterson and Schuldt's
scheme [26] and it is based on Boneh and Boyen's scheme [6] and Waters' scheme
[33]. Our construction is as follows.

Setup: Choose groups $\mathbb{G}$ and $\mathbb{G}_T$ of prime order p such that a pairing $e : \mathbb{G} \times \mathbb{G} \rightarrow$
$\mathbb{G}_T$ can be constructed and pick a generator g of $\mathbb{G}$. Now, pick $\alpha \leftarrow_R \mathbb{Z}_p$,
compute $g_1 = g^\alpha$ and pick $g_2 \leftarrow_R \mathbb{G}$. Furthermore, pick $u', x' \leftarrow_R \mathbb{G}$ and
vector $\mathbf{X} = (x_i)$ of length n_m, whose entries are random elements from
$\mathbb{G}$. Here, n_m denotes the bit length of a message. The verification key is
$vkey = (\mathbb{G}, \mathbb{G}_T, e, g, g_1, g_2, u', x', \mathbf{X})$ and the master key is $msk = g_2^\alpha$.

IExt: Let v be an identity which is selected from a small ID space $\{1, 2, ..., n\}$
(n is a positive integer). To construct the private key, sk_v, of the identity v,
pick $r_u \leftarrow_R \mathbb{Z}_p$ and compute $sk_v = (g_2^\alpha (u' g_1^v)^{r_u}, g^{r_u})$.

ISig: Let $\mathbf{m}$ be a bit string representing a message and let $\mathcal{X} \subset \{1, ..., n_m\}$ be
the set of indicies j such that $\mathbf{m}[j] = 1$, where $\mathbf{m}[j]$ is the jth bit of $\mathbf{m}$. A
signature of v on $\mathbf{m}$ is constructed by picking $r_m \leftarrow_R \mathbb{Z}_p$ and computing
$$\sigma = \left(g_2^\alpha (u' g_1^v)^{r_u} \left(x' \prod_{j \in \mathcal{X}} x_j \right)^{r_m}, g^{r_u}, g^{r_m} \right) \in \mathbb{G}^3.$$

IVf: Given a signature $\sigma = (V, R_u, R_m) \in \mathbb{G}^3$ of an identity v on a message
$\mathbf{m}$, a verifier accepts σ if the following equality holds: $e(V, g) = e(g_2, g_1)$
$e(u' g_1^v, R_u) e \left(x' \prod_{j \in \mathcal{X}} x_j, R_m \right).$

The above scheme employs small identity space, and thus, in order to prove
that the scheme is wID-CMA secure, it is sufficient to prove that the scheme is
sID-CMA secure. Concretely, we will prove the following theorem.

Theorem 7. *The above scheme is* $(\tau', q_s, q_e, \epsilon')$-*sID-CMA secure assuming that
the* (τ, ϵ)-*CDH assumption holds in* $\mathbb{G}$, *where* $\epsilon' = (4q_s(n_m + 1))\epsilon$, $\tau' = \tau - O((q_e + q_s n_m)\rho_1 + (q_e + q_s)\rho_2)$, *and* ρ_1 *and* ρ_2 *are the time for a multiplication
and an exponentiation in* $\mathbb{G}$, *respectively.*

We will give the proof of Theorem 7 in the full version of this paper.

6.2   An Efficient Quasi Group Signature

In Sect. 3.2, we showed that there exists a black-box construction of a QA-GS
scheme from any W-IBS scheme. Therefore, we can easily construct an efficient
QA-GS scheme from the W-IBS scheme described in Sect. 6.1.

6.3 Performance

Table 1 shows comparison between the proposed QA-GS scheme and the conventional FA-GS schemes [12] and [19] in terms of secret signing key size, signature size, and computational cost for signing and verification. The computational cost means the number of multiplications and that of pairing computations.[3] The FA-GS schemes in [12] is provably secure in the random oracle model, and the other schemes are provably secure in the standard model. We assume that the representation of $\mathbb{G}$, $\mathbb{G}_1$, $\mathbb{G}_2$, $\mathbb{G}_T$ are 172, 172, 1020, and 1020 bits, respectively, and $|p| = 171$ bits in the FA-GS scheme [12].[4] We then assume that both of the representation of $\mathbb{G}$ and its subgroup $\mathbb{G}_q$ of order q are 172 bits, and $|n| = 1024$ bits in the FA-GS scheme [19]. Finally, we assume that the representation of $\mathbb{G}$ is 172 bits, the length of message and that of identity are 160[5] and 80 bits, respectively, in our scheme.

Table 1. Comparisons on size and cost in group signature schemes (M: multiplication, P: pairing computation)

	[12]	[19]	Ours
Secret signing key size (bits)	857	1196	424
Signature size (bits)	1714	860	596
Cost (Sign)	$140M$	$1974M$	$404M$
Cost (Verify)	$140M+P$	$961M+4P$	$242M+3P$
Assumption	SDH,DDH	SDH,MOMSDH	CDH
Security	FA	FA	QA
Model	RO	w/o RO	w/o RO

Our scheme is more efficient than the other schemes in terms of secret signing key size and signature size, and our scheme is more efficient than the scheme in [19] in terms of computational cost for signing and verification. It shows that our scheme is a QA-GS scheme which is the most efficient and provably secure in the standard model.[6]

Remark 1. *The QA-GS scheme based on [26] is also an efficient scheme. The secret signing key size is 424 bits, the signature size is 596 bits, the computational cost for signing is 465M, and the computational cost for verification is 240M+3P. On the other hand, in our scheme, those are 424 bits, 596 bits, 404M, and 242M+3P, respectively. Therefore, our scheme is more efficient than the QA-GS scheme based on [26].*

---

[3] We assume that the number of multiplications for $(g_{11}^{a_{11}} g_{21}^{a_{21}} \cdots g_{p1}^{a_{p1}})$, ..., $(g_{1q}^{a_{1q}} g_{2q}^{a_{2q}} \cdots g_{pq}^{a_{pq}})$ is about $(\min\{p,q\} + pq/\log(pq \log B)) \log B$ if each exponent is below B (by using Pippenger's algorithm [28]).

[4] An asymmetric pairing $e : \mathbb{G}_1 \times \mathbb{G}_2 \to \mathbb{G}_T$ is used in the FA-GS scheme [12].

[5] The length of message can be arbitrary size by using any collision-resistant hash function.

[6] The public key size in our scheme is larger than the other schemes. However, that is no problem since, unlike a secret signing key, a public key does not need to be stored a small-capacity device, such as a smart card, in many cases.

Remark 2. *As mentioned in a footnote in Sect. 3.1, we can also directly construct a* QA-*GS scheme from standard digital signatures (and this can be considered as a special case of our generic construction in Sect. 3.2). This scheme is also fairly efficient, but unfortunately its signature length is still larger than that of our scheme. More specifically, a signature in this scheme consists of two signatures and a verification key of the underlying scheme, and therefore, even if we use the most practical schemes (in the standard model), i.e. [7] or [33], as an underlying signature scheme, signature length of the resulting scheme significantly exceeds that of our scheme since a signature length of [7] and [33] are at least 344 bits for the same security level.*

Acknowledgment

The authors would like to thank anonymous reviewers of Africacrypt'09 for their invaluable comments.

References

1. Abdalla, M., Warinschi, B.: On the Minimal Assumptions of Group Signature Schemes. In: López, J., Qing, S., Okamoto, E. (eds.) ICICS 2004. LNCS, vol. 3269, pp. 1–13. Springer, Heidelberg (2004)
2. Ateniese, G., Camenisch, J., Joye, M., Tsudik, G.: A Practical and Provably Secure Coalition-Resistant Group Signature Scheme. In: Bellare, M. (ed.) CRYPTO 2000. LNCS, vol. 1880, pp. 255–270. Springer, Heidelberg (2000)
3. Bellare, M., Micciancio, D., Warinschi, B.: Foundations of Group Signatures: Formal Definitions, Simplified Requirements, and a Construction Based on General Assumptions. In: Biham, E. (ed.) EUROCRYPT 2003. LNCS, vol. 2656, pp. 614–629. Springer, Heidelberg (2003)
4. Bellare, M., Shi, H., Zhang, C.: Foundations of Group Signatures. In: Menezes, A. (ed.) CT-RSA 2005. LNCS, vol. 3376, pp. 136–153. Springer, Heidelberg (2005)
5. Bellare, M., Namprempre, C., Neven, G.: Security proofs for identity-based identification and signature schemes. In: Cachin, C., Camenisch, J.L. (eds.) EUROCRYPT 2004. LNCS, vol. 3027, pp. 268–286. Springer, Heidelberg (2004)
6. Boneh, D., Boyen, X.: Efficient Selective-ID Secure Identity Based Encryption Without Random Oracles. In: Cachin, C., Camenisch, J.L. (eds.) EUROCRYPT 2004. LNCS, vol. 3027, pp. 223–238. Springer, Heidelberg (2004)
7. Boneh, D., Boyen, X.: Short Signatures Without Random Oracles. In: Cachin, C., Camenisch, J.L. (eds.) EUROCRYPT 2004. LNCS, vol. 3027, pp. 56–73. Springer, Heidelberg (2004)
8. Camenisch, J., Stadler, M.: Efficient Group Signatures Schemes for Large Groups. In: Kaliski Jr., B.S. (ed.) CRYPTO 1997. LNCS, vol. 1294, pp. 410–424. Springer, Heidelberg (1997)
9. Castelluccia, C.: How to Convert Any ID-Based Signature Schemes into A Group Signature Scheme, http://eprint.iacr.org/2002/116
10. Chaum, D., van Heyst, E.: Group signatures. In: Davies, D.W. (ed.) EUROCRYPT 1991. LNCS, vol. 547, pp. 257–265. Springer, Heidelberg (1991)
11. Dolev, D., Dwork, C., Naor, M.: Non-Malleable Cryptography. In: Proc. of ACM STOC 1991, pp. 542–552 (1991)
12. Furukawa, J., Imai, H.: An Efficient Group Signature Scheme from Bilinear Maps. IEICE Trans. Fundamentals E89-A(5), 1328–1338 (2006)

13. Galindo, D., Herranz, J., Kiltz, E.: On the Generic Construction of Identity-Based Signatures with Additional Properties. In: Lai, X., Chen, K. (eds.) ASIACRYPT 2006. LNCS, vol. 4284, pp. 178–193. Springer, Heidelberg (2006)
14. Gentry, C., Silverberg, A.: Hierarchical ID-Based Cryptography. In: Zheng, Y. (ed.) ASIACRYPT 2002. LNCS, vol. 2501, pp. 548–566. Springer, Heidelberg (2002)
15. Gertner, Y., Malkin, T., Myers, S.: Towards a Separation of Semantic and CCA Security for Public Key Encryption. In: Vadhan, S.P. (ed.) TCC 2007. LNCS, vol. 4392, pp. 434–455. Springer, Heidelberg (2007)
16. Goldwasser, S., Micali, S.: Probabilistic Encryption. JCSS 28(2), 270–299 (1984)
17. Impagliazzo, R., Rudich, S.: Limits on the Provable Consequences of One-Way Permutations. In: Proc. of ACM STOC 1989, pp. 44–61 (1989)
18. Lamport, L.: Constructing Digital Signatures from a One Way Function. In: SRI Intl. CSL 1998 (October 1979)
19. Liang, X., Cao, Z., Shao, J., Lin, H.: Short Group Signature without Random Oracles. In: Qing, S., Imai, H., Wang, G. (eds.) ICICS 2007. LNCS, vol. 4861, pp. 69–82. Springer, Heidelberg (2007)
20. Liu, J.K., Wei, V.K., Wong, D.S.: Linkable Spontaneous Anonymous Group Signature for Ad Hoc Groups. In: Wang, H., Pieprzyk, J., Varadharajan, V. (eds.) ACISP 2004. LNCS, vol. 3108, pp. 325–335. Springer, Heidelberg (2004)
21. Manulis, M., Sadeghi, A.R., Schwenk, J.: Linkable Democratic Group Signatures. In: Chen, K., Deng, R., Lai, X., Zhou, J. (eds.) ISPEC 2006. LNCS, vol. 3903, pp. 187–201. Springer, Heidelberg (2006)
22. Maurer, U., Yacobi, Y.: Non-interactive Public-Key Cryptography. In: Davies, D.W. (ed.) EUROCRYPT 1991. LNCS, vol. 547, pp. 498–507. Springer, Heidelberg (1991)
23. Maurer, U., Yacobi, Y.: A Non-interactive Public-Key Distribution System. Designs, Code and Cryptography 9(3), 305–316 (1996); full version of [22]
24. Naor, M., Yung, M.: Universal One-Way Hash Functions and Their Cryptographic Applications. In: Proc. of ACM STOC 1989, pp. 33–43 (1989)
25. Nguyen, L., Safavi-Naini, R.: Efficient and Provably Secure Trapdoor-free Group Signature Schemes from Bilinear Pairings. In: Lee, P.J. (ed.) ASIACRYPT 2004. LNCS, vol. 3329, pp. 372–386. Springer, Heidelberg (2004)
26. Paterson, K., Schuldt, J.: Efficient Identity-based Signatures Secure in the Standard Model. In: Batten, L.M., Safavi-Naini, R. (eds.) ACISP 2006. LNCS, vol. 4058, pp. 207–222. Springer, Heidelberg (2006)
27. Petersen, H.: How to Convert Any Digital Signature Scheme into A Group Signature Scheme. In: Proc. of Security Protocols Workshop 1997, pp. 177–190 (1997)
28. Pippenger, N.: On the Evaluation of Powers and Related Problems. In: Proc. of FOCS 1976, pp. 258–263 (1976)
29. Popescu, C.: An Efficient ID-Based Group Signature Scheme, http://www.cs.ubbcluj.ro/~studia-i/2002-2/4-Popescu.pdf
30. Rackoff, C., Simon, D.R.: Non-Interactive Zero-Knowledge Proof of Knowledge and Chosen Ciphertext Attack. In: Feigenbaum, J. (ed.) CRYPTO 1991. LNCS, vol. 576, pp. 433–444. Springer, Heidelberg (1992)
31. Reingold, O., Trevisan, L., Vadhan, S.: Notions of Reducibility between Cryptographic Primitives. In: Naor, M. (ed.) TCC 2004. LNCS, vol. 2951, pp. 1–20. Springer, Heidelberg (2004)
32. Rompel, J.: One-Way Functions are Necessary and Sufficient for Secure Signatures. In: Proc. of ACM STOC 1990, pp. 387–394 (1990)
33. Waters, B.: Efficient identity-based encryption without random oracles. In: Cramer, R. (ed.) EUROCRYPT 2005. LNCS, vol. 3494, pp. 114–127. Springer, Heidelberg (2005)

Practical Threshold Signatures with Linear Secret Sharing Schemes*

İlker Nadi Bozkurt, Kamer Kaya**, and Ali Aydın Selçuk

Department of Computer Engineering
Bilkent University
Ankara 06800, Turkey
{bozkurti,kamer,selcuk}@cs.bilkent.edu.tr

Abstract. Function sharing deals with the problem of distribution of the computation of a function (such as decryption or signature) among several parties. The necessary values for the computation are distributed to the participating parties using a secret sharing scheme (SSS). Several function sharing schemes have been proposed in the literature, with most of them using Shamir secret sharing as the underlying SSS. In this paper, we investigate how threshold cryptography can be conducted with any linear secret sharing scheme and present a function sharing scheme for the RSA cryptosystem. The challenge is that constructing the secret in a linear SSS requires the solution of a linear system, which normally involves computing inverses, while computing an inverse modulo $\varphi(N)$ cannot be tolerated in a threshold RSA system in any way. The threshold RSA scheme we propose is a generalization of Shoup's Shamir-based scheme. It is similarly robust and provably secure under the static adversary model. At the end of the paper, we show how this scheme can be extended to other public key cryptosystems and give an example on the Paillier cryptosystem.

Keywords: Linear secret sharing, threshold cryptography, function sharing.

1 Introduction

The secure storage of the private keys of a cryptosystem is an important problem. Possession of a highly sensitive key by an individual may not be desirable as the key can easily be lost or as the individual may not be fully trusted. Giving copies of the key to more than one individual increases the risk of compromise. A solution to this problem is to give shares of the key to several individuals, forcing them to cooperate to find the secret key. This not only reduces the risk of losing the key but also makes compromising the key more difficult. In threshold

---

* This work is supported in part by the Turkish Scientific and Technological Research Agency (TÜBİTAK), under grant number 108E150.
** Supported by the Turkish Scientific and Technological Research Agency (TÜBİTAK) Ph.D. scholarship.

B. Preneel (Ed.): AFRICACRYPT 2009, LNCS 5580, pp. 167–178, 2009.

cryptography, secret sharing deals with this problem, namely, sharing a highly sensitive secret among a group of n users such that only when a sufficient number t of them come together can the secret be reconstructed. Well-known secret sharing schemes (SSS) in the literature include Shamir [17] based on polynomial interpolation, Blakley [2] based on hyperplane geometry, and Asmuth-Bloom [1] based on the Chinese Remainder Theorem.

A shortcoming of secret sharing schemes is the need to reveal the secret shares during the reconstruction phase. The system would be more secure if the subject function can be computed without revealing the secret shares or reconstructing the secret. This is known as the function sharing problem. A function sharing scheme requires distributing the function's computation according to the underlying SSS such that each part of the computation can be carried out by a different user and then the partial results can be combined to yield the function's value without disclosing the individual secrets. Several protocols for function sharing have been proposed in the literature [4,5,6,7,18,10,16]. Nearly all these protocols use Shamir secret sharing as the underlying SSS.

1.1   Secret Sharing Schemes

The problem of secret sharing and the first solutions were introduced in 1979 independently by Shamir [17] and Blakley [2]. A (t, n)-secret sharing scheme is used to distribute a secret d among n people such that any coalition of size t or more can construct d but smaller coalitions cannot.

Shamir secret sharing is based on polynomial interpolation over a finite field. It uses the fact that we can find a secret polynomial of degree $t - 1$ given t data points. To generate a polynomial $f(x) = \sum_{i=0}^{t-1} a_i x^i$, a_0 is set to the secret value and the coefficients a_1 to a_{t-1} are assigned random values in the field. The value $f(i)$ is given to user i. When t out of n users come together, they can construct the polynomial using Lagrange interpolation and can find the secret.

Blakley secret sharing scheme has a different approach based on hyperplane geometry: To implement a (t, n) threshold scheme, each of the n users is given a hyperplane equation in a t dimensional space over a finite field such that each hyperplane passes through a certain point. The intersection point of the hyperplanes is the secret. When t users come together, they can solve the system of equations to find the secret.

Both Shamir and Blakley are linear threshold secret sharing schemes: As Karnin et al. [11] observed, Shamir SSS is a subclass of a broader class of linear secret sharing. The polynomial share computation can be represented as a matrix multiplication by using a Vandermonde matrix. Similarly, the secret and the shares of the Blakley SSS can be represented as a linear system $Ax = y$ where the matrix A and the vector y are obtained from the hyperplane equations.

1.2   Function Sharing Schemes

Function sharing is the concept of distributing the computation of a function such that when a sufficient number of users come together they can compute the value of the function without revealing their secret shares. This problem is related to secret sharing as the secret values needed for partial computations are distributed using secret sharing.

Several solutions for sharing the RSA, ElGamal, and Paillier private key operations have been proposed in the literature [4,5,6,7,9,12,13,16,18]. Almost all of these schemes have been based on the Shamir SSS.

The additive nature of the Lagrange's interpolation formula used in the combining phase of Shamir's scheme makes it an attractive choice for function sharing, but it also provides several challenges. One of the most significant challenges is the computation of inverses in $\mathbb{Z}_{\varphi(N)}$ for the division operations in Lagrange's formula, while $\varphi(N)$ should not be known by the users. There are two main difficulties in this respect:

1. An inverse x^{-1} will not exist modulo $\varphi(N)$ if $\gcd(x, \varphi(N)) \neq 1$.
2. Even when x^{-1} exists it should not be computable by a user, since that would enable computing $\varphi(N)$.

Early solutions to this problem, albeit not very efficient, were given in [4,16]. Afterwards an ingenious solution was given by Shoup [18] where he removed the need of taking inverses in Lagrange interpolation altogether.

Shoup's practical RSA scheme has inspired similar works on different cryptosystems. Fouque et al. [9] proposed a similar Shamir-based threshold solution for the Paillier cryptosystem and used it in e-voting and lottery protocols. Later, Lysyanskaya and Peikert [13] improved this work and obtained a threshold Paillier encryption scheme secure under the adaptive security model. The current paper is also inspired by Shoup's work.

1.3   Our Contribution

In this work, we show how to generalize Shoup's ideas to do function sharing with any linear SSS, and we give a robust threshold RSA signature scheme. A linear SSS, where the solution is based on solving a linear system, naturally requires computing inverses for reconstructing the secret. We show how to utilize such a system for function sharing while avoiding computation of inverses modulo $\varphi(N)$ completely.

We also discuss how this approach can be applied to other public key cryptosystems and show an example on the Paillier decryption function.

2   Linear Secret Sharing Schemes

A linear (t, n) threshold secret sharing scheme can be defined as follows: Let A be a full-rank public $n \times t$ matrix with entries chosen from $\mathcal{F} = \mathbb{Z}_m^*$ for a prime m. Let $x = (x_1, x_2, \ldots, x_t)^T$ be a secret vector from $\mathcal{F}^t$. Let a_{ij} denote the entry at the ith row and jth column of the matrix A.

2.1 Dealing Phase

The dealer chooses a secret vector $x \in \mathcal{F}^t$ where the first entry x_1 is set to the secret value (the RSA private key d in our case) and the values of the other coordinates are set randomly from the field $\mathcal{F}$. The ith user will get a his share $y_i \in \mathcal{F}$,

$$y_i = a_{i1}x_1 + a_{i2}x_2 + \ldots + a_{it}x_t. \tag{1}$$

For a (t, n) threshold scheme there will be n such shares, and hence we will have an $n \times t$ linear system

$$Ax = y. \tag{2}$$

The dealer then sends the secret value of y_i to user i for $1 \leq i \leq n$ and makes the matrix A public.

2.2 Share Combining Phase

Share combining step is simply finding the solution of a linear system of equations. Suppose that a coalition $\mathcal{S} = \{i_1, \ldots, i_t\}$ of users come together. They form a matrix $A_\mathcal{S}$ using their equations and solve

$$A_\mathcal{S}x = y_\mathcal{S}, \tag{3}$$

where $y_\mathcal{S}$ is the vector of the secret shares of the users. The secret is found as the first coordinate of the solution.

3 Sharing RSA Signature Computation

In this section, we describe our threshold RSA signature scheme which works with any linear SSS in general.

3.1 Setup

In the RSA setup phase, choose the RSA primes $p = 2p' + 1$ and $q = 2q' + 1$, where p' and q' are large primes. Compute the RSA modulus as $N = pq$. Let $m = p'q'$. The public key e is chosen as a prime number, details of which will be explained in the next section. After choosing e, the private key d is computed such that $ed \equiv 1 \pmod{m}$. Then the dealer shares the private key d among n users using a linear threshold SSS described in Section 2.

The dealer also chooses v as a generator of Q_N, where Q_N is the subgroup of squares in $\mathbb{Z}_N^*$. He computes and broadcasts

$$v_i = v^{y_i} \in Q_N, \tag{4}$$

for $1 \leq i \leq n$, which are the verification keys to be used in the proofs of correctness of the partial signatures, where y_i is the secret share of user i.

3.2   Signing

Let $H(.)$ be a hash function mapping input messages to $\mathbb{Z}_N^*$ and let $w = H(M) \in \mathbb{Z}_N^*$ be the hashed message to be signed. Assume a coalition $\mathcal{S}$ of size t wants to obtain the signature $s = w^d \bmod N$.

Generating partial signatures. Let $\mathcal{S} = \{i_1, \ldots, i_t\}$ be the coalition of t users, forming the linear system

$$A_\mathcal{S} x = y_\mathcal{S}.$$

Let c_{ij} be the ij-th cofactor of matrix $A_\mathcal{S}$ and let $C_\mathcal{S}$ be the adjugate matrix,

$$C_\mathcal{S} = \begin{pmatrix} c_{11} & c_{21} & \cdots & c_{t1} \\ c_{12} & c_{22} & \cdots & c_{t2} \\ \vdots & \vdots & \ddots & \vdots \\ c_{1t} & c_{2t} & \cdots & c_{tt} \end{pmatrix}.$$

If we denote the determinant of $A_\mathcal{S}$ by $\Delta_\mathcal{S}$ we have,

$$A_\mathcal{S} C_\mathcal{S} = C_\mathcal{S} A_\mathcal{S} = \Delta_\mathcal{S} I_t, \tag{5}$$

where I_t denotes the $t \times t$ identity matrix.

For our scheme, each user $i \in \mathcal{S}$ computes his partial signature as

$$s_i = w^{2c_{i1} y_i} \bmod N. \tag{6}$$

Verifying partial signatures. Each user computes and publishes a proof of correctness for the verification of his partial signature. The proof of correctness of the partial signature of user i is a proof that the discrete logarithm of s_i^2 to the base

$$\tilde{s}_i = w^{4c_{i1}} \bmod N \tag{7}$$

is the same as the discrete logarithm of v_i to the base v. To prove this, a protocol by Shoup [18] which is a non-interactive version of Chaum and Pedersen's [3] interactive protocol is used:

Let $L(n)$ be the bit-length of n. Let H' be a hash function, whose output is an L_1-bit integer, where L_1 is a secondary security parameter. To construct the proof of correctness, user i chooses a random number $r \in \{0, 1, \ldots, 2^{L(N)+2L_1} - 1\}$, computes

$$v' = v^r \bmod N,$$
$$s' = \tilde{s}_i^r \bmod N,$$
$$D = H'(v, \tilde{s}_i, v_i, s_i^2, v', s'),$$
$$\sigma = y_i D + r.$$

Then user i publishes his proof of correctness as (σ, D).

To verify this proof of correctness, one checks whether

$$D \overset{?}{=} H'(v, \tilde{s}, v_i, s_i^2, v^\sigma v_i^{-D}, \tilde{s}_i^\sigma s_i^{-2D}).$$

Combining partial signatures. To combine the partial signatures, we simply compute

$$\overline{s} = \prod_{i \in S} s_i \bmod N. \tag{8}$$

Note that, by equation (5), we have

$$\overline{s} = w^{d\,\delta} \bmod N, \tag{9}$$

where

$$\delta = 2\,\Delta_S. \tag{10}$$

Given that e is a prime number relatively prime to Δ_S, it is easy to compute the signature $s = w^d \bmod N$ from $\overline{s}$. Take

$$s = \overline{s}^a w^b \bmod N, \tag{11}$$

where a and b are integers such that

$$\delta a + e b = 1, \tag{12}$$

which can be obtained by the extended Euclidean algorithm on δ and e.

4   Choosing e

The choice of e is critical in the setup phase because the solution depends on e and Δ_S being relatively prime. To achieve this, we can either choose a special matrix whose determinant is known to be relatively prime to e, or choose e as a sufficiently large prime according to t and n so that the probability that Δ_S is divisible by e will be negligible for any coalition S.

4.1   Choosing e probabilistically

The probability of a random integer's being divisible by a prime e is $1/e$. So, if we have a (t, n) threshold scheme, the probability that the determinant of none of the $\binom{n}{t}$ A_S matrices will be divisible by e is $\left(1 - \frac{1}{e}\right)^{\binom{n}{t}}$. If we take $e \gg \binom{n}{t}$, we have

$$\left(1 - \frac{1}{e}\right)^{\binom{n}{t}} \approx 1. \tag{13}$$

4.2   Choosing a Vandermonde Matrix as the Coefficient Matrix

A simple choice for the matrix A that enables us to guarantee that e will be relatively prime to the determinant of the coefficient matrix is to choose the rows of the matrix A as the rows of a Vandermonde matrix. Note that this is exactly the case for Shamir secret sharing. Then A_S will have the following form for a coalition S of size t:

$$A_\mathcal{S} = \begin{pmatrix} 1 & a_1 & a_1^2 & \dots & a_1^{t-1} \\ 1 & a_2 & a_2^2 & \dots & a_2^{t-1} \\ \vdots & \vdots & \vdots & \ddots & \vdots \\ 1 & a_t & a_t^2 & \dots & a_t^{t-1} \end{pmatrix}$$

The determinant of the Vandermonde matrix is nonzero, provided that no two rows are identical, and is given by the following formula:

$$|A_\mathcal{S}| = \prod_{i,j=1,i<j}^{t} (a_i - a_j) \tag{14}$$

Without loss of generality take $(a_1, a_2, \dots, a_n) = (1, 2, \dots, n)$. Obviously,

$$\prod_{i,j=1,i<j}^{t} (a_i - a_j) \mid \prod_{i,j=1,i<j}^{n} (a_i - a_j).$$

We also have,

$$\prod_{i,j=1,i<j}^{n} (a_i - a_j) = 1^{\alpha_1} 2^{\alpha_2} \dots (n-1)^{\alpha_{n-1}} \tag{15}$$

for some $\alpha_1, \alpha_2, \dots, \alpha_{n-1}$. Hence by choosing e as a prime greater than or equal to n we can guarantee that the determinant of any $A_\mathcal{S}$ will be relatively prime to e.

5   Security Analysis

Now we will prove that the proposed threshold RSA signature scheme is secure provided that the standard RSA signature is secure. We assume a static adversary model in the sense that the adversary controls exactly $t - 1$ users and chooses them at the beginning of the attack. The adversary obtains all secret information of the corrupted users along with the public parameters of the system. She can control the actions of the corrupted users, asking for partial signatures of messages of her choice but cannot corrupt any other user in due course.

First we will analyze the proof of correctness. Then using this analysis we will prove that the proposed threshold signature scheme is secure.

5.1   Analysis of the Proof of Correctness

For generating and verifying the proof of correctness, the following properties hold:

Completeness. If the ith user is honest then the proof succeeds since

$$v^\sigma v_i^{-D} = v^{y_i D} v^r v_i^{-D} = v^r = v'$$

and

$$\tilde{s}_i^\sigma s_i^{-2D} = w^{4c_{i1}(y_i D + r)} w^{-4c_{i1} y_i D} = s^r = s'.$$

Soundness. To prove the soundness of the proof of correctness, we have to show that the adversary cannot construct a valid proof of correctness for an incorrect share, except with negligible probability. Let (σ, D) be a valid proof of correctness for a message w and partial signature s_i. We have $D = H'(v, \tilde{s}_i, v_i, s_i^2, v', s')$, where

$$\tilde{s}_i = w^{4c_{i1}}, v' = v^\sigma v_i^{-D}, s' = \tilde{s}_i^\sigma s_i^{-2D}.$$

Obviously $\tilde{s}_i$, v_i, s_i^2, v' and s' all lie in Q_n and we know that v is a generator of Q_n. So we have

$$\tilde{s}_i = v^\alpha, v_i = v^{y_i}, s_i^2 = v^\beta, v' = v^\gamma, s' = v^\mu,$$

for some integers $\alpha, \beta, \gamma, \mu$. From this we have,

$$\sigma - Dy_i \equiv \gamma \pmod{m} \tag{16}$$
$$\sigma\alpha - D\beta \equiv \mu \pmod{m}. \tag{17}$$

From equations (16) and (17) we get,

$$D(\beta - y_i\alpha) \equiv \alpha\gamma - \mu \pmod{m}. \tag{18}$$

A share is correct, if and only if,

$$\beta \equiv y_i\alpha \pmod{m}. \tag{19}$$

If (19) does not hold, then it does not hold either mod p' or mod q' and so (18) uniquely determines D mod p' or D mod q'. But the distribution of D is uniform in the random oracle model, so this happens with negligible probability.

Zero knowledge simulatability. To prove zero knowledge simulatability, we will use the random oracle model for the hash function and construct a simple simulator that simulates the adversary's view without knowing the value y_i. When an uncorrupted user wants to create a proof (σ, D) for a message w and partial signature s_i, the simulator chooses $D \in \{0, \ldots, 2^{L_1} - 1\}$ and $\sigma \in \{0, \ldots, 2^{L(N)+2L_1} - 1\}$ at random and defines the value of the random oracle at $(v, \tilde{s}_i, v_i, s_i^2, v^\sigma v_i^{-D}, \tilde{s}_i^\sigma s_i^{-2D})$ to be D. Note that, the value of the random oracle is not defined at this point with all but negligible probability. When the adversary queries the oracle, if the value of the oracle was already set the simulator returns that value, otherwise it returns a random value. It is obvious that the output of this simulator is statistically indistinguishable from real output.

5.2   Security of the Proposed Signature Scheme

To reduce the problem of the security of the proposed threshold signature scheme to that of the standard RSA signature, the following proof constructs another simulator.

Theorem 1. *In the random oracle model for H', the proposed threshold signature scheme is a secure threshold signature scheme (robust and non-forgeable) under the static adversary model given that the standard RSA signature scheme is secure.*

Proof. We will simulate the threshold protocol with no information on the secret where the output of the simulator is indistinguishable in the adversary's view. Afterwards, we will show that the secrecy of the private key d is not disrupted by the values obtained by the adversary. Thus, if the threshold RSA scheme is not secure, i.e. an adversary who controls $t - 1$ users can forge signatures in the threshold scheme, one can use this simulator to forge a signature in the standard RSA signature scheme.

Let $i_1, \ldots, i_{t-1}$ be the set of corrupted players. To simulate the adversary's view, we simply choose the y_{i_j} values belonging to the set of corrupted players at random from the set $\{0, \ldots, \lfloor N/4 \rfloor - 1\}$. The corrupted players' secret key shares are random numbers in the set $\{0, \ldots, m - 1\}$. Once these values are chosen, the values y_i for the uncorrupted players are completely determined modulo m, but cannot easily be computed. However, given $w, s \in \mathbb{Z}_N^*$ with $s^e = w$, we can easily compute s_{i_t} for an uncorrupted user i_t as

$$s_{i_t} = w^{2c_{t1}y_{i_t}} = s^{2\Delta_S} w^{-2\sum_{j=1}^{t-1} c_{j1}y_{i_j}}. \tag{20}$$

Note the dependence of Δ_S and c_{j1} values on the coalition $\{i_1, \ldots, i_{t-1}, i_t\}$.

Using this technique, we can generate the values $v, v_1, \ldots, v_n$, and also generate any share s_i of a signature, given the standard RSA signature. These values produced by the simulator and the proof of correctness given in this section are computationally indistinguishable from the real ones. Hence, the threshold RSA signature scheme based on a linear SSS is secure given that the standard RSA signature scheme is secure.   □

6 Application to Other PKCs

So far, we investigated only how to share the RSA signature function by using a linear SSS. The same approach can also be used to share the RSA decryption function since the signature and decryption functions are mostly identical. Besides RSA, the proposed approach can also be used to share other public key cryptosystems where the private key is used in the exponent, such as the ElGamal [8], Naccache-Stern [14] and the Paillier [15] decryption functions.

Below, as an example, we describe how our approach can be utilized for sharing the Paillier decryption function. The scheme works along the same lines as Fouque et al.'s extension [9] of Shoup's work to the Paillier cryptosystem.

6.1 The Paillier Cryptosystem

The Paillier PKC is based on the properties of Carmichael function over $\mathbb{Z}_{N^2}$ where N is an RSA composite. Security of the cryptosystem is based on the intractability of computing discrete logarithms in $\mathbb{Z}_{N^2}$ without the Carmichael number $\lambda(N)$.

Key Generation. Let $N = pq$ where p and q are large prime integers. Let g be an arbitrary element from $\mathbb{Z}_{N^2}$ such that its order is a multiple of N. Let $\lambda = (p-1)(q-1)/2$ denote the the Carmichael function for N. The public and private keys are (N, g) and λ, respectively.

Encryption. Let w be the message to be encrypted. Choose a random $r \in \mathbb{Z}_{N^2}$ and compute the ciphertext as

$$s = g^w r^N \bmod N^2.$$

Decryption. The plaintext is obtained by

$$w = \frac{L(s^\lambda \bmod \mathbb{Z}_{N^2})}{L(g^\lambda \bmod \mathbb{Z}_{N^2})} \bmod N$$

where $L(x) = (x-1)/N$ for $x \equiv 1 \pmod{N}$.

Paillier proved that this scheme is semantically secure under the assumption that it is hard to detect whether a given random element in $\mathbb{Z}_{N^2}$ is an N-residue. The cryptosystem possesses the following homomorphic properties:

$$E(w_1 + w_2) = E(w_1).E(w_2)$$
$$E(k.w) = E(w)^k.$$

6.2   Sharing the Paillier Decryption Function

Since $\lambda(N)$ must be kept secret, the inverse computation problem is similar to the one we encountered while sharing the RSA signature function. Our threshold Paillier scheme is given below:

Key Generation. In the Paillier setup phase, choose two safe primes $p = 2p' + 1$ and $q = 2q' + 1$, where p' and q' are large primes and $\gcd(N, \varphi(N)) = 1$ for $N = pq$. Let $m = p'q'$. Let $\beta \in_R \mathbb{Z}_N^*$ and $(a, b) \in_R \mathbb{Z}_N \times \mathbb{Z}_N^*$. Compute

$$g = (1 + N)^a \times b^N \bmod N^2.$$

Share the private key $d = \beta m$ among n users with modulo Nm by using the linear SSS. Let

$$\theta = L\left(g^{\beta m}\right) = a\beta m \bmod N.$$

Set the public key as (g, N, θ). Choose v as a generator of Q_{N^2}, where Q_{N^2} is the cyclic group of squares in $\mathbb{Z}_{N^2}$. Compute the verification keys

$$v_i = v^{y_i} \in Q_{N^2}$$

for $1 \le i \le n$ as before.

Encryption. Let w be the message to be encrypted. Choose a random $r \in \mathbb{Z}_{N^2}$ and compute the ciphertext as $s = g^w r^N \bmod N^2$. Let $m = p'q'$.

Decryption. Let s be the ciphertext to be decrypted and $\mathcal{S} = \{i_1, \ldots, i_t\}$ denote a coalition of t users that will compute the plaintext together. Let $A_{\mathcal{S}}$ be the coalition matrix and $C_{\mathcal{S}}$ be the corresponding adjugate matrix, respectively, as in Section 3. Each member $i \in \mathcal{S}$ computes his partial value as

$$s_i = s^{2c_{i1}y_i} \bmod N^2$$

where c_{i1} is the ith element of the first row of $C_{\mathcal{S}}$. He also generates a proof of correctness which is used to prove that the discrete logarithm of s_i^2 to the base $\tilde{s} = w^{4c_{i1}}$ is the same as the discrete logarithm of v_i to the base v. Note that the proof is now working on a cyclic group of unknown order mN.

After the partial decryptions are obtained, the combining algorithm computes the plaintext

$$w = \frac{L\left(\prod_{i \in \mathcal{S}} s_i \bmod N^2\right)}{2\Delta_{\mathcal{S}}\theta} \bmod N.$$

Note that

$$\prod_{i \in \mathcal{S}} s_i \equiv s^{2\Delta_{\mathcal{S}}\beta m}$$

$$\equiv g^{2\Delta_{\mathcal{S}}\beta m w}$$

$$\equiv (1 + N)^{2\Delta_{\mathcal{S}}a\beta m w}$$

$$\equiv 1 + 2\Delta_{\mathcal{S}}a\beta m w N$$

$$\equiv 1 + 2\Delta_{\mathcal{S}}\theta w N \pmod{N^2}.$$

7   Conclusion

We showed how to do threshold cryptography with linear secret sharing in general. We presented a robust RSA threshold signature scheme based on a linear SSS. The proposed signature scheme generalizes Shoup's threshold RSA signature based on Shamir secret sharing, and is as efficient and practical as Shoup's scheme.

Besides RSA, this approach can be extended to other public key cryptosystems where the private key is used in the exponent. As an example we demonstrated how Paillier decryption function can be shared by this approach. ElGamal and Naccache-Stern knapsack cryptosystems are some other systems that can benefit from the proposed solution.

Acknowledgements

We would like to thank Ahmet Güloğlu for informative discussions and his comments on this paper. We would also like to thank anonymous AfricaCrypt referees for their valuable comments which significantly helped to improve the paper.

References

1. Asmuth, C., Bloom, J.: A modular approach to key safeguarding. IEEE Trans. Information Theory 29(2), 208–210 (1983)
2. Blakley, G.: Safeguarding cryptographic keys. In: Proc. of AFIPS National Computer Conference (1979)
3. Chaum, D., Pedersen, T.P.: Wallet databases with observers. In: Brickell, E.F. (ed.) CRYPTO 1992. LNCS, vol. 740, pp. 89–105. Springer, Heidelberg (1993)
4. Desmedt, Y.: Some recent research aspects of threshold cryptography. In: Okamoto, E. (ed.) ISW 1997. LNCS, vol. 1396, pp. 158–173. Springer, Heidelberg (1998)
5. Desmedt, Y., Frankel, Y.: Threshold cryptosystems. In: Brassard, G. (ed.) CRYPTO 1989. LNCS, vol. 435, pp. 307–315. Springer, Heidelberg (1990)
6. Desmedt, Y., Frankel, Y.: Shared generation of authenticators and signatures. In: Feigenbaum, J. (ed.) CRYPTO 1991. LNCS, vol. 576, pp. 457–469. Springer, Heidelberg (1992)
7. Desmedt, Y., Frankel, Y.: Homomorphic zero-knowledge threshold schemes over any finite abelian group. SIAM Journal on Discrete Mathematics 7(4), 667–679 (1994)
8. ElGamal, T.: A public key cryptosystem and a signature scheme based on discrete logarithms. IEEE Trans. Information Theory 31(4), 469–472 (1985)
9. Fouque, P.A., Poupard, G., Stern, J.: Sharing decryption in the context of voting or lotteries. In: Frankel, Y. (ed.) FC 2000. LNCS, vol. 1962, pp. 90–104. Springer, Heidelberg (2001)
10. Huang, H.F., Chang, C.C.: A novel efficient (t,n) threshold proxy signature scheme. Information Sciences 176(10), 1338–1349 (2006)
11. Karnin, E.D., Greene, J.W., Hellman, M.E.: On secret sharing systems. IEEE Transactions on Information Theory 29, 35–41 (1983)
12. Kaya, K., Selçuk, A.A.: Threshold cryptography based on Asmuth–Bloom secret sharing. Information Sciences 177(19), 4148–4160 (2007)
13. Lysyanskaya, A., Peikert, C.: Adaptive security in the threshold setting: From cryptosystems to signature schemes. In: Boyd, C. (ed.) ASIACRYPT 2001. LNCS, vol. 2248, pp. 331–350. Springer, Heidelberg (2001)
14. Naccache, D., Stern, J.: A new public key cryptosystem. In: Fumy, W. (ed.) EUROCRYPT 1997. LNCS, vol. 1233, pp. 27–36. Springer, Heidelberg (1997)
15. Paillier, P.: Public key cryptosystems based on composite degree residuosity classes. In: Stern, J. (ed.) EUROCRYPT 1999. LNCS, vol. 1592, pp. 223–238. Springer, Heidelberg (1999)
16. De Santis, A., Desmedt, Y., Frankel, Y., Yung, M.: How to share a function securely? In: Proc. of STOC 1994, pp. 522–533 (1994)
17. Shamir, A.: How to share a secret? Comm. ACM 22(11), 612–613 (1979)
18. Shoup, V.: Practical threshold signatures. In: Preneel, B. (ed.) EUROCRYPT 2000. LNCS, vol. 1807, pp. 207–220. Springer, Heidelberg (2000)

Certified Encryption Revisited

Pooya Farshim[1] and Bogdan Warinschi[2]

[1] Departamento de Informática, Universidade do Minho,
Campus de Gualtar, 4710-057 Braga, Portugal
farshim@di.uminho.pt
[2] Department of Computer Science, University of Bristol,
Merchant Venturers Building, Woodland Road,
Bristol BS8 1UB, United Kingdom
bogdan@cs.bris.ac.uk

Abstract. The notion of certified encryption had recently been suggested as a suitable setting for analyzing the security of encryption against adversaries that tamper with the key-registration process. The flexible syntax afforded by certified encryption suggests that identity-based and certificateless encryption schemes can be analyzed using the models for certified encryption. In this paper we explore the relationships between security models for these two primitives and that for certified encryption. We obtain the following results.

We show that an identity-based encryption scheme is secure if and only if it is secure when viewed as a certified encryption scheme. This result holds under the (unavoidable) restriction that registration occurs over private channels. In the case of certificateless encryption we observe that a similar result cannot hold. The reason is that existent models explicitly account for attacks against the non-monolithic structure of the secret keys whereas certified encryption models treat secret keys as whole entities. We propose an extension for certified encryption where the adversary is allowed to partially modify the secret keys of honest parties. The extension that we propose is very general and may lead to unsatisfiable notions. Nevertheless, we exhibit one instantiation for which we can prove the desired result: a certificateless encryption is secure if and only if its associated certified encryption scheme is secure.

As part of our analysis, and a result of separate interest we confirm the folklore belief that for both IBE and CLE, security in the single-user setting (as captured by existent models) is equivalent to security in the multi-user setting.

Keywords: Identity-Based Encryption, Certificateless Encryption, Certified Encryption, Security Models, Corrupt Decryption.

1 Introduction

Background. Research on public-key encryption has tacitly assumed that important preconditions for widespread use of the primitive can somehow be achieved. These conditions include the existence of trusted third parties that

B. Preneel (Ed.): AFRICACRYPT 2009, LNCS 5580, pp. 179–197, 2009.

register keys, the existence of secure public directories and, of course, social acceptance. The typical envisioned solution is the use of digital certificates which can be somehow made transparent to the user. An immediate benefit of this separation of concerns is that research can concentrate on the more important/technical/difficult aspects of the primitive without the burden of explicitly considering the use of certificates. The downside is that attacks on the primitive that take advantage somehow by the steps of the protocol used to register keys are not directly captured. In turn, this may lead to insecurities in protocols where encryption is used as a building block [19].

Recent proposals that aim to address the difficulties associated to certificate management are identity-based encryption (IBE) [18,8] and certificateless encryption (CLE) [1,9,3,11]. For both of the primitives, encryption can be carried out without actually verifying a certificate: unless party ID carries out a registration process with the certification authorities, he/she would not be able to decrypt ciphertexts addressed to ID. The importance of the registration protocol to the security of the primitives is in this case clear, and indeed, attacks that use registration are captured by the models for these primitives.

A systematic way of analyzing the security of public-key encryption schemes in the presence of adversaries that can interfere with the key-registration process has recently been put forth by Boldyreva *et al.* [7]. They give a general and flexible syntax for encryption schemes and a rigorous security model that captures a variety of attacks that involve the registration process. The authors call schemes secure in their model *certified encryption* (CE) schemes. An interesting observation is that the syntax of certified encryption is so general that (with small modifications) both identity-based and certificateless encryption can be viewed as certified encryption schemes. Unfortunately this syntactic fit does not shed any light on the relation between the security models for certified encryption and those for IBE/CLE. In an ideal situation, security for an IBE/CLE scheme in the sense defined by certified encryption would be equivalent to security in some standard security model for IBE/CLE, but no formal result in this sense is known. The goal of this paper is to clarify the relations between certified encryption and standard models for IBEs/CLEs. Our results are as follows.

IBE as certified encryption. We first explore the connection between IBE and certified encryption. We show how to transform an arbitrary IBE scheme IBE into a certified encryption one IBE-2-CE(IBE), by providing an appropriate registration protocol. Such a protocol is explicitly required by the syntax of certified encryption but is only implicitly defined in IBE literature. To avoid trivial attacks, the secret key which the certification authority produces to register a user needs to be sent over private channels.

We then show that IBE satisfies indistinguishability against adaptively chosen identity and ciphertext attacks if IBE-2-CE(IBE) is a secure certified encryption scheme, if registration takes place over private channels. If registration took place over public channels, the resulting certified encryption scheme would be trivially insecure. This result shows that the security model for certified encryption captures all attacks against an IBE scheme that are also captured by existing literature.

A useful step in our analysis, and a result of independent interest, is a proof that confirms in the IBE setting, single-user security and multi-user security are equivalent notions. This result had been known for standard public key encryption [4] but the theorems there do not immediately extend to IBEs.

CLE as certified encryption with corrupt decryption. Next, we concentrate on certificateless encryption. Results for this case are complicated by the multiple deeply related but subtly different security models that have been developed for this primitive [1,9,3,11]. For our analysis we have settled on the stable models of Al-Riyami and Paterson [1] as formalized by Dent [11]. Our results extend via appropriate modifications to the other models as well.

Recall that to avoid the key-escrow problem specific to the IBE settings, in CLE schemes the secret key of parties is obtained from two basic components: one component is produced by the certification authority and is tied to the identity of the party, the second component is usually an actual secret key created by the user himself. Since secret keys are not monolithic, security models for certificateless encryption have to account for a variety of attacks which target part of the key. For example, the adversary is allowed to learn half of the key (say the component given by the CA) but should not learn any information as long as the rest of the secret key remains secret.

In the certified encryption model of [7] secret keys are viewed as monolithic entities which can be either corrupt or non-corrupt, but no partial corruption is possible. In consequence, not all attacks against CLE schemes are captured by the certified encryption models and a result similar to that for IBE is not immediately possible. Since the above attacks reflect real-life practical concerns, it is important to explore extensions of the CE model that takes these attacks into consideration. An important contribution of this paper is one such extension.

Since we work in a very general setting (in particular there is no required form for the secret keys) we provide the adversary with *indirect* access to the secret keys of parties via an abstract class of functions $\mathcal{F}$ that parameterizes the security experiment. The adversary may use these functions to modify the value of a secret key when used for decryptions. That is, the adversary can see decryptions under $f(sk)$ where $f \in \mathcal{F}$ and sk is a key of some honest party. The notion resembles that of security under related keys attacks for symmetric primitives [6]. We call the resulting model *certified encryption with corrupt decryption*.

Although this is not the most general extension one could imagine, for example $f(sk)$ could be provided directly to the adversary, the resulting model suffices to capture all attacks against certificateless encryption. In fact, the model is so strong that no certificateless scheme would be secure in this model for rather trivial reasons (which we discuss later in the paper). We therefore provide a slight relaxation which allows us to prove the desired implications: scheme CLE is secure in a standard CLE model if and only if the associated scheme CLE-2-CE(CLE) is a secure certified encryption scheme. Our result relies on a lemma that states that for certificateless encryption, security in single-user settings is equivalent to that in multi-user settings.

Related work. The importance of including the key-registration process in the analysis has been first pointed out by Shoup [19] and exemplified by Kaliski [15]. The registration protocol is explicitly modeled in several papers where security relies on parties possessing the secret keys of registered public keys, e.g. [14,5].

The first efficient and provably-secure identity-based encryption scheme, was introduced by Boneh and Franklin in [8]. An alternative scheme was given by Sakai and Kasahara [17]. The security of both of these schemes relies on random oracles. Waters [20], and later Gentry [13], proposed practical IBE schemes which are secure in the standard model of computation.

Following the original work of Al-Ryiami and Paterson [1], who proposed the concept of certificateless encryption, many other constructions and several variations of the primitive have been proposed. Cheng and Comley (CC) in [9] simplified the syntax of the primitive by integrating the full secret key algorithm into the decryption procedure. They also extended the security model which allows an adversary to extract the locally computed secret values of users. Baek, Safavi-Naini and Susilo (BSS) [3] further simplified the CLE definition by letting the users generate their public keys based on their partial private keys obtained from the certification authority. This allows for CLE schemes that do not rely on bilinear maps on elliptic curves. Furthermore, a notion of security known as Denial-of-Decryption [16], can only be achieved in the BSS formulation. In this attack, an adversary replaces the public key of a user with a value which results in valid ciphertexts, but the receiver upon decryption recovers invalid or different plaintext than those which where intended to be communicated. A good survey of certificateless public-key encryption schemes and security models has recently been provided by Dent [11].

Paper organization. In Section 2 we recall the notion of certified encryption model and motivate and explain our extension. In Section 3 we recall the relevant security notions for identity-based encryption, and prove that certified encryption models can be used to analyze IBE schemes. In Section 4 we recall the definitions and security models for certificateless encryption, motivate a relaxation for certified encryption and prove the equivalence between certified and certificateless encryption models.

2   Certified Encryption

In this section we present our extension to the notion of certified encryption of Boldyreva *et al.* [7]. We start with recalling that model.

Certified Encryption with Honest Decryption

As discussed in the introduction, this security notion for encryption explicitly takes into account the registration process and captures secrecy of plaintexts even when the adversary can tamper with the registration process.

Syntax. Formally, a certified encryption scheme is defined via a five-tuple of polynomial-time algorithms as follows.

1. $\mathsf{Setup_{CE}}(1^k)$ is a probabilistic *parameter-generation* algorithm. It takes input 1^k, where k is the security parameter, and outputs some parameters I, available to all parties. For the sake of readability we omit I from the input of the parties.
2. $\mathbb{G}_{CE}(I)$ is a probabilistic *key-generation* algorithm. It takes input a set of parameters I, and outputs a pair $(\mathsf{SK_{CA}}, \mathsf{PK_{CA}})$ consisting of a secret key and a matching public key for the certification authority.
3. $(\mathbb{C}_{CE}, \mathbb{U}_{CE})$ is a pair of interactive probabilistic algorithms forming the (two-party) public-key *registration protocol*. Algorithm $\mathbb{C}_{CE}$ takes input a secret key $\mathsf{SK_{CA}}$. Algorithm $\mathbb{U}_{CE}$ takes input the identity ID of a user and the public key $\mathsf{PK_{CA}}$ corresponding to $\mathsf{SK_{CA}}$. As result of the interaction, the output of $\mathbb{C}_{CE}$ is $(\mathsf{ID}, \mathsf{PK}, \mathsf{cert})$, where PK is a public key and cert is an issued certificate. The local output of $\mathbb{U}_{CE}$ is $(\mathsf{ID}, \mathsf{PK}, \mathsf{SK}, \mathsf{cert})$, where SK is a secret key that user uses to decrypt ciphertexts. We write

$$((\mathsf{ID}, \mathsf{PK}, \mathsf{cert}), (\mathsf{ID}, \mathsf{PK}, \mathsf{SK}, \mathsf{cert})) \leftarrow (\mathbb{C}_{CE}(\mathsf{SK_{CA}}), \mathbb{U}_{CE}(\mathsf{ID}, \mathsf{PK_{CA}}))$$

   for the result of this interaction. Either party can quit the execution prematurely, in which case the output of the party is set to $\bot$.
4. $\mathbb{E}_{CE}(\mathsf{m}, \mathsf{ID}, \mathsf{PK}, \mathsf{cert}, \mathsf{PK_{CA}})$ is a probabilistic *encryption* algorithm that takes input a message $\mathsf{m} \in \mathbb{M}_{CE}(I)$, a user's identity ID, a public encryption key PK, a certificate cert, and the authority's public key $\mathsf{PK_{CA}}$, and outputs a ciphertext $\mathsf{c} \in \{0,1\}^* \cup \{\bot\}$.
5. $\mathbb{D}_{CE}(\mathsf{c}, \mathsf{ID}, \mathsf{PK}, \mathsf{SK}, \mathsf{cert}, \mathsf{PK_{CA}})$ is a *deterministic* decryption algorithm which takes as input a ciphertext c, a user's identity ID, a secret decryption key SK, a certificate cert, the authority's public key $\mathsf{PK_{CA}}$, and outputs $\mathsf{m} \in \mathbb{M}_{CE}(I) \cup \{\bot\}$.

Correctness. A certified encryption scheme is *correct* if the decryption algorithm is the inverse of the encryption algorithm. In other words for any $\mathsf{ID} \in \{0,1\}^*$ and any $\mathsf{m} \in \mathbb{M}_{CE}(I)$ if we compute $\leftarrow \mathsf{Setup_{CE}}(1^k)$; $(\mathsf{SK_{CA}}, \mathsf{PK_{CA}}) \leftarrow \mathbb{G}_{CE}(I)$; $((\mathsf{ID}, \mathsf{PK}, \mathsf{cert}), (\mathsf{ID}, \mathsf{PK}, \mathsf{SK}, \mathsf{cert})) \leftarrow (\mathbb{C}_{CE}(\mathsf{SK_{CA}}), \mathbb{U}_{CE}(\mathsf{ID}, \mathsf{PK_{CA}}))$; $\mathsf{c} \leftarrow \mathbb{E}_{CE}(\mathsf{m}, \mathsf{ID}, \mathsf{PK}, \mathsf{cert}, \mathsf{PK_{CA}})$; $\mathsf{m}' \leftarrow \mathbb{D}_{CE}(\mathsf{c}, \mathsf{ID}, \mathsf{PK}, \mathsf{SK}, \mathsf{cert}, \mathsf{PK_{CA}})$ then $\mathsf{m}' = \mathsf{m}$

Security. The security of a certified encryption scheme CE is defined through the experiments in Figure 1. Experiment $\mathbf{Exp}_{CE,\mathcal{A},b}^{mCE\text{-}CCA\text{-}I}(1^k)$ models the situation where the certification authority is honest. Experiment $\mathbf{Exp}_{CE,\mathcal{A},b}^{mCE\text{-}CCA\text{-}M}(1^k)$ models the situation where the authority is corrupt. Both experiments involve an adversary $\mathcal{A}$ (which may run in multiple stages) and is parameterized by bit b and maintain two lists $\mathsf{RegListPub}$ and $\mathsf{RegListSec}$ used to store public and secret information pertaining to users. The adversary can read the content of $\mathsf{RegListPub}$. The adversaries in the experiments have access to a set of oracles $\mathcal{O}$ that provides the adversaries with the following capabilities. Each oracle corresponds to a type of queries that the adversary can issue. These queries are the following:

- Register(ID, L): When this query is issued, with parameters some identity ID and a label L ∈ {honest, corrupt} the oracle answers as follows. If L = honest then the registration protocol is executed internally by the oracle, i.e. $((\text{ID}', \text{PK}', \text{cert}'), (\text{ID}, \text{PK}, \text{SK}, \text{cert})) \leftarrow (\mathbb{C}_{CE}(\text{SK}_{CA}), \mathbb{U}_{CE}(\text{ID}, \text{PK}_{CA}),$ the entry (ID, PK, cert) is stored in RegListPub and (ID, PK, SK, cert) is stored in RegListSec. If L = corrupt then the registration protocol is executed with the adversary playing the role of the ID (i.e. running a corrupt version of $\mathbb{U}_{CE}$). At the end of the execution, when $\mathbb{C}_{CE}$ outputs some values (ID, PK, cert), the tuple (ID, PK, cert) in stored in RegListPub.
- Encrypt(m_0, m_1, ID, PK, cert): On input two messages m_0 and m_1 and an a tuple (ID, PK, cert), this oracle returns the ciphertext $\mathbb{E}_{CLE}(m_b, \text{ID}, \text{PK}, \text{cert}, \text{PK}_{CA})$.
- Decrypt(c, ID, PK, cert): If the entry (ID, PK, SK, cert) for some SK occurs in RegListSec, then the oracle returns $\mathbb{D}_{CE}(c, \text{ID}, \text{PK}, \text{SK}, \text{cert}, \text{PK}_{CA})$. Otherwise, it returns ⊥.

$$\textbf{Exp}_{CE,\mathcal{A},b}^{mCE\text{-}CCA\text{-}I}(1^k)$$
1. $I \leftarrow \text{Setup}_{CE}(1^k)$
2. $(\text{SK}_{CA}, \text{PK}_{CA}) \leftarrow \mathbb{G}_{CE}(I)$
3. $b' \leftarrow \mathcal{A}_1^{\mathcal{O}}(I, \text{PK}_{CA}, \text{st})$
4. Return b'

$$\textbf{Exp}_{CE,\mathcal{A},b}^{mCE\text{-}CCA\text{-}M}(1^k)$$
1. $I \leftarrow \text{Setup}_{CE}(1^k)$
2. $(\text{PK}_{CA}, \text{st}) \leftarrow \mathcal{A}_0(I)$
3. $b' \leftarrow \mathcal{A}_1^{\mathcal{O}}(\text{st})$
4. Return b'

Fig. 1. Experiments for defining security for a certified encryption scheme CE against chosen-ciphertext attacks, in the private channels model. Here st is some state information. Adversary $\mathcal{A}$ is required not to cause any of the following events. 1) Calling Register with L = corrupt if CA was corrupted; 2) Calling Encrypt on two messages with unequal lengths, or using a tuple (ID, PK, cert) ∈ RegListPub if it also placed a Register(ID, corrupt) query, or if CA is corrupt, with a tuple (ID, PK, cert) ∉ RegListPub; 3) Placing Decrypt(c, ID, PK, cert) with the ciphertext c previously received from the left-right encryption oracle on a query involving (ID, PK, cert).

We say that a certified encryption scheme CE is secure against type x adversaries (x = I is for the case when the CA is honest throughout the attack, and x = M indicates that the CA is corrupted) with chosen-ciphertext capabilities, if the advantage of any probabilistic polynomial-time (ppt) adversary $\mathcal{A}$, defined by:

$$\textbf{Adv}_{CE,\mathcal{A}}^{mCE\text{-}CCA\text{-}x}(k) := \Pr\left[\textbf{Exp}_{CE,\mathcal{A},1}^{mCE\text{-}CCA\text{-}x}(1^k) = 1\right] - \Pr\left[\textbf{Exp}_{CE,\mathcal{A},0}^{mCE\text{-}CCA\text{-}x}(1^k) = 1\right]$$

is a negligible function.

Remark1. All of the security notions regarding encryption schemes in this paper use the indistinguishability definitional paradigm. To unclutter notation we do not explicitly indicate this in our notation. Also, in the experiment above we use mCE in our notation (as opposed to CE) to indicate that our model for certified encryption schemes is multi-user (the adversary has multiple left-right queries for potentially different identities/public keys).

Remark2. Our model is concerned specifically with schemes where registration takes place over secure channels: the transcript of the registration protocol executed by an honest party is not returned to the adversary. A model for public channels can then be obtained by giving these transcripts to the adversary.

Remark3. The original model of Boldyreva *et al.* only considered one experiment where the adversary chooses in the beginning to corrupt the CA or not. For clarity, we chose to have two separate experiments, one for each possibility. A scheme is secure in the sense of [7] iff it is secure in type I and type M models.

Certified Encryption with Corrupted Decryption

In this section we describe the extension that we propose to the certified encryption model. We are motivated by the observation that an important class of attacks is not captured by the model above. Indeed, the adversary lacks the ability to tamper the secret keys of honest parties since once an honest party is registered the adversary can only involve that party's secret key in decryptions (via the decryption oracle) but can do nothing more. Yet, scenarios of practical interest where such attacks may occur can be easily envisioned. Consider for example the setting of certificateless encryption where the secret keys of party ID has two components SK and D. The first component is generated by the user and the second one by a trusted authority. One can imagine a setting where the two parts of the key are stored separately (e.g. D a trusted card and SK stored on a computer) and the adversary may tamper with one of the two (by gaining access to only one of the two devices). Decryptions performed by the user now involve the modified key.

Next we present an extension of the certified encryption model which includes extra power for the adversary. We are trying to be as general as possible (and not concentrate for example on the case where secret keys are as in certificateless encryption). We allow the adversary to apply any function f from a set $\mathcal{F}$ to the secret keys of parties, before the keys are used for decryption. Formally, we replace the query $\mathsf{Decrypt}(\mathsf{c}, \mathsf{ID}, \mathsf{PK}, \mathsf{cert})$ that adversary $\mathcal{A}$ is allowed to issue in with a new query $\mathsf{MalDecrypt}(f, \mathsf{c}, \mathsf{ID}, \mathsf{PK}, \mathsf{cert})$ with $f \in \mathcal{F}$. When this query is issued, the oracle searches the list $\mathsf{RegListSec}$ for an entry $(\mathsf{ID}, \mathsf{PK}, \mathsf{cert}, \mathsf{SK})$. If no such entry exists the oracle returns $\perp$. Otherwise the oracle returns m computed as: $\mathsf{m} \leftarrow \mathbb{D}_{\mathsf{CE}}(\mathsf{c}, f(\mathsf{ID}, \mathsf{PK}, \mathsf{SK}, \mathsf{cert}), \mathsf{PK}_{\mathsf{CA}})$.

The resulting model is quite flexible in the abilities that the adversary can gain. However, due to the high level of abstraction it is difficult to describe which are the trivial attacks that the adversary is not allowed to perform. In this paper, and for the specific function sets $\mathcal{F}$ that we use, the following restrictions on this oracle is imposed.

- $\mathcal{A}$ is not allowed to query the $\mathsf{MalDecrypt}$ oracle on $(f, \mathsf{c}, \mathsf{ID}, \mathsf{PK}, \mathsf{cert})$ with $f = \mathrm{Id}$ and a ciphertext c which was previously received from the left-right encryption oracle on a query involving $(\mathsf{ID}, \mathsf{PK}, \mathsf{cert})$.

- $\mathcal{A}$ is not allowed to query the MalDecrypt oracle on $(f, \mathsf{c}, \mathsf{ID}, \mathsf{PK}, \mathsf{cert})$ with an f such that $f(\mathsf{ID}, \mathsf{PK}, \mathsf{SK}, \mathsf{cert}) = (\mathsf{ID}, \mathsf{PK}, \mathsf{SK}', \mathsf{cert})$ with $\mathsf{SK} \neq \mathsf{SK}'$ and ciphertext c previously received from the left-right encryption oracle on a query involving $(\mathsf{ID}, \mathsf{PK}, \mathsf{cert})$.

The second condition is imposed as the CLE model does not allow replacing the public key of the challenge value with the *same* value as that used to generate the challenge ciphertext but providing a new SK.

We say that scheme CE is secure against chosen-ciphertext attacks with $\mathcal{F}$-decryptions if for any ppt adversary $\mathcal{A}$, its advantage defined in the usual way is a negligible function.

Remark1. Our model generalizes the one of the previous section: if $\mathcal{F}$ contains only the identity function, the extended model is not more stringent than the previous one. Similarly, if we set $\mathcal{F}$ to be the empty set we obtain CPA attacks.

Remark2. The functions in $\mathcal{F}$ are fixed, and do not adaptively change. One interesting direction is to consider a stronger model where the functions have as parameter the local state of the adversary. For example, this would allow f to depend on the secret key of the CA when the CA is corrupt.

Remark3. The extended model allows the adversary to modify the keys that honest parties use for decryption, but does not allow it to learn these keys (or parts of these keys) adaptively. Another plausible way to strengthen the model that we propose would be to add adversarial capabilities that account for these possibilities. We do not pursue this direction in this paper since the model that we have is sufficient to capture security for both IBE and CLE schemes.

3   Identity-Based Encryption as Certified Encryption

In this section we recall the syntax and the relevant security models for identity-based encryption. Our presentation mainly follows [8]. We also introduce a multi-user security model and prove the folklore result that the single-user security is equivalent to multi-user security. Then, we show that an IBE scheme secure in the certified encryption sense is secure in a standard model for IBEs.

In the identity-based setting, parties use their unstructured identities as their own public keys, and have associated secret keys generated by a trusted third party. The formal definition of an identity-based encryption scheme IBE can be found in [8].

3.1   Single-User Security for IBE schemes

Security. Security of the identity-based encryption scheme IBE is defined, as usual, via indistinguishability games. In Figure 2 we describe experiments for chosen-ciphertext attacks. The adversary in those experiments have access to a set of oracles $\mathcal{O}$ that formalizes its various capabilities. Both experiments involve an adversary $\mathcal{A}$ and are parameterized by a bit $b \in \{0, 1\}$ which the adversary aims to determine. The adversary has access to a key extraction oracle Extract and a decryption oracle Decrypt, both parameterized by $(I, \mathsf{Mpk}, \mathsf{Msk})$. The oracle process the following queries:

- Extract(ID): This oracle on input an identity ID returns $\mathbb{X}_{\mathsf{IBE}}(\mathsf{ID}, \mathsf{Msk})$.
- Decrypt(c, ID): This oracle on input a ciphertext/identity pair $(\mathsf{c}, \mathsf{ID})$ answers with $\mathbb{D}_{\mathsf{IBE}}(\mathsf{c}, \mathsf{ID}, \mathsf{D})$ where $\mathsf{D} \leftarrow \mathbb{X}_{\mathsf{IBE}}(\mathsf{ID}, \mathsf{Msk})$.

$\mathbf{Exp}_{\mathsf{IBE},\mathcal{A},b}^{\mathsf{ID\text{-}CCA}}(1^k)$
1. $I \leftarrow \mathsf{Setup}_{\mathsf{IBE}}(1^k)$
2. $(\mathsf{Msk}, \mathsf{Mpk}) \leftarrow \mathbb{G}_{\mathsf{IBE}}(I)$
3. $(\mathsf{m}_0, \mathsf{m}_1, \mathsf{ID}^*, \mathsf{st}) \leftarrow \mathcal{A}_1^{\mathcal{O}}(I, \mathsf{Mpk})$
4. $\mathsf{c}^* \leftarrow \mathbb{E}_{\mathsf{IBE}}(\mathsf{m}_b, \mathsf{ID}^*, \mathsf{Mpk})$
5. $b' \leftarrow \mathcal{A}_2^{\mathcal{O}}(\mathsf{c}^*, \mathsf{st})$
6. Return b'

$\mathbf{Exp}_{\mathsf{IBE},\mathcal{A},b}^{\mathsf{mID\text{-}CCA}}(1^k)$
1. $I \leftarrow \mathsf{Setup}_{\mathsf{IBE}}(1^k)$
2. $(\mathsf{Msk}, \mathsf{Mpk}) \leftarrow \mathbb{G}_{\mathsf{IBE}}(I)$
3. $b' \leftarrow \mathcal{A}^{\mathcal{O}}(I, \mathsf{Mpk})$
4. Return b'

Fig. 2. Experiments for defining security of identity-based scheme IBE against chosen-ciphertext attacks. In both experiments, st is some state information. We require that messages m_0 and m_1 be of equal length. The model on the left if for the single-user setting. Here, we require that adversary $\mathcal{A}$ does not cause any of the following events. 1) Querying Extract on ID^*; 2) $\mathcal{A}_2$ querying Decrypt$(\mathsf{c}^*, \mathsf{ID}^*)$. The model on the right is for the multi-user setting. Here we require that adversary $\mathcal{A}$ does not cause any of the following events. 1) Calling Encrypt on two messages with unequal lengths; 2) Querying Extract on any identity sent to the left-right encryption oracle; 3) $\mathcal{A}$ querying Decrypt on a pair $(\mathsf{c}, \mathsf{ID})$ with c previously received from the left-right encryption oracle on a query involving ID.

We say that scheme IBE is secure against chosen-ciphertext attacks in the single-user setting (or ID-CCA secure) if the advantage of any ppt adversary $\mathcal{A}$ defined in the usual way is a negligible function.

3.2 Multi-user Security for IBE Schemes

The security model for IBE schemes that we presented in above was a single-user one: the adversary targets only a single identity to attack and receives only one challenge ciphertext under this identity. It is more realistic to analyze the security of a scheme in the *multi-user* setting, where many challenge ciphertexts on many adversarially-chosen identities are available. We formalize this model via the experiments shown in Figure 2 on the right.

The experiment involves an adversary $\mathcal{A}$ and is parameterized by a bit $b \in \{0, 1\}$ which the adversary aims to determine. In addition to the oracles Extract and Decrypt (as defined for the single-user setting) the adversary also has access to a left-right encryption oracle with access to $(I, \mathsf{Mpk}, \mathsf{Msk})$. This oracle behaves as follows. On a query Encrypt$(\mathsf{m}_0, \mathsf{m}_1, \mathsf{ID})$, with m_0 and m_1 bit-strings of equal length, and identity ID, the oracle returns the ciphertext $\mathbb{E}_{\mathsf{IBE}}(\mathsf{m}_b, \mathsf{ID}, \mathsf{Mpk})$.

We say that an identity-based encryption scheme IBE is secure against chosen-ciphertext attacks in the multi-user setting (or mID-CCA secure) if the advantage of any ppt adversary $\mathcal{A}$, defined in the usual way is negligible. Although security in the multi-user setting seems a stronger requirement we prove that, as for the case of standard public-key encryption, it is equivalent to security in the single-user setting.

Lemma 1. *Identity base encryption scheme* IBE *is secure in the multi-user setting if and only if it is secure in the single-user setting.*

The proof is by a standard hybrid argument and is given in the full version of the paper.

We show that for any ppt adversary $\mathcal{A}$ against an IBE scheme IBE in the multi-user sense, there exists a ppt adversary $\mathcal{B}$ against the scheme in the single-user setting such that:

$$\mathbf{Adv}_{\mathsf{IBE},\mathcal{A}}^{\mathsf{mID\text{-}CCA}}(k) \leq Q_{\mathsf{ID}}(k) \cdot Q_{\mathsf{E}}(k) \cdot \mathbf{Adv}_{\mathsf{IBE},\mathcal{B}}^{\mathsf{ID\text{-}CCA}}(k).$$

Here $Q_{\mathsf{ID}}(k)$ denotes the number of different identities sent to the left-right encryption oracle, and $Q_{\mathsf{E}}(k)$ denotes the maximum number of left-right encryption queries per identity.

3.3   The IBE-2-CE Transformation

In this section we explain how any identity-based encryption scheme naturally gives rise to an *associated* identity-based certified encryption scheme. Then, we prove that if the resulting scheme is a secure certified encryption scheme, then the original scheme was a secure identity-based encryption scheme.

Fix an arbitrary identity-based encryption scheme IBE. The associated certified encryption scheme IBE-2-CE(IBE) is given as follows. $\mathsf{Setup}_{\mathsf{CE}}$, $\mathbb{G}_{\mathsf{CE}}$, $\mathbb{E}_{\mathsf{CE}}$ and $\mathbb{D}_{\mathsf{CE}}$ are identical to $\mathsf{Setup}_{\mathsf{IBE}}$, $\mathbb{G}_{\mathsf{IBE}}$, $\mathbb{E}_{\mathsf{IBE}}$ and $\mathbb{D}_{\mathsf{IBE}}$, respectively, and the registration protocol is defined as follows.

1. $\mathbb{U}_{\mathsf{CE}}(\mathsf{ID}, \mathsf{PK}_{\mathsf{CA}})$: Sends ID to the CA;
2. $\mathbb{C}_{\mathsf{CE}}(\mathsf{SK}_{\mathsf{CA}})$: Receives ID, runs $\mathsf{D} \leftarrow \mathbb{X}_{\mathsf{IBE}}(\mathsf{ID}, \mathsf{SK}_{\mathsf{CA}})$, and sends D to user ID. It outputs $(\mathsf{ID}, \epsilon, \epsilon)$ locally and terminates;
3. $\mathbb{U}_{\mathsf{CE}}(\mathsf{ID}, \mathsf{PK}_{\mathsf{CA}})$: Receives D, outputs $(\mathsf{ID}, \epsilon, \mathsf{D}, \epsilon)$ locally and terminates.

The registration process consists in the user sending his identity ID to the certification authority who extracts the secret key associated to ID and sends it to the user. The first main result of this paper is that identity-based encryption scheme can be analyzed in the models for certified encryption.

Theorem 1. *Let* IBE *be an arbitrary identity-based encryption scheme and set* CE := IBE-2-CE(IBE). *Then* IBE *is secure if and only if* CE *is a secure certified-encryption scheme with CA honest.*

The proof of the theorem can be found in the full version of the paper where we show that for any ppt adversary $\mathcal{A}$ against IBE in the single-user setting there exists a ppt type I adversary $\mathcal{B}$ against CE such that:

$$\mathbf{Adv}_{\mathsf{IBE},\mathcal{A}}^{\mathsf{ID\text{-}CCA}}(k) \leq Q_{\mathsf{ID}}(k) \cdot \mathbf{Adv}_{\mathsf{CE},\mathcal{B}}^{\mathsf{mCE\text{-}CCA\text{-}I}}(k).$$

Here $Q_{\mathsf{ID}}(k)$ denotes the number of distinct identities queried during the ID-CCA experiment, that is the number of distinct identities sent to the Extract or the Decrypt oracle, together with ID^*. Conversely, for any ppt adversary $\mathcal{A}$ against CE there exists ppt adversary $\mathcal{B}$ against IBE in the multi-user setting such that:

$$\mathbf{Adv}_{\mathsf{CE},\mathcal{A}}^{\mathsf{mCE\text{-}CCA\text{-}I}}(k) \leq \mathbf{Adv}_{\mathsf{IBE},\mathcal{B}}^{\mathsf{mID\text{-}CCA}}(k).$$

We note that the security model for IBE schemes that we have presented has an extraction oracle which computes a secret key with fresh random coins for an identity which is re-submitted to this oracle. This is in line with the certified security model where a user can invoke multiple runs of the register protocol on an identity. In the setting where only one private key can be extracted, the CE should be modified so that it only allows the users to register once.

4   Certificateless Encryption as Certified Encryption

In this section we recall the syntax and the relevant security models for certificateless encryption. The security models that we use are those of Al-Ryiami and Paterson [1], as described by Dent [11]. We also introduce a multi-user security model for the primitive, and prove the folklore theorem that single-user security is equivalent to multi-user security. Towards proving that certificateless encryption can be viewed and analyzed using the certified encryption model we introduce a slight weakening of the latter. We then prove that a certificateless scheme is secure if and only if it is secure as a certified encryption scheme.

Syntax. In a certificateless encryption scheme users' public keys consist of their identity and a user-generated public value. A recipient uses two partial secret values, corresponding to it identity and public value, to decrypt ciphertexts. Formally, a certificateless encryption scheme CLE is specified by six polynomial-time algorithms as follows.

1. $\mathsf{Setup}_{\mathsf{CLE}}(1^k)$. A probabilistic *setup* algorithm, which takes as input the security parameter 1^k and returns the descriptions of underlying groups, message space $\mathbb{M}_{\mathsf{CLE}}(I)$ and ciphertext space $\mathbb{C}_{\mathsf{CLE}}(I)$. This algorithm is executed by the key-generation center (KGC), which publishes I. We assume that I is available to all parties and do not include it explicitly in the various algorithms that define the scheme.
2. $\mathbb{G}_{\mathsf{CLE}}(I)$. A probabilistic algorithm for *KGC key-generation* which outputs a master secret key, master public key pair (Msk, Mpk).
3. $\mathbb{X}_{\mathsf{CLE}}(\mathsf{ID}, \mathsf{Msk})$. A probabilistic algorithm for *partial private key extraction* which takes as input an identifier string $\mathsf{ID} \in \{0,1\}^*$, the master secret key Msk, and returns a partial secret key D. This algorithm is run by the KGC, after verifying the user's identity.

4. $\mathbb{U}_{\mathsf{CLE}}(\mathsf{ID}, \mathsf{Mpk})$ A probabilistic algorithm for *user key-generation* which takes an identity and the master public key, and outputs a secret value SK and a public key PK. This algorithm is run by a user to obtain a public key and a secret value which can be used to construct a full private key. The public key is published without certification.

5. $\mathbb{S}_{\mathsf{CLE}}(\mathsf{SK}, \mathsf{D}, \mathsf{Mpk})$. A probabilistic algorithm for *full secret key extraction* which takes a secret value SK and a partial private key D as well as the master public key and returns a full secret key S.

6. $\mathbb{E}_{\mathsf{CLE}}(\mathsf{m}, \mathsf{ID}, \mathsf{PK}, \mathsf{Mpk})$. This is the probabilistic *encryption* algorithm. On input of a message $\mathsf{m} \in \mathbb{M}_{\mathsf{CLE}}(I)$, receiver's identifier ID, the receiver's public key PK, and the master public key Mpk, this algorithm outputs a ciphertext $\mathsf{c} \in \mathbb{C}_{\mathsf{CLE}}(I)$ or an error symbol $\bot$.

7. $\mathbb{D}_{\mathsf{CLE}}(\mathsf{c}, \mathsf{ID}, \mathsf{PK}, \mathsf{S}, \mathsf{Mpk})$. This is the deterministic decryption algorithm. On input of a ciphertext c, an identity ID, a public key PK, the receiver's full private key S, and Mpk this algorithm outputs a message m or $\bot$.

The correctness of a CLE scheme is defined in the usual way.

4.1   Single-User Security for CLE Schemes

We next recall two different security models for the privacy of encrypted plaintexts. In line with previous literature we classify the attackers as of type I, and M. Intuitively, the attackers in these models correspond to the following usage scenarios for a certificateless encryption scheme. A type I attack corresponds to the case where the key-generation center is honest; a type M attack corresponds to the case where the key-generation center is dishonest, and in particular, may generate its public/secret key pair not following the prescribed key-generation algorithm. In both models the adversaries are allowed to obtain partial secret keys (i.e. D that correspond to identity ID), secret values (i.e. SK that correspond to identity ID), replace public keys of parties, etc. For a more elaborate discussion on the various existent models for certificateless encryption we refer the reader to [11].

Experiments that define security against these attackers are given in Figure 3. Both experiments are parameterized by bit $b \in \{0, 1\}$ and involve an adversary $\mathcal{A}$. The adversaries in those experiments have access to a set of oracle $\mathcal{O}$ that formalizes its various capabilities. The oracle maintains internally two lists Real and Fake. The entries of both lists are of the form $(\mathsf{ID}, \mathsf{PK}, \mathsf{SK})$ and maintain the secret keys that correspond to the public key of identity ID. List Real keeps track of the "real" public/secret keys of parties and its entries are not modified during the execution. The list Fake maintains keys associated to identities, but we allow the adversary to modify them. In all experiments the oracles are parameterized by $(I, \mathsf{Mpk}, \mathsf{Msk})$, where we assume $\mathsf{Msk} = \bot$ in type M models. The queries that the adversary can make to the oracle are as follows.

- ReqPK$(\mathsf{ID}, \mathsf{D})$: This query has as parameter an identity ID and a partial private key D and is processed as follows. In experiment against a type I adversary the query is processed as follows: The oracle checks the list Real

$\mathbf{Exp}_{CLE,\mathcal{A},b}^{CL\text{-}CCA\text{-}I}(1^k)$

  1. $I \leftarrow \mathsf{Setup}(1^k)$
  2. $(\mathsf{Msk}, \mathsf{Mpk}) \leftarrow \mathbb{G}_{CLE}(I)$
  3. $(\mathsf{m}_0, \mathsf{m}_1, \mathsf{ID}^*, \mathsf{st}) \leftarrow \mathcal{A}_1^{\mathcal{O}}(I, \mathsf{Mpk})$
  4. $\mathsf{c}^* \leftarrow \mathbb{E}_{CLE}(\mathsf{m}_b, \mathsf{ID}^*, \mathsf{PK}^*, \mathsf{Mpk})$
  5. $b' \leftarrow \mathcal{A}_2^{\mathcal{O}}(\mathsf{c}^*, \mathsf{st})$
  6. Return b'

$\mathbf{Exp}_{CLE,\mathcal{A},b}^{CL\text{-}CCA\text{-}M}(1^k)$

  1. $(I) \leftarrow \mathsf{Setup}_{CLE}(1^k)$
  2. $(\mathsf{Mpk}, \mathsf{st}) \leftarrow \mathcal{A}_0(I)$
  3. $(\mathsf{m}_0, \mathsf{m}_1, \mathsf{ID}^*, \mathsf{st}) \leftarrow \mathcal{A}_1^{\mathcal{O}}(\mathsf{st})$
  4. $\mathsf{c}^* \leftarrow \mathbb{E}_{CLE}(\mathsf{m}_b, \mathsf{ID}^*, \mathsf{PK}^*, \mathsf{Mpk})$
  5. $b' \leftarrow \mathcal{A}_2^{\mathcal{O}}(\mathsf{c}^*, \mathsf{st})$
  6. Return b'

Fig. 3. Experiments for defining type I and M security for certificateless encryption scheme CLE. Here st is some state information and we require the messages m_0 and m_1 to be of equal length. Encryption in stage 5 is performed with respect to the public key PK^* associated with ID^* on the Fake list. In all of the above experiments the adversary $\mathcal{A}$ is required not to cause any of the following events. 1) It places both an ExtractFSK query on ID^*; 2) $\mathcal{A}_2$ placing a Decrypt query on $(\mathsf{L}, \mathsf{c}^*, \mathsf{ID}^*)$ and the current public key of ID^* retrieved from the list associated to L is PK^*.

for an entry of the form $(\mathsf{ID}, \mathsf{PK}, \mathsf{SK}, \mathsf{D}', \mathsf{S})$. If such an entry exists it returns PK. Otherwise, it 1) Executes $(\mathsf{PK}, \mathsf{SK}) \leftarrow \mathbb{U}_{CLE}(\mathsf{ID}, \mathsf{Mpk})$; 2) Sets $\mathsf{D} \leftarrow \mathbb{X}_{CLE}(\mathsf{ID}, \mathsf{Msk})$; 3) Computes $\mathsf{S} \leftarrow \mathbb{S}_{CLE}(\mathsf{SK}, \mathsf{D}, \mathsf{Mpk})$; and 4) Adds $(\mathsf{ID}, \mathsf{PK}, \mathsf{SK}, \mathsf{D}, \mathsf{S})$ to both the Real and Fake lists. It then returns PK to the adversary. Note that the provided D is not used in this case.

In the experiment against a type M adversary the query is processed as follows: The oracle checks the list Real for an entry of the form $(\mathsf{ID}, \mathsf{PK}, \mathsf{SK}, \mathsf{D}', \mathsf{S})$. If such an entry exists, it replaces D' with D and returns PK. Otherwise, it 1) Executes $(\mathsf{PK}, \mathsf{SK}) \leftarrow \mathbb{U}_{CLE}(\mathsf{ID}, \mathsf{Mpk})$; 2) Computes $\mathsf{S} \leftarrow \mathbb{S}_{CLE}(\mathsf{SK}, \mathsf{D}, \mathsf{Mpk})$; and 3) Adds $(\mathsf{ID}, \mathsf{PK}, \mathsf{SK}, \mathsf{D}, \mathsf{S})$ to the Real and Fake lists and returns PK.

- ReplacePK$(\mathsf{ID}, \mathsf{PK}, \mathsf{SK})$: This query has as parameters an identity ID, a public key PK, and a secret key SK. When such a query is issued, the oracle searches the Fake lit for an entry $(\mathsf{ID}, \mathsf{PK}', \mathsf{SK}', \mathsf{D}, \mathsf{S})$ and replaces it with $(\mathsf{ID}, \mathsf{PK}, \mathsf{SK}, \mathsf{D}, \mathsf{S})$. We assume, wlog, that the adversary has previously performed a ReqPK query involving ID so that such an entry always exits.
- ExtractPSK(ID): The oracle returns the D component from the Real list. We assume ReqPK has already been called on ID.
- ExtractFSK(ID): The oracle returns the S component from the Real list. We assume ReqPK has already been called on ID.
- Decrypt$(\mathsf{L}, \mathsf{c}, \mathsf{ID})$: This oracle has as parameters $\mathsf{L} \in \{\mathsf{fake}, \mathsf{real}\}$, an identity ID and a ciphertext c. On input $(\mathsf{real}, \mathsf{c}, \mathsf{ID})$ searches the list Real for an entry $(\mathsf{ID}, \mathsf{PK}, \mathsf{SK}, \mathsf{D}, \mathsf{S})$. and on input $(\mathsf{fake}, \mathsf{c}, \mathsf{ID})$ the oracle finds an entry $(\mathsf{ID}, \mathsf{PK}, \mathsf{SK}, \mathsf{D}, \mathsf{S})$ in Fake. It outputs $\mathsf{m} \leftarrow \mathbb{D}_{CLE}(\mathsf{c}, \mathsf{ID}, \mathsf{PK}, \mathsf{S}, \mathsf{Mpk})$. We assume that prior to any query to the decryption oracle that contains ID, the adversary made at some point a query ReqPK$(\mathsf{ID}, \mathsf{D})$.

The experiments that we consider are summarized in Figure 3. In addition to the restrictions that we outline there, we define three specific classes of adversaries. Type I and M adversaries correspond to adversaries in the literature. Type I* adversary is a variant of type I which is useful in deriving our later results. These classes are defined by the following additional restrictions:

- *Type* I: Adversary Both $\mathcal{A}_1$ does not issue a ReplacePK(ID*, PK, SK) query and $\mathcal{A}$ does not place an ExtractPSK(ID*) query;
- *Type* I*: Adversary $\mathcal{A}$ does not issue an ExtractPSK(ID*) query;
- *Type* M: Adversary $\mathcal{A}$ does not issue a ReplacePK or an ExtractPSK query.

For x $\in$ {I, I*, M}, we say that a certificateless encryption scheme CLE is type x secure in the single-user CCA setting if for all ppt adversaries $\mathcal{A}$ its advantage defined below is a negligible function.

$$\mathbf{Adv}_{\mathsf{CLE},\mathcal{A}}^{\mathsf{CL-CCA-x}}(k) := \Pr\left[\mathbf{Exp}_{\mathsf{CLE},\mathcal{A},1}^{\mathsf{CL-CCA-x}}(1^k) = 1\right] - \Pr\left[\mathbf{Exp}_{\mathsf{CLE},\mathcal{A},0}^{\mathsf{CL-CCA-x}}(1^k) = 1\right].$$

4.2   Multi-user Security for CLE Schemes

Analogously to the identity-based setting, we extend the security models for certificateless encryption schemes to multi-user scenario in two different dimensions. First, we consider a setting where parties may possess more than one public key, and second, the adversary may receive multiple challenge ciphertexts. The experiments defining security are give in Figure 4.

$\mathbf{Exp}_{\mathsf{CLE},\mathcal{A},b}^{\mathsf{mCL-CCA-I}}(1^k)$
1. $I \leftarrow \mathsf{Setup}_{\mathsf{CLE}}(1^k)$
2. $(\mathsf{Msk}, \mathsf{Mpk}) \leftarrow \mathbb{G}_{\mathsf{CLE}}(I)$
3. $b' \leftarrow \mathcal{A}^{\mathcal{O}}(I, \mathsf{Mpk})$
4. Return b'

$\mathbf{Exp}_{\mathsf{CLE},\mathcal{A},b}^{\mathsf{mCL-CCA-M}}(1^k)$
1. $I \leftarrow \mathsf{Setup}_{\mathsf{CLE}}(1^k)$
2. $(\mathsf{Mpk}, \mathsf{st}) \leftarrow \mathcal{A}_0(I)$
3. $b' \leftarrow \mathcal{A}_1^{\mathcal{O}}(\mathsf{st})$
4. Return b'

Fig. 4. Experiments for defining type I and M security for a certificateless encryption scheme CLE in the multi-user setting. Here st is some state information. For the experiment against type I adversary $\mathcal{A}$, the adversary is required not to cause any of the following events. 1) Calling Encrypt on two messages with unequal lengths; 2) Calling ExtractFSK on an identity submitted to the Encrypt oracle; 3) Calling Decrypt query on (L, c, ID, PK) with c previously received from a left-right encryption query involving the pair (ID, PK) and such that the public key associated to ID on the list characterized by L is still PK.

We give separate experiments for type I and type M attackers. Both experiments involve an adversary $\mathcal{A}$ and are parameterized by bit $b \in \{0, 1\}$ that the adversary aims to determine. The oracles available to the adversary include those in the single-user setting. In addition, the adversary has access to a left-right encryption oracle, which is parameterized by b and has access to $(I, \mathsf{Mpk}, \mathsf{Msk})$. When the left-right oracle receives a query Encrypt(m_0, m_1, ID, PK) with m_0 and m_1 equal length messages, this oracle returns the ciphertext $\mathbb{E}_{\mathsf{CLE}}(m_b, \mathsf{ID}, \mathsf{PK}, \mathsf{Mpk})$. Also, the queries to the ReqPK oracle trigger the generation of a new secret value/private key each time an identity is submitted, even if this identity had been previously submitted as a query.

We say that a certificateless encryption scheme CLE is type x secure for x $\in$ {I, I*, M} in the multi-user CCA setting if for all ppt adversaries $\mathcal{A}$ of type x, its advantage defined below is a negligible function.

$$\mathbf{Adv}_{\mathsf{CLE},\mathcal{A}}^{\mathsf{mCL\text{-}CCA\text{-}x}}(k) := \Pr\left[\mathbf{Exp}_{\mathsf{CLE},\mathcal{A},1}^{\mathsf{mCL\text{-}CCA\text{-}x}}(1^k) = 1\right] - \Pr\left[\mathbf{Exp}_{\mathsf{CLE},\mathcal{A},0}^{\mathsf{mCL\text{-}CCA\text{-}x}}(1^k) = 1\right].$$

The power afforded to the adversary in the multi-user extension seems quite extensive. Nevertheless, we can still show that security in the single-user setting is equivalent to security in the multi-user setting. Since one implication is obvious, we only state and prove the more difficult direction.

Lemma 2. *Let* CLE *be an arbitrary certificateless encryption scheme secure against type* x *adversaries in the single-user* CCA *setting (for some* x $\in$ {I, I*, M}*). Then,* CLE *is secure against type* x *adversaries in the multi-user setting* CCA.

We prove the lemma in the full version of the paper where we show that for any ppt type x adversary $\mathcal{A}$ against a CLE scheme CLE in the multi-user setting, there exists a ppt type x adversary $\mathcal{B}$ against the scheme in the single-user setting such that:

$$\mathbf{Adv}_{\mathsf{CLE},\mathcal{A}}^{\mathsf{mCL\text{-}CCA\text{-}x}}(k) \leq Q_{\mathsf{ID}}(k) \cdot Q_{\mathsf{PK}}(k) \cdot Q_{\mathsf{E}}(k) \cdot \mathbf{Adv}_{\mathsf{CLE},\mathcal{B}}^{\mathsf{CL\text{-}CCA\text{-}x}}(k).$$

Here $Q_{\mathsf{ID}}(k)$ is the number of distinct identities sent to the left-right encryption oracle, $Q_{\mathsf{PK}}(k)$ the maximum number of public key replacement query on an identity, and $Q_{\mathsf{E}}(k)$ the maximum number of left-right encryption queries on an identity/public key pair.

4.3   The CLE-2-CE Transformation

We now move towards proving that CLE schemes can be analyzed using CE models. First, we provide a syntactic transformation that associates a CE scheme to a CLE scheme. Just as for IBE encryption, the only needed change is a registration protocol; the rest of the algorithms of the scheme remain essentially the same.

Given a CLE scheme we define its associated CE scheme CE = CLE-2-CE(CLE) by setting $\mathsf{Setup}_{\mathsf{CE}}$, $\mathbb{G}_{\mathsf{CE}}$, $\mathbb{E}_{\mathsf{CE}}$ and $\mathbb{D}_{\mathsf{CE}}$ to be $\mathsf{Setup}_{\mathsf{CLE}}$, $\mathbb{G}_{\mathsf{CLE}}$, $\mathbb{E}_{\mathsf{CLE}}$ and $\mathbb{D}_{\mathsf{CLE}}$ algorithms of CLE respectively. The registration protocol is defined as follows. In the registration protocol the user sends his identity to the CA who computes the partial key that corresponds to ID which he then sends back to ID. The user then generates a public key/secret key pair and sends the public key back to the CA. More formally, the registration protocol is as follows:

1. $\mathbb{U}_{\mathsf{CE}}(\mathsf{ID}, \mathsf{PK}_{\mathsf{CA}})$: Sends ID to the CA;
2. $\mathbb{C}_{\mathsf{CE}}(\mathsf{SK}_{\mathsf{CA}})$: Receives ID, runs $\mathsf{D} \leftarrow \mathbb{X}_{\mathsf{CLE}}(\mathsf{ID}, \mathsf{SK}_{\mathsf{CA}})$ and sends D to user ID;
3. $\mathbb{U}_{\mathsf{CE}}(\mathsf{ID}, \mathsf{PK}_{\mathsf{CA}})$: Receives D, runs $(\mathsf{SK}, \mathsf{PK}) \leftarrow \mathbb{U}_{\mathsf{CLE}}(\mathsf{ID}, \mathsf{PK}_{\mathsf{CA}})$ and then $\mathsf{S} \leftarrow \mathbb{S}_{\mathsf{CLE}}(\mathsf{SK}, \mathsf{D}, \mathsf{Mpk})$. It sends PK to CA and outputs $(\mathsf{ID}, \mathsf{PK}, \mathsf{S}, \epsilon)$.
4. $\mathbb{C}_{\mathsf{CE}}(\mathsf{SK}_{\mathsf{CA}})$: Receives PK and outputs $(\mathsf{ID}, \mathsf{PK}, \epsilon)$.

The main results of this section relate the security of CLE to that of the transformed scheme CLE-2-CE(CLE). We start with the case when the attacker is of type M, as in this case we get perfect equivalence between the models:

Theorem 2. *Let* CLE *be an arbitrary certificateless encryption scheme and set* CE := CLE-2-CE(CLE). *Then* CLE *is secure against type* M *adversaries in the single-user* CCA *model if and only if* CE *is secure certified encryption scheme against adversaries in the* CCA *model which corrupt the CA.*

The proof, which we give in the full version of the paper shows that for any ppt type M adversary $\mathcal{A}$ against CLE there exists a type M adversary $\mathcal{B}$ against CE such that type

$$\mathbf{Adv}_{\mathsf{CLE},\mathcal{A}}^{\mathsf{CL\text{-}CCA\text{-}M}}(k) \leq Q_{\mathsf{ID}}(k) \cdot \mathbf{Adv}_{\mathsf{CE},\mathcal{B}}^{\mathsf{mCE\text{-}CCA\text{-}M}}(k).$$

Conversely, we have that for any adversary $\mathcal{A}$ of type M against CE there exists an type M adversary $\mathcal{B}$ against CLE such that:

$$\mathbf{Adv}_{\mathsf{CE},\mathcal{A}}^{\mathsf{mCE\text{-}CCA\text{-}M}}(k) \leq \mathbf{Adv}_{\mathsf{CLE},\mathcal{B}}^{\mathsf{mCL\text{-}CCA\text{-}M}}(k).$$

Unfortunately, for the case of honest CA (that is, type I adversaries) the situation is more complex and a similar theorem does not immediately hold. To clarify some of the difficulties, notice that via the transformation CLE-2-CE (and in fact via any other similar transformation), the resulting CE scheme can be trivially attacked using the powers that the adversary has in the CE model. The adversary, interacting with an honest CA does the following. It starts the execution of the registration protocol for some identity ID and stops (prematurely) before executing step (4) of the protocol. That is, the adversary computes (PK, SK) as prescribed, but does not send PK to the certification authority. The adversary then calls the left-right encryption oracle on a message pair and (ID, PK). Notice that since the CA did not complete the protocol, the tuple (ID, PK) does not occur in the list RegListSec (which would have rendered the query invalid). However, the adversary has the right decryption key and thus can immediately win the game.

What happens here is that although from the point of view of the adversary, the key PK had been registered, the experiment does not (and cannot) capture this situation since it does not get access to PK. Since encryptions under such keys can be trivially decrypted by the adversary, this attack motivates a weaker, but still reasonable security model for certified encryption. The class of attackers that we consider have the right to issue left-right encryption queries only for identities that had not been corrupt. We call these kind of adversaries *weak*.

An additional important observation is that to simulate the queries of the CLE setting a CE adversary needs somehow access to the internal structure of the secret keys of parties. We provide such access using corrupt decryption oracles (as defined in Section 2). We show that we can the obtain a characterization of security against type I adversaries in the CLE sense via security in CE models.

Recall that an adversary with corrupted decryption capabilities can choose functions in a set $\mathcal{F}$ to be applied to the secret keys of parties, before these secret keys are used to decrypt ciphertexts of the adversary's choice. For our purposes, we consider the set:

$$\mathcal{F} := \{(\mathsf{Id}, \mathsf{Id}, f, \mathsf{Id}) : f = \mathsf{Id} \text{ or } f = f_{\mathsf{SK}} : (\mathsf{SK}', \mathsf{D}') \mapsto (\mathsf{SK}, \mathsf{D}') \text{ for some } \mathsf{SK} \in \{0,1\}^* \}.$$

which consists essentially of functions that allow changing the first component SK' of the secret key of a party with any other secret key SK that the adversary chooses. We can the use adversary with $\mathcal{F}$-decryption to obtain the desired link. We start with a relation between type I* adversaries against CLE and weak adversaries with $\mathcal{F}$-decryptions.

Proposition 1. *Let* CLE *be a certificateless encryption scheme and set* CE := CLE-2-CE(CLE). *Then* CLE *is secure against type* I* *adversaries in the single-user* CCA *model if and only if* CE *is secure against weak type* I *adversaries with* $\mathcal{F}$*-decryptions.*

The proof, which we give in the full version of the paper shows that for any ppt type I* (resp. type M) adversary $\mathcal{A}$ against CLE in the single-user setting there exists a ppt adversary $\mathcal{B}$ against CE in $\mathcal{F}$-extended model which does not corrupt (resp. corrupts) the CA such that:

$$\mathbf{Adv}_{\mathsf{CLE},\mathcal{A}}^{\mathsf{CL\text{-}CCA\text{-}I}^*}(k) \leq Q_{\mathsf{ID}}(k) \cdot \mathbf{Adv}_{\mathsf{CE},\mathcal{B}}^{\mathsf{mCE\text{-}}\mathcal{F}\text{-}\mathsf{CCA\text{-}I}^-}(k),$$

Here $Q_{\mathsf{ID}}(k)$ denotes the maximum number of IDs queried to the experiment.

Conversely, for any ppt weak CE adversary $\mathcal{A}$ against CE with $\mathcal{F}$-decryptions there exists a ppt type I* adversary $\mathcal{B}$ against CLE in the multi-user setting such that:

$$\mathbf{Adv}_{\mathsf{CE},\mathcal{A}}^{\mathsf{mCE\text{-}}\mathcal{F}\text{-}\mathsf{CCA\text{-}I}^-}(k) \leq \mathbf{Adv}_{\mathsf{CLE},\mathcal{B}}^{\mathsf{mCL\text{-}CCA\text{-}I}^*}(k).$$

To obtain a relation between security against type I adversaries in the CLE model and security in the sense of certified encryption we first establish a link between models where the adversary can obtain information on D using corruption capabilities and models where the adversary corrupts the master secret key. The proof of the following is given in the full version of the paper.

Lemma 3. *For any ppt type* I *adversary* $\mathcal{A}$*, there exists ppt adversaries* $\mathcal{B}_1$ *and* $\mathcal{B}_2$ *in type* I* *and type* M *models respectively such that:*

$$\mathbf{Adv}_{\mathsf{CLE},\mathcal{A}}^{\mathsf{CL\text{-}CCA\text{-}I}}(k) \leq \mathbf{Adv}_{\mathsf{CLE},\mathcal{B}_1}^{\mathsf{CL\text{-}CCA\text{-}I}^*}(k) + \mathbf{Adv}_{\mathsf{CLE},\mathcal{B}_2}^{\mathsf{CL\text{-}CCA\text{-}M}}(k),$$

The following theorem is the main result of this section. Informally it says that security of certificateless schemes can be analyzed using certified encryption models, extended with corrupt decryptions. The proof of the theorem follows from Proposition 1 and Lemma 3.

Theorem 3. CLE *scheme is secure against type* I *and type* M *attackers if and only if* CLE-2-CE(CLE) *is secure against type* I *weak CE adversaries with* $\mathcal{F}$*-decryption and also secure against type* M *adversaries.*

Acknowledgments

The work carried out by the first author was supported in part by the Scientific and Technological Research Council of Turkey (TÜBİTAK) while at Middle East Technical University. The second author has been supported in part by the European Commission through the IST Programme under Contract IST-2007-216646 ECRYPT II grant. The information in this document reflects only the author's views, is provided as is and no guarantee or warranty is given that the information is fit for any particular purpose. The user thereof uses the information at its sole risk and liability.

References

1. Al-Riyami, S.S., Paterson, K.G.: Certificateless Public-Key Cryptography. In: Laih, C.-S. (ed.) ASIACRYPT 2003. LNCS, vol. 2894, pp. 452–473. Springer, Heidelberg (2003)
2. Au, M.H., Chen, J., Liu, J.K., Mu, Y., Wong, D.S., Yang, G.: Malicious KGC Attacks in Certificateless Cryptography. In: ACM Symposium on Information, Computer and Communications Security, March 2007, pp. 302–311 (2007)
3. Baek, J., Safavi-Naini, R., Susilo, W.: Certificateless Public Key Encryption Without Pairing. In: Zhou, J., López, J., Deng, R.H., Bao, F. (eds.) ISC 2005. LNCS, vol. 3650, pp. 134–148. Springer, Heidelberg (2005)
4. Bellare, M., Boldyreva, A., Micali, S.: Public-Key Encryption in a Multi-User Setting: Security Proofs and Improvements. In: Preneel, B. (ed.) EUROCRYPT 2000. LNCS, vol. 1807, pp. 259–274. Springer, Heidelberg (2000)
5. Bellare, M., Boldyreva, A., Staddon, J.: Multi-Recipient Encryption Schemes: Security Notions and Randomness Re-Use. In: Desmedt, Y.G. (ed.) PKC 2003. LNCS, vol. 2567, pp. 85–99. Springer, Heidelberg (2002)
6. Bellare, M., Kohno, T.: A Theoretical Treatment of Related-Key Attacks: RKA-PRPs, RKA-PRFs, and Applications. In: Biham, E. (ed.) EUROCRYPT 2003. LNCS, vol. 2656, pp. 491–506. Springer, Heidelberg (2003)
7. Boldyreva, A., Fischlin, M., Palacio, A., Warinschi, B.: A Closer Look at PKI: Security and Efficiency. In: Okamoto, T., Wang, X. (eds.) PKC 2007. LNCS, vol. 4450, pp. 458–475. Springer, Heidelberg (2007)
8. Boneh, D., Franklin, M.: Identity-Based Encryption from the Weil Pairing. SIAM Journal on Computing 32, 586–615 (2003)
9. Cheng, Z., Comley, R.: Efficient Certificateless Public Key Encryption. Cryptology ePrint Archive, Report 2005/012 (2005)
10. Dent, A.W.: A Note On Game-Hopping Proofs. Cryptology ePrint Archive, Report 2006/260 (2006)
11. Dent, A.W.: A Survey of Certificateless Encryption Schemes and Security Models. International J. of Information Security 7(5), 349–377 (2008)
12. Gentry, C.: Certificate-Based Encryption and the Certificate Revocation Problem. In: Biham, E. (ed.) EUROCRYPT 2003. LNCS, vol. 2656, pp. 272–293. Springer, Heidelberg (2003)
13. Gentry, C.: Practical Identity-Based Encryption without Random Oracles. In: Vaudenay, S. (ed.) EUROCRYPT 2006. LNCS, vol. 4004, pp. 445–464. Springer, Heidelberg (2006)

14. Herzog, J., Liskov, M., Micali, S.: Plaintext Awareness via Key Registration. In: Boneh, D. (ed.) CRYPTO 2003. LNCS, vol. 2729, pp. 548–564. Springer, Heidelberg (2003)
15. Kaliski, B.: An Unknown Key-Share Attack on the MQV Key Agreement Protocol. ACM Transactions on Information and System Security – TISSEC 4(3), 275–288 (2001)
16. Liu, J.K., Au, M.H., Susilo, W.: Self-Generated-Certificate Public Key Cryptography and Certificateless Signature/Encryption Scheme in the Standard Model. In: Proceedings of the 2nd ACM Symposium on Information, Computer and Communications Security, pp. 273–283. ACM Press, New York (2007)
17. Sakai, R., Kasahara, M.: ID-Based Cryptosystems with Pairing on Elliptic Curve. In: Symposium on Cryptography and Information Security – SCIS 2003 (2003)
18. Shamir, A.: Identity-Based Cryptosystems and Signature Schemes. In: Blakely, G.R., Chaum, D. (eds.) CRYPTO 1984. LNCS, vol. 196, pp. 47–53. Springer, Heidelberg (1985)
19. Shoup, V.: On Formal Models for Secure Key Exchange. IBM Research Report
20. Waters, B.: Efficient Identity-Based Encryption Without Random Oracles. In: Cramer, R. (ed.) EUROCRYPT 2005. LNCS, vol. 3494, pp. 114–127. Springer, Heidelberg (2005)

Threshold Attribute-Based Signatures and Their Application to Anonymous Credential Systems

Siamak F. Shahandashti[1] and Reihaneh Safavi-Naini[2]

[1] School of Computer Science and Software Engineering
University of Wollongong, Australia
http://www.uow.edu.au/~sfs166
[2] Department of Computer Science
University of Calgary, Canada
http://www.cpsc.ucalgary.ca/~rei

Abstract. In this paper we propose *threshold attribute-based signatures* (t-ABS). A t-ABS scheme enables a signature holder to prove possession of signatures by revealing only the relevant attributes of the signer, hence providing *signer-attribute privacy* for the signature holder. We define t-ABS schemes, formalize their security and propose two t-ABS schemes: a basic scheme secure against selective forgery and a second one secure against existential forgery, both provable in the standard model, assuming hardness of the CDH problem. We show that our basic t-ABS scheme can be augmented with two extra protocols that are used for efficiently issuing and verifying t-ABS signatures on committed values. We call the augmented scheme a threshold attribute based c-signature scheme (t-ABCS). We show how a t-ABCS scheme can be used to realize a secure *threshold attribute-based anonymous credential system* (t-ABACS) providing issuer-attribute privacy. We propose a security model for t-ABACS, give a concrete scheme using t-ABCS scheme, and prove that the credential system is secure if the t-ABCS scheme is secure.

1 Introduction

Inspired by the recent works on attribute based encryption, we introduce a new signature scheme that we call a *threshold attribute-based signature*, or t-ABS for short. In a t-ABS scheme, a central authority issues secret keys to the signers according to their *attributes*. A signer with an attribute set Att can use his secret key to sign using any subset $A \subset Att$ of attributes. A signature can be verified against a verification attribute set B and verification succeeds if $A \cap B$ has at least size d. Threshold attribute-based signatures have attractive applications that are outlined below.

A t-ABS scheme provides *threshold attribute-based verification*. This can be used for applications such as identity-based signatures that use biometric information as user identities. Since different measurements by different entities will not be exactly the same, *fuzzy verification* is required. Using a threshold

B. Preneel (Ed.): AFRICACRYPT 2009, LNCS 5580, pp. 198–216, 2009.

attribute-based signature scheme with suitable choices of mapping and threshold, "close" biometric measurements map to attribute sets for signing and verification that have sufficient overlap and result in successful verification of signature.

A t-ABS scheme enables the signer to choose their "signing identity" (relevant attributes) at the time of signing, and also enables *signature holders* to present for verification a signature that corresponds to subset of signer's attributes. John Smith, a member of 2008 Executive Committee of the Workers' Union (WU), can sign documents as, and a document signed by him can be presented as signed by an Executive Committee member of the WU, a 2008 Executive Committee member of the WU, or simply John Smith. In general, to provide the above functionalities for signers with n "atomic" identities, a t-ABS scheme can be implemented that has signing keys and signatures, both of size $O(n)$. Providing the same functionalities using standard or identity-based signatures requires $O(2^n)$ signing keys, one for each "combined" identity, and $O(2^n)$ signatures, each produced using one of the keys. Note that the naïve solution of $O(n)$ signing keys of a standard or an identity-based signature, is not secure because signatures can be "mixed and matched" with each other. A group of signers, for example, can collude to produce a signature corresponding to the union of their identities. A t-ABS scheme provides *collusion resistance*: that is, no group of colluding signers can produce a signature that could not be generated by a single member of the group.

A t-ABS scheme supports *signer-attribute privacy* for users. Consider a scenario where participation in a poll requires proof of residency in one of the 27 counties in a state. Assume these proofs are cards that are signed by either the local government or police authorities of the counties and are given to the long-time residents and foreign workers, respectively. For privacy reasons, card holders may want to be able to prove that they own a card but protect their other details. Proving possession of such a card is equivalent to proving possession of a card issued by one of the 2×27 possible entities. Using standard or identity-based signatures, a voter can prove possession of a valid signature with respect to one of the 2×27 public keys or identities through a proof of size 2×27 (see *proofs of partial knowledge* [9]). A t-ABS can provide a proof of size $2+27$ and ensures that the attributes of the signer (and hence unnecessary information of the card holder) are not revealed. Signers in such schemes will each have two attributes: one corresponding to 'police' or 'local government', and the other representing one of the 27 counties. Verification of residency is then equivalent to requiring that a card is verifiable with a threshold of $d = 2$ attributes from the set of all $2+27$ attributes above. Generally, if the universal set of attributes can be partitioned such that each signer has at most one attribute from each partition, a t-ABS enables threshold verification with additive cost in terms of the sizes of partitions, while the cost of proofs of partial knowledge is multiplicative. The functionality becomes the only solution in cases where, because of the size and spread of the system, it is practically infeasible for the signature holder and the verifier to know all possible signers.

As a prime example to show their application range, we show that t-ABS schemes, when augmented with additional protocols for signing and proof of signature ownership, can be used to construct *attribute-based anonymous credential systems*, providing features including *attribute privacy*, which is an essential property in anonymous credential systems, since verifying credentials against specific public keys reveals unnecessary information about the users and contradicts the "raison d'être" of such systems.

1.1   Our Contributions

We first define *threshold attribute-based signatures* that support threshold attribute verification. Then, we construct a t-ABS scheme and prove its security against selective forgery in the standard model based on the CDH assumption. We discuss how security and efficiency of the scheme can be improved and give a secure construction against existential forgery in the standard model.

We then introduce an algorithm for *converting* a signature and an interactive protocol for signature verification. The convert algorithm enables the signature holder to construct a signature that is verifiable with the attribute set $A \cap B$ and prevents the verifier from learning the signer's attributes that are outside $A \cap B$. We call this level of privacy *weak signer-attribute privacy*. The interactive verification protocol enables the signature holder to prove possession of a valid signature without allowing the verifier to learn even $A \cap B$. The verifier learns nothing more than the fact that $|A \cap B| \geq d$ We call this level of privacy *(strong) signer-attribute privacy*. We denote a t-ABS with an efficient conversion algorithm and an efficient interactive verification protocol by t-ABS$^+$. We provide efficient conversion algorithms and interactive verification protocols for both our proposed schemes and prove that they provide both levels of signer-attribute privacy.

A t-ABCS scheme consists of a t-ABS$^+$ scheme, a commitment scheme, and three additional protocols for (i) obtaining a signature on a committed value, (ii) proving knowledge of a signature, and (iii) proving knowledge of a signature on a committed value. We give security definitions for t-ABCS schemes and a concrete efficient construction that is based on our basic t-ABS scheme and prove security of the construction.

We give a definition for threshold attribute-based anonymous credential systems (t-ABACS) and formalize their security using the *simulation paradigm*. The approach uses a comparison of a real system model with an ideal system model. The ideal model for the t-ABACS captures the security and privacy properties of an anonymous credential system, including the *issuer-attribute privacy* property. We show how t-ABCS schemes can be used to realize this model. That is, we give a concrete t-ABACS system and prove its security in the sense that it remains indistinguishable from an ideal system from the viewpoint of any polynomial-time adversary. This results in a concrete t-ABCS scheme with security based on the CDH problem.

1.2   Related Work

Attribute-based encryption (ABE) was introduced by Sahai and Waters as an extension of identity-based encryption where each identity is considered as a set of descriptive attributes [21]. The scheme, called "fuzzy identity-based encryption", allows a threshold attribute-based decryption of encrypted data. Our attribute-based signature scheme can be viewed as the signature counterpart of their encryption scheme.

Attribute-based signatures extend the *identity-based signature* of Shamir [23] by allowing identity of a signer to be a set of descriptive attributes rather than a single string. Identity-based signature can be seen as a specific case of our schemes with identity size and threshold both equal to one.

Independent of our work, there have been other attempts to define and realize *attribute-based signatures*. The schemes of [25] and [13] are direct applications of the known transform from identity-based encryptions to identity-based signatures [10]. In both works, authors do not consider any notion of privacy. The works of [15,14] and [17] capture weaker notions of anonymity where signers only remain anonymous within the group of entities possessing the same attributes and the verifiers must know which attributes are used to sign a message to be able to verify. Khader [16] and Maji et al. [20] treat attribute-privacy as a fundamental requirement of attribute-based signatures. However, both schemes require the signer to know the verification policy at the time of signing. This is a major limitation for their proposed scheme, particularly in applications where other than the signer and the verifier, there are also 'signature holders'. An example of such applications is a credential system where a credential holder needs to satisfy different verification policies depending on the occasion, and it is important for efficiency and useability purposes not to require different credentials for each verification policy. Also, security proofs of Khader and Maji et al. are in the random oracle or generic group models, while all our proofs are in the standard model and use the well-known assumption of computational Diffie-Hellman. Finally, none of the previous attribute-based signatures had been extended to credential systems.

Anonymous credential systems (a.k.a. *pseudonym systems*) were introduced by Chaum [6,7] and more recently further formalized and studied in [19,18]. A credential in such systems is issued and verified on a user pseudonym, which in turn is bound to the user's secret identity. Users remain anonymous since their pseudonyms hide their secret identity. Besides, transactions involving the same user remain unlinkable.

2   Notation and Preliminaries

We use different font families to denote *variables*, algorithms, strings, and SE-CURITY NOTIONS, respectively. By "$x \leftarrow \mathsf{X}(a)$" we denote that X is run on input a and the output is assigned to x. We also use "$A \dashv(X)\!\rightarrow B$ if C" to denote that A sends X to B if condition C holds. The condition can be complex and include logical connectors. We use the proof of knowledge notation originated in [4]. For instance, "ZK-PoK$\{x : a = g^x\}$" denotes a zero knowledge (ZK [12]) proof of knowledge (PoK [1]) of x such that $a = g^x$, with a and g being public inputs to the protocol. We use zero knowledge proofs of knowledge for conjunctive statements about discrete logarithms. Efficient versions of such proofs can be found in the literature. e.g., in the work by Cramer et al. [8].

Lagrange interpolation for a polynomial $q(\cdot)$ over $\mathbb{Z}_p$ of order $d - 1$ and a set $S \subset \mathbb{Z}_p$ with size $|S| = d$ is calculated as $q(x) = \sum_{i \in S} q(i)\Delta_{i,S}(x)$, where

$$\Delta_{i,S}(x) \triangleq 1 \cdot \prod_{j \in S, j \neq i} \frac{x - j}{i - j} \quad \text{for all } i \in S \text{ (and extendedly for all } i \in \mathbb{Z}_p).$$

Camenisch and Lysyanskaya have shown that a tuple consisting of a signature, a commitment scheme, and efficient protocols for issuing and verifying signatures on committed values, is sufficient for realizing anonymous credential schemes [3,18]. We denote such tuples by *C-signatures*.

A well-known paradigm that is used to define security of cryptographic protocols is the simulation (a.k.a. "ideal vs. real") paradigm, originating from [11]. The intuition behind such definitions is that a real system is secure if it emulates a certain ideal system designed to trivially guarantee security properties expected from the system. We use this paradigm to define a framework for analyzing security of our credential system.

3   Threshold Attribute-Based Signatures

We assume there is a universal set of attributes $\mathbb{U}$ that is publicly known. Each signer is associated with a subset $Att \subset \mathbb{U}$ of attributes that is verified by a central authority. In the following, we assume that the signing attribute set A is equal to the signer attribute set Att and we use the terms interchangeably. This has minimal impact in our definitions of the schemes and their security. We define the signature for a fixed threshold d and discuss flexible thresholds later.

Definition: A threshold attribute-based signature (t-ABS) is a quadruple of algorithms as follows:

Setup is the algorithm run by a central authority on input the security parameter k and outputs a master secret key msk and a master public key mpk.

KeyGen is the algorithm run by the central authority on inputs msk and a set of signer attributes A and generates a secret signing key ssk for the signer.

Sign is the algorithm run by a signer on inputs ssk and a message m and generates a signature σ on the message.

Verify is the algorithm run by a verifier on inputs mpk, a message signature pair (m, σ), and a verification attribute set B and outputs 1 if σ is a valid signature by a signer who has at least d of the attributes in B, i.e., $|A \cap B| \geq d$.

Correctness: A signature generated by a signer with attributes A must pass the verification test for any B if $|A \cap B| \geq d$.

Unforgeability: For a t-ABS scheme defined as above we require that it is *existentially unforgeable* against *chosen message and attribute set attacks*. In particular, we define the following game between a challenger and the adversary.

Setup Phase: The challenger runs the Setup algorithm and gives mpk to the adversary.

Query Phase: The adversary is allowed to ask queries for the following:
- a secret key of a signer with attributes of its choice α, and
- a signature of a signer with any attribute set of its choice α on a message of its choice m.

Forgery Phase: The adversary outputs a triplet (μ, σ, β), consisting of a message μ, a forged signature σ, and a verification attribute set β, and wins if σ is a valid signature with respect to (mpk, μ, β) and
- for all queried sets of attributes α, we have $|\alpha \cap \beta| < d$, and
- for all queried pairs (α, m), we have $m \neq \mu$ or $|\alpha \cap \beta| < d$.

If no polynomial adversary has a considerable advantage in the above game, we say that the t-ABS scheme is existentially unforgeable against chosen message and attribute set attacks, or EUF-CMAA-secure for short. We also consider a weaker notion of security, *selective unforgeability* against chosen message and attribute set attacks (SUF-CMAA-security) in which the adversary should commit to the target forgery message and verification attribute set in the beginning of the attack. We define this notion in the full version of this paper [22] and discuss how an existentially unforgeable scheme can be constructed given a selectively unforgeable scheme.

Collusion Resistance: It is important to note that the above definition of unforgeability guarantees *collusion resistance* in the sense that no colluding group of users can generate a signature that is not generable by one of the colluders. This is because if a group of signers can construct a signature that none of them could individually produce, then this is a forgery as per the above definition.

3.1   Additional Protocols

A signature holder can always check a signature σ against possible verification attribute sets to deduce information about the signer's attributes. To preserve privacy of signers we equip our t-ABS scheme with an additional algorithm for *converting* the signature to another signature that is verifiable against B and only reveals the d chosen attributes of the signer. The *converted signature* can be seen as a B-designated signature that contains a minimal subset of attributes from the original set of signer attributes, that allows the verification to succeed. Attribute privacy is obtained by using an *interactive verification* protocol iVerify that allows the signature holder to prove possession of a valid converted signature without revealing the chosen d attributes in common between A and B.

We call our t-ABS scheme equipped with both the above conversion algorithm and interactive verification protocol a t-ABS$^+$ scheme, which formally contains the Setup, KeyGen, and Sign algorithms as defined in the t-ABS scheme plus the following:

Convert is the algorithm run by a signature holder on inputs mpk, a message signature pair (m, σ), and a verification attribute set B and generates a converted signature $\tilde{\sigma}$ on the message.

CvtVerify is the algorithm run by a verifier on inputs mpk, a message converted-signature pair $(m, \tilde{\sigma})$, and a verification attribute set B and outputs 1 if $\tilde{\sigma}$ is a valid converted signature by a signer who has at least d of the attributes in B, i.e., if $|A \cap B| \geq d$.

iVerify is an interactive verification protocol for proving knowledge of a converted signature on the *prover* side and verifying a converted signature on the *verifier* side. The public inputs are the t-ABS master public key mpk, a message m, and verifier's verification attribute set B. Prover's private input is a converted signature $\tilde{\sigma}$. The verifier has no private input. At the end of the protocol execution the verifier will output a binary value reflecting prover's converted signature validity against B.

Correctness: Any converted signature calculated from a valid signature must (i) pass the CvtVerify test, and (ii) make the verifier in the iVerify protocol accept.

Unforgeability: We require that converted signatures in our t-ABS$^+$ scheme are also existentially unforgeable under chosen message and attribute set attacks. Since knowledge of a signature is sufficient for producing a converted signature (using algorithm Convert), converted signature unforgeability implies signature unforgeability, and hence, is a stronger notion of security. Converted signature existential unforgeability under chosen message and attribute set attacks (C-EUF-CMAA-security) is defined through a game, in which the setup and query phases are the same as the EUF-CMAA-security game above, but σ in the forgery phase is replaced with $\tilde{\sigma}$. That is, the adversary is given the same resources, but is expected to forge a nontrivial valid *converted* signature instead of a nontrivial valid signature. The full game is transcribed in the full version of this paper [22] for completion.

Weak Signer-Attribute Privacy: A converted signature should not reveal any attribute of the signer other than the d of them common with B chosen by the signature holder at the time of conversion. Thus, we require that whatever a verifier can deduce about other attributes of the signer given a converted signature, can also be deduced given merely the d attributes as well. This ensures that only the d attributes of the signer that are chosen by the signature holder are revealed to the verifier given a converted signature. We call this property *weak signer-attribute privacy*.

Signer-Attribute Privacy: We also require that the iVerify protocol is a zero knowledge proof of knowledge of a valid converted signature with respect to the public inputs (mpk, m, β). This ensures that (i) only provers in possession of a valid converted signature are indeed successful in proving so, and (ii) the proof reveals no information other than the validity of the prover's converted signature to the verifier. We call this property *(full) signer-attribute privacy*. Note that property (ii) guarantees that proofs of possession of signatures from different signers satisfying the verification policy remain indistinguishable for the verifier. Furthermore, it guarantees that multiple proofs of possession of even the same signature (from the same signer) remain unlinkable for the verifier.

Flexible Threshold: To achieve a flexible threshold, one can use either or a combination of the following two techniques based on the application at hand: (i) designing multiple schemes with different thresholds and (ii) using *dummy* attributes. We discuss the latter in detail in the full version of this paper [22].

3.2   Constructions

We propose a threshold attribute-based signature based on bilinear maps. We make use of some design techniques of earlier works of [21] and [2].

The Scheme: Signer attributes are assumed to be sets of at most n elements of $\mathbb{Z}_p$. Although generally, identities can be sets of at most n arbitrary strings and a collision resistant hash function is used to map the strings to elements of $\mathbb{Z}_p$. We use $N = \{1, 2, \ldots, n+1\}$ to denote the set of possible attributes. In the following, a basic scheme with a fixed threshold d is introduced. We will later discuss how to extend our scheme for verifiers with different thresholds. Let $\mathbb{G}_1 = \langle g \rangle$ be a group of prime order p and a bilinear map $e : \mathbb{G}_1 \times \mathbb{G}_1 \to \mathbb{G}_2$ be defined. Let $(e, \mathbb{G}_1, \mathbb{G}_2)$ be of public knowledge. We present the scheme for signing messages in $\mathbb{Z}_p$. Although, the message space can be expanded to contain arbitrary messages using a collision resistant hash function to map strings to $\mathbb{Z}_p$.

Setup(1^k): Pick y randomly from $\mathbb{Z}_p$ and set $g_1 = g^y$. Pick random elements g_2, $h, t_1, t_2, \ldots, t_{n+1}$ from $\mathbb{G}_1$. Define and output the following:

$$T(x) \overset{\triangle}{=} g_2^{x^n} \prod_{i=1}^{n+1} t_i^{\Delta_{i,N}(x)}, \quad msk = y, \quad mpk = (g, g_1, g_2, t_1, t_2, \ldots, t_{n+1}, h)$$

KeyGen(msk, A): Choose a random $d - 1$ degree polynomial $q(x)$ such that $q(0) = y$, choose random elements r_i in $\mathbb{Z}_p$ for $i \in A$, and output

$$ssk = \langle \ \{ \ g_2^{q(i)} T(i)^{r_i}, \quad g^{r_i} \ \}_{i \in A} \ \rangle$$

Sign(ssk, m): Parse the signing key as $ssk = \langle \{ssk_{1i}, ssk_{2i}\}_{i \in A} \rangle$, pick random elements s_i in $\mathbb{Z}_p$ for all $i \in A$, and output

$$\sigma = \langle \ A, \ \{ \ ssk_{1i}(g_1^m \cdot h)^{s_i}, \quad ssk_{2i}, \quad g^{s_i} \ \}_{i \in A} \ \rangle$$

Verify(mpk, m, σ, B): Parse the signature as $\sigma = \langle A, \{\sigma_{1i}, \sigma_{2i}, \sigma_{3i}\}_{i \in A} \rangle$. Select an $S \subseteq A \cap B$ such that $|S| = d$ and check if the following equation holds:

$$\prod_{i \in S} \left(\frac{e(\sigma_{1i}, g)}{e(T(i), \sigma_{2i}) \cdot e(g_1^m \cdot h, \sigma_{3i})} \right)^{\Delta_{i,S}(0)} = e(g_2, g_1) \tag{1}$$

Correctness and Unforgeability: The correctness proof is straightforward and can be found in the full version of this paper [22] for completion. Unforgeability is implied by converted-signature unforgeability that we prove later in Theorem 1.

Additional Protocols: The concrete conversion and converted-signature verification algorithms are as follows:

Convert(mpk, m, σ, B): Parse the signature as $\sigma = \langle A, \{\sigma_{1i}, \sigma_{2i}, \sigma_{3i}\}_{i \in A} \rangle$. Select an $S \subseteq A \cap B$ such that $|S| = d$. Calculate the converted signature components as follows and output $\tilde{\sigma} = \langle \{\tilde{\sigma}_{1i}, \tilde{\sigma}_{2i}, \tilde{\sigma}_{3i}\}_{i \in B} \rangle$.

for all $i \in S$:
$$\tilde{\sigma}_{1i} \leftarrow \sigma_{1i}^{1/\Delta_{i,B \setminus S}(0)} \qquad \tilde{\sigma}_{2i} \leftarrow \sigma_{2i}^{1/\Delta_{i,B \setminus S}(0)} \qquad \tilde{\sigma}_{3i} \leftarrow \sigma_{3i}^{1/\Delta_{i,B \setminus S}(0)}$$

for all $i \in B \setminus S$:
$$\tilde{\sigma}_{1i} \leftarrow (T(i)g_1^m h)^{1/\Delta_{i,B \setminus S}(0)} \qquad \tilde{\sigma}_{2i} \leftarrow g^{1/\Delta_{i,B \setminus S}(0)} \qquad \tilde{\sigma}_{3i} \leftarrow g^{1/\Delta_{i,B \setminus S}(0)}$$

CvtVerify($mpk, m, \tilde{\sigma}, B$): Parse the converted signature as $\tilde{\sigma} = \langle \{\tilde{\sigma}_{1i}, \tilde{\sigma}_{2i}, \tilde{\sigma}_{3i}\}_{i \in B} \rangle$. Check if the following equation holds:

$$\prod_{i \in B} \left(\frac{e(\tilde{\sigma}_{1i}, g)}{e(T(i), \tilde{\sigma}_{2i}) \cdot e(g_1^m \cdot h, \tilde{\sigma}_{3i})} \right)^{\Delta_{i,B}(0)} = e(g_2, g_1) \qquad (2)$$

Furthermore, the iVerify protocol flow is as follows:

1. The signature holder randomizes the converted signature by first choosing random elements s_i' and r_i' for $i \in B$ and then calculating the following. Note that the resulting randomized converted signature is a valid converted signature itself.

$$\breve{\sigma}_{1i} = \tilde{\sigma}_{1i} \cdot T(i)^{r_i'} (g_1^m \cdot h)^{s_i'}, \quad \breve{\sigma}_{2i} = \tilde{\sigma}_{2i} \cdot g^{r_i'}, \quad \breve{\sigma}_{3i} = \tilde{\sigma}_{3i} \cdot g^{s_i'}$$

2. The signature holder chooses random values τ_i for all $i \in B$ and sets $\hat{\sigma}_{1i} \leftarrow \breve{\sigma}_{1i}^{1/\tau_i}$ and sends $\langle \{\hat{\sigma}_{1i}, \breve{\sigma}_{2i}, \breve{\sigma}_{3i}\}_{i \in B} \rangle$ to the verifier.
3. Both the signature holder and the verifier calculate the following for all $i \in B$:

$$u_0 \leftarrow e(g_2, g_1) \qquad\qquad u_{1i} \leftarrow e(\hat{\sigma}_{1i}, g)^{\Delta_{i,B}(0)}$$
$$u_2 \leftarrow \prod_{i \in B} e(T(i), \breve{\sigma}_{2i})^{\Delta_{i,B}(0)} \qquad u_3 \leftarrow \prod_{i \in B} e(g_1^m h, \breve{\sigma}_{3i})^{\Delta_{i,B}(0)}$$

4. The signature holder performs the following ZK-PoK for the verifier:

$$\text{ZK-PoK}\{ \ (\{\tau_i\}_{i \in B}) : \ \prod_{i \in B} u_{1i}^{\tau_i} = u_0 u_2 u_3 \ \}$$

Correctness and Unforgeability: The correctness proof is straightforward. We discuss unforgeability of our scheme, and in particular, the proof of the following theorem, in the full version of this paper [22].

Theorem 1. *The above t-ABS⁺ scheme is* C-SUF-CMAA-*secure if the CDH problem is hard. As a direct corollary, the underlying t-ABS scheme is* SUF-CMAA-*secure if the CDH problem is hard.*

Full Security: Existential unforgeability can be achieved in either of the following techniques:

General Reduction: Any selectively unforgeable t-ABS can be proved to be existentially unforgeable. This general reduction is discussed in the full version of this paper [22]. This reduction is not efficient and introduces a large penalty factor to the security of the scheme, but it is carried out in the standard model.

Random Oracles: An alternative method is use random oracles to hash messages and signer attributes at the time of signing. This method is discussed in the full version of this paper [22] and provides an efficient reduction. The penalty factor here is substantially lower than that of the general reduction.

Waters' Technique: Waters proposed a technique [24] that can be used here to achieve a *tight* existential unforgeability reduction in the standard model at the price of large public keys. To use this technique, we needs to (i) add ℓ random elements $h_1, h_2, \ldots, h_\ell$ from $\mathbb{G}_1$ to mpk in Setup algorithm, where ℓ is the (maximum) bit length of messages, and (ii) replace all instances of $g_1^m h$ in the above scheme with $W(m)$, where Waters function $W(\cdot)$ is defined as $W(m) = h \prod h_i^{m_i}$, where m_i denotes the i-th bit of m. The concrete scheme is transcribed in the full version of this paper [22] for completeness.

We use the algebraic properties of our basic scheme to construct C-signatures in the next section and we do not know if C-signatures can be constructed based on the Waters modification of our basic t-ABS scheme. Thus, our construction of C-signatures admits only to the first two techniques.

Signer-Attribute Privacy: One can see that since $B \setminus S$ components of the converted signature are publicly simulatable, weak signer-attribute privacy is achieved. Furthermore, the iVerify protocol can be proved to be both a proof of knowledge and zero knowledge and hence full signer-attribute privacy is also achieved. We prove the following theorem in the full version of this paper [22]:

Theorem 2. *The above t-ABS⁺ scheme achieves both weak and full signer-attribute privacy.*

Efficiency: The iVerify protocol is of size linear in the size of the verification attribute set. This is in contrast with the discussed partial proofs of knowledge, which require proofs of size linear in the number of possible signers.

4   Threshold Attribute-Based C-Signatures

We define threshold attribute-based C-signatures to accommodate for construction of an anonymous credential scheme where users' pseudonyms are in the form of committed values. Hence, we extend t-ABS$^+$ schemes to schemes supporting efficient protocols for signing and verifying signatures on committed values (i.e., pseudonyms) in the following.

Definition: A t-ABCS scheme consists of a t-ABS$^+$ scheme defined on messages in the form $\tilde{m} = (m, r)$, a commitment scheme, and the following additional protocols:

iCSign: An *interactive signing protocol* for signing a committed value on the *signer* side and obtaining a signature on a committed value on the *user* side. The public inputs are the t-ABS master public key mpk, the commitment public key cpk, and a commitment M. User's private inputs are the message to be signed (m, r) containing a random value r such that $M = \mathsf{Commit}(cpk, m, r)$. Signer's private input is its signing key ssk. At the end of the protocol execution the user will output signer's signature σ on m.

iHVerify: An *interactive verification protocol* for proving knowledge of a converted signature on the *prover* side and verifying a converted signature on the *verifier* side. The public inputs are the t-ABS master public key mpk and verifier's verification attribute set B. Prover's private input is the tuple $(m, r, \tilde{\sigma})$ such that $\tilde{\sigma}$ is a converted signature on (m, r). The verifier has no private input. At the end of the protocol execution the verifier will output a binary value reflecting prover's converted signature validity against B.

iCVerify: An *interactive verification protocol* for proving knowledge of a converted signature on a committed value on the *prover* side and verifying a converted signature on a committed value on the *verifier* side. The public inputs are the t-ABS master public key mpk, the commitment public key cpk, a commitment M', and verifier's verification attribute set B. Prover's private input is the tuple $(m, r, r', \tilde{\sigma})$ such that $M' = \mathsf{Commit}(cpk, m, r')$ and $\tilde{\sigma}$ is a converted signature on (m, r). The verifier has no private input. At the end of the protocol execution the verifier will output a binary value reflecting prover's converted signature validity against B.

Correctness: For iCSign, if σ is the output the user gets from running the protocol with a signer with attribute set A, then $\mathsf{Convert}(mpk, (m, r), \sigma, B)$ must pass the verification test for any B if $|A \cap B| \geq d$. For iHVerify and iCVerify, if the prover and the verifier follow the protocol, then the verifier's outputs must be the same as $\mathsf{CvtVerify}(mpk, (m, r), \tilde{\sigma}, B)$.

Security: Besides security of the underlying t-ABS$^+$ scheme, we require the following security properties from the additional protocols. We require that the iCSign protocol is secure in the following senses:

Security for the user: Users with different private inputs m should remain indistinguishable for the signer, even if the signer acts maliciously. In particular, we require that for any mpk and cpk, there is no malicious signer that can distinguish if it is interacting with a user with private input containing m_0 or with a user with private input containing m_1.

Security for the signer: The protocol should reveal (almost) no information other than a single signature on a known committed value to a user, even though the user acts maliciously. In particular, we require that for any mpk and cpk, and for any (possibly malicious) user, there exists a *simulator*, that with only a *one-time* access to the signing oracle, can simulate a signer's interaction with the user.

We require that iHVerify is (i) zero knowledge (*security for the prover*) and (ii) a proof of knowledge of a triplet $(m, r, \tilde{\sigma})$ s.t. $\mathsf{CvtVerify}(mpk, (m, r), \tilde{\sigma}, B) = 1$ (*security for the verifier*). We require that iCVerify is (i) zero knowledge (*security for the prover*) and (ii) a proof of knowledge of a quadruple $(m, r, r', \tilde{\sigma})$ s.t. $M' = \mathsf{Commit}(cpk, m, r')$ and $\mathsf{CvtVerify}(mpk, (m, r), \tilde{\sigma}, B) = 1$ (*security for the verifier*).

A t-ABCS scheme is said to be correct and secure if the underlying t-ABS$^+$ scheme defined on messages in the form $\tilde{m} = (m, r)$, is correct and C-EUF-CMAA-secure, the commitment scheme is correct and secure, and the associated iCSign, iHVerify and iCVerify protocols are correct and secure for the user, signer, prover and verifier in the above senses.

4.1   Construction

Our t-ABS$^+$ scheme can be modified to be defined on messages in the form of $\tilde{m} = (m, r)$ as follows: (i) in the Setup algorithm, add a random element g_3 from $\mathbb{G}_1$ to the master public key and set $mpk = (g, g_1, g_2, g_3, t_1, t_2, \ldots, t_{n+1}, h)$, and (ii) in the Sign, Verify, Convert, and CvtVerify algorithms, replace g_1^m with $g_1^m g_3^r$ everywhere. We will not use the iVerify protocol. The scheme is provided in the full version of this paper [22] for completion.

For the commitment scheme, consider the scheme with $cpk = (g_1, g_3)$ and $\mathsf{Commit}(cpk, m, r) = g_1^m g_3^r$. This scheme is (unconditionally) hiding and (computationally) binding if the discrete logarithm problem is hard. For this commitment scheme and the above t-ABS scheme, we introduce the following additional protocols to construct a t-ABCS scheme altogether. The iCSign protocol is as follows:

1. the user gives a ZK-PoK of (m, r) such that $M = g_1^m \cdot g_3^r$.
2. the signer picks random elements s_i in $\mathbb{Z}_p$ for all $i \in A$, calculates the signature as below, and sends it to the user.

$$\sigma = \Big\langle \quad A, \quad \{ \quad ssk_{1i}(Mh)^{s_i}, \quad ssk_{2i}, \quad g^{s_i} \quad \}_{i \in A} \quad \Big\rangle$$

The iHVerify and iCVerify protocols are as follows:

1. The signature holder randomizes the converted signature by first choosing random elements s_i' and r_i' for $i \in B$ and then calculating the following. Note that the resulting randomized converted signature is a valid converted signature itself.

$$\check{\sigma}_{1i} = \tilde{\sigma}_{1i} \cdot T(i)^{r_i'} \left(g_1^m g_3^r \cdot h\right)^{s_i'}, \quad \check{\sigma}_{2i} = \tilde{\sigma}_{2i} \cdot g^{r_i'}, \quad \check{\sigma}_{3i} = \tilde{\sigma}_{3i} \cdot g^{s_i'}$$

2. The signature holder chooses random values τ_i for all $i \in B$ and sets $\hat{\sigma}_{1i} \leftarrow \check{\sigma}_{1i}^{1/\tau_i}$ and sends $\langle \{\hat{\sigma}_{1i}, \check{\sigma}_{2i}, \check{\sigma}_{3i}\}_{i \in B} \rangle$ to the verifier.
3. Both the signature holder and the verifier calculate the following for all $i \in B$:

$$u_0 \leftarrow e(g_2, g_1) \qquad\qquad u_{1i} \leftarrow e(\hat{\sigma}_{1i}, g)^{\Delta_{i,B}(0)}$$
$$u_2 \leftarrow \prod_{i \in B} e(T(i), \check{\sigma}_{2i})^{\Delta_{i,B}(0)} \qquad u_{31} \leftarrow \prod_{i \in B} e(g_1, \check{\sigma}_{3i})^{\Delta_{i,B}(0)}$$
$$u_{32} \leftarrow \prod_{i \in B} e(g_3, \check{\sigma}_{3i})^{\Delta_{i,B}(0)} \qquad u_{33} \leftarrow \prod_{i \in B} e(h, \check{\sigma}_{3i})^{\Delta_{i,B}(0)}$$

4. The signature holder performs the following ZK-PoK for the verifier based on the protocol:
   for iHVerify:
   ZK-PoK$\{\ (m, r, \{\tau_i\}_{i \in B}) : \quad \prod_{i \in B} u_{1i}^{\tau_i} = u_0 u_2 u_{31}^m u_{32}^r u_{33}\ \}$
   for iCVerify:
   ZK-PoK$\{\ (m, r, r', \{\tau_i\}_{i \in B}) : M' = g_1^m g_3^{r'} \wedge \prod_{i \in B} u_{1i}^{\tau_i} = u_0 u_2 u_{31}^m u_{32}^r u_{33}\ \}$

Correctness: In the full version of this paper [22] we briefly show that the above protocols are correct.

Security: On unforgeability of our scheme, in the full version of this paper [22] we prove Theorem 3 that comes in the following. As discussed under un-forgeability of our t-ABS scheme, the theorem implies that the scheme is also EUF-CMAA-secure with a loose reduction or is EUF-CMAA-secure with an efficient reduction in the random oracle model. On security of our additional protocols, we prove the Theorem 4 that comes in the following in the full version of this paper [22]. Furthermore, hardness of the CDH problem implies hardness of the discrete logarithm problem, which, in turn, is sufficient for the commitment scheme to be secure. Thus our t-ABCS scheme is secure as per our definition if CDH is hard.

Theorem 3. *The above t-ABCS scheme, i.e. our t-ABS$^+$ scheme defined on messages in the form $\tilde{m} = (m, r)$, is* C-SUF-CMAA-*secure if the CDH problem is hard.*

Theorem 4. *The above protocols* iCSign, iHVerify, *and* iCVerify *are secure for the user, signer, prover and verifier.*

5   Threshold Attribute-Based Anonymous Credential Systems

A credential system involves users and organizations. Organizations issue *cre-dentials* to users and users prove possession of their credentials to organizations to get a service or be granted new credentials. An *anonymous credential system* is one in which users are known to organizations only by their *pseudonyms*. Such pseudonyms should be well-defined, i.e., each pseudonyms must belong to only one user.

A basic anonymous credential system must support a minimal of three basic protocols, respectively for *forming* pseudonyms, *granting* credentials, and *verifying* credentials. All these protocols are between a user and an organization. An optional property is that the system supports two types of verification protocols: one for verifying possession of a credential on a formed pseudonym, and another for simply verifying possession of a credential, which implements more efficiently than the former.

An attribute-based anonymous credential system is one with attribute-based organizations. Each organization is given a signing key based on its attributes by a trusted authority. A threshold attribute-based anonymous credential system is an attribute-based anonymous credential system that supports threshold attribute-based verification. We assume that the "credential types" in the system are fixed and each organization is given extra attributes for the credential types they can issue. Hence, a credential only consists of the identity of the credential holder and the signature of the organization on it, using the appropriate signing attribute for the credential type.

5.1   Security Framework

To formalize a security definition for a threshold attribute-based anonymous credential system, we use the simulation paradigm. Two models of a credential system, an ideal model and a real model, are presented. We consider *static active* adversaries, i.e., the adversary can corrupt a fixed set of entities in the beginning of the system execution. In the real model, the entities communicate with each other directly. However, in the ideal model, they communicate via a *trusted party*, which locally performs all calculations needed to carry out a transaction and just reports the corresponding outputs to the entities involved. To be able to compare the two systems, a scheduler entity is introduced to the system, called the *environment*. The environment schedules transactions in the system and gathers output information. In particular, it tells each entity which protocol to carry out and with whom. At the end of the protocol, the entities involved report back the outcome of the protocol. The environment proceeds in periods and in each period it schedules only one transaction. Concurrent scheduling is not considered[1]. This framework of security is the same as Camenisch and Lysyanskaya's framework in [3,18].

A real system is said to be secure if it *emulates* a secure ideal model. Emulation means that for any arbitrary scheduling and any arbitrary real-model adversary, an ideal-model adversary can be found such that the two systems are indistinguishable in the eyes of the environment. An implication of this property is that whatever the real-model adversary can extract from the real-model system, the ideal-model adversary can extract from the ideal-model system. Now, since the ideal system is defined in a way that its security against ideal-model adversaries is guaranteed, the security of the real-model system follows.

---

[1] Our security notion is weaker than Canetti's *Universal Composability* [5]. We do not address a *composable* notion of security, but rather a *stand-alone* one.

5.2   Ideal Model for t-ABACS

The Model: In the ideal, model users and organizations interact through a trusted party T who makes sure the system remains anonymous and secure. There is a public universal set of attributes $\mathbb{U}$. Each organization O has a set of attributes $A_O \in \mathbb{U}$, which is assumed to be public and fixed during the life of the system. Besides, each organization O, based on its policy at a certain time, might be interested to verify users' credentials against a set of attributes. We denote this set by $B_O \in \mathbb{U}$. B_O is assumed to be made public by the organization and fixed during each period. T maintains three lists: a list of user passwords L_P, a list of user pseudonyms L_N, and a list of issued credentials L_C. L_P contains pairs of user names U and their passwords K_U and is used to authenticate (and hence identify) users. We assume that initially this list contains all user names and passwords in the system. L_N contains triplets of user names U, organization names O, and their pseudonyms N_{UO}, and is initially empty and is filled by T as users form pseudonyms with organizations. L_C contains pairs of user pseudonyms N_{UI} and issuer organization attributes A_I, and is initially empty and is filled by T as organizations grant credentials to users. User pseudonyms are assumed to be chosen randomly and independent of user identities. We transcribe the model briefly in the following. For a full transcript see the full version of this paper [22]. There are four basic operations as follows.

FormNym $[U \leftrightarrow \mathsf{T} \leftrightarrow O] (N_{UO})$:   This is a protocol between a user U and an organization O. U wants to form a pseudonym N_{UO} with O. The protocol flow is as follows.

$U -\!(\,\mathsf{FormNym}, (U, K_U), N_{UO}, O\,)\!\rightarrowtail \mathsf{T}$
$\mathsf{T} -\!(\,\mathsf{FormNym}, N_{UO}\,)\!\rightarrowtail O$ if $(U, K_U) \in L_P$
$O -\!(\,d\,)\!\rightarrowtail \mathsf{T}$
$\mathsf{T} -\!(\,d\,)\!\rightarrowtail U$ (If $d = 1$, T also stores the triple (U, O, N_{UO}) in L_N.)

GrantCred $[U \leftrightarrow \mathsf{T} \leftrightarrow I] (N_{UI})$:   This is a protocol between a user U and an organization I. U, who has already formed a pseudonym N_{UI} with I, wants to be granted a credential by I on N_{UI}. I has set of attributes A_I. The protocol flow is as follows.

$U -\!(\,\mathsf{GrantCred}, (U, K_U), N_{UI}, I\,)\!\rightarrowtail \mathsf{T}$
$\mathsf{T} -\!(\,\mathsf{CrantCred}, N_{UI}\,)\!\rightarrowtail I$ if $(U, K_U) \in L_P \,\wedge\, (U, I, N_{UI}) \in L_N$
$I -\!(\,d\,)\!\rightarrowtail \mathsf{T}$
$\mathsf{T} -\!(\,d\,)\!\rightarrowtail U$ (If $d = 1$, T also stores the pair (N_{UI}, A_I) in L_C.)

VerifyCred $[U \leftrightarrow \mathsf{T} \leftrightarrow V] (N_{UI})$:   This is a protocol between a user U and an organization V. U has a credential issued by I on N_{UI}. A_I is the attribute set of I and B_V is the verification attribute set of V. U wants V to verify that she has been issued a credential by an organization whose attributes A_I has at least d elements in common with V's verification attribute set B_V. The protocol flow is as follows.

$U \multimap (\texttt{VerifyCred}, (U, K_U), N_{UI}, V) \rightarrowtail \mathsf{T}$
$\mathsf{T} \multimap (\texttt{VerifyCred}) \rightarrowtail V$ if $(U, K_U) \in L_P \wedge [\exists I : (U, I, N_{UI}) \in L_N \wedge$
    $(N_{UI}, A_I) \in L_C \wedge |A_I \cap B_V| \geq d]$
$V \multimap (\texttt{Ack}) \rightarrowtail \mathsf{T}$
$\mathsf{T} \multimap (\texttt{Ack}) \rightarrowtail U$

VerifyCredOnNym $[U \leftrightarrow \mathsf{T} \leftrightarrow V] (N_{UI}, N_{UV})$:   This is a protocol between a user U and an organization V. U has a credential issued by I on N_{UI} and has already formed a pseudonym N_{UV} with V. A_I is the attribute set of I and B_V is the verification attribute set of V. U wants V to verify that she has been issued a credential by an organization whose attributes A_I has at least d elements in common with V's verification attribute set B_V. The protocol flow is as follows.

$U \multimap (\texttt{VerifyCredOnNym}, (U, K_U), N_{UI}, V, N_{UV}) \rightarrowtail \mathsf{T}$
$\mathsf{T} \multimap (\texttt{VerifyCredOnNym}, N_{UV}) \rightarrowtail V$ if $(U, K_U) \in L_P \wedge (U, V, N_{UV}) \in L_N \wedge$
    $[\exists I : (U, I, N_{UI}) \in L_N \wedge (N_{UI}, A_I) \in L_C \wedge |A_I \cap B_V| \geq d]$
$V \multimap (\texttt{Ack}) \rightarrowtail \mathsf{T}$
$\mathsf{T} \multimap (\texttt{Ack}) \rightarrowtail U$

Properties of the Model: The above model ensures a number of security and privacy properties, including attribute-based unforgeability of credentials (since records can be added to L_C only when a credential is issued), anonymity of users (since users are known to organizations merely by their pseudonyms), unlinkability of transactions (since the user uses different pseudonyms in the transactions), non-transferability of credentials (since L_N is checked during both issuing and verifying the credentials), and issuer-attribute privacy (since Verifiers do not find out anything other than the fact that the issuer attributes satisfy the verification policy).

5.3   A Concrete t-ABACS System

Consider a t-ABACS scheme as defined above. We propose the following t-ABACS system based on this scheme. We assume that there is a trusted signing key generator authority outside the system that issues signing keys for organizations based on their attributes. Organizations' attribute sets are all subsets of a universal public attribute set $\mathbb{U}$. There is also a trusted pseudonym consistency authority that issues zeroth credentials to users and makes sure that pseudonyms remain well-defined.

Init(1^k): During the initiation phase, each user U in the system picks a secret SK_U and each organization O with attributes A_O contacts the system signing key generator to get a signing secret key ssk_O.

FormNym $[U \leftrightarrow O] (N_{UO})$: User U picks a random r_O and forms a commitment to her secret $N_{UO} = \texttt{Commit}(cpk, SK_U, r_O)$. She then sends N_{UO} to O and proves that she knows the pair (SK_U, r_O) using a ZK-PoK protocol. U and O save N_{UO} as the pseudonym of U with O.

GrantCred $[U \leftrightarrow I] (N_{UI})$: User U and credential issuer I who have already formed a pseudonym $N_{UI} = \texttt{Commit}(cpk, SK_U, r_I)$, carry out the iCSign protocol on public inputs (mpk, cpk, N_{UI}), user's private input (SK_U, r_I), and issuer's private input ssk_I. User stores her output σ.

VerifyCred $[U \leftrightarrow V]$ (N_{UI}): User U calculates $\tilde{\sigma} = \mathsf{Convert}(mpk, (SK_U, r_I), \sigma, B_V)$. Then, user U and verifier V carry out the iHVerify protocol on public inputs (mpk, B_V) and user's private inputs $(SK_U, r_I, \tilde{\sigma})$.

VerifyCredOnNym $[U \leftrightarrow V]$ (N_{UI}, N_{UV}): User U calculates the converted signature $\tilde{\sigma} = \mathsf{Convert}(mpk, (SK_U, r_I), \sigma, B_V)$. Then, user U and verifier V who have already formed a pseudonym $N_{UV} = \mathsf{Commit}(cpk, SK_U, r_V)$, carry out the iCVerify protocol on public inputs (mpk, cpk, N_{UV}, B_V) and user's private inputs $(SK_U, r_I, r_V, \tilde{\sigma})$.

We assume that pseudonyms are well-defined, i.e., each pseudonym belongs to only one user. This can be guaranteed by requiring all users to reveal their identity and pseudonyms to a trusted authority and get a *zeroth credential* for each of their pseudonyms. Possession of this zeroth credential is proved to an organization at the time of forming the pseudonym.

We prove that the above system is a secure implementation of the t-ABACS system. In particular, in the full version of this paper [22] we prove the following theorem. In our proof, we provide an ideal-model adversary for any real-model adversary and briefly show how they remain indistinguishable in the eyes of the environment as long as the underlying t-ABCS scheme is secure.

Theorem 5. *The above concrete t-ABACS system emulates the ideal model for a t-ABACS if the underlying t-ABCS scheme is* EUF-CMAA-*secure.*

6   Conclusions and Open Problems

We introduced a new scheme called a threshold attribute-based signature, which allows verification of signatures as originating from a fuzzy signer. We proposed a basic t-ABS scheme and provided a tight security reduction for its selective unforgeability based on the CDH assumption and discussed how to achieve existential unforgeability in both the standard and the random oracle models. We showed that our basic t-ABS scheme admits to an efficient threshold attribute-based C-signature that can be used as a building block to realize privacy-enhanced anonymous credential systems. However, existential unforgeability of our t-ABCS scheme is based on a loose generic reduction. Designing t-ABCS schemes with tight existential unforgeability remains an open problem.

Our attribute-based signature scheme and hence our attribute-based anonymous credential system only support simple threshold verification policies. A possible future enhancement is to design schemes that support more complex verification policies. That is, to generalize our schemes to ones in which the verification algorithm gets as input a (complex) policy, rather than a verification attribute set, and the verification goes through if the signer attributes satisfy the verification policy. Systems based on such schemes can accommodate for a larger application spectrum.

References

1. Bellare, M., Goldreich, O.: On Defining Proofs of Knowledge. In: Brickell, E.F. (ed.) CRYPTO 1992. LNCS, vol. 740, pp. 390–420. Springer, Heidelberg (1993)
2. Boneh, D., Boyen, X.: Efficient Selective-ID Secure Identity-Based Encryption Without Random Oracles. In: Cachin, C., Camenisch, J.L. (eds.) EUROCRYPT 2004. LNCS, vol. 3027, pp. 223–238. Springer, Heidelberg (2004)
3. Camenisch, J., Lysyanskaya, A.: An Efficient System for Non-transferable Anonymous Credentials with Optional Anonymity Revocation. In: Pfitzmann, B. (ed.) EUROCRYPT 2001. LNCS, vol. 2045, pp. 93–118. Springer, Heidelberg (2001)
4. Camenisch, J., Stadler, M.: Efficient Group Signature Schemes for Large Groups. In: Kaliski Jr., B.S. (ed.) CRYPTO 1997. LNCS, vol. 1294, pp. 410–424. Springer, Heidelberg (1997)
5. Canetti, R.: Universally Composable Security: A New Paradigm for Cryptographic Protocols. In: FOCS 2001, pp. 136–145. IEEE, Los Alamitos (2001)
6. Chaum, D.: Security Without Identification: Transaction Systems to Make Big Brother Obsolete. Commun. ACM 28(10), 1030–1044 (1985)
7. Chaum, D., Evertse, J.-H.: A Secure and Privacy-protecting Protocol for Transmitting Personal Information Between Organizations. In: Odlyzko, A.M. (ed.) CRYPTO 1986. LNCS, vol. 263, pp. 118–167. Springer, Heidelberg (1987)
8. Cramer, R., Damgård, I., MacKenzie, P.D.: Efficient Zero-Knowledge Proofs of Knowledge Without Intractability Assumptions. In: Imai, H., Zheng, Y. (eds.) PKC 2000. LNCS, vol. 1751, pp. 354–373. Springer, Heidelberg (2000)
9. Cramer, R., Damgård, I., Schoenmakers, B.: Proofs of Partial Knowledge and Simplified Design of Witness Hiding Protocols. In: Desmedt, Y.G. (ed.) CRYPTO 1994. LNCS, vol. 839, pp. 174–187. Springer, Heidelberg (1994)
10. Gentry, C., Silverberg, A.: Hierarchical ID-Based Cryptography. In: Zheng, Y. (ed.) ASIACRYPT 2002. LNCS, vol. 2501, pp. 548–566. Springer, Heidelberg (2002)
11. Goldreich, O., Micali, S., Wigderson, A.: Proofs that Yield Nothing But Their Validity for All Languages in NP Have Zero-Knowledge Proof Systems. J. ACM 38(3), 691–729 (1991)
12. Goldwasser, S., Micali, S., Rackoff, C.: The Knowledge Complexity of Interactive Proof Systems. SIAM J. Comput. 18(1), 186–208 (1989)
13. Guo, S., Zeng, Y.: Attribute-based Signature Scheme. In: Int'l Conf. on Information Security and Assurance (ISA 2008), pp. 509–511. IEEE, Los Alamitos (2008)
14. Khader, D.: Attribute Based Group Signature with Revocation. Cryptology ePrint Archive, Report 2007/241 (2007), http://eprint.iacr.org/2007/241
15. Khader, D.: Attribute Based Group Signatures. Cryptology ePrint Archive, Report 2007/159 (2007), http://eprint.iacr.org/2007/159
16. Khader, D.: Authenticating with Attributes. Cryptology ePrint Archive, Report 2008/031 (2008), http://eprint.iacr.org/2008/031
17. Li, J., Kim, K.: Attribute-Based Ring Signatures. Cryptology ePrint Archive, Report 2008/394 (2008), http://eprint.iacr.org/2008/394
18. Lysyanskaya, A.: Signature Schemes and Applications to Cryptographic Protocol Design. Ph.D thesis, Massachusetts Institute of Technology (2002)
19. Lysyanskaya, A., Rivest, R.L., Sahai, A., Wolf, S.: Pseudonym Systems. In: Heys, H.M., Adams, C.M. (eds.) SAC 1999. LNCS, vol. 1758, pp. 184–199. Springer, Heidelberg (2000)
20. Maji, H., Prabhakaran, M., Rosulek, M.: Attribute-Based Signatures: Achieving Attribute-Privacy and Collusion-Resistance. Cryptology ePrint Archive, Report 2008/328 (2008), http://eprint.iacr.org/2008/328

21. Sahai, A., Waters, B.: Fuzzy Identity-Based Encryption. In: Cramer, R. (ed.) EU-ROCRYPT 2005. LNCS, vol. 3494, pp. 457–473. Springer, Heidelberg (2005)
22. Shahandashti, S.F., Safavi-Naini, R.: Threshold Attribute-Based Signatures and Their Application to Anonymous Credential Systems. Cryptology ePrint Archive, Report 2009/126 (2009), http://eprint.iacr.org/2009/126
23. Shamir, A.: Identity-Based Cryptosystems and Signature Schemes. In: Blakely, G.R., Chaum, D. (eds.) CRYPTO 1984. LNCS, vol. 196, pp. 47–53. Springer, Heidelberg (1985)
24. Waters, B.: Efficient Identity-Based Encryption Without Random Oracles. In: Cramer, R. (ed.) EUROCRYPT 2005. LNCS, vol. 3494, pp. 114–127. Springer, Heidelberg (2005)
25. Yang, P., Cao, Z., Dong, X.: Fuzzy Identity Based Signature. Cryptology ePrint Archive, Report 2008/002 (2008), http://eprint.iacr.org/2008/002

Anonymity from Public Key Encryption to Undeniable Signatures

Laila El Aimani

b-it, Dahlmannstr. 2, Universität Bonn, 53113 Bonn, Germany
elaimani@bit.uni-bonn.de

Abstract. Anonymity or "key privacy" was introduced in [1] as a new security notion a cryptosystem must fulfill, in some settings, in addition to the traditional indistinguishability property. It requires an adversary not be able to distinguish pairs of ciphertexts based on the keys under which they are created. Anonymity for undeniable signatures is defined along the same lines, and is considered a relevant requirement for such signatures.

Our results in this paper are twofold. First, we show that anonymity and indistinguishability are not as orthogonal to each other (i.e., independent) as previously believed. In fact, they are equivalent under certain circumstances. Consequently, we confirm the results of [1] on the anonymity of ElGamal's and of Cramer-Shoup's schemes, based on existing work about their indistinguishability. Next, we constructively use anonymous encryption together with secure digital signature schemes to build anonymous convertible undeniable signatures. In this context, we revisit a well known undeniable signature scheme, whose security remained an open problem for over than a decade, and prove that it is not anonymous. Moreover, we repair this scheme so that it provides the anonymity feature and analyze its security in our proposed framework. Finally, we analyze an efficient undeniable signature scheme, which was proposed recently, in our framework; we confirm its security results and show that it also enjoys the selective conversion feature.

Keywords: Encryption schemes, Anonymity, KEM/DEM, Convertible undeniable signatures, Generic construction.

1 Introduction

The classical security notion a cryptosystem must fulfill is data privacy or indistinguishability. It captures the inability of an attacker to distinguish pairs of ciphertexts based on the messages they encrypt. In [1], the authors propose an additional notion called anonymity, which formalizes the property of key privacy. As a matter of fact, an adversary, in possession of two public keys and a ciphertext formed by encrypting some data under one of the two keys, should not be able to tell under which key the ciphertext was created. The formalization of this new security notion was motivated by the numerous applications

B. Preneel (Ed.): AFRICACRYPT 2009, LNCS 5580, pp. 217–234, 2009.
© Springer-Verlag Berlin Heidelberg 2009

in which surfaced anonymity, such as undeniable signatures. In fact, undeniable signatures were introduced in [3] to limit the self-authenticating property of digital signatures. In these signatures, the verifier has no means to check the validity/invalidity of a purported signature without interacting with the signer via the confirmation/denial protocols. Undeniable signatures proved critical in situations where privacy or anonymity is an important issue. A typical example is this real-life scenario; An employer issues a job offer to a certain candidate. Naturally, the employer needs to compete with the other job offers in order to attract the good candidate. Therefore, he does not wish the offer to be revealed to his competitors. At the same time, the candidate needs more than a verbal or unsigned agreement in order to protect himself from the employer not keeping his promise. In such a situation, undeniable signatures come to rescue since they are : 1. only verified with the help of the signer, 2. non transferable, 3. binding in the sense that a signer cannot deny a signature he has actually issued.

1.1   Related Work

Exploring the relationship between data privacy and key privacy in public key encryption schemes came very natural to researchers. Indeed, in their seminal work [1], the authors observe that the new notion is totally different from data privacy, as there exist encryption schemes that satisfy one notion but not the other. They also claimed that "it is not hard to see that the goals of data privacy and key privacy are orthogonal". Recently, this claim was proven in [15] by exhibiting a technique that upgrades the key privacy (in a cryptosystem already enjoying this property) but destroys the data privacy, and vice versa. Such a result can be considered as negative, since it only shows how to build a cryptosystem which has one property but not the other. But what about the opposite? Can one specify simple assumptions to hold in an encryption scheme so that key privacy yields data privacy and vice versa? Such an approach has been considered in the literature for a different primitive, namely undeniable signatures. In fact, invisibility and anonymity are two security properties that are closely related in undeniable signatures. The first one requires an adversary not be able to distinguish a valid signature on a certain message from any uniformly chosen bit-string from the signatures space, whereas the second notion refers to the hardness of, in possession of a signature and two public keys, telling under which key the signature was created. Since the introduction of undeniable signatures, these two notions were treated separately and many schemes emerged which either meet the first notion or the second, until 2003 where a comprehensive study [8] led to the conclusion that anonymity and invisibility are essentially the same under certain conditions. With such a result, one can seek only one notion when designing undeniable signatures.

Besides, since the notion of anonymity is formalized for encryption schemes, it would be good to see whether one could constructively use anonymous encryption to build anonymous undeniable signatures using the famous "encryption of a signature" paradigm. In fact, this paradigm was successfully used to design generic constructions for designated confirmer signatures [2]. It consists in generating a digital signature on the message to be signed, then encrypting the given signature using an appropriate cryptosystem. Confirmation or denial of a given signature are achieved via proofs of knowledge of NP (co-NP) statements (inherent to the relations defined by the decryption and the verification algorithms in the underlying cryptosystem and signature scheme respectively). Such a resort to proofs of knowledge of general NP (co-NP) statements has constituted the main weakness of the method since these proofs are not necessarily efficient enough for practice. This problem was solved by allowing homomorphic cryptosystems in the design of [6], which impacts positively the efficiency of the confirmation/denial protocols, without compromising the security of the resulting undeniable signatures.

1.2   Our Contributions

Our results are twofold. In an attempt to bridge the gap between anonymity and indistinguishability in encryption schemes, we specify simple conditions to hold in the given cryptosystem so that anonymity implies indistinguishability and vice versa. This will allow a direct use of existing results about data/key privacy of asymmetric encryption schemes rather than "doing the work" from scratch as claimed in [1]. As a consequence, we confirm the results in [1] that prove the anonymity under chosen plaintext attacks of ElGamal's cryptosystem and the anonymity under chosen ciphertext attacks of Cramer-Shoup's encryption, assuming the intractability of the Decisional Diffie-Hellman problem (DDH). It is worth noting that Halevi [9] has already done the first direction, namely, observed a simple condition to hold in a cryptosystem so that indistinguishability yields anonymity. In the present work, we express this condition analogously to the condition defined in [8], which allows to derive, in undeniable signatures, anonymity from a variant of invisibility. Moreover, we define another sufficient condition to derive indistinguishability from anonymity.

Next, we extend the panel of anonymous encryption applications by a generic construction of anonymous undeniable signatures. To be specific, we use as building blocks anonymous cryptosystems under *chosen plaintext attacks* obtained from the "hybrid encryption paradigm", thus allowing also homomorphic schemes as in [6], and secure digital signatures. In this context, we define the notion of anonymity for Key Encapsulation Mechanisms (KEMs) and provide a similar study to that of public key encryption on the equivalence between anonymity and indistinguishability for KEMs. As an application to our generic construction, we revisit a well known undeniable signature [5], whose security remained an open problem since 1996 [1] and which was reported recently in [11] [2]. We show that the scheme does not provide the anonymity and thus the invisibility feature. Moreover, we modify it so that it becomes anonymous and we analyze its security in our proposed framework. Finally, we confirm the security

---

[1] In Sec.5.2 of [5], the authors wrote that "We therfore conjecture that..." on the invisibility of their scheme.

[2] In Sec.1.1 of [11], the authors wrote "However, invisibility is not proved in these schemes..." referring to the schemes proposed in [5].

results of a very efficient scheme that was proposed recently and show that it also enjoys the selective conversion feature.

2 Preliminaries

2.1 Public Key Encryption Schemes

An asymmetric encryption scheme Γ consists of the following algorithms:

1. Γ.keygen is a probabilistic key generation algorithm which returns pairs of public and private keys (pk, sk) depending on the security parameter κ,
2. Γ.encrypt is a probabilistic encryption algorithm which takes as input a public key pk and a plaintext m, runs on a random tape u and returns a ciphertext c, and
3. Γ.decrypt is a deterministic decryption algorithm which takes on input a private key sk and a ciphertext c, and returns the corresponding plaintext m or the symbol $\perp$. We require that if (pk, sk) is a valid key pair, then $\forall m\colon \mathsf{decrypt_{sk}}\left(\mathsf{encrypt_{pk}}(m)\right) = m$.

The classical security goal a cryptosystem should fulfill is indistinguishability, denoted IND. Like any security goal, when combined with the three attack models, namely, CPA (Chosen Plaintext Attack), PCA (Plaintext Checking Attack) and CCA (Chosen Ciphertext Attack), it yields three security notions IND-ATK, for ATK $\in \{\mathrm{CPA}, \mathrm{PCA}, \mathrm{CCA}\}$:

Definition 1. *A cryptosystem Γ is IND-ATK if no distinguisher $\mathcal{A}$ has a non negligible advantage in the following game:*

- **Phase 1.** *The challenger runs Γ.keygen to obtain a key pair (pk, sk). The adversary $\mathcal{A}$ is given the public key pk and is allowed to query a plaintext checking oracle[3] in case ATK = PCA, or a decryption oracle[4] in case ATK = CCA, for any input of his choice.*
- **Challenge.** *Eventually, $\mathcal{A}$ outputs two messages m_0 and m_1. The challenger picks uniformly at random a bit b from $\{0,1\}$, encrypts m_b and feeds the result $c^\star$ to $\mathcal{A}$.*
- **Phase 2.** *$\mathcal{A}$ is again given access to the oracles he had access to in **Phase 1**, which accept now any query except when it is on the challenge ciphertext $c^\star$. At the end of this phase, $\mathcal{A}$ outputs a bit b'. $\mathcal{A}$ wins the game if $b = b'$.*

We define $\mathcal{A}$'s advantage as $\mathsf{Adv}(\mathcal{A}) = |\Pr[b = b'] - \frac{1}{2}|$, where the probability is taken over the random choices of the adversary $\mathcal{A}$ and of the challenger.
An encryption scheme is (t, ϵ, q_o)-IND-ATK-secure if no adversary $\mathcal{A}$, operating in time t and issuing q_o queries to the available oracles, wins the game with probability greater than ϵ.

---

[3] A plaintext checking oracle gets as input a pair (m, c) and decides whether the ciphertext c encrypts the message m.

[4] A decryption oracle gets a ciphertext c as input and outputs the corresponding plaintext.

An encryption scheme is said to provide anonymity if it is difficult for an adversary to distinguish pairs of ciphertexts based on the keys under which they are created. The formal definition as introduced in [1] is defined as follows.

Definition 2. *A cryptosystem Γ is ANO-ATK-secure (ATK $\in$ {CPA, PCA, CCA}) if no distinguisher $\mathcal{A}$ has a non negligible advantage in the following game:*

- **Phase 1.** *The challenger runs twice the algorithm Γ.keygen to obtain two public keys pk_0 and pk_1 with the two corresponding private keys sk_0 and sk_1. The adversary $\mathcal{A}$ is given the public keys pk_0 and pk_1, and is allowed to query the plaintext checking oracles (corresponding to both keys pk_0 and pk_1) in case ATK = PCA, or the given decryption oracles in case ATK = CCA, for any message of his choice.*
- **Challenge.** *Eventually, $\mathcal{A}$ outputs a message m. The challenger picks uniformly at random a bit b from {0, 1}, encrypts m under pk_b, and feeds the result $c^\star$ to $\mathcal{A}$.*
- **Phase 2.** *$\mathcal{A}$ is again given access to the previous oracles defined in **Phase 1**, which accept now all queries except those on the challenge ciphertext $c^\star$. At the end of this phase, $\mathcal{A}$ outputs a bit b'. $\mathcal{A}$ wins the game if $b = b'$.*

We define $\mathcal{A}$'s advantage as $\mathsf{Adv}(\mathcal{A}) = |\Pr[b = b'] - \frac{1}{2}|$, where the probability is taken over the random choices of the adversary $\mathcal{A}$ and of the challenger.
An encryption scheme is (t, ϵ, q_o)-ANO-ATK-secure if no adversary $\mathcal{A}$, operating in time t and issuing q_o queries to the allowed oracles, wins the game with probability greater than ϵ.

2.2   Convertible Undeniable Signatures (CUS)

Syntax

Setup. On input the security parameter κ, outputs the public parameters.
Key generation. Generates probabilistically a key pair (sk, pk) consisting of the private and of the public keys respectively.
Signature. On input the public parameters, the private key sk and a message m outputs an undeniable signature μ.
Verification. This is an algorithm run by the signer to check the validity of an undeniable signature μ issued on m, using his private key sk.
Confirmation/Denial protocol. These are interactive protocols between a prover and a verifier. Their common input consists of the public parameters of the scheme, the signature μ and the message m in question. The prover, that is the signer, uses his private key sk to convince the verifier with the validity (invalidity) of the signature μ on m. At the end of the protocol, the verifier either accepts the proof or rejects it.
Selective conversion. Takes as input the undeniable signature to be converted, the message on which was issued the signature and the private key sk. The result is either $\perp$ or a string which can be universally verified as a valid digital signature.

Universal conversion. Releases a universal receipt, using sk, that makes all undeniable signatures universally verifiable.

Selective/Universal verification. On input a signature, a message, a receipt and the public key pk, outputs 1 if the signature is valid and 0 otherwise.

Security model. In addition to the completeness of the above algorithms/protocols, the proofs inherent to the confirmation/denial protocols must be sound and non-transferable. Moreover, a convertible undeniable signature scheme requires two further properties, that are unforgeability and anonymity.

Unforgeability. The standard security requirement that a convertible signature scheme should fulfill is the existential unforgeability against a chosen message attack (EUF-CMA). It is defined through the following game.

- **Setup.** The adversary $\mathcal{A}$ is given the public parameters of the scheme he trying to attack in addition to the universal receipt allowing the conversion of all undeniable signatures into ordinary ones.
- **Queries.** $\mathcal{A}$ queries the signing oracle adaptively on at most q_s messages.
- **Output.** At the end, $\mathcal{A}$ outputs a pair consisting of a message m, that has not been queried yet, and a string μ. $\mathcal{A}$ wins the game if μ is a valid undeniable signature on m.

We say that a convertible undeniable signature scheme is (t, ϵ, q_s)-EUF-CMA secure if there is no adversary, operating in time t, that wins the above game with probability greater than ϵ.

Anonymity. As in public key encryption, anonymity refers to the hardness of distinguishing signatures based on the keys under which they are created. It is defined as follows.

- $\mathcal{A}$ gets the public parameters in addition to two public key pk_0 and pk_1 from his challenger $\mathcal{R}$.
- **Phase 1.** $\mathcal{A}$ queries the signing, confirmation/denial and selective conversion oracles, corresponding to both keys pk_0 and pk_1, in an adaptive way.
- **Challenge.** Eventually, $\mathcal{A}$ outputs a message $m^\star$ that has not been queried before to both signing oracles and requests a challenge signature $\mu^\star$. $\mathcal{R}$ picks uniformly at random a bit $b \in_R \{0,1\}$, signs $m^\star$ using pk_b and feeds the result $\mu^\star$ to $\mathcal{A}$.
- **Phase 2.** $\mathcal{A}$ can adaptively query the previous oracles with the exception of not querying $m^\star$ to the signing oracles or $(m^\star, \mu^\star)$ to the confirmation/denial and selective conversion oracles.
- **Output.** $\mathcal{A}$ outputs a bit b'. He wins the game if $b = b'$. We define $\mathcal{A}$'s advantage as $\mathsf{Adv}(\mathcal{A}) = |\Pr[b = b'] - \frac{1}{2}|$

We say that a convertible undeniable signature scheme is $(t, \epsilon, q_s, q_v, q_{sc})$-ANO-CMA secure if no adversary operating in time t, issuing q_s queries to the signing oracles, q_v queries to the confirmation/denial oracles and q_{sc} queries to the selective conversion oracles wins the above game with advantage greater than ϵ.

3   Key Privacy versus Data Privacy for Encryption Schemes

We are now able to state our first result, namely, present conditions that suffice to conclude on the anonymity of an encryption scheme given existing results about its indistinguishability and vice versa. Our result builds from the work of [8] on undeniable signatures and extends it to public key encryption.

In this section, we stress that every choice of the security parameter κ defines a key space $\mathsf{PK} - \mathsf{SK}$ (corresponding to the space of key pairs $(\mathsf{pk}, \mathsf{sk})$ generated by the keygen algorithm), a message space M and a ciphertext space C. In particular, the ciphertext space C depends merely on κ and not on a specific key.

3.1   Key Privacy versus Data Privacy

Let Γ be a public key encryption scheme given by its three algorithms: Γ.keygen, Γ.encrypt and Γ.decrypt. The following are the properties needed to prove the relationship between key privacy and data privacy.

Property A. Let κ be a security parameter and let $(\mathsf{pk}, \mathsf{sk})$ be an output of Γ.keygen. Consider the uniform distribution on M. Then, the distribution on C corresponding to the random variable $\Gamma.\mathsf{encrypt}_{\mathsf{pk}}(m)$ $(m \xleftarrow{R} \mathsf{M})$ is computationally indistinguishable from uniform.

Property B. Let κ be a security parameter and let $m \in \mathsf{M}$ be an arbitrary message. Consider the distribution induced by the probabilistic algorithm Γ.keygen on the key space $\mathsf{PK} - \mathsf{SK}$. Then, from a key pair $(\mathsf{pk}, \mathsf{sk})$ sampled according to this distribution, the distribution on C corresponding to the random variable $\Gamma.\mathsf{encrypt}_{\mathsf{pk}}(m)$ is computationally indistinguishable from uniform.

Intuitively, Property A means basically the following: for a fixed key and varying messages, encryptions look random. As previously mentioned, the same property has been formulated differently in [9], where it requires encryptions of random messages to be independent of the public key used for encryption. Property B suggests that, for a fixed message and varying keys, encryptions look random.

We get now to the relation between anonymity and indistinguishability. Theorem 1 says that if Property A holds in an encryption scheme Γ, then indistinguishability implies anonymity. Theorem 2 requires Property B for anonymity to yield indistinguishability in a given cryptosystem. Both theorems stand in all attack models ($\mathrm{ATK} \in \{\mathrm{CPA}, \mathrm{PCA}, \mathrm{CCA}\}$).

Theorem 1. *Let Γ be a public key encryption scheme that has Property A. If Γ is IND-ATK-secure then it is ANO-ATK-secure, where $ATK \in \{CPA, PCA, CCA\}$.*

Proof. Given an anonymity adversary $\mathcal{A}^{\mathsf{ano-atk}}$, we will create an indistinguishability adversary $\mathcal{A}^{\mathsf{ind-atk}}$ in the same attack model ATK. Let pk_0 be the input to

$\mathcal{A}^{\mathsf{ind-atk}}$. $\mathcal{A}^{\mathsf{ind-atk}}$ will run Γ.keygen to generate a public key pk_1 together with its corresponding private key sk_1.

Queries made by $\mathcal{A}^{\mathsf{ano-atk}}$ are answered in the following way: if they are with respect to the key pk_0, they are forwarded to $\mathcal{A}^{\mathsf{ind-atk}}$'s own challenger. Otherwise, in case they are with respect to pk_1, they are answered by $\mathcal{A}^{\mathsf{ind-atk}}$ using the private key sk_1.

When $\mathcal{A}^{\mathsf{ano-atk}}$ outputs a message m_0 and requests a challenge, $\mathcal{A}^{\mathsf{ind-atk}}$ chooses a message m_1 uniformly at random from M that he will pass, together with m_0 to his challenger. $\mathcal{A}^{\mathsf{ind-atk}}$ will get an encryption Γ.encrypt$_{\mathsf{pk}_0}(m_b)$ of either m_0 or m_1 ($b \xleftarrow{R} \{0,1\}$) which he will forward to $\mathcal{A}^{\mathsf{ano-atk}}$. Queries by $\mathcal{A}^{\mathsf{ano-atk}}$ continue to be handled as before.

If Γ.encrypt$_{\mathsf{pk}_0}(m_b)$ corresponds to the encryption of m_0 (under pk_0), then with overwhelming probability it is not an encryption of m_0 under pk_1. Otherwise, if Γ.encrypt$_{\mathsf{pk}_0}(m_b)$ is the encryption of m_1 (under pk_0), then by virtue of Property A, Γ.encrypt$_{\mathsf{pk}_0}(m_1)$ is a random element in C and with overwhelming probability it is not an encryption of m_0 under either key.

At the end of the game, $\mathcal{A}^{\mathsf{ano-atk}}$ outputs a guess b' on the key under which Γ.encrypt$_{\mathsf{pk}_0}(m_b)$ was created. $\mathcal{A}^{\mathsf{ind-atk}}$ will then output the same guess b'.

Let ϵ be the advantage of $\mathcal{A}^{\mathsf{ano-atk}}$. We have $\epsilon = \Pr(b' = 0|b = 0) - \frac{1}{2}$. In fact, $\mathcal{A}^{\mathsf{ano-atk}}$ is expected to work only when $b = 0$ (proper simulation), which explains the conditional probability. In this case, $\mathcal{A}^{\mathsf{ano-atk}}$ is considered successful when he recognizes the challenge to be an encryption under pk_0 of the message m_0.

The advantage of $\mathcal{A}^{\mathsf{ind-atk}}$ is, according to Definition 1, $\Pr(b' = b) - \frac{1}{2}$ and we have:

$$\mathsf{Adv}(\mathcal{A}^{\mathsf{ind-atk}}) = \Pr(b' = b) - \frac{1}{2} = \Pr(b' = 0, b = 0) + \Pr(b' = 1, b = 1) - \frac{1}{2}$$

$$= \Pr(b' = 0|b = 0)\Pr(b = 0) + \Pr(b' = 1|b = 1)\Pr(b = 1) - \frac{1}{2}$$

$$\approx (\epsilon + \frac{1}{2})\frac{1}{2} + \frac{1}{2}\frac{1}{2} - \frac{1}{2} = \frac{1}{2}\epsilon$$

The last inequality, due to $\Pr(b' = 1|b = 1) \approx \frac{1}{2}$, is explained by the fact that in case $b = 1$, there is a negligible chance for Γ.encrypt$_{\mathsf{pk}_0}(m_1)$ to be also an encryption of m_0 under pk_1. $\square$

Theorem 2. *Let Γ be a public key encryption scheme that has Property B. If Γ is ANO-ATK-secure then it is IND-ATK-secure, where $ATK \in \{CPA, PCA, CCA\}$.*

Proof. From an indistinguishability adversary $\mathcal{A}^{\mathsf{ind-atk}}$ with advantage ϵ, we will construct an anonymity adversary $\mathcal{A}^{\mathsf{ano-atk}}$ as follows.

Let $(\mathsf{pk}_0, \mathsf{pk}_1)$ be the input to $\mathcal{A}^{\mathsf{ano-atk}}$. $\mathcal{A}^{\mathsf{ano-atk}}$ will run $\mathcal{A}^{\mathsf{ind-atk}}$ on pk_0. Queries made by $\mathcal{A}^{\mathsf{ind-atk}}$ will be simply passed to $\mathcal{A}^{\mathsf{ano-atk}}$'s own challenger.

At some time, $\mathcal{A}^{\mathsf{ind-atk}}$ outputs two messages m_0, m_1. $\mathcal{A}^{\mathsf{ano-atk}}$ will forward m_0 to his challenger and obtain the challenge Γ.encrypt$_{\mathsf{pk}_b}(m_0)$ where $b \xleftarrow{R} \{0,1\}$.

$\mathcal{A}^{ano-atk}$ will then pass the challenge to $\mathcal{A}^{ind-atk}$ and continue to handle queries as previously.

In case $b = 0$, the challenge encryption is a valid encryption of m_0 and an invalid encryption of m_1 under pk_0. In the other case, since pk_1 (together with sk_1) is sampled from $PK - SK$ and Property B holds, then $\Gamma.\text{encrypt}_{pk_1}(m_0)$ is a random element in C and with overwhelming probability it is not a encryption of m_1 under pk_0. Therefore, when $\mathcal{A}^{ind-atk}$ outputs his guess b', $\mathcal{A}^{ano-atk}$ will forward the same guess to his own challenger.

The advantage of $\mathcal{A}^{ind-atk}$ in such an attack is defined by: $\text{Adv}(\mathcal{A}^{ind-atk}) = \epsilon = \Pr(b' = 0|b = 0) - \frac{1}{2}$. In fact, $\mathcal{A}^{ind-atk}$ is expected to work only when $b = 0$. In this case, $\mathcal{A}^{ind-atk}$ is considered successful when he recognizes the challenge to be an encryption of m_0 under pk_0.

The overall advantage of $\mathcal{A}^{ano-atk}$ is according to Definition 2:

$$\text{Adv}(\mathcal{A}^{ano-atk}) = \Pr(b' = b) - \frac{1}{2} = \Pr(b' = 0, b = 0) + \Pr(b' = 1, b = 1) - \frac{1}{2}$$

$$= \Pr(b' = 0|b = 0)\Pr(b = 0) + \Pr(b' = 1|b = 1)\Pr(b = 1) - \frac{1}{2}$$

$$\approx (\epsilon + \frac{1}{2})\frac{1}{2} + \frac{1}{2}\frac{1}{2} - \frac{1}{2} = \frac{1}{2}\epsilon$$

In fact, $\Pr(b' = 1|b = 1) \approx \frac{1}{2}$, because in the case where $b = 1$, there is a negligible chance for $\Gamma.\text{encrypt}_{pk_1}(m_0)$ to be also an encryption of m_1 under pk_0. $\qquad\square$

3.2 On the Orthogonality between Key Privacy and Data Privacy

In [15], the authors propose a technique that turns an anonymous cryptosystem into a distinguishable anonymous cryptosystem, and vice versa. The idea consists in considering the augmented scheme which appends the message to its encryption (using the original scheme). Since the new ciphertext does not reveal more information about the public key than the original scheme does, it is still anonymous. Concerning the other part, from an indistinguishable scheme one can consider the cryptosystem consisting of appending the public key to the encryption of the message. The new scheme does not reveal more information about the message than the original scheme does. Therefore, it is still indistinguishable. However, it is not anonymous since it discloses the public key.

Theorem 2 complies with this result since the first cryptosystem (obtained by appending the message to the ciphertext) does not have Property B; for a fixed message m, the distribution considered in Property B is easily distinguished from uniform. In fact, the probability that an ciphertext sampled according to this distribution equals a ciphertext whose suffix is different from m is exactly zero. Similarly, Theorem 1 is in accordance with this result since the cryptosystem obtained by appending the public key to the ciphertext does not have Property A. Indeed, for a fixed key pk, the probability that an ciphertext sampled from the distribution considered in Property A equals another ciphertext whose suffix differs from pk is exactly zero.

3.3   ElGamal's Encryption Revisited

In the previous paragraph, we showed that our results consent to the negative results in [15] concerning the independence of key privacy from data privacy. In fact, as Properties B and A do not hold in the augmented cryptosystems respectively, one cannot deduce one security notion from the other. In this section, we confirm the positive results in [1] concerning the anonymity of ElGamal's encryption[7]. The scheme uses a prime order group $(\mathbb{G}, \cdot)$ generated by g $(|\mathbb{G}| = d)$:

keygen	encrypt$_{\mathsf{pk}}(m)$	decrypt$_{\mathsf{sk}}(c_1, c_2)$
$x \xleftarrow{R} \mathbb{Z}_d$	$t \xleftarrow{R} \mathbb{Z}_d$	$m \leftarrow c_2 c_1^{-x}$
$y \leftarrow g^x$	$c_1 \leftarrow g^t$	Return(m)
$\mathsf{pk} \leftarrow (d, g, y)$	$c_2 \leftarrow my^t$	
$\mathsf{sk} \leftarrow (d, g, x)$	Return(c_1, c_2)	
Return$(\mathsf{pk}, \mathsf{sk})$		

ElGamal's encryption is IND-CPA-secure under the hardness of the Decisional Diffie-Hellman problem (DDH)[5]. Actually, the following holds: $\mathsf{Adv}(\mathcal{A}_{\mathsf{ElGamal}}^{\mathsf{ind-cpa}} = \mathsf{Adv}(\mathcal{R}^{\mathsf{ddh}})$. To analyze the ANO-CPA property of ElGamal, it suffices to check whether Property A holds.

The ciphertext space consists of all elements $(g^t, my^t) \in \mathbb{G} \times \mathbb{G}$ such that:

$$\mathsf{C} = \{(g^t, my^t) \in \mathbb{G} \times \mathbb{G} : t \xleftarrow{R} \mathbb{Z}_d, \ m \in \mathsf{M}, \ (y = g^x, x) \in \mathsf{PK} - \mathsf{SK}\} = \mathbb{G} \times \mathbb{G}.$$

We herewith show that the distribution on C, corresponding to the random variable $\mathsf{ElGamal.encrypt}_y(m)$ for a fixed key y and a message m sampled uniformly at random from M, is exactly the uniform distribution. Let $(a_1, a_2) \in \mathsf{C}$ be a fixed value from $\mathbb{G} \times \mathbb{G}$.

$$\Pr[(g^t, my^t) = (a_1, a_2)] = \Pr[g^t = a_1] \Pr[my^t = a_2 | g^t = a_1]$$
$$= \frac{1}{d} \Pr[my^t = a_2 | y^t = a_1^x] = \frac{1}{d} \Pr[m = a_2 a_1^{-x}] = \frac{1}{d}\frac{1}{d}.$$

The last equality is due to the fact that m was sampled uniformly at random from $\mathsf{M} = \mathbb{G}$. We conclude with Theorem 1 that ElGamal's encryption is ANO-CPA secure under the DDH assumption and we have: $\mathsf{Adv}(\mathcal{R}^{\mathsf{ddh}}) \geq \frac{1}{2}\mathsf{Adv}(\mathcal{A}_{\mathsf{ElGamal}}^{\mathsf{ano-cpa}})$, which complies with Theorem 1 in [1]. Similarly, one can analyze the anonymity of Cramer-Shoup's cryptosystem [4] and confirm again Theorem 2 of [1] that states the scheme to be ANO-CCA secure under the DDH assumption.

Before concluding this section, it is worth noting that Property A highlights a strength of the discrete-log-based world in contrast to the RSA-based world. Concretely, let Γ be an RSA-based encryption scheme where the public key comprises the RSA modulus N to be used. If the ciphertext c (seen as a set) contains an element $e \in \mathbb{Z}_N$, then the scheme will never have Property A. In fact, for a fixed key pk (where $N \in \mathsf{pk}$) and a message m chosen uniformly at

---

[5] The DDH problem consists in, given a group $(\mathbb{G}, \cdot)$ generated by g, the group elements g^x, y and z, decide whether $z = y^x$.

random from M, the probability that Γ.encrypt$_{pk}(m)$ equals an element $c' \in C$ with the component $e' \geq N$, is exactly zero. Therefore, it is easy to distinguish the distribution on C, defined in Property A, from the uniform distribution. This argument conforms again to the result in [1], namely, the fact that RSA-OAEP is not anonymous thought it is indistinguishable in the most powerful attack model.

4   Construction of CUS from Anonymous Cryptosystems

In this section, we illustrate the need for anonymity of public key encryption in the area of undeniable signatures. First, we recall the "hybrid encryption paradigm" along with the underlying building blocks, namely Key and Data Encapsulation Mechanisms (KEMs and DEMs). Moreover, we define the anonymity notion for these mechanisms and provide a study on the relation between anonymity and indistinguishability in KEMs. Finally, we proceed to the description of the application.

4.1   Hybrid Encryption Paradigm

Key Encapsulation Mechanisms (KEMs). A KEM is a tuple of algorithms $\mathcal{K} = (\text{keygen}, \text{encap}, \text{decap})$ where

- $\mathcal{K}$.keygen probabilistically generates a key pair (pk, sk),
- $\mathcal{K}$.encap, or the *encapsulation* algorithm which, on input the public key pk, runs on a random tape u and generates a *session key* denoted k and its *encapsulation* e, and
- $\mathcal{K}$.decap, or the *decapsulation* algorithm. Given the private key sk and the element e, this algorithm computes the decapsulation k of e, or returns $\perp$ if e is invalid.

We define the anonymity security notion for such a mechanism as follows.

Definition 3. $\mathcal{K}$ *is said to be* (t, ϵ)-ANO-CPA *secure if no adversary* $\mathcal{A}$, *operating in time* t, *wins the following game with probability greater than* ϵ:

- **Phase 1.** *The challenger runs twice the algorithm* $\mathcal{K}$.keygen *to obtain two public keys* pk_0 *and* pk_1 *with the two corresponding private keys* sk_0 *and* sk_1. *The adversary* $\mathcal{A}$ *is given the public keys* pk_0 *and* pk_1.
- **Challenge.** *The challenger picks uniformly at random a bit b from* $\{0, 1\}$. *Then, he computes a given key* $k^\star$ *together with its encapsulation* $e^\star$ *with regard to the public key* pk_b. *The challenge is* $(e^\star, k^\star)$.
- **Phase 2.** $\mathcal{A}$ *outputs a bit* b', *representing his guess of the pair* $(e^\star, k^\star)$ *being created under* $\text{pk}_{b'}$, *and wins the game if* $b = b'$. *We define* $\mathcal{A}$'s *advantage as* $\text{Adv}(\mathcal{A}) = |\Pr[b = b'] - \frac{1}{2}|$, *where the probability is taken over the random choices of the adversary* $\mathcal{A}$ *and of the challenger.*

Similarly to the study provided in the previous section, we formulate a further property which is sufficient for anonymity to induce indistinguishability. Informally speaking, this property suggests that for a fixed encapsulation e and varying public keys pk (with the corresponding private keys sk), the resulting decapsulations $\mathsf{decap}_{\mathsf{sk}}(e)$ look random.

Again, we stress that every choice of the security parameter κ defines a key space $\mathsf{PK} - \mathsf{SK}$ (corresponding to the space of key pairs $(\mathsf{pk}, \mathsf{sk})$), an encapsulation space E (corresponding to the encapsulations generated by the KEM encapsulation algorithm) and a session key space K (corresponding to the session keys generated by the KEM decapsulation algorithm).

Property C. Let κ be a security parameter. Let further e be an arbitrary encapsulation value from E. Consider the distribution induced by the probabilistic algorithm keygen on the key space $\mathsf{PK} - \mathsf{SK}$. Then, from a key $(\mathsf{pk}, \mathsf{sk})$ sampled according to this distribution, the distribution on K, corresponding to the random variable $\mathsf{decap}_{\mathsf{sk}}(e)$, is computationally indistinguishable from uniform.

Theorem 3. *Let $\mathcal{K}$ be a key encapsulation mechanism that has Property C. If $\mathcal{K}$ is ANO-ATK-secure then it is IND-ATK-secure, where $ATK \in \{CPA, PCA, CCA\}$.*

Proof. First assume that the distribution on the session key space K is *exactly* the uniform distribution. From an indistinguishability adversary $\mathcal{A}^{\mathsf{ind-atk}}$ with advantage ϵ, we will construct an anonymity adversary $\mathcal{A}^{\mathsf{ano-atk}}$ as follows.

Let $(\mathsf{pk}_0, \mathsf{pk}_1)$ be the input to $\mathcal{A}^{\mathsf{ano-atk}}$. $\mathcal{A}^{\mathsf{ano-atk}}$ will run $\mathcal{A}^{\mathsf{ind-atk}}$ on pk_0. Queries made by $\mathcal{A}^{\mathsf{ind-atk}}$ will be simply passed to $\mathcal{A}^{\mathsf{ano-atk}}$'s own challenger. Note that pk_1 is independent of the view of $\mathcal{A}^{\mathsf{ind-atk}}$.

At some time, $\mathcal{A}^{\mathsf{ano-atk}}$ gets from his challenger a challenge (e, k) and is asked to tell the key (pk_0 or pk_1) under which it was created. $\mathcal{A}^{\mathsf{ano-atk}}$ will forward this challenge to $\mathcal{A}^{\mathsf{ind-atk}}$. In case it was created under pk_1, since pk_1 (together with the corresponding private key) is sampled from $\mathsf{PK} - \mathsf{SK}$, Property C implies that $k = \mathcal{K}.\mathsf{decap}_{\mathsf{sk}_1}(e)$ is a uniformly random element of K. Therefore, the value k is either the decapsulation of e under pk_0, or a uniformly random element in K, and thus compatible with the game $\mathcal{A}^{\mathsf{ind-atk}}$ is designed to play.

Further queries by $\mathcal{A}^{\mathsf{ind-atk}}$ continue to be handled as before. At the end, $\mathcal{A}^{\mathsf{ind-atk}}$ will output a bit representing his guess for k being the decapsulation of e under the public key pk_0 or not. $\mathcal{A}^{\mathsf{ano-atk}}$ will use this bit as his guess for the key under which k was created. It is clear that:

$$\mathsf{Adv}(\mathcal{A}^{\mathsf{ano-atk}}) = \mathsf{Adv}(\mathcal{A}^{\mathsf{ind-atk}})$$

Now assume that the distribution on K is *only indistinguishable* from uniform. Let $\mathcal{A}^{\mathsf{ind-atk}}$ be an indistinguishability distinguisher. If the advantage of $\mathcal{A}^{\mathsf{ind-atk}}$ in the reduction described above is non-negligibly different from the advantage of $\mathcal{A}^{\mathsf{ind-atk}}$ in a real attack, then $\mathcal{A}^{\mathsf{ind-atk}}$ can be easily used as a distinguisher for the distribution considered by Property C. As a consequence:

$$\mathsf{Adv}(\mathcal{A}^{\mathsf{ano-atk}}) \approx \mathsf{Adv}(\mathcal{A}^{\mathsf{ind-atk}})$$

where $\approx$ means "equal up to negligible terms". $\qquad\qquad\square$

Theorem 4. *Let $\mathcal{K}$ be a key encapsulation mechanism. If $\mathcal{K}$ is IND-ATK-secure then it is ANO-ATK-secure, where $ATK \in \{CPA, PCA, CCA\}$.*   □

The proof is similar to that of Theorem 1 and will appear in the full version of the paper.

Data Encapsulation Mechanisms (DEMs). DEMs are secret key encryption algorithms. They are, similarly to public key encryption (See Section 2.1), given by the same three algorithms (keygen, encrypt and decrypt), with the exception of generating only one key in the keygen algorithm which will serve for encryption as well as for decryption.

The security notion for DEMs, that corresponds to the ANO-CPA notion for public key encryption, can be defined as follows.

Definition 4. *A DEM $\mathcal{D}$ is said to be (t, ϵ)-ANO-OT-secure (Anonymous under a One Time Attack) if no adversary $\mathcal{A}$, operating in time t, wins the following game with probability greater than ϵ:*

- **Phase 1.** *The challenger runs twice the algorithm $\mathcal{D}$.keygen to obtain two keys $\mathcal{D}.k_0$ and $\mathcal{D}.k_1$.*
- **Challenge.** *The adversary outputs eventually a message $m^\star$. The challenger picks uniformly at random a bit b from $\{0,1\}$ and encrypts $m^\star$, in $c^\star$, under $\mathcal{D}.k_b$. The challenge is $(m^\star, c^\star)$.*
- **Phase 2.** *$\mathcal{A}$ outputs a bit b', representing his guess of $c^\star$ being the encryption of $m^\star$, under $\mathcal{D}.k_{b'}$ and wins the game if $b = b'$. We define $\mathcal{A}$'s advantage as $\mathsf{Adv}(\mathcal{A}) = |\Pr[b = b'] - \frac{1}{2}|$, where the probability is taken over the random choices of the adversary $\mathcal{A}$ and of the challenger.*

Note that the above notion corresponds to the ANO-CPA notion in the public key world because the adversary does not have any oracle access. In fact, in the secret key scenario, the adversary cannot even encrypt messages of his choice (Chosen Plaintext Attack) since he does not have the key at his disposal.

The hybrid encryption paradigm. It consists in combining KEMs with secure secret key encryption algorithms or Data Encapsulation Mechanisms (DEMs) to build encryption schemes. In fact, one can fix a session key k using the KEM, then uses it to encrypt a message using an efficient DEM. Decryption is achieved by first recovering the key from the encapsulation (part of the ciphertext) then applying the DEM decryption algorithm. It can be shown that one can obtain an ANO-CPA-secure cryptosystem from an ANO-CPA-secure KEM combined with an ANO-OT-secure DEM. The proof is similar the one of the indistinguishability notion, which is given in [10].

4.2   The Application

It consists of a generic construction of anonymous convertible undeniable signatures from certain anonymous cryptosystems and secure digital signatures. Our construction is very similar to the construction in [6] in the sense that it is also a variant of the "the encryption of a signature" paradigm where the components are secure digital signatures and weakly secure cryptosystems obtained from the "hybrid encryption" paradigm. However, our construction differs from the one in [6] in the following points:

1. Our construction considers ANO-CPA-secure KEMs in order to achieve ANO-CMA-secure signatures, whereas the construction in [6] considers IND-CPA-secure KEMs to get INV-CMA-secure signatures.
2. Our construction encrypts only a part of the signature, oppositely to the construction in [6], and thus allows more flexibility without compromising the overall security of the resulting signatures.
3. Finally, our construction enjoys the selective conversion feature and so does the construction in [6], though it is claimed to support only the universal conversion.

4.3   The Construction

Let Σ be a digital signature scheme given by Σ.keygen which generates a key pair (Σ.sk, Σ.pk), Σ.sign and Σ.verify. Let furthermore $\mathcal{K}$ be a KEM given by $\mathcal{K}$.keygen which generates a key pair ($\mathcal{K}$.pk, $\mathcal{K}$.sk), $\mathcal{K}$.encap and $\mathcal{K}$.decap. Finally, we consider a DEM $\mathcal{D}$ given by $\mathcal{D}$.encrypt and $\mathcal{D}$.decrypt.

For a message m, we write the signature on m generated using Σ as $\sigma = (s, r)$, where r reveals no information about m nor about (Σ.sk, Σ.pk). I.e., there exists an algorithm that inputs a message m and a key pair (Σ.sk, Σ.pk) and outputs a string statistically indistinguishable from r, where the probability is taken over the message and the key pair spaces considered by Σ. Note that every signature scheme produces signatures of the given form, since a signature can be always written as the concatenation of itself and the empty string (the message-key-independent part). We assume that s belongs to the message space of $\mathcal{D}$.

Let $m \in \{0, 1\}^{\star}$ be a message, we propose the following construction.

Setup. Call $\mathcal{K}$.setup and Σ.setup.

Key generation. Call Σ.keygen and $\mathcal{K}$.keygen to generate Σ.sk, Σ.pk, $\mathcal{K}$.pk and $\mathcal{K}$.sk respectively. Set the public key to (Σ.pk, $\mathcal{K}$.pk) and the private key to (Σ.sk, $\mathcal{K}$.sk).

Signature. Fix a key k together with its encapsulation e, then compute a (digital) signature $\sigma = \Sigma.\text{sign}_{\Sigma.\text{sk}}(m\|e) = (s, r)$ on $m\|e$. Output $\mu = (e, \mathcal{D}.\text{encrypt}_k(s), r)$.

Verification (by the signer). To check the validity of an undeniable signature $\mu = (\mu_1, \mu_2, \mu_3)$, issued on a certain message m, the signer first computes $k = \mathcal{K}.\text{decap}_{\mathcal{K}.\text{sk}}(\mu_1)$ then calls Σ.verify on $(\mathcal{D}.\text{decrypt}_k(\mu_2), \mu_3)$ and $m\|\mu_1$ using Σ.pk. μ is valid if and only if the output of the latter item is 1.

Confirmation/Denial protocol. To confirm (deny) a purported signature μ, the signer checks its validity using the verification algorithm described above. According to the result, the signer issues a zero knowledge proof of knowledge of the decryption of (μ_1, μ_2) that, together with μ_3, passes (does not pass) the verification algorithm Σ.verify.

Selective conversion. To convert a given signature $\mu = (\mu_1, \mu_2, \mu_3)$ issued on a certain message m, the signer first checks its validity. In case it is valid, the signer computes $k = \mathcal{K}.\mathsf{decap}_{\mathcal{K}.\mathsf{sk}}(\mu_1)$ and outputs $(\mathcal{D}.\mathsf{decrypt}_k(\mu_2), \mu_3)$ otherwise he outputs $\bot$.

Universal conversion. Release $\mathcal{K}.\mathsf{sk}$.

4.4 Analysis

We first observe that completeness of the above algorithms follows from the completeness of the underlying building blocks, namely the KEM/DEM mechanisms and the digital signature scheme. Completeness, soundness and non-transferability of the confirmation/denial protocols are met by our construction as a direct consequence of the zero-knowledge proofs of knowledge. Moreover, we state that the construction resists existential forgeries and that signatures are anonymous.

Theorem 5. *Our generic construction is (t, ϵ, q_s)-EUF-CMA secure if the underlying digital signature scheme is (t, ϵ, q_s)-EUF-CMA secure.* □

Theorem 6. *Our generic construction is $(t, \epsilon, q_s, q_v, q_{sc})$-ANO-CMA secure if it is (t, ϵ', q_s)-EUF-CMA secure and the underlying DEM and KEM are ANO-OT secure and $(t + q_s(q_v + q_{sc}), \epsilon \cdot (1 - \epsilon')^{q_v + q_{sc}})$-ANO-CPA secure respectively.* □

The proofs are similar to those of Theorems 1 and 2 in [6], and will appear in the full version of the paper.

5 Application: Damgård-Pedersen Scheme [5]

5.1 The Scheme

It consists of first generating a provably secure variant of ElGamal's signature on the given message, then encrypting the message-key-dependent part using ElGamal's encryption. For example, one can use the Modified ElGamal signature scheme [13] to first generate a digital signature (s, r) on the message, then encrypt s (message-key-dependent part) using ElGamal's encryption. The result $(\mathsf{ElGamal.encrypt}(s), r) = (c_1, c_2, r)$ forms the undeniable signature.

5.2 Security Analysis

Since the publication of the scheme [5], its authors conjectured that unforgeability rests on the hardness of the discrete logarithm problem, whilst anonymity is based on the DDH problem. This conjecture was reported recently in [11]. In this paragraph, we show that the above scheme is not anonymous and thus not invisible either.

Lemma 1. *The scheme in [5] is not ANO-CMA-secure.*

Proof. The proof will investigate the weakness of the ElGamal encryption which consists in the possibility of, given an encryption of a certain message, creating a different ciphertext for the same message.

Let $\mathcal{A}$ be an ANO-CMA adversary against the construction. Let furthermore pk_0 and pk_1 be the challenge public keys for the undeniable signature scheme. Eventually $\mathcal{A}$ outputs a message $m \xleftarrow{R} \{0,1\}^\star$ and gets a challenge signature $\mu = (\mu_1, \mu_2, \mu_3)$ with respect to a given key. He is then requested to tell under which key the signature was created. By construction, (μ_1, μ_2) is an ElGamal encryption of a certain value s with respect to a key pk_b, $b \in \{0,1\}$. In fact, We know that the ElGamal public key, say $y_b = g^{x_b}$ (the group considered is $(\langle g \rangle, \cdot)$ with prime order d), is part of the undeniable signature public key pk_b. Therefore $\mathcal{A}$ picks uniformly at random a bit $b' \xleftarrow{R} \{0,1\}$ and a value $\ell \xleftarrow{R} \mathbb{Z}_d^\times$. He then forms the new signature $\mu' = (g^\ell \mu_1, y_{b'}^\ell \mu_2, \mu_3)$ on the message m. Finally, he queries this new signature to the confirmation/denial oracle corresponding to the public key $pk_{b'}$. If the outcome of this query is the confirmation protocol, then $\mathcal{A}$ will output the bit b' (μ was indeed a signature on m under $pk_{b'}$, thus the new signature μ' is also a signature on m with respect to the same key), otherwise he outputs the bit $1 - b'$.   □

Remark 1. From the study in [8] (Theorem 4), it follows that the scheme is not invisible under a chosen message attack.

5.3   Repairing the Scheme

With our generic construction, we are able to repair the scheme so that it becomes anonymous. In fact, instead of first producing an ElGamal signature on the message m to be signed, one first fixes an ElGamal encryption key $k = y^t$ together with its encapsulation $e = g^t$, then produces the ElGamal signature (s, r) on the message to be signed concatenated with the encapsulation e. Finally, the signer encrypts s in μ_2, by multiplying it with the key y^t. The undeniable signature consists of the triple (e, μ_2, r).

It is easy to see that this new scheme is an instantiation of our framework with the Modified ElGamal signature and the ElGamal encryption (KEM/DEM). Therefore, our security analysis applies to it. Since the Modified ElGamal signatures are proven EUF-CMA secure in the random oracle under the discrete logarithm assumption. Also, we know that the KEM corresponding to the ElGamal encryption is IND-CPA-secure under the DDH assumption. Thus, according to Theorem 4, it is ANO-CPA-secure under the same assumption. Moreover, the DEM associated to the ElGamal cryptosystem is trivially ANO-OT-secure. Therefore, we conclude with Theorem 5 and Theorem 6 on the unforgeability and on the anonymity of the scheme respectively. To prove its invisibility, we can use Theorem 3 of [8] in addition to the fact that the scheme enjoys Property B for undeniable signatures (described also in [8]).

Unfortunately, this scheme suffers the weakened proof methodology that uses the random oracle, which makes the security questionable. In fact, techniques from [12] suggest that it is highly improbable to reduce the discrete logarithm problem to the security of ElGamal's signatures in the standard model. Therefore, we consider the scheme proposed in [6], whose security is investigated in the standard model. It consists of first fixing a key k and its encapsulation e using the linear Diffie-Hellman KEM, then producing a Waters' [14] signature (s, r) on $m \| e$, where m is the message to be signed. The output undeniable signature is $(e, k + s, R)$ (k and s are elements of a group $(\mathbb{G}, +)$). This scheme can be seen as a special instantiation of our framework. Moreover, according to Theorem 4, the linear Diffie-Hellman KEM is ANO-CPA secure under the Decision Linear Assumption. As a result, we are able to confirm the security results of the resulting undeniable signature and prove that it also offers the selective conversion feature, which entitles it to be the most efficient *convertible* (both selectively and universally) undeniable signature scheme with regard to the signature length and cost, the security model and the underlying assumptions.

6   Conclusion

In this paper, we proved that key privacy and data privacy in encryption schemes are related to a certain extent. In fact, under some conditions, we showed that one notion yields the other. This allows to use existing work on the data privacy of some schemes in order to derive their anonymity. Moreover, we illustrated the need for anonymity of public key encryption in the area of undeniable signatures. More precisely, we proposed a generic construction of anonymous convertible undeniable signatures using anonymous cryptosystems in the weakest attack model, obtained from the "hybrid encryption paradigm", and secure digital signature schemes. In this context, we defined the anonymity notion for Key Encapsulation Mechanisms and provided a study on the equivalence between this notion and the indistinguishability notion. Appropriate instantiations of our construction result in well known signatures whose security remained unknown since 1996, or in signatures with appealing features (efficient, standard model, popular assumptions).

Acknowledgments

I thank the anonymous reviewers of AfricaCypt'09 for their careful reading and helpful comments. I also appreciate the remarks of Joachim von zur Gathen and Daniel Loebenberger. This work was funded by the B-IT Foundation and the Land Nordrhein-Westfalen.

References

1. Bellare, M., Boldyreva, A., Desai, A., Pointcheval, D.: Key-Privacy in Public-Key Encryption. In: Boyd, C. (ed.) ASIACRYPT 2001. LNCS, vol. 2248, pp. 566–582. Springer, Heidelberg (2001)

2. Camenisch, J., Michels, M.: Confirmer Signature Schemes Secure against Adaptative Adversaries. In: Preneel, B. (ed.) EUROCRYPT 2000. LNCS, vol. 1807, pp. 243–258. Springer, Heidelberg (2000)
3. Chaum, D., van Antwerpen, H.: Undeniable Signatures. In: Brassard, G. (ed.) CRYPTO 1989. LNCS, vol. 435, pp. 212–216. Springer, Heidelberg (1990)
4. Cramer, R., Shoup, V.: Design and Analysis of Practical Public-Key Encryption Schemes Secure Against Adaptive Chosen Ciphertext Attack. SIAM J. Comput. 33(1), 167–226 (2003)
5. Damgård, I.B., Pedersen, T.P.: New Convertible Undeniable Signature Schemes. In: Maurer, U.M. (ed.) EUROCRYPT 1996. LNCS, vol. 1070, pp. 372–386. Springer, Heidelberg (1996)
6. Aimani, L.E.: Toward a Generic Construction of Universally Convertible Undeniable Signatures from Pairing-Based Signatures. In: Roy Chowdhury, D., Rijmen, V., Das, A. (eds.) Progress in Cryptology - INDOCRYPT 2008. LNCS, vol. 5365, pp. 145–157. Springer, Heidelberg (2008)
7. Gamal, T.E.: A Public Key Cryptosystem and a Signature Scheme based on Discrete Logarithms.. IEEE Trans. Inf. Theory 31, 469–472 (1985)
8. Galbraith, S.D., Mao, W.: Invisibility and Anonymity of Undeniable and Confirmer Signatures. In: Joye, M. (ed.) CT-RSA 2003. LNCS, vol. 2612, pp. 80–97. Springer, Heidelberg (2003)
9. Halevi, S.: A sufficient condition for key-privacy (2005), http://eprint.iacr.org/2005/005
10. Herranz, J., Hofheinz, D., Kiltz, E.: KEM/DEM: Necessary and Sufficient Conditions for secure Hybrid Encryption (August 2006), http://eprint.iacr.org/2006/265.pdf
11. Kurosawa, K., Takagi, T.: New Approach for Selectively Convertible Undeniable Signature Schemes. In: Lai, X., Chen, K. (eds.) ASIACRYPT 2006. LNCS, vol. 4284, pp. 428–443. Springer, Heidelberg (2006)
12. Paillier, P., Vergnaud, D.: Discrete-Log Based Signatures May Not Be Equivalent to Discrete-Log. In: Roy, B. (ed.) ASIACRYPT 2005. LNCS, vol. 3788, pp. 1–20. Springer, Heidelberg (2005)
13. Pointcheval, D., Stern, J.: Security Arguments for Digital Signatures and Blind Signatures.. J. Cryptology 13(3), 361–396 (2000)
14. Waters, B.: Efficient Identity-Based Encryption Without Random Oracles. In: Cramer, R. (ed.) EUROCRYPT 2005. LNCS, vol. 3494, pp. 114–127. Springer, Heidelberg (2005)
15. Zhang, R., Hanaoka, G., Imai, H.: Orthogonality between Key Privacy and Data Privacy, Revisited. In: Pei, D., Yung, M., Lin, D., Wu, C. (eds.) INSCRYPT 2007. LNCS, vol. 4990, pp. 313–327. Springer, Heidelberg (2008)

Security Analysis of Standard Authentication and Key Agreement Protocols Utilising Timestamps

Manuel Barbosa and Pooya Farshim

Departamento de Informática, Universidade do Minho,
Campus de Gualtar, 4710-057 Braga, Portugal
{mbb,farshim}@di.uminho.pt

Abstract. We propose a generic modelling technique that can be used to extend existing frameworks for theoretical security analysis in order to capture the use of timestamps. We apply this technique to two of the most popular models adopted in literature (Bellare-Rogaway and Canetti-Krawczyk). We analyse previous results obtained using these models in light of the proposed extensions, and demonstrate their application to a new class of protocols. In the timed CK model we concentrate on modular design and analysis of protocols, and propose a more efficient timed authenticator relying on timestamps. The structure of this new authenticator implies that an authentication mechanism standardised in ISO-9798 is secure. Finally, we use our timed extension to the BR model to establish the security of an efficient ISO protocol for key transport and unilateral entity authentication.

Keywords: Timestamp, Key Agreement, Entity Authentication.

1 Introduction

The analysis of key agreement protocols has received a lot of attention within the cryptographic community, as they are central components in secure communication systems. Theoretical treatment of these protocols has been performed under computational models of security, under symbolic models and, more recently, under hybrid models which bridge the gap between these two approaches. However, a common trait to all previous work in this area is the abstraction of time, even when key agreement protocols are explicitly synchronous and resort to representations of local time in their definitions. The use of timestamps in key distribution protocols was suggested by Denning and Sacco [13]. Nowadays, protocols such as Kerberos [19], the entity authentication protocols in ISO-9798 [16,17], and the key agreement protocols in ISO-11770 [15] rely on timestamps. In this paper we are concerned with the formal security analysis of such protocols.

Perhaps the most common use of timestamps in cryptographic protocols is to counteract replay and interleaving attacks, and to provide uniqueness or timeliness guarantees [18, Section 10.3.1]. In this sense, timestamps are an alternative to challenge-response mechanisms using fresh random nonces and to message

B. Preneel (Ed.): AFRICACRYPT 2009, LNCS 5580, pp. 235–253, 2009.

sequence numbers. In comparison to challenge-response mechanisms, protocols using timestamps will typically require one less message to complete and will not require parties to generate random numbers. On the downside, the receiver must keep a small amount of ephemeral local state to detect the replay of valid messages within an acceptance window. The amount of state that must be kept when using timestamps can also be seen as an advantage when compared, for example, with solutions using sequence numbers where the receiver must keep static long-term state for each possible peer. In other application scenarios, there is no real alternative to the use of timestamps. Examples of this are the implementation of time-limited privileges, such as those awarded by Kerberos tickets, or the legal validity of authenticated documents, such as X.509 public key certificates.

In short, timestamps are extensively used in cryptographic protocols and they are adopted in mainstream (de facto) cryptographic standards, because they have interesting security properties that can be advantageous in many real-world scenarios. However, to the best of our knowledge, the use of timestamps has not been addressed in previously published work on the theoretical security analysis of cryptographic protocols. In particular, the current formal security models for the analysis of cryptographic protocols do not allow capturing this sort of mechanism in any reasonable way.

The security of this sort of mechanism relies on the use of a common time reference. This means that each party must have a local clock and that these must be synchronised to an extent that accommodates the acceptance window that is used. The local clocks must also be secure to prevent adversarial modification: if an adversary is able to reset a clock backwards, then it might be able to restore the validity of old messages; conversely, by setting a clock forward, the adversary might have advantage in preparing a message for some future point in time. These assumptions on the security and synchronisation of local clocks may be seen as disadvantages of using timestamps, since in many environments they may not be realistic. For example, it is common that the synchronisation of local clocks in a distributed environment is enforced by communication protocols that must themselves be secure in order for this assumption to be valid.

Our contribution. In this paper, we propose a general approach to the enrichment of said models to permit analysing protocols relying on timestamps. Our focus is on a generic modelling technique, which can be applied to virtually any framework for the analysis of cryptographic protocols. For concreteness, we apply this technique to two of the most popular models adopted in literature (the family of models stemming from the work of Bellare and Rogaway [4] and the model proposed by Canetti and Krawczyk in [2]), analyse previous results obtained using these models in light of the proposed extensions, and demonstrate their application to a new class of protocols.

An additional contribution of this paper is that the examples we use to demonstrate our approach are standardised protocols that lacked a formal proof of security until now. In particular, the timestamped authenticator we present in Section 4 was described in a footnote in the original paper by Canetti and Krawczyk in [2], but no proof of security was provided to support the claim.

Furthermore, the structure of this new authenticator (and the security proof we provide) imply that a signature-based unilateral authentication mechanism standardised in ISO-9798-3 is secure for message authentication. Similarly, to the best of our knowledge, the ISO-11770-2 key transport protocol we analyse in Section 5 previously lacked a formal proof of security to validate the informal security guarantees described in the standard.

Structure of the paper. In Section 2 we briefly review the related work. In Section 3 we introduce our modelling approach and then use it propose extensions to the BR model and the CK model that permit capturing timestamping techniques, and discuss the implications for previous results. Finally, we present two examples of how the extended models can be used to analyse concrete protocols: an efficient authenticator in the CK model in Section 4, and a one-pass key exchange protocol from ISO-11770-2 in Section 5. We conclude the paper with a discussion on directions for future work in Section 6.

2   Related Work

Bellare and Rogaway [4] gave the first formal model of security for the analysis of authentication and key agreement protocols. It is a game-based definition, in which the adversary is allowed to interact with a set of oracles that model communicating parties in a network, and where the adversary's goal is to distinguish whether the challenge it is given is a correctly shared key or is a randomly generated value. This seminal paper also provided the first computational proof of security for a cryptographic protocol. Subsequent work by the same authors [5] corrected a flaw in the original formulation and a considerable number of publications since have brought the model to maturity. This evolution included the simplification and refinement of the concept of matching conversations, the introduction of session identifiers and the observation that these are most naturally defined as message traces, the adaptation of the model to different scenarios, and the use of the model to capture different security goals [6]. In the remainder of this paper we will refer to the security models that follow this approach of Bellare and Rogaway as BR models.

In [2] Bellare, Canetti and Krawczyk proposed a modular approach to the design and analysis of authentication and key agreement protocols. This work adapted the concept of simulatability, and showed how one could analyse the security of a protocol in an ideally authenticated world and then use an authenticator to compile it into a new protocol providing the same security guarantees in a more realistic model. Canetti and Krawczyk [11] later corrected some problems with the original formulation of this model by merging their simulation-based approach (in particular they maintained the notions of emulation and compilation that enable the modular construction of protocols) with an indistinguishability-based security definition for key exchange protocols. This enabled the authors to prove a composition theorem, whereby combining a key agreement protocol with an authenticated encryption scheme for message transfer yields a two-stage secure message exchange system. In this paper we will refer to the security models that follow this approach of Canetti and Krawczyk as CK models.

Handling of time in related work. Cryptographic protocols are analysed in abstract models, where participants and adversaries are represented by processes exchanging messages through communication channels. Central to these models is the way they capture the timeliness of physical communication networks. In particular, it is possible to split these models into two categories by looking at the way they handle the activation of processes and the delivery of sent messages. In synchronous models, time is captured as a sequence of rounds. In each round all processes are activated simultaneously, and messages are exchanged instantly. In asynchronous models, there is no explicit assumption on the global passing of time. Process activation is usually message-driven and the adversary controls message delivery and participant activation.

Synchronous models are usually adapted when the focus is on a timeliness guarantee, such as termination of a process. However, asynchronous models are taken as better abstraction of real communication systems, as they make no assumptions about network delays and the relative execution speed of the parties, and they nicely capture the view that communications networks are hostile environments controlled by malicious agents [1]. For this reason, asynchronous models, such as the ones described earlier in this section, are much more widely used. This trend, however, comes at the cost of abstracting away many of the practical uses of time-variant parameters in cryptographic protocols, which rely on explicit representations of time. For example, it is common practice to treat timestamps as random nonces, or to assume that all transmitted messages are different. This is an understandable strategy to simplify analyses, but misses security-critical protocol implementation aspects such as buffering previously received messages to avoid replay attacks, or the use of timestamps and windows of acceptance to reduce the size of said message buffers [18, Section 10.3.1].

3   Adding Time Awareness to BR and CK Models

3.1   General Approach

The objective is to obtain a framework for the analysis of key agreement protocols relying on timestamps, where one can argue that they satisfy a formal security definition. We do not introduce an all-new time-aware analysis framework, which would mean our findings might break away from the current state-of-the-art and might not be easily comparable to previously published results. Instead, we propose to extend the existing models for the analysis of key agreement protocols in a natural way, taking care to preserve an acceptable degree of backward-compatibility. The basic idea of our approach is applicable to several competing analysis frameworks that are currently used by researchers in this area, and it does not imply the adoption of any particular one.

To demonstrate this principle, we propose to extend the BR and CK models referred in Section 2 in very similar terms. The most important change that we introduce is that we provide the communicating parties with internal clocks. These clocks are the only means available to each party to determine the current (local) time. To preserve the common asynchronous trait in these models, where the adversary controls the entire sequence of events occurring during an execution, we do not allow the clocks to progress independently. Instead, we leave it to the adversary to control the individual clocks of parties: we allow it to perform a `Tick` (or activation) query through which it can increment the internal clock of an honest party (of course it has complete control of the clocks of corrupted parties). The adversary is not allowed to reset or cause the internal clocks to regress in any way. This restriction captures the real-world assumption we described in Section 1 that the internal clocks of honest parties must be, to some extent, secure.

The addition of these elements to the BR and CK models allows us to capture the notion of time and internal clock drifts. We preserve the asynchronous nature of the model by allowing the adversary to freely control the perception of time passing at the different parties. Through the `Tick` mechanism, the adversary is able to induce any conceivable pattern in the relative speed of local clocks, and may try to use this capability to obtain advantage in attacking protocols that rely on local time measurements to construct and/or validate timestamps. Of course by giving this power to the adversary, we are enabling it to drive internal clocks significantly out of synchrony with respect to each other. However, a secure protocol using explicit representations of time should make it infeasible for an adversary to take advantage of such a strategy, or at least should permit formally stating the amount of drift that can tolerated. At this point, it is important to distinguish two types of security guarantees that may be obtained from timestamps and that we aim to capture using this modelling strategy.

Resistance against replay attacks. Recall that, in protocols that use timestamps to prevent replay attacks, the receiver defines an acceptance window and temporarily stores received messages until their timestamps expire. The width of the acceptance window must be defined as a trade-off between the required amount of storage space, the expected message transmission frequency, speed and processing time; and the required synchronisation between the clocks of sender and receiver. Received messages are discarded if they have invalid timestamps, or if they are repeats within the acceptance window.

In this setting, the local clocks are not explicitly used to keep track of elapsed time, but simply to ensure that the receiver does not have to store all previously received messages to prevent accepting duplicates. In fact, for this purpose, timestamps are essentially equivalent to sequence numbers. Furthermore, synchronisation of clocks between sender and receiver is less of a timeliness issue, and more of an interoperability problem. For example, two honest parties using this mechanism might not be able to communicate at all, even without the active intervention of any adversary, should their clocks values be sufficiently apart. In our extended model, this is reminiscent of a Denial-of-Service attack, which is usually out of the scope of cryptographic security analyses. Consistently with this view and with the original models, the security definitions for cryptographic protocols using timestamps in this context remain unchanged: it is accepted that the adversary may be able to prevent successful completions of protocols (e.g. by driving internal clocks significantly out of synchronisation, or simply by not delivering messages) but it should not be able to break the security requirements in any useful way.

Timeliness guarantees. For protocols that use timestamps to obtain timeliness guarantees on messages, the local clock values are taken for what they really mean: time measurements. In this context, timestamped messages are typically valid for a longer period of time, and timeliness guarantees can be provided to either the sender or the receiver, or to both. For example, the sender may want to be sure that a message will not be accepted by an honest receiver outside its validity period, which is defined with respect to the sender's own internal clock. Conversely, the receiver may require assurance that an accepted message was generated *recently* with respect to its own local clock, where *recently* is quantifiable as a time interval.

To deal with these guarantees we need to capture accuracy assumptions on the internal clocks of the honest parties in the system. We can do this by imposing limits on the maximum pair-wise drift that the adversary can induce between the internal clocks of different parties. In our modelling approach, we capture this sort of security requirement by stating that a protocol enforcing such a timeliness property must guarantee that any adversary breaking this requirement must be overstepping its maximum drift allowance with overwhelming probability.

3.2   Extending the CK Model

Brief review of the CK model [2,11]. An n-party message-driven protocol is a collection of n programs. Each program is run by a different party with some initial input that includes the party's identity, random input and the security parameter. The program waits for an *activation*: (1) the arrival of an *incoming message* from the network, or (2) an *action request* coming from other programs run by the party. Upon activation, the program processes the incoming data, starting from its current internal state, and as a result it can generate outgoing messages to the network and action requests to other programs run by the party. In addition, a *local output* value is generated and appended to a cumulative output tape, which is initially empty. The protocol definition includes an *initialisation* function I that models an initial phase of out-of-band and authenticated information exchange between the parties. Function I takes a random input r and the security parameter κ, and outputs a vector $I(r, \kappa) = I(r, \kappa)_0, \ldots, I(r, \kappa)_n$. The component $I(r, \kappa)_0$ is the public information that becomes known to all parties and to the adversary. For $i > 0$, $I(r, \kappa)_i$ becomes known only to P_i.

The *Unauthenticated-Links Adversarial Model* (UM) defines the capabilities of an active man-in-the-middle attacker and its interaction with a protocol [11]. The participants are parties $P_1, \ldots, P_n$ running an n-party protocol π on inputs $x_1, \ldots, x_n$, respectively, and an adversary $\mathcal{U}$. For initialisation, each party

P_i invokes π on local input x_i, security parameter κ and random input; the initialisation function of π is executed as described above. Then, while $\mathcal{U}$ has not terminated do:

1. $\mathcal{U}$ may **activate** π within some party, P_i. An activation can take two forms:
   (a) An **action request** q. This activation models requests or invocations coming from other programs run by the party.
   (b) An **incoming message** m with a specified sender P_j. This activation models messages coming from the network. We assume that every message specifies the sender of the message and its intended recipient.
   If an activation occurred then the activated party P_i runs its program and hands $\mathcal{U}$ the resulting outgoing messages and action requests. Local outputs produced by the protocol are known to $\mathcal{U}$ except for those labeled **secret**.
2. $\mathcal{U}$ may **corrupt** a party P_i. Upon corruption, $\mathcal{U}$ learns the current internal state of P_i, and a special message is added to P_i's local output. From this point on, P_i is no longer activated and does not generate further local output.
3. $\mathcal{U}$ may issue a **session-state reveal** for a specified session within some party P_i. In this case, $\mathcal{U}$ learns the current internal state of the specified session within P_i. This event is recorded through a special note in P_i's local output.
4. $\mathcal{U}$ may issue a **session-output query** for a specified session within some party P_i. In this case, $\mathcal{U}$ learns any output from the specified session that was labeled **secret**. This event is recorded through a special note in P_i's local output.

The *global output* of running a protocol in the UM is the concatenation of the cumulative local outputs of all the parties, together with the output of the adversary. The global output resulting from adversary $\mathcal{U}$ interacting with parties running protocol π is seen as an ensemble of probability distributions parameterised by security parameter $k \in \mathbb{N}$ and the input to the system[1] $\mathbf{x} \in \{0,1\}^*$, and where the probability space is defined by the combined coin tosses of the adversary and the communicating parties. Following the original notation, we denote this ensemble by $\text{UNAUTH}_{\mathcal{U},\pi}$.

The *Authenticated-Links Adversarial Model* (AM) is identical to the UM, with the exception that the adversary is constrained to model an ideal authenticated communications system. The AM-adversary, denoted $\mathcal{A}$ cannot inject or modify messages, except if the specified sender is a corrupted party or if the message belongs to an exposed session. Analogously to $\text{UNAUTH}_{\mathcal{U},\pi}$, we have that $\text{AUTH}_{\mathcal{A},\pi}$ is the ensemble of random variables representing the global output for a computation carried out in the authenticated-links model.

Due to space limitations, we refer the reader to [11] for the definitions related to Key-Exchange protocols and their security in the CK model.

Introducing local clocks. Our modification to the previous model is based on a special program that we call LocalTime.

---

[1] The concatenation of global public data with individual local inputs for each party.

Definition 1 (LocalTime Program). *The* LocalTime *program follows the syntax of message-driven protocols. The program does not accept messages from the network or transmit messages to the network. The program is deterministic and it is invoked with the empty input. It maintains a* clock *variable as internal state, which is initialised to* 0. *The program accepts a single external request, with no parameters, which is called* Tick. *When activated by the* Tick *request, the program increments the counter and outputs* Local Time: <clock> , *where* <clock> *denotes the value of the* clock *variable.*

We introduce the *timed* variants of the UM and AM, which we refer to as TUM and TAM, and we require that each party in the TUM and in the TAM runs a single instance of LocalTime. Note that in the timed models, the adversary may control the value of the internal clock variables at will, by sending the Tick request to any party. Consistently with the original models, we assume that the local output at a given party P_i is readable only by the adversary and the programs running in the same party. Alternatively, the internal clock variable can be seen as part of the local state of each party, which is read-only to other programs and protocols running in the same environment. This means, in particular, that a program which enables a party P_i to participate in a given protocol may use the local clock value at that party, but is otherwise unaware of any other time references. We disallow protocols from issuing the Tick request to their local clock themselves.

Remark. The approach we followed to integrate the local clocks into the communicating parties in the CK model deserves a few words of explanation. Firstly, the adversary's interactions with parties in the CK model are either external requests to protocols, or message deliveries. Our choice of modelling the local clock as a separate program that accepts the Tick activation as an external request is consistent with this principle, and allows the adversary to control the local clock as desired. Secondly, by causing the LocalTime program to produce local output after each tick, protocol outputs do not need to include information about the time at which a certain event occurred in order to make timeliness properties explicit: this follows directly from the cumulative nature of the local output. Finally, our approach makes the concept of protocol emulation time-aware: the fact that the local clock progression is observable in the local output of each party also implies that any protocol π' that emulates a protocol π (see Definition 2 below) is guaranteed to preserve any timeliness properties formulated over the global output when the original protocol is run in the TAM.

Modular protocol design in the timed models. Central to the methodology of [2] are the concepts of *protocol emulation, compiler,* and *authenticator*, which we directly adapt for protocol translations between the TAM and the TUM.

Definition 2 (Emulation). *A protocol π' emulates a protocol π in the TUM if, for any TUM adversary $\mathcal{U}_T$, there exists TAM adversary $\mathcal{A}_T$ such that, for all input vectors, the global output resulting from running $\mathcal{A}_T$ against π in the TAM is computationally indistinguishable from that obtained when running $\mathcal{U}_T$ against π' in the TUM.*

We emphasise that the global outputs resulting from running a protocol in the timed models include the local outputs produced by the LocalTime program, which reflect the sequence of Tick queries performed by the adversary at each party, and that these outputs are captured by the emulation definition above.

Definition 3 (Timed-Authenticator). *A compiler $\mathcal{C}$ is an algorithm that takes for input descriptions of protocols and outputs descriptions of protocols. A* timed-authenticator *is a compiler $\mathcal{C}$ that, for any TAM protocol π, the protocol $\mathcal{C}(\pi)$ emulates π in the TUM.*

One can show establish, in an almost identical way to Theorem 6 in [11], that:

Theorem 1. *Let π be an SK-secure (see [11] for definition) key exchange protocol in the TAM and let $\mathcal{C}$ be a timed-authenticator. Then $\pi' := \mathcal{C}(\pi)$ is an SK-secure key exchange protocol in the TUM.*

In Section 4 we prove that, not only the original AM-to-UM authenticators proposed in [2] are also timed-authenticators, but also that through the use of timestamps one can obtain more efficient timed-authenticators. However, in order to argue that these results are meaningful, we need to revisit the modular approach to the development of cryptographic protocols introduced in [2]. With the introduction of the timed models, we have now four options for the design and analysis of protocols. For convenience, one would like to carry out the design in the authenticated models (AM and TAM), where adversaries are more limited in their capabilities and security goals are easier to achieve. The choice of whether or not to use a timed model should depend only on whether or not the protocol relies on time-dependent parameters to achieve security. On the other hand, and without loss of generality, we will assume that the overall goal is to translate these protocols into the TUM, which is the most general of the more realistic unauthenticated models, given that it accommodates protocols which may or may not take advantage of the local clock feature. To support this methodology, we first formalise a class of protocols for which the timed models are not particularly relevant.

Definition 4 (Time-Independence). *A protocol π is* time-independent *if its behaviour is oblivious of the LocalTime protocol, i.e. if protocol π does not use the outputs of the LocalTime protocol in any way.*

One would expect that, for time-independent protocols, the TUM (resp. TAM) would be identical to the UM (resp. AM). In particular, all of the results obtained in [11] for specific time-independent protocols should carry across to the timed models we have introduced. Unfortunately, proving a general theorem establishing that, for any time-independent protocol, in the UM (resp. AM) one can simply recast it in the TUM (resp. TAM) to obtain a protocol which emulates the original one (and satisfying the same security definitions) is not possible given our definition of the LocalTime program. This is because it is, by definition, impossible to recreate local time outputs by individual parties in the UM (resp. AM), and hence a simulation-based proof does not go through. However, for the specific case of SK-security, we can prove the following theorem establishing that, for time-independent protocols, one can perform the analysis in the UM (resp. AM) and the results will still apply in the TUM (resp. TAM).

Theorem 2. *If a time-independent UM-protocol (resp. AM-protocol) π is SK-secure, then it is also SK-secure when run in the TUM (resp. TAM).*

Remark. We emphasise that, although we are able to show that the newly proposed timed models are a coherent extension to the work in [2,11] for the design and analysis of key exchange protocols relying on time-dependent parameters, we are not able to establish a general theorem that carries through all of the previous results in the CK model. In particular, we cannot prove a theorem stating that AM-to-UM emulation implies TAM-to-TUM emulation for time-independent protocols. This would automatically imply that all authenticators are also timed-authenticators (we will return to this discussion in Section 4). However, the proof for such a theorem does not seem to go through because the definition of emulation is not strong enough to guarantee that, using the existence of suitable AM adversary for all TUM-adversaries, one is able to construct the required TAM-adversary that produces an indistinguishable sequence of Tick queries.

Theorem 2, combined with the concrete time-dependent and time-independent timed-authenticators in Section 4, provides the desired degree of flexibility in designing SK-secure KE protocols, as shown in the table below.

Lower Layer	Time-independent authenticator	Time-dependent authenticator
Time-independent in the AM	Use the original CK modular approach to obtain an SK-secure protocol in the UM. Apply Theorem 2 to move to the TUM.	Use Theorem 2 to move result to the TAM. Apply the timed-authenticator in Section 4 to obtain an SK-secure KE protocol in the TUM.
Time-dependent in the TAM	Apply one of the original authenticators in [2], which are also timed-authenticators by Theorem 3, to obtain an SK-secure KE in the TUM.	Apply the timed-authenticator in Section 4 to obtain an SK-secure KE protocol in the TUM.

3.3   Extending the BR Model

Brief review of the BR model [4,6]. Protocol participants are the elements of a non-empty set $\mathcal{ID}$ of principals. Each principal $A \in \mathcal{ID}$ is named by a fixed-length string, and they all hold public information and long-lived cryptographic private keys. Everybody's private key and public information is determined by running a key generator. During the execution of a protocol, there may be many running instances of each principal $A \in \mathcal{ID}$. We call instance i of principal A an oracle, and we denote it Π_A^i. Each instance of a principal might be embodied as a process (running on some machine) which is controlled by that principal.

Intuitively, protocol execution proceeds as follows. An initiator-instance speaks first, producing a first message. A responder-instance may reply with a message of its own, intended for the initiator-instance. This process is intended to continue for some fixed number of flows, until both instances have *terminated*, by which time each instance should also have *accepted*. Acceptance may occur at any time, and it means that the party holds a session key sk, a session identifier sid (that can be used to uniquely name the ensuing session), and a partner identifier pid (that names the principal with which the instance believes it has just exchanged a key). The session key is secret, but the other two parameters are considered public. An instance can accept at most once.

Adversary $\mathcal{A}$ is defined as a probabilistic algorithm which has access to an arbitrary number of instance oracles, as described above, to which he can place the following queries:

- Send(A, B, i, m): This delivers message m, which is claimed to originate in party B, to oracle Π_A^i. The oracle computes what the protocol says to, and returns back the response. Should the oracle accept, this fact, as well as the session and partner identifiers will be made visible to the adversary. Should the oracle terminate, this too will be made visible to the adversary. To initiate the protocol with an instance of A as initiator, and an instance of B as responder, the adversary should call Send(A, B, i, λ) on an unused instance i of A.
- Reveal(A, i): If oracle Π_A^i has accepted, holding some session key sk, then this query returns sk to the adversary.
- Corrupt(A): This oracle returns the private key corresponding to party A[2].
- Test(A, i): If oracle Π_A^i has terminated, holding some session key sk and $pid = B$, then the following happens. A coin b is flipped. If $b = 1$, then sk is returned to the adversary. Otherwise, a random session key, drawn from the appropriate distribution, is returned.

To capture the security of authenticated key agreement protocols (AKE), we require the following definitions.

Definition 5 (Partnering). *We say that Π_A^i is the partner of $\Pi_{A'}^{i'}$ if (1) Both oracles has accepted and hold (sk, sid, pid) and (sk', sid', pid') respectively; (2) $sk = sk'$, $sid = sid'$, $pid = A'$, and $pid' = A$; and (3) No oracle besides Π_A^i and $\Pi_{A'}^{i'}$ has accepted with session identity sid. Note that partnership is symmetric.*

Definition 6 (Freshness). *Π_A^i is fresh if no reveal or corrupt queries are placed on Π_A^i or its partner Π_B^j.*

Definition 7 (AKE Security). *We say that a key exchange protocol is AKE secure if for any probabilistic polynomial-time adversary $\mathcal{A}$, the probability that*

---

[2] For simplicity we adopt the weak corruption model, where Corrupt does not return the states of all instances of A.

$\mathcal{A}$ *guesses the bit b chosen in a fresh test session is negligibly different from* $1/2$. *The advantage of the adversary, which returns a bit* b', *is defined to be:*

$$\mathrm{Adv}_{\mathrm{KE}}^{\mathrm{AKE}}(\mathcal{A}) := |2\Pr[b = b'] - 1|.$$

Definition 8 (Entity Authentication (EA)). *We say that a key exchange protocol provides initiator-to-responder authentication if, for any probabilistic polynomial-time adversary* $\mathcal{A}$ *attacking the protocol in the above model, the probability,* $\mathrm{Adv}_{\mathrm{KE}}^{\mathrm{I2R}}(\mathcal{A})$, *that some honest responder oracle* Π_B^j *terminates with pid* $=$ *A, an honest party, but has no partner oracle is negligible.*

Remark. The restriction of being honest that we have imposed above, is introduced to model the setting where authentication relies on symmetric keys. This is the case for the protocol we analyse in Section 5. In the asymmetric setting, however, only the authenticated party (initiator in the above) needs to be honest.

Introducing local clocks. To ensure consistency with the structure of the BR model, we provide each party with a `clock` variable, which is initially set to zero. This variable is read-only state, which is accessible to all the instances of a protocol running at a given party (very much like the private keys). In order to model the adversarial control of clocks at different parties we enhance its capabilities by providing access to the following oracle:

– `Tick(A)`: increment the `clock` variable at party A, and return it.

It is interesting to note that the relation between the timed version of the BR model and the original one is identical to that we established in the previous section between the TUM and the UM in the CK model. Specifically, one can formulate the notions of AKE security and entity authentication without change in the timed BR model. It is also straightforward to adapt the definition of time-independence to protocols specified in the BR model and prove that, for all time-independent protocols, AKE security and entity authentication are preserved when we move from the original to the timed version of the model. We omit the equivalent of Theorem 2 for BR models due to space limitations. However, the observation that such a theorem holds is important to support our claim that the extension we propose to the BR model is a natural one.

Capturing timeliness guarantees. The definition of entity authentication formulated over the timed BR model is a good case study for capturing timeliness guarantees in security definitions. The existential guarantee stated in the definition implicitly refers to two events: (1) the termination of the protocol-instance that obtains the authentication guarantee; and (2) the acceptance of the partner protocol-instance that is authenticated. It seems natural to extend this definition with additional information relating the points in time at which the two events occur. To achieve this, we must first isolate a category of adversaries for which making such claims is possible.

Definition 9 (δ-synchronisation). *An adversary in the timed BR model satisfies δ-synchronisation if it never causes the* clock *variables of any two (honest) parties to differ by more than δ.*

The previous definition captures the notion that clocks must be synchronised in order to achieve any sort of timeliness guarantee, as described in Section 3.1. We are now in a position to state an alternative version of the entity authentication definition. Let Π_A^i and Π_B^j be two partner oracles where the latter has terminated. Also, let $t_B(E)$ be the function returning the value of the local clock at B when event E occurred. Finally, let $\mathtt{acc}(A, i)$ denote the event that Π_A^i accepted, and let $\mathtt{term}(B, j)$ denote the event that Π_B^j terminated.

Definition 10 (β-Recent Entity Authentication (β-REA)). *We say that a key exchange protocol provides β-recent initiator-to-responder authentication if it provides initiator-to-responder authentication, and furthermore for any honest responder oracle Π_B^j which has terminated with partner Π_A^i, with A honest, we have that: $|t_B(\mathtt{term}(B, j)) - t_B(\mathtt{acc}(A, i))| \leq \beta$.*

The above definition captures attacks such as that described in [14], where an adversary uses a post-dated clock at a client to impersonate as him later, when correct time is reached at the server side. In Section 5 we will prove that a concrete key agreement protocol using timestamps satisfies the previous definition, as long as the adversary is guaranteed to comply with δ-synchronisation.

4   An Example in the CK Model: Timed-Authenticators

The concept of *authenticator* is central to the modular approach to the analysis of cryptographic protocols proposed in [2]. Authenticators are compilers that take protocols shown to satisfy a set of properties in the AM, and produce protocols which satisfy equivalent properties in the UM. Bellare et al. [2] propose a method to construct authenticators based on the simple message transfer (MT) protocol: they show that any protocol which emulates the MT protocol in the UM can be used as an authenticator. Authenticators constructed using in this way are called *MT-authenticators*.

In this section we show that this method can be easily adapted to the timed versions of the CK model introduced in the previous section. We start by recalling the definition of the MT-protocol and note that, when run in the timed models, the local output at each party permits reading the local time at which the MT-protocol signalled the reception and transmission of messages.

Definition 11 (The MT-Protocol). *The protocol takes empty input. Upon activation within P_i on action request* $\mathtt{Send}(P_i, P_j, m)$*, party P_i sends the message (P_i, P_j, m) to party P_j, and outputs "P_i* sent *m* to *P_j". Upon receipt of a message (P_i, Pj, m), P_j outputs "P_j* received *m* from *P_i".*

Now, let λ be a protocol that emulates the MT-protocol in the TUM and, similarly to the modular construction in [2], define a compiler $\mathcal{C}_\lambda$ that on input a protocol π produces a protocol $\pi' = \mathcal{C}_\lambda(\pi)$ defined as follows.

- When π' is activated at a party P_i it first invokes λ.
- Then, for each message sent in protocol π, protocol π' activates λ with the action request for sending the same message to the same specified recipient.
- Whenever π' is activated with some incoming message, it activates λ with the same incoming message.
- When λ outputs "P_i received m from P_j", protocol π is activated with incoming message m from P_j.

We complete this discussion with two theorems. Theorem 3 is the equivalent of Theorem 3 in [2]. Theorem 4 is the equivalent of Propositions 4 and 5 in [2].

Theorem 3. *Let π be a protocol in the TAM, and let λ be protocol which emulates the MT-protocol in the TUM, then $\pi' := \mathcal{C}_\lambda(\pi)$ emulates π in the TUM.*

Theorem 4. *The signature-based and the encryption-based MT-authenticators proposed in [2] both emulate the MT-protocol in the TUM.*

The proofs are identical to the original ones, with the following exception: when a TUM-adversary activates the LocalTime protocol of a (dummy) TUM-party by a Tick request, the simulating TAM-adversary invokes the LocalTime protocol of the corresponding party in the TAM, and passes back the output, without change, to the TUM-adversary.

Theorem 4 establishes that the original compilers proposed in [2] can also be used to translate protocols from the TAM to the TUM, i.e. they are also timed-authenticators. Intuitively, this is possible because these constructions are oblivious of time and of the LocalTime programs added to the timed models, and the MT-protocol is not claimed to provide concrete timeliness guarantees.

To complete this section, we present a more efficient one-round timed authenticator, which uses timestamps to eliminate the challenge-response construction used in the original authenticators. The protocol is shown in Figure 1, and is an adaptation of the signature-based MT-authenticator in [2]. It is parameterised with a positive integer δ which defines the width of the timestamp acceptance window. We observe that this protocol is structurally equivalent to the signature-based unilateral authentication protocol in the ISO-9798-3 standard [17] when one uses message m in place of the optional text-fields allowed by the standard. This implies that Theorem 5 below establishes the validity of the claim in standard ISO-9798-3 that this protocol can be used for message authentication.

The following theorem, whose proof can be found in the full version of the paper, formally establishes the security properties of the protocol in Figure 1.

Theorem 5. *Assume that the signature scheme in use is secure against the standard notion of chosen message attacks (UF-CMA). Then protocol $\lambda_{Sig}(\delta)$ emulates the MT-protocol in the TUM.*

Remark. There is a subtlety involving adversaries who can forge signatures but do not disrupt the simulation needed by a TAM-adversary. Consider an adversary who activates party P^* to send message m at local time t^*, but does not deliver the message to the intended recipient. Instead, it forges a signature on the same

---

Protocol $\lambda_{Sig}(\delta)$

- The initialisation function I first invokes, once for each party, the key generation algorithm of a signature scheme secure against chosen message attacks with security parameter κ. Let Sig and Ver denote the signing and verification algorithms. Let s_i and v_i denote the signing and verification keys associated with party P_i. The public information includes all public keys: $I_0 = v_1, \ldots, v_n$. P_i's private information is $I_i = s_i$.
- Each party keeps as protocol state a list L where it stores message/timestamp pairs (m, t) corresponding to previously received and accepted messages.
- When activated within party P_i and with external request to send message m to party P_j, protocol $\lambda_{Sig}(\delta)$ invokes a two-party protocol that proceeds as follows:
  - First, P_i checks the local time value t and constructs a message $(m\|t\|\mathsf{Sig}(m\|t\|P_j, s_i))$ and sends it to P_j.
  - Then, P_i outputs "P_i sent m to P_j".
  - Upon receipt of $(m\|t\|\sigma)$ from P_i, party P_j accepts m and outputs "P_j received m from P_i" if:
    * the signature σ is successfully verified by $\mathsf{Ver}(m\|t\|P_j, v_i)$.
    * $t \in [t' - \delta, t' + \delta]$, where t' is the value of the local time at P_j when the message is received.
    * list L does not contain the pair (m, t).
  - Finally, P_j updates list L adding the pair (m, t) and deleting all pairs $(\hat{m}, \hat{t})$ where $\hat{t} \notin [t' - \delta, t' + \delta]$.

Fig. 1. A signature-based timed-authenticator in the TUM

message, but with a later timestamp, and delivers this message to the intended recipient much later in the simulation run, taking care that the timestamp in the forged message is valid at that time. This adversary does not cause a problem in the proof of the above theorem, since the message is delivered only once. In fact, this is an attack on the timeliness properties of the authenticator, which are not captured in the formulation of the MT-protocol. This attack would be an important part of the proof that protocol $\lambda_{Sig}(\delta)$ emulates a version of the MT-protocol with timeliness guarantees, where messages are only accepted on the receiver's side if they are delivered within some specific time interval after they are added to the set M.

5   An Example in the BR Model: A Standard AKE Protocol

In this section we use the timed BR model to analyse the security of a one-pass key agreement protocol offering unilateral authentication, as defined in the ISO-11770-2 standard. The protocol is formalised in Figure 2. It is a key transport protocol that uses an authenticated symmetric encryption scheme to carry a fresh session key between the initiator and the responder. The use of timestamps permits achieving AKE security in one-pass, and the reception of the single message in the protocol effectively allows the responder to authenticate the initiator. In fact, this protocol is presented in the ISO-11770-2 standard as a particular use of a unilateral authentication protocol presented in ISO-9798-2, where the session key is transmitted in place of a generic text field. As explained in Section 3.3, the security proof we present here can be easily adapted to show that the underlying ISO-9798-2 protocol is a secure unilateral EA protocol.

ISO-11770-2 informally states the following security properties for the protocol in Figure 2. The session key is supplied by the initiator party, and AKE security is guaranteed by the confidentiality property of the underlying authenticated encryption scheme. The protocol provides unilateral authentication: the mechanism enables the responder to authenticate the initiator. Entity authentication is achieved by demonstrating knowledge of a secret authentication key, i.e. the entity using its secret key to encipher specific data. For this reason, the protocol requires an authenticated encryption algorithm which provides, not only data confidentiality, but also data integrity and data origin authentication. Uniqueness and timeliness is controlled by timestamps: the protocol uses timestamps to prevent valid messages (authentication information) from being accepted at a later time or more than once.

The protocol requires that parties are able to maintain mechanisms for generating or verifying the validity of timestamps: the deciphered data includes a timestamp that must be validated by the recipient. Parties maintains a list L to detect replay attacks. In relation to forward secrecy, note that if an adversary gets hold of a ciphertext stored in L, and furthermore at some point it corrupts

---

Protocol $\pi_{\texttt{AuthEnc}}(\delta)$

- The initialisation function I first invokes, once for each pair of parties, the key generation algorithm of an authenticated symmetric encryption scheme with security parameter κ and sets the secret information of party A with pair B to be $K_{A,B}$. I_A is set to be the list of the keys A shares with B for all parties B.
- All parties keep as protocol state a list L where it stores ciphertext/timestamp pairs (c, t) corresponding to previously received and accepted messages.
- When activated within party A to act as initiator, and establish a session with party B, the protocol proceeds as follows.
    - A checks the local time value t. It generates a random session key sk and sets $c \leftarrow \texttt{AuthEnc}(sk||t||B, K_{A,B})$. It then sends (A, B, c) to B.
    - A accepts sk as the session key, (A, B, c) as sid, B as pid, and terminates.
- Upon receipt of (A, B, c), the responder accepts a key sk as the session key, (A, B, c) as sid, A as pid and terminates if:
    - B is the identity of responder.
    - c successfully decrypts to $(sk||t||B)$ under $K_{A,B}$.
    - $t \in [t' - \delta, t' + \delta]$, where t' is local time at B when the message is received.
    - List L does not contain the pair (c, t).
- Finally, B updates the list L, adding the pair (c, t) and deleting all pairs $(\hat{c}, \hat{t})$ where $\hat{t} \notin [t' - \delta, t' + \delta]$.

Fig. 2. One-pass key agreement with unilateral authentication from ISO-11770-2

the owner of the list, it can compute the secret key for the corresponding past session. Identifier B is included in the ciphertext to prevent a substitution attack, i.e. the re-use of this message by an adversary masquerading as B to A. Where such attacks cannot occur, the identifier may be omitted [15].

The following theorem, whose proof can be found in the full version of the paper, formally establishes the security properties of the protocol in Figure 2.

Theorem 6. *The protocol $\pi_{\mathtt{AuthEnc}}(\delta)$ in Figure 2 is an AKE secure key exchange protocol in the timed BR model if the underlying authenticated encryption scheme is secure in the IND-CPA and INT-CTXT senses. This protocol also provides initiator-to-responder authentication if the authenticated encryption scheme is INT-CTXT secure. More precisely, we have:*

$$\mathrm{Adv}_{\mathrm{KE}}^{\mathrm{I2R}}(\mathcal{A}) \leq 2q^2 \cdot \mathrm{Adv}_{\mathtt{AuthEnc}}^{\mathrm{INT-CTXT}}(\mathcal{B}_1) + q^2 q_s / |\mathcal{K}|,$$

$$\mathrm{Adv}_{\mathrm{KE}}^{\mathrm{AKE}}(\mathcal{A}) \leq q^2(2 + q_s) \cdot \mathrm{Adv}_{\mathtt{AuthEnc}}^{\mathrm{INT-CTXT}}(\mathcal{B}_1) + q^2 q_s \cdot \mathrm{Adv}_{\mathtt{AuthEnc}}^{\mathrm{IND-CPA}}(\mathcal{B}_2) + q^2 q_s / |\mathcal{K}|.$$

Here a uniform distribution on the key space $\mathcal{K}$ is assumed, q is the maximum number of parties involved in the attack, and q_s is the maximum number of sessions held at any party.

Furthermore, if the adversary respects β-synchronisation, then the protocol guarantees $(\beta + \delta)$-recent initiator-to-responder authentication.

6   Conclusion

In this paper we proposed a general modelling technique that can be used to extend current models for the analysis of key agreement protocols, so that they capture the use of timestamps. We have shown that two popular analysis frameworks (CK and BR models) can be extended in a natural way using this technique, and that this permits addressing a new class of real-world protocols that, until now, lacked a complete formal treatment. The paper also leaves many open problems that can be addressed in future work. We conclude the paper by referring some of these topics. The approach we introduced can be applied to extend other theoretical models, the most interesting of which is perhaps the Universal Composability framework of Canetti [10]. Orthogonally, there are many key agreement and authentication protocols which rely on timestamps and that could benefit from a security analysis in a time-aware framework. Kerberos [19] is an example of such a protocol, which utilises timestamps in a setting where a server is available. In order to rigorously analyse the security of this protocol, one would need to define a timed version of three-party key agreement security models. Moving away from key agreement and authentication protocols, our approach opens the way for the formal analysis of time-related cryptographic protocols such as those aiming to provide secure message timestamping and clock-synchronisation. Finally, it would be interesting to see how one could apply a similar approach to security models that try to capture public key infrastructures, where the temporal validity of certificates is usually ignored.

Acknowledgments

The authors would like to thank Alex Dent for proposing and discussing the original ideas that led to this work. We would also like to thank Bogdan Warinschi and Paul Morrissey for important discussions. The authors were funded in part by the WITS project (FCT - PTDC/EIA/71362/2006) and the eCrypt II project (EU FP7 - ICT-2007-216646). The second author was also supported in part by the Scientific and Technological Research Council of Turkey (TÜBİTAK) while at Middle East Technical University.

References

1. Backes, M.: Unifying Simulatability Definitions in Cryptographic Systems under Different Timing Assumptions. In: Amadio, R., Lugiez, D. (eds.) CONCUR 2003. LNCS, vol. 2761, pp. 350–365. Springer, Heidelberg (2003)
2. Bellare, M., Canetti, R., Krawczyk, H.: A Modular Approach to the Design and Analysis of Authentication and Key Exchange Protocols. In: The 30th Annual ACM Symposium on Theory of Computing, pp. 419–428. ACM Press, New York (1998)
3. Bellare, M., Namprempre, C.: Authenticated Encryption: Relations Among Notions and Analysis of the Generic Composition Paradigm. In: Okamoto, T. (ed.) ASIACRYPT 2000. LNCS, vol. 1976, pp. 531–545. Springer, Heidelberg (2000)
4. Bellare, M., Rogaway, P.: Entity Authentication and Key Distribution. In: Stinson, D.R. (ed.) CRYPTO 1993. LNCS, vol. 773, pp. 232–249. Springer, Heidelberg (1994)
5. Bellare, M., Rogaway, P.: Provably Secure Session Key Distribution: The Three Party Case. In: Proceedings of the 27th Annual Symposium on the Theory of Computing, pp. 57–66. ACM Press, New York (1995)
6. Bellare, M., Pointcheval, D., Rogaway, P.: Authenticated Key Exchange Secure Against Dictionary Attacks. In: Preneel, B. (ed.) EUROCRYPT 2000. LNCS, vol. 1807, pp. 139–155. Springer, Heidelberg (2000)
7. Blanchet, B., Jaggard, A.D., Scedrov, A., Tsay, J.-K.: Computationally Sound Mechanised Proofs for Basic and Public-Key Kerberos. In: 2008 ACM Symposium on Information, Computer and Communications Security, pp. 87–99. ACM Press, New York (2008)
8. Blanchet, B., Pointcheval, D.: Automated Security Proofs with Sequences of Games. In: Dwork, C. (ed.) CRYPTO 2006. LNCS, vol. 4117, pp. 537–554. Springer, Heidelberg (2006)
9. Boldyreva, A., Kumar, V.: Provable-Security Analysis of Authenticated Encryption in Kerberos. In: 2007 IEEE Symposium on Security and Privacy, pp. 92–100. IEEE Computer Society, Los Alamitos (2007)
10. Canetti, R.: Universally Composable Security: A New Paradigm for Cryptographic Protocols. In: 42nd IEEE Symposium on Foundations of Computer Science, pp. 136–145. IEEE Computer Society Press, Los Alamitos (2001)
11. Canetti, R., Krawczyk, H.: Analysis of Key-Exchange Protocols and Their Use for Building Secure Channels. In: Pfitzmann, B. (ed.) EUROCRYPT 2001. LNCS, vol. 2045, pp. 453–474. Springer, Heidelberg (2001)

12. Canetti, R., Krawczyk, H.: Universally Composable Notions of Key-Exchange and Secure Channels. In: Knudsen, L.R. (ed.) EUROCRYPT 2002. LNCS, vol. 2332, pp. 337–351. Springer, Heidelberg (2002)
13. Denning, D.E., Sacco, G.M.: Timestamps in Key Distribution Protocols. Communications of the ACM 24(8), 533–536 (1981)
14. Gong, L.: A Security Risk of Depending on Synchronized Clocks. ACM SIGOPS Operating Systems Review 26(1), 49–53 (1992)
15. ISO/IEC 11770-2: Information Technology – Security Techniques – Key Management – Part 2: Mechanisms Using Symmetric Techniques (2008)
16. ISO/IEC 9798-2: Information Technology – Security Techniques – Entity Authentication – Part 2: Mechanisms Using Symmetric Encipherment Algorithms (1999)
17. ISO/IEC 9798-3: Information Technology – Security Techniques – Entity Authentication – Part 3: Mechanisms Using Digital Signature Techniques (1998)
18. Menezes, A.J., van Oorschot, P.C., Vanstone, S.A.: Handbook of Applied Cryptography. CRC Press, Boca Raton (2001)
19. Newman, C., Yu, T., Hartman, S., Raeburn, K.: The Kerberos Network Authentication Service, V5 (2005), http://www.ietf.org/rfc/rfc4120

Password-Authenticated Group Key Agreement with Adaptive Security and Contributiveness

Michel Abdalla[1], Dario Catalano[2], Céline Chevalier[1], and David Pointcheval[1]

[1] École Normale Supérieure, CNRS-INRIA, Paris, France
[2] Università di Catania, Catania, Italy

Abstract. Adaptively-secure key exchange allows the establishment of secure channels even in the presence of an adversary that can corrupt parties adaptively and obtain their internal states. In this paper, we give a formal definition of contributory protocols and define an ideal functionality for password-based group key exchange with explicit authentication and contributiveness in the UC framework. As with previous definitions in the same framework, our definitions do not assume any particular distribution on passwords or independence between passwords of different parties. We also provide the first steps toward realizing this functionality in the above strong adaptive setting by analyzing an efficient existing protocol and showing that it realizes the ideal functionality in the random-oracle and ideal-cipher models based on the CDH assumption.

1 Introduction

Motivation. The main goal of an authenticated key exchange (AKE) protocol is to allow users to establish a common key over a public channel, even in the presence of adversaries. The most common way to achieve this goal is to rely either on a public-key infrastructure (PKI) or on a common high-entropy secret key [38]. Unfortunately, these methods require the existence of trusted hardware capable of storing high-entropy secret keys. In this paper, we focus on a different and perhaps more realistic scenario where secret keys are assumed to be short passwords.

Since the seminal work by Bellovin and Merritt [11], password-based key exchange has become quite popular. Due to their low entropy, passwords are easily memorizable by humans and avoid the need for trusted hardware. On the other hand, the low entropy of passwords makes them vulnerable to exhaustive search: perfect forward secrecy thus becomes quite important. This notion means that, even if the password of a user is later guessed or leaked, keys established before the leakage of the password remain private. Depending on the security model, the privacy of the common key has been modeled via either "semantic security" [9] or the indistinguishability of the actual protocol and an ideal one [26]. In both cases, the leakage of the long-term secret (password in our case) has been modeled by "corruption" queries. Unfortunately, the long-term secret may not be the only information leaked during a corruption. Ephemeral secret leakage has also been shown to cause severe damages in some contexts [36,35], and

B. Preneel (Ed.): AFRICACRYPT 2009, LNCS 5580, pp. 254–271, 2009.

thus "strong corruptions" should also be considered [39]. However, it may not be very realistic to allow the designer of a protocol to decide which information is revealed by such a *strong corruption* query. The universal composability (UC) framework [24], on the other hand, allows for a different approach: whenever a strong corruption occurs, the adversary breaks into the corrupted players, learns whatever information is required to complete the session, and controls the player thereafter. This seems to be a more realistic scenario.

In this paper, we consider stronger corruptions for password-based group key exchange protocols in the adaptive setting, in which adversaries are allowed to corrupt parties adaptively based on the information they have gathered so far. Despite numerous works on group key exchange protocols, very few schemes have been proven to withstand strong corruptions in case of adaptive adversaries. In the context of group AKE, Katz and Shin [34] proposed a compiler that converts any protocol that is secure in a weak corruption model into one that is secure in a strong corruption model, but their protocol relies on signatures and does not work in the password-based scenario. In the case of password-based group AKE protocols, to the best of our knowledge, the only scheme shown to withstand adaptive corruptions is due to Barak *et al.* [6] using general techniques from multi-party computation. Unfortunately, their scheme is not practical.

Another issue being considered here is contributiveness [13,28,21]. In the security model of Katz and Shin [34], one just wants to prevent the adversary from fully determining the key, unless it has corrupted one player in the group. However, it would be better to ensure the randomness of the key, unless it has compromised the security of a sufficient number of players: If the adversary has not compromised too many players, it should not be able to bias the key. There are several advantages in adopting a stronger notion of contributiveness for group key exchange protocols. First, it creates a clear distinction between key distribution and key agreement protocols by describing in concrete terms the notion that, in a group key agreement protocol, each player should contribute equally to the session key. Second, it renders the protocol more robust to failures in that the session key is guaranteed to be uniformly distributed even if some of the players do not choose their contributions correctly (due to hardware malfunction, for instance). Third, it avoids scenarios in which malicious insiders secretly leak the value of a session key to a third party by either imposing specific values for the session key or biasing the distribution of the latter, so that the key can be quickly derived, and then the communication eavesdropped in real-time. For instance, we can imagine the third party to be an intelligence agency, such as the CIA, and the malicious insiders to be malicious pieces of hardware and software (pseudo-random generator) installed by the intelligence agency. Finally, if the absence of subliminal channels [40] can be guaranteed during the lifetime of the session key starting from the moment at which the players have gathered enough information to compute this key (a property that needs to be studied independently), then no malicious insider would be able to help an outsider to eavesdrop on the communication in real-time. Interestingly, since all the functions used in the later rounds are deterministic, this property seems to be satisfied by our scheme. Of course, one cannot keep an insider from later revealing this key, but by then it might already be too late for this information to be useful. On the other hand, physical protections or network surveillance can ensure that no stream of data can be sent out by the players during a confidential discussion. Contributiveness and the absence of subliminal channels therefore guarantee no real-time eavesdropping, thus yielding a new security property. However, the study of subliminal channels is out of the scope of the present paper.

Contributions. There are three main contributions in this paper. First, we investigate a stronger notion of security for password-based group AKE. This is done by combining recent results on the topic of AKE in the UC framework, including the work of Canetti *et al.* [25], Katz and Shin [34], and Barak *et al.* [6]. The first one described the ideal functionality for the password-based AKE, and proved that a variant of the KOY/GL scheme [33,31] securely realizes the functionality, only against static adversaries (no strong corruptions available during the protocol, but at the beginning only). Katz and Shin [34] provided the ideal functionality for the group AKE, and proved that the new, derived, security notion is actually stronger than the usual Bresson *et al.* one [20]. Furthermore they formalized a strong corruption model, where honest players may be compromised at any point during the execution of the protocol, and leak their long-term and ephemeral secrets (the entire internal state). Barak *et al.* [6] considered protocols for general multi-party computation in the absence of authenticated channels. In particular, they provided a general and conceptually simple solution (though inefficient) to a number of problems including password-based group AKE.

In Section 2, we propose an ideal functionality for password-based group AKE. Our new functionality guarantees mutual authentication, which explicitly ensures that, if a player accepts the session key, then all his intended partners have been involved in the protocol and obtained the key material. Note however that the adversary may modify subsequent flows and eventually make some of the players reject while others accept. The protocol also inevitably leaks some information to the adversary, who ends up learning whether the parties share a common secret key or not. Following the approach suggested in [25], we assume that the passwords are chosen by the environment, who then hands them to the parties as input. This is the strongest security model, since it does not assume any distribution on passwords. Furthermore, it allows the environment to force players to run the protocol using different (possibly related) passwords. This is useful, for example, to model the situation where users mistype their passwords.

Vulnerability of the passwords (whose entropy may be low) is modeled via split functionalities [6] and TestPwd queries. The use of split functionalities captures the fact that the adversary can always partition the players into disjoint subgroups and engage in separate executions of the protocols with each of these subgroups, playing the role of the other players. This is because, in the absence of strong authentication mechanisms such as signatures, honest parties cannot distinguish the case in which they interact with each other from the case in which they interact with the adversary. The use of TestPwd queries captures the fact that, within a particular subgroup, the adversary may be able to further divide the set of honest users into smaller groups and locally test the passwords of these users. In fact, in most password-based protocols based on the Burmester-Desmedt group key exchange [23] such as the one by Abdalla *et al.* [2], the adversary can test the value of the password held by a user by simply playing the role of the neighbors of this user.

A second contribution of this paper is to strengthen the original security model of Katz and Shin [34] by incorporating the notion of contributiveness in the functionality described above. A protocol is said to be (t, n)-contributory if no adversary can bias the key as long as (strictly) less than t players in the group of n players have been corrupted. This is stronger than the initial Katz-Shin model, which prevents the adversary from choosing the key, but as long as there is no corrupted player in the game ($t = 1$), only. Of course, one cannot prevent an insider adversary from learning the key, and possibly revealing it or the communication itself to outside. But we may hope to prevent "subliminal" leakage of information that would allow eavesdropping in real-time.

Our last contribution is to show that a slight variant of the password-based group AKE protocol by Abdalla *et al.* [2] based on the Burmester-Desmedt protocol [22,23] and briefly described in Section 3, securely realizes the new functionality, even against adaptive adversaries. The proof is given in Section 4 (the details can be found in the full version) and is in the ideal-tweakable-cipher and random-oracle models [8]. Even though, from a mathematical point of view, it would be preferable to have a proof in the standard model, we point out that the protocol presented here is the first group key exchange protocol realizing such strong security notions, namely adaptive security against strong corruptions in the UC framework and $(n/2, n)$-contributiveness. In addition, our protocol is quite efficient. We also provide a modification to achieve $(n - 1, n)$-contributiveness.

Related work. Since the seminal Diffie-Hellman key exchange protocol [29], several extensions of that protocol to the group setting were proposed in the literature [22,5] without a formal security model. The first security model for group key exchange was proposed by Bresson *et al.* [20], who later extended it to dynamic and concurrent setting [15,16], using the same framework as Bellare *et al.* [9,10]. In the UC framework [24], the first security model in the group setting was proposed by Katz and Shin [34]. Their model is quite strong, allowing the adversary to corrupt players adaptively and learn their internal state. In the password-based scenario, most of the previous work focused on the 2-party case. The first security models to appear [7,14] were based on the frameworks by Bellare *et al.* [9,10] and by Shoup [39]. In the UC framework, the first ones to propose an ideal functionality for password-based AKE were Canetti *et al.* [25]. More recently, Abdalla *et al.* [3] showed that the 2-party password-based authenticated key exchange protocol in [18] is also secure in the UC model against adaptive adversaries assuming the random-oracle and ideal-cipher models. In the group password-based AKE scenario, there has been a few protocols proposed in the literature, from the initial work by Bresson *et al.* [17,19] to the more recent proposals by Dutta and Barua [30], Abdalla *et al.* [1,2,4], and Bohli *et al.* [12]. However, none of them appear to satisfy the security requirements being considered in this paper.

2 Definition of Security

Notations. We denote by k the security parameter. An event is said to be negligible if it happens with probability less than the inverse of any polynomial in k. If G is a finite set, $x \xleftarrow{R} G$ indicates the process of selecting x uniformly and at random in G (thus we implicitly assume that G can be sampled efficiently).

The UC Framework. Throughout this paper we assume basic familiarity with the universal composability framework. The interested reader is referred to [24,25] for details. The model considered in this paper is the UC framework with joint state proposed by Canetti and Rabin [27].

Adaptive Adversaries. In this paper, we consider adaptive adversaries which are allowed to arbitrarily corrupt players at any moment during the execution of the protocol, thus getting complete access to their internal memory. In a real execution of the protocol, this is modeled by letting the adversary $\mathcal{A}$ obtain the password and the internal state of the corrupted player. Moreover, $\mathcal{A}$ can arbitrarily modify the player's strategy. In an ideal execution of the protocol, the simulator $\mathcal{S}$ gets the corrupted player's password and has to simulate its internal state, in a way that remains consistent to what was already provided to the environment.

Contributory Protocols. In addition, we consider a stronger corruption model against insiders than the one proposed by Katz and Shin in [34], where one allows the adversary to choose the session key as soon as there is a corruption. On the contrary, we define here a notion of contributory protocol which guarantees the distribution of the session keys to be random as long as there are enough honest participants in the session: the adversary cannot bias the distribution unless it controls a large number of players. More precisely, we say that a protocol is (t, n)-contributory if the group consists of n people and if the adversary cannot bias the key as long as it has corrupted (strictly) less than t players. More concretely, we claim that our proposed protocol is $(n/2, n)$-contributory, which means that the adversary cannot bias the key as long as there are at least half honest players. We even show in the full version that $(n-1, n)$-contributiveness can be fulfilled by running parallel executions of our protocol.

The Random Oracle and Ideal Tweakable Cipher. In [25], Canetti *et al.* showed that there doesn't exist any protocol that UC-emulates the two-party password-based key-exchange functionality in the plain model (i.e. without additional setup assumptions). Here we show how to securely realize a similar functionality without setup assumption but working in the random-oracle and ideal-tweakable-cipher models instead. The random oracle [8] ideal functionality was already defined by Hofheinz and Müller-Quade in [32] (see the full version).

Similarly, it is straightforward to derive the functionality for the ideal tweakable cipher primitive [37], see the full version. Note that for both, since the session identifier sid will be included either in the input for the random oracle, or in the tweak for the ideal-tweakable cipher, we can have one instantiation of each only, in the joint state, and not different instantiations for each session, which is more realistic. Note however that in some cases (in the first flow, see Figure 3), only some part of the session identifier will be included, which could lead to collisions with other sessions sharing this part of the sid (which by construction of the protocol are the sessions generated by the split functionality, see below). It could lead to a problem in the simulation, and more precisely in case of programming the random oracle or the ideal cipher. But programming is used in the proof only for honest players in a session. And with the split functionality, honest players are separated into various sets that make a *partition* (they are all disjoint): since a player cannot belong to two distinct sessions, this means that the simulator will have to program the random oracle or the decryption only once and that no problem occurs (except, of course, if the adversary has already asked the critical query, e.g. an encryption leading to the particular ciphertext, but this happens with negligible probability only).

Split Functionalities. Without any strong authentication mechanisms, the adversary can always partition the players into disjoint subgroups and execute independent sessions of the protocol with each subgroup, playing the role of the other players. Such an attack is unavoidable since players cannot distinguish the case in which they interact with each other from the case where they interact with the adversary. The authors of [6] addressed this issue by proposing a new model based on *split functionalities* which guarantees that this attack is the only one available to the adversary.

The split functionality is a generic construction based upon an ideal functionality: Its description can be found on Figure 1. In the initialization stage, the adversary adaptively chooses disjoint subsets of the honest parties (with a unique session identifier that is fixed for the duration of the protocol). More precisely, the protocol starts with a session identifier sid. Then, the initialization stage generates some random values which, combined together and with sid, create the new session identifier sid', shared by all parties which have received the same values – that is, the parties of the disjoint subsets. The important point here is that the subsets create a *partition* of the players, thus forbidding communication among the subsets. During the computation, each subset H activates a separate instance of the functionality $\mathcal{F}$. All these functionality instances are independent: The executions of the protocol for each subset H can only be related in the way the adversary chooses the inputs of the players it controls. The parties $P_i \in H$ provide their own inputs and receive their own outputs (see the first item of "computation" in Figure 1), whereas the adversary plays the role of all the parties $P_j \notin H$ (see the second item).

The Group Password-Based Key Exchange Functionality with Mutual Authentication. In this section, we discuss the $\mathcal{F}_{GPAKE}$ functionality (see Figure 2). The multi-session extension of our functionality would be similar

Given a functionality $\mathcal{F}$, the split functionality $s\mathcal{F}$ proceeds as follows:

Initialization:

- Upon receiving $(\texttt{Init}, sid)$ from party P_i, send $(\texttt{Init}, sid, P_i)$ to the adversary.
- Upon receiving a message $(\texttt{Init}, sid, P_i, H, sid_H)$ from $\mathcal{A}$, where H is a set of party identities, check that P_i has already sent $(\texttt{Init}, sid)$ and that for all recorded $(H', sid_{H'})$, either $H = H'$ and $sid_H = sid_{H'}$ or H and H' are disjoint and $sid_H \neq sid_{H'}$. If so, record the pair (H, sid_H), send $(\texttt{Init}, sid, sid_H)$ to P_i, and invoke a new functionality $(\mathcal{F}, sid_H)$ denoted as $\mathcal{F}_H$ and with set of honest parties H.

Computation:

- Upon receiving $(\texttt{Input}, sid, m)$ from party P_i, find the set H such that $P_i \in H$ and forward m to $\mathcal{F}_H$.
- Upon receiving $(\texttt{Input}, sid, P_j, H, m)$ from $\mathcal{A}$, such that $P_j \notin H$, forward m to $\mathcal{F}_H$ as if coming from P_j.
- When $\mathcal{F}_H$ generates an output m for party $P_i \in H$, send m to P_i. If the output is for $P_j \notin H$ or for the adversary, send m to the adversary.

Fig. 1. Split Functionality $s\mathcal{F}$

to the one proposed by Canetti and Rabin [27]. Our starting points are the group key exchange functionality described in [34] and the (two party) password-based key exchange functionality given in [25]. Our aim is to combine the two of them and to add mutual authentication and (t, n)-contributiveness. The new definition still remains very general: letting $t = 1$, we get back the case in which the adversary may manage to set the key when it controls at least a player, as in [25].

First, notice that the functionality is not in charge of providing the passwords to the participants. Rather we let the environment do this. As already pointed out in [25], such an approach allows to model, for example, the case where some users may use the same password for different protocols and, more generally, the case where passwords are chosen according to some arbitrary distribution (i.e. not necessarily the uniform one). Moreover, notice that allowing the environment to choose the passwords guarantees forward secrecy, basically for free. More generally, this approach allows to preserve security[1] even in those situations where the password is used (by the same environment) for other purposes.

In the following we denote by n the number of players involved in a given execution of the protocol. The functionality starts with an initialization step during which it basically waits for each player to notify its interest in participating to the protocol. More precisely, we assume that every player starts a new session of the protocol with input $(\texttt{NewSession}, sid, P_i, Pid, pw_i)$, where P_i is the identity of the player, pw_i is its password and Pid represents the set of (identities of)

---

[1] By "preserved" here we mean that the probability of breaking the scheme is basically the same as the probability of guessing the password.

The functionality $\mathcal{F}_{GPAKE}$ is parameterized by a security parameter k and the parameter t of the contributiveness. It interacts with an adversary $\mathcal{S}$ and a set of parties $P_1,\ldots,P_n$ via the following queries:

- **Initialization.** Upon receiving (NewSession, sid, P_i, Pid, pw_i) from player P_i for the first time, where Pid is a set of at least two distinct identities containing P_i, record (sid, P_i, Pid, pw_i), mark it **fresh**, and send (sid, P_i, Pid) to $\mathcal{S}$. Ignore any subsequent query (NewSession, sid, P_j, Pid', pw_j) where $Pid' \neq Pid$.
  If there are already $|Pid| - 1$ recorded tuples (sid, P_j, Pid, pw_j) for players $P_j \in Pid \setminus \{P_i\}$, then record (sid, Pid, ready) and send it to $\mathcal{S}$.

- **Password tests.** Upon receiving a query (TestPwd, sid, P_i, Pid, pw') from the adversary $\mathcal{S}$, if there exists a record of the form (sid, P_i, Pid, pw_i) which is **fresh**:
  - If $pw_i = pw'$, mark the record **compromised** and reply to $\mathcal{S}$ with "correct guess".
  - If $pw_i \neq pw'$, mark the record **interrupted** and reply to $\mathcal{S}$ with "wrong guess".

- **Key Generation.** Upon receiving a message $(sid, Pid, \text{ok}, sk)$ from $\mathcal{S}$ where there exists a recorded tuple (sid, Pid, ready), then, denote by n_c the number of corrupted players, and
  - If all $P_i \in Pid$ have the same passwords and $n_c < t$, choose $sk' \in \{0,1\}^k$ uniformly at random and store (sid, Pid, sk'). Next, for all $P_i \in Pid$ mark the record (sid, P_i, Pid, pw_i) **complete**.
  - If all $P_i \in Pid$ have the same passwords and $n_c \geq t$, store (sid, Pid, sk). Next, for all $P_i \in Pid$ mark the record (sid, P_i, Pid, pw_i) **complete**.
  - In any other case, store (sid, Pid, error). For all $P_i \in Pid$ mark the record (sid, P_i, Pid, pw_i) **error**.

  When the key is set, report the result (either **error** or **complete**) to $\mathcal{S}$.

- **Key Delivery.** Upon receiving a message (deliver, b, sid, P_i) from $\mathcal{S}$, then if $P_i \in Pid$ and there is a recorded tuple (sid, Pid, α) where $\alpha \in \{0,1\}^k \cup \{\text{error}\}$, send (sid, Pid, α) to P_i if b equals **yes** or (sid, Pid, error) if b equals **no**.

- **Player Corruption.** If $\mathcal{S}$ corrupts $P_i \in Pid$ where there is a recorded tuple (sid, P_i, Pid, pw_i), then reveal pw_i to $\mathcal{S}$. If there also is a recorded tuple (sid, Pid, sk), that has not yet been sent to P_i, then send (sid, Pid, sk) to $\mathcal{S}$.

Fig. 2. Functionality $\mathcal{F}_{GPAKE}$

players with whom it intends to share a session key. Once all the players (sharing the same sid and Pid) have sent their notification message, $\mathcal{F}_{GPAKE}$ informs the adversary that it is ready to start a new session of the protocol.

In principle, after the initialization stage is over, all the players are ready to receive the session key. However the functionality waits for $\mathcal{S}$ to send an "ok" message before proceeding. This allows $\mathcal{S}$ to decide the exact moment when the key should be sent to the players and, in particular, it allows $\mathcal{S}$ to choose the exact moment when corruptions should occur (for instance $\mathcal{S}$ may decide to corrupt some party P_i before the key is sent but after P_i decided to participate to a given session of the protocol, see [34]).

Once the functionality receives a message $(sid, Pid, \texttt{ok}, sk)$ from $\mathcal{S}$, it proceeds to the key generation phase. This is done as follows. If all players in Pid share the same password and less than t players are corrupted, the functionality chooses a key sk' uniformly and at random in the appropriate key space. If all players in Pid share the same password but t or more players are corrupted, then the functionality allows $\mathcal{S}$ to fully determine the key by letting $sk' = sk$. In all the remaining cases no key is established.

Remark. For sake of simplicity, we chose to integrate the contributiveness in the UC-functionality. However, one could say that this is not necessary and that one could have kept the original functionality for group password-based key exchange, and then studying the contributiveness as an extra property. But then, two security proofs would be needed.

Our definition of the $\mathcal{F}_{GPAKE}$ functionality deals with corruptions of players in a way quite similar to that of $\mathcal{F}_{GPAKE}$ in [34], in the sense that if the adversary has corrupted some participants, it may determine the session key, but here only if there are enough corrupted players. Notice however that $\mathcal{S}$ is given such power only before the key is actually established. Once the key is set, corruptions allow the adversary to know the key but not to choose it.

Following [25], a correct password test is captured by marking the corresponding record as **compromised**, and a failed attempt by marking it as **interrupted**. Records that are neither **compromised** nor **interrupted** are initially marked as **fresh**. Once a key is established, all records in Pid are marked as **complete**. Changing the fresh status of a record whenever a password test occurs or a (valid) key is established, is aimed at limiting the number of password tests to at most one per player. However, the **TestPwd** queries seem unnecessary in the context of split functionalities, since by splitting the whole group in subgroups of one player each, the adversary can already test one password per player. However, adding this query just allows $\mathcal{A}$ to test an additional password per player, which does not change significantly its power in practice. Besides, these queries are needed in the security analysis of our protocol, which we tried to keep as efficient as possible. Designing an efficient protocol while getting rid of theses queries is an interesting open problem.

In any case, after the key generation, the functionality informs the adversary about the result, meaning with this that the adversary is informed on whether a key was actually established or not. In particular, this means that the adversary is also informed on whether the players share the same password or not. At first glance this may seem like a dangerous information to provide to the adversary. We argue, however, that this is not the case in our setting. Indeed, being all the passwords chosen by the environment, such an information could be available to the adversary anyway. Moreover, it does not seem critical to hide the status of the protocol (i.e. if it completed correctly or not), as in practice this information is often easily obtained by simply monitoring its execution (if the players suddenly stop their communications, there must have been some problem).

Finally the key is sent to the players according to the schedule chosen by $\mathcal{S}$. This is formally modeled by means of key delivery queries. We assume that, once $\mathcal{S}$ asks to deliver the key to a player, the key is sent immediately.

Notice that, the mutual authentication indeed means that if one of the players accepts, then all players share the key material; but, it doesn't mean that they all accept. Indeed, we cannot assume that all the flows are correctly forwarded by the adversary: it can modify just one flow, or at least omit to deliver one flow. This attack, called *denial of service*, is modeled in the functionality by the key delivery: the adversary can choose whether it wants the player to receive or not the good key/messages simply with the help of the keyword b set to yes or no.

3   Our Scheme

Description. Our solution builds on an earlier protocol by Abdalla *et al.* [2] and is described in Figure 3. Let $\mathcal{E}$ and $\mathcal{D}$ be the encryption and decryption schemes of an ideal tweakable cipher scheme. We denote by $\mathcal{E}_{pw}^{\ell}(m)$ an encryption of the message m using the ideal tweakable cipher, label ℓ and password pw. Similarly, the decryption is denoted as $\mathcal{D}_{pw}^{\ell}(c)$. The protocol uses five different random oracles, denoted by $\mathcal{H}_i$ for all $i = 0, \ldots, 4$. We denote by q_{h_i} the number of queries made to the oracle $\mathcal{H}_i$ ($q_h = q_{h_0} + \cdots + q_{h_4}$), and by $k_i = 2^{\ell_i}$ the output size: $\mathcal{H}_i : \{0,1\}^k \to \{0,1\}^{\ell_i}$. For an optimal instantiation, we will assume that for all i, $\ell_i = 2k$ (collisions), where k is the security parameter. Finally let (SKG, Sign, Ver) be a one-time signature scheme, SKG being the signature key generation, Sign the signing algorithm and Ver the verifying algorithm. Note that we do not require a *strong* one-time signature: Here, the adversary is allowed to query the signing oracle at most once, and should not be able to forge a signature of *another* authenticator.

Informally, and omitting the details, the algorithm can be described as follows: First, each player chooses a random exponent x_i and computes $z_i = g^{x_i}$ and an encryption z_i^* of z_i. It then applies SKG to generate a pair (SK_i, VK_i) of signature keys, and commits to the values VK_i and z_i^*. In the second round, it reveals these values (the use of the commitment will be explained later in this section). We stress that the second round does not begin until all commitments have been received. At this point, the session identifier becomes $ssid' = ssid\|c_1\|\ldots\|c_n$. It will be included, and verified, in all the subsequent hash values. Then, after verifying the commitments of the others, each couple (P_i, P_{i-1}) of players computes a common Diffie-Hellman value $Z_i = g^{x_i x_{i-1}}$, leading to a hash value X_i for each player. Each player then commits to this X_i, and once all these commitments have been received, it reveals the value X_i. In the next round, the players check this second round of commitments and compute an authenticator and the associated signature. Finally, the players check these authenticators and signatures, and if they are all correct, they compute the session key and mark their session as complete.

As soon as a value received by P_i doesn't match with the expected value, it aborts, setting the key $sk_i = $ error. In particular, every player checks that $c_i = \mathcal{H}_3(ssid, z_i^*, VK_i, i)$, $c_i' = \mathcal{H}_4(ssid', X_i, i)$ and $\text{Ver}(VK_i, Auth_i, \sigma_i) = 1$.

$$(1a)\ x_i \xleftarrow{R} \mathbb{Z}_q^* \qquad z_i = g^{x_i} \qquad z_i^* = \mathcal{E}_{pw_i}^{ssid,i}(z_i)$$
$$(\mathrm{VK}_i, \mathrm{SK}_i) \leftarrow \mathsf{SKG} \qquad c_i = \mathcal{H}_3(ssid, z_i^*, \mathrm{VK}_i, i) \qquad \xrightarrow{\quad c_i \quad}$$

....

After this point, the session identifier becomes $ssid' = ssid\|c_1\|\dots\|c_n$.

$(1b)$ sends z_i^* and VK_i $\qquad\qquad\qquad\qquad\qquad\qquad \xrightarrow{\quad z_i^*, \mathrm{VK}_i \quad}$

$(2a)$ checks $c_j = \mathcal{H}_3(ssid, z_j^*, \mathrm{VK}_j, j) \quad \forall j \neq i$
  and aborts if one of these values is incorrect

$$z_{i-1} = \mathcal{D}_{pw_i}^{ssid,i-1}(z_{i-1}^*) \qquad Z_i = (z_{i-1})^{x_i}$$
$$z_{i+1} = \mathcal{D}_{pw_i}^{ssid,i+1}(z_{i+1}^*) \qquad Z_{i+1} = (z_{i+1})^{x_i}$$
$$h_i = \mathcal{H}_2(Z_i) \qquad\qquad h_{i+1} = \mathcal{H}_2(Z_{i+1})$$
$$X_i = h_i \oplus h_{i+1} \qquad\qquad c_i' = \mathcal{H}_4(ssid', X_i, i) \qquad \xrightarrow{\quad c_i' \quad}$$

$(2b)$ sends X_i $\qquad\qquad\qquad\qquad\qquad\qquad\qquad\qquad \xrightarrow{\quad X_i \quad}$

(3) checks $c_j' = \mathcal{H}_4(ssid', X_j, j) \quad \forall j \neq i$
  and aborts if one of these values is incorrect
$$h_{j+1} = X_j \oplus h_j \qquad \forall j = i, \dots, n+i-1 \pmod n$$
$$\mathcal{A}uth_i = \mathcal{H}_1(ssid', (z_1^*, X_1, h_1), \dots, (z_n^*, X_n, h_n), i)$$
$$\sigma_i = \mathsf{Sign}(\mathrm{SK}_i, \mathcal{A}uth_i) \qquad\qquad\qquad \xrightarrow{\quad \sigma_i \quad}$$

(4) computes all the $\mathcal{A}uth_j$ and checks $\mathsf{Ver}(\mathrm{VK}_j, \mathcal{A}uth_j, \sigma_j)$. If they are
  correct, then marks the session as complete and
  sets $sk_i = \mathcal{H}_0(ssid', h_1, \dots, h_n)$. Otherwise, sets $sk_i = \mathtt{error}$.

Fig. 3. Description of the protocol for player P_i, with index i and password pw_i

We now highlight some of the differences between the scheme of [2] and our scheme, described in Figure 3. First, our construction does not require any random nonce in the first round (the session id constructed after the first flow is enough). Moreover, to properly implement the functionality, we return an error message to the players whenever they don't share the same password (mutual authentication). For sake of simplicity, two additional modifications are not directly related to the functionality. First, we use an ideal tweakable cipher rather than a different symmetric key for each player. Second, the values X_i's are here computed as the xor of two hashes.

Due to the split functionality, the players are partitioned according to the values they received during the first round (*i.e.* before the dotted line in Figure 3). All the c_i are shared among them – and thus the z_i^* and VK_i due to the random oracle $\mathcal{H}_3$ – and their session identifier becomes $ssid' = ssid\|c_1\|\dots\|c_n$. In round 3, the signature added to the authentication flow prevents the adversary from being able to change an authenticator to another value. At the beginning of each flow, the players wait until they have received all the other values of the previous flow before sending their new one. This is particularly important between flow(1a) and flow(1b) and similarly between flow(2a) and flow(2b). Since the session identifier $ssid'$ is included in all the hash values, and in the latter signature, only players in the same subset can accept and conclude with a common key.

Finally, the contributory property is ensured by the following modification: In the first and second rounds, each player starts by sending a commitment of the value it has just computed (using a random oracle), denoted as c_i and c'_i. Due to this commitment, it is impossible for a player to compute its z_i^* (or X_i) once it has seen the others: Every player has to commit its z_i^* (or X_i) at the same time as the others, and this value cannot depend on the other values sent by the players.

Finally we point out that, in our proof of security, we don't need to assume that the players erase any ephemeral value before the end of the computation of the session key.

Computational Diffie-Hellman Assumption. Denote by $G = \langle g \rangle$ a finite cyclic (multiplicative) group of prime order q. If x, y are chosen uniformly at random in $\mathbb{Z}_q^*$, the CDH assumption states it is computationally intractable to output g^{xy} given g, g^x and g^y.

It is easy to see that if P_i and P_{i+1} have the same passwords, they will compute in round 2 the same $Z_{i+1} = (z_{i+1})^{x_i} = g^{x_i x_{i+1}} = (z_i)^{x_{i+1}}$. If the passwords are different, we denote by Z_i^R a value computed by P_i on its right side, and Z_{i+1}^L a value computed by P_{i+1} on its left side. We have:

$$Z_i^R = CDH_g(\mathcal{D}_{pw_i}^{ssid,i-1}(z_{i-1}^*), \mathcal{D}_{pw_i}^{ssid,i}(z_i^*)) \qquad h_i^R = \mathcal{H}_2(Z_i^R)$$
$$Z_{i+1}^L = CDH_g(\mathcal{D}_{pw_i}^{ssid,i}(z_i^*), \mathcal{D}_{pw_i}^{ssid,i+1}(z_{i+1}^*)) \qquad h_{i+1}^L = \mathcal{H}_2(Z_{i+1}^L)$$

(here CDH_g denotes the Diffie-Hellman function in base g that given on input g^a, g^b outputs g^{ab}), and then Z_{i+1}^L and Z_{i+1}^R are likely different.

Pictorially, the situation can be summarized as follows

$$P_{i-1} \xleftarrow{\quad Z_i^L \neq Z_i^R \quad} P_i \xleftarrow{\quad Z_{i+1}^L \neq Z_{i+1}^R \quad} P_{i+1}$$

Each P_i computes $X_i = h_i^R \oplus h_{i+1}^L$ and thus, once the values X_j's are published, all the players can iteratively compute all the h_j's required to compute the authenticators $\mathcal{A}uth_i$ and later the session key sk_i.

Our Main Theorem. Let $\widehat{s\mathcal{F}}_{GPAKE}$ be the multi-session extension of the split functionality $s\mathcal{F}_{GPAKE}$ and let $\mathcal{F}_{RO}$ and $\mathcal{F}_{ITC}$ be the ideal functionalities that provide a random oracle and an ideal tweakable cipher to all parties.

Theorem 1. *The protocol presented in Figure 3 securely realizes $\widehat{s\mathcal{F}}_{GPAKE}$ in the $(\mathcal{F}_{RO}, \mathcal{F}_{ITC})$-hybrid model, in the presence of adaptive adversaries, and is $(n/2, n)$-contributory.*

Corollary 1. *This protocol, along with small modifications described in   the full version (with $n/2$ parallel executions of the original scheme), securely realizes $\widehat{s\mathcal{F}}_{GPAKE}$ in the $(\mathcal{F}_{RO}, \mathcal{F}_{ITC})$-hybrid model, in the presence of adaptive adversaries, and is $(n-1, n)$-contributory.*

Note that if the signature scheme is subliminal channel free [40] (*e.g.* only one signature σ is valid for a given pair $(VK, \mathcal{A}uth)$), then after the first round ($1a$), everything is deterministic, and thus no subliminal channel is available for an adversary to leak any information to an eavesdropper.

4   Proof of Theorem 1

We prove the protocol to be $(n/2, n)$-contributory, and show in the full version how to get the $(n-1, n)$-contributiveness, with some parallel executions. Note that $n/2$ here implicitly means $\lfloor n/2 \rfloor$ when n is odd.

We start by giving an attack showing that the $(n/2+1)$-contributory cannot be satisfied. For sake of simplicity, assume that n is odd and consider the following situation in which there are $\lfloor n/2 \rfloor$ honest players (denoted as P_i) and $\lfloor n/2 \rfloor + 1$ corrupted players (denoted as $\mathcal{A}_i$, since they are under the control of the adversary $\mathcal{A}$):

$$\mathcal{A}_1 \quad P_1 \quad \mathcal{A}_2 \quad P_2 \quad \mathcal{A}_3 \ldots P_{\lfloor n/2 \rfloor} \quad \mathcal{A}_{\lfloor n/2 \rfloor} \quad \mathcal{A}_{\lfloor n/2 \rfloor + 1}$$

Since $\mathcal{A}$ knows the random values of the corrupted players, it learns, from flows $(1b)$, the values Z_i, and thus $h_i = \mathcal{H}_2(Z_i)$ for $i = 1, \ldots, n-1$, before it plays on behalf of $\mathcal{A}_{\lfloor n/2 \rfloor + 1}$. Even if it cannot modify anymore z_n in flow $(1b)$ (already committed to in c_n), it can choose Z_n to bias the value h_n, and thus the final key, that is defined as $sk = \mathcal{H}_0(ssid', h_1, \ldots, h_n)$: this is introduced in flow $(2a)$. Such an inconsistent value is possible since no honest player can verify the value of Z_n: this is the Diffie-Hellman value between two corrupted players.

More generally, if $\mathcal{A}$ controls enough players so that each honest player is between two corrupted players, then it can learn, from flows $(1b)$, the values X_i that will be sent in flows $(2b)$. If two corrupted players are neighbors, they can send a value X_i of their choice, since it comes from a Diffie-Hellman value between these two corrupted players. In the attack above, the adversary could learn all the h_i early enough, so that its control on h_n could bias the key. If it can control an h_i, without knowing the other values, there is still enough entropy in the key derivation: the final key is uniformly distributed: contributiveness.

We now prove the $(n/2, n)$-contributiveness, using the above intuition. We need to construct, for any real-world adversary $\mathcal{A}$ (interacting with real parties running the protocol), an ideal-world adversary $\mathcal{S}$ (interacting with dummy parties and the functionality $\widehat{s\mathcal{F}}_{GPAKE}$) such that no environment $\mathcal{Z}$ can distinguish between an execution with $\mathcal{A}$ in the real world and $\mathcal{S}$ in the ideal world with non-negligible probability.

We incrementally define a sequence of games starting from the one describing a real execution of the protocol in the real world, and ending up with game $\mathbf{G_8}$ which we prove to be indistinguishable with respect to the ideal experiment. The key point will be $\mathbf{G_7}$. $\mathbf{G_0}$ is the real-world game. In $\mathbf{G_1}$, we start by simulating the encryption, decryption and hash queries, canceling some unlikely events (such as collisions). Granted the ideal tweakable cipher model (see details in the full version), we can extract the passwords used by $\mathcal{A}$ for players corrupted from the beginning of the session. $\mathbf{G_2}$ and $\mathbf{G_3}$ allow $\mathcal{S}$ to be sure that the authenticators for non-corrupted players are always oracle-generated. In $\mathbf{G_4}$, we show how to deal with the simulation of the first flows. In $\mathbf{G_5}$, we deal with only oracle-generated flows. In $\mathbf{G_6}$, we deal with (possibly) non-oracle-generated flows from round 2. $\mathbf{G_7}$ is the crucial game, where we show how to simulate the non-corrupted players without the knowledge of their passwords, even in the case of corruptions before round 2. Finally, we show that $\mathbf{G_8}$, in which we only replace the hybrid queries by the real ones, is indistinguishable from the ideal game.

Following [25], we say that a flow is *oracle-generated* if it was sent by an honest player (our simulation) and arrives without any alteration to the player it was meant to. We say it is *non-oracle-generated* otherwise, that is either if it was sent by an honest player and modified by the adversary, or if it was sent by a corrupted player or a player impersonated by the adversary: in all these cases, we say that the sender is an *attacked* player. In brief, our simulation controls the random coins of oracle-generated flows, whereas the adversary may control them in non-oracle-generated flows.

Note that since we consider the split functionality, the players have been partitioned in sets according to what they received during the very first flow flow(1a). In the following, we can thus assume that all the players have received the same flow(1a) and flow(1b) (under the binding property of the commitment $\mathcal{H}_3$). Oracle-generated flows flow(1a) have been sent by players that will be considered honest in this session, whereas non-oracle-generated flows have been sent by the adversary, the corresponding players are thus assumed corrupted from the beginning of the session, since the adversary has chosen the password. Note that an advantage of this model (using both a random oracle for the commitment, and an ideal tweakable cipher for the encryption) is that we know the passwords used by the adversary (and thus corrupted players) in this first round, by simply looking in the tables that will be defined in $\mathbf{G_2}$, and they are the same in the view of any honest player. The extraction of the password may fail, if the adversary has computed either the ciphertext or the commitment at random, but then it has no chance to make the protocol conclude successfully. Also note that if flow(2a) is oracle-generated, then flow(2b) must be oracle-generated also with overwhelming probability, due to $\mathcal{H}_4$, as above. As a result, we set $sk_i = \texttt{error}$ whenever an inconsistency is noted by a player (incorrect commitment opening, or invalid signature). The latter then aborts its execution.

Adaptive Corruptions and Connected Components. For simplicity, we consider that the simulator maintains, for each honest player, a list symbolizing its internal state:

$$\Lambda_i = (pw_i, SK_i, x_i, z_i, z_i^*, c_i, Z_i^R, Z_{i+1}^L, h_i^R, h_{i+1}^L, X_i, c_i'),$$

where the superscripts L and R denote the neighbor with whom the value is shared (the left or the right neighbor). When a player gets corrupted, the simulator has to provide the adversary with this list. Most of the fields will be chosen at random during the simulation, with the remaining values initially set to $\perp$. As soon as a player gets corrupted, the simulator recovers its password, which will help to fill the other fields in a consistent way, granted the programmability of the random oracle and the ideal tweakable cipher.

The knowledge of P_i's password indeed helps the simulator to fill in the internal state (see below). Note that this allows $\mathcal{S}$ to send the values (in particular the X_i) in a way that remains consistent with respect to the view of the adversary, and even the environment. Informally, $\mathcal{S}$ does this by partitioning the set of players into a number of connected components. Each component consists of all the connected players sharing the same password (the one used to generate the first flow). Below, we show that all S has to do for the simulation to work is to make sure the produced values are consistent only for the players belonging to the same components. Indeed, for neighbor players belonging to different components, S can basically produce completely unrelated values, without worrying about being caught, since the decrypted values z_i are unrelated, from the beginning of the protocol.

Simulator: Session Initialization. The aim of the first flow is to create the subsets H of players involved in the same protocol execution (see the split functionality, Section 2 and Figure 1). S chooses the values $(\mathrm{SK}_i, \mathrm{VK}_i)$ on behalf of the honest players and the value z_i^* at random, rather than asking an encryption query (it does not know the passwords of the players), then computes and sends the commitments c_i to A. The environment initializes a session for each honest (dummy) player, which is modeled by the Init queries sent to the split functionality. The adversary (from the view of the c_i of the honest players) makes its decision about the subgroups it wants to make: it sends c_i on behalf of the players it wants to impersonate (they will become corrupted from the beginning of the session). We then define the H sets according to the received $\{c_j\}$: the honest players that have received the same $\{c_j\}$ (possibly modified by the adversary) are in the same subgroup H. The simulator forwards these sets H (which make a partition of all the honest players) to the split functionality. The latter then initializes ideal functionalities with sid_H, for each subgroup H: all the players in the same session received and thus use the same $\{c_j\}$. The environment gets back this split, via the dummy players, and then sends the NewSession queries on behalf of the latter, according to the appropriate sid_H. The simulator uses the commitments sent from the adversary to extract the password used (granted the ideal tweakable cipher), and thus sends the appropriate NewSession queries on behalf of the corrupted players (note that in case that no password can be extracted, a random one is used). Then, we can focus on a specific session $ssid' = sid_H$ for some set H.

Simulator: Main Idea. In a nutshell, the simulation of the remaining of the protocol depends on the knowledge of the passwords by the simulator. First, if the simulator knows the password of a player, it does everything honestly for every player belonging to its connected component. Otherwise, it sets everything at random. In case of corruption, S learns the password, and can program the oracles (random oracle and ideal tweakable cipher) and fill in the internal state of the player (and of all the players in its connected component) in a consistent way. This last phase, which consists in programming the oracle, may fail, but only if the adversary can solve an intractable problem (computational Diffie-Hellman).

   More precisely, in most of the cases, the simulator S just follows the protocol on behalf of all the honest players. The main difference between the simulated players and the real honest players is that S does not engage on a particular password on their behalf. However, if A generates/modifies a flow(1a) or a flow(2a) message that is delivered to player P in session $ssid'$, then $\mathcal{S}$ extracts the password pw, granted the ideal tweakable cipher, and uses it in a TestPwd query to the functionality. If this is a correct guess, then $\mathcal{S}$ uses this password on behalf of P, and proceeds with the simulation.

The key point of the simulation consists in sending coherent X_i's, whose values completely determine the session key. To this aim, we consider two cases. First, if the players are all honest and share the same password, the h_i's are chosen at random, but identically between two neighbors. If they do not share the same passwords, they are simply set at random. Second, if there are corrupted players among the players, the simulator determines the connected components, as described above, and the trick consists in making the simulation coherent within those components since the adversary has no means of guessing what happens between two components (the passwords are different).

If a session aborts or terminates, $\mathcal{S}$ reports it to $\mathcal{A}$. If the session terminates with a session key sk, then $\mathcal{S}$ makes a Key Delivery call to $\widehat{\mathcal{F}}_{GPAKE}$, specifying the session key. But recall that unless enough players are corrupted, $\widehat{\mathcal{F}}_{GPAKE}$ will ignore the key specified by $\mathcal{S}$, and thus we do not have to bother with the key in these cases.

5   Conclusion

This paper investigates a stronger security notion against insider adversaries, for password-based group AKE. The protocol as presented in Section 3 achieves $(n/2, n)$-contributiveness; We also show how to achieve $(n-1, n)$-contributiveness in the full version.

Acknowledgments

This work was supported in part by the French ANR-07-SESU-008-01 PAMPA Project and the European Commission through the IST Program under Contract ICT-2007-216646 ECRYPT II.

References

1. Abdalla, M., Bohli, J.-M., Gonzalez Vasco, M.I., Steinwandt, R. (Password) authenticated key establishment: From 2-party to group. In: Vadhan, S.P. (ed.) TCC 2007. LNCS, vol. 4392, pp. 499–514. Springer, Heidelberg (2007)
2. Abdalla, M., Bresson, E., Chevassut, O., Pointcheval, D.: Password-based group key exchange in a constant number of rounds. In: Yung, M., Dodis, Y., Kiayias, A., Malkin, T.G. (eds.) PKC 2006. LNCS, vol. 3958, pp. 427–442. Springer, Heidelberg (2006)
3. Abdalla, M., Catalano, D., Chevalier, C., Pointcheval, D.: Efficient two-party password-based key exchange protocols in the UC framework. In: Malkin, T.G. (ed.) CT-RSA 2008. LNCS, vol. 4964, pp. 335–351. Springer, Heidelberg (2008)

4. Abdalla, M., Pointcheval, D.: A scalable password-based group key exchange protocol in the standard model. In: Lai, X., Chen, K. (eds.) ASIACRYPT 2006. LNCS, vol. 4284, pp. 332–347. Springer, Heidelberg (2006)
5. Ateniese, G., Steiner, M., Tsudik, G.: Authenticated group key agreement and friends. In: ACM CCS 1998 Conference on Computer and Communications Security, pp. 17–26. ACM Press, New York (1998)
6. Barak, B., Canetti, R., Lindell, Y., Pass, R., Rabin, T.: Secure computation without authentication. In: Shoup, V. (ed.) CRYPTO 2005. LNCS, vol. 3621, pp. 361–377. Springer, Heidelberg (2005)
7. Bellare, M., Pointcheval, D., Rogaway, P.: Authenticated key exchange secure against dictionary attacks. In: Preneel, B. (ed.) EUROCRYPT 2000. LNCS, vol. 1807, pp. 139–155. Springer, Heidelberg (2000)
8. Bellare, M., Rogaway, P.: Random oracles are practical: A paradigm for designing efficient protocols. In: ACM CCS 1993, pp. 62–73. ACM Press, New York (1993)
9. Bellare, M., Rogaway, P.: Entity authentication and key distribution. In: Stinson, D.R. (ed.) CRYPTO 1993. LNCS, vol. 773, pp. 232–249. Springer, Heidelberg (1994)
10. Bellare, M., Rogaway, P.: Provably secure session key distribution: The three party case. In: 27th ACM STOC, pp. 57–66. ACM Press, New York (1995)
11. Bellovin, S.M., Merritt, M.: Encrypted key exchange: Password-based protocols secure against dictionary attacks. In: 1992 IEEE Symposium on Security and Privacy, pp. 72–84. IEEE Computer Society Press, Los Alamitos (1992)
12. Bohli, J.-M., Gonzalez Vasco, M.I., Steinwandt, R.: Password-authenticated constant-round group key establishment with a common reference string. Cryptology ePrint Archive, Report 2006/214 (2006)
13. Bohli, J.-M., Vasco, M.I.G., Steinwandt, R.: Secure group key establishment revisited. Int. J. Inf. Secur. 6(4), 243–254 (2007)
14. Boyko, V., MacKenzie, P.D., Patel, S.: Provably secure password-authenticated key exchange using Diffie-Hellman. In: Preneel, B. (ed.) EUROCRYPT 2000. LNCS, vol. 1807, pp. 156–171. Springer, Heidelberg (2000)
15. Bresson, E., Chevassut, O., Pointcheval, D.: Provably authenticated group Diffie-Hellman key exchange – the dynamic case. In: Boyd, C. (ed.) ASIACRYPT 2001. LNCS, vol. 2248, pp. 290–309. Springer, Heidelberg (2001)
16. Bresson, E., Chevassut, O., Pointcheval, D.: Dynamic group Diffie-Hellman key exchange under standard assumptions. In: Knudsen, L.R. (ed.) EUROCRYPT 2002. LNCS, vol. 2332, pp. 321–336. Springer, Heidelberg (2002)
17. Bresson, E., Chevassut, O., Pointcheval, D.: Group Diffie-Hellman key exchange secure against dictionary attacks. In: Zheng, Y. (ed.) ASIACRYPT 2002. LNCS, vol. 2501, pp. 497–514. Springer, Heidelberg (2002)
18. Bresson, E., Chevassut, O., Pointcheval, D.: Security proofs for an efficient password-based key exchange. In: ACM CCS 2003, pp. 241–250. ACM Press, New York (2003)
19. Bresson, E., Chevassut, O., Pointcheval, D.: A security solution for IEEE 802.11's ad-hoc mode: Password authentication and group Diffie-Hellman key exchange. International Journal of Wireless and Mobile Computing 2(1), 4–13 (2007)
20. Bresson, E., Chevassut, O., Pointcheval, D., Quisquater, J.-J.: Provably authenticated group Diffie-Hellman key exchange. In: ACM CCS 2001, pp. 255–264. ACM Press, New York (2001)
21. Bresson, E., Manulis, M.: Securing group key exchange against strong corruptions and key registration attacks. Int. J. Appl. Cryptol. 1(2), 91–107 (2008)

22. Burmester, M., Desmedt, Y.: A secure and efficient conference key distribution system (extended abstract). In: De Santis, A. (ed.) EUROCRYPT 1994. LNCS, vol. 950, pp. 275–286. Springer, Heidelberg (1995)
23. Burmester, M., Desmedt, Y.: A secure and scalable group key exchange system. Information Processing Letters 94(3), 137–143 (2005)
24. Canetti, R.: Universally composable security: A new paradigm for cryptographic protocols. In: 42nd Annual Symposium on Foundations of Computer Science, pp. 136–145. IEEE Computer Society Press, Los Alamitos (2001)
25. Canetti, R., Halevi, S., Katz, J., Lindell, Y., MacKenzie, P.D.: Universally composable password-based key exchange. In: Cramer, R. (ed.) EUROCRYPT 2005. LNCS, vol. 3494, pp. 404–421. Springer, Heidelberg (2005)
26. Canetti, R., Krawczyk, H.: Analysis of key-exchange protocols and their use for building secure channels. In: Pfitzmann, B. (ed.) EUROCRYPT 2001. LNCS, vol. 2045, pp. 453–474. Springer, Heidelberg (2001)
27. Canetti, R., Rabin, T.: Universal composition with joint state. In: Boneh, D. (ed.) CRYPTO 2003. LNCS, vol. 2729, pp. 265–281. Springer, Heidelberg (2003)
28. Desmedt, Y., Pieprzyk, J., Steinfeld, R., Wang, H.: A non-malleable group key exchange protocol robust against active insiders. In: Katsikas, S.K., López, J., Backes, M., Gritzalis, S., Preneel, B. (eds.) ISC 2006. LNCS, vol. 4176, pp. 459–475. Springer, Heidelberg (2006)
29. Diffie, W., Hellman, M.E.: New directions in cryptography. IEEE Transactions on Information Theory 22(6), 644–654 (1976)
30. Dutta, R., Barua, R.: Password-based encrypted group key agreement. International Journal of Network Security 3(1), 30–41 (2006)
31. Gennaro, R., Lindell, Y.: A framework for password-based authenticated key exchange. In: Biham, E. (ed.) EUROCRYPT 2003. LNCS, vol. 2656, pp. 524–543. Springer, Heidelberg (2003)
32. Hofheinz, D., Müller-Quade, J.: Universally composable commitments using random oracles. In: Naor, M. (ed.) TCC 2004. LNCS, vol. 2951, pp. 58–76. Springer, Heidelberg (2004)
33. Katz, J., Ostrovsky, R., Yung, M.: Efficient password-authenticated key exchange using human-memorable passwords. In: Pfitzmann, B. (ed.) EUROCRYPT 2001. LNCS, vol. 2045, pp. 475–494. Springer, Heidelberg (2001)
34. Katz, J., Shin, J.S.: Modeling insider attacks on group key-exchange protocols. In: ACM CCS 2005, pp. 180–189. ACM Press, New York (2005)
35. Koblitz, N., Menezes, A.: Another look at "Provable security". II. In: Barua, R., Lange, T. (eds.) INDOCRYPT 2006. LNCS, vol. 4329, pp. 148–175. Springer, Heidelberg (2006) (invited talk)
36. Krawczyk, H.: HMQV: A high-performance secure diffie-hellman protocol. In: Shoup, V. (ed.) CRYPTO 2005. LNCS, vol. 3621, pp. 546–566. Springer, Heidelberg (2005)
37. Liskov, M., Rivest, R.L., Wagner, D.: Tweakable block ciphers. In: Yung, M. (ed.) CRYPTO 2002. LNCS, vol. 2442, pp. 31–46. Springer, Heidelberg (2002)
38. Needham, R.M., Schroeder, M.D.: Using encryption for authentication in large networks of computers. Communications of the Association for Computing Machinery 21(21), 993–999 (1978)
39. Shoup, V.: On formal models for secure key exchange. Technical Report RZ 3120, IBM (1999)
40. Simmons, G.J.: The prisoners' problem and the subliminal channel. In: CRYPTO 1983, pp. 51–67. Plenum Press, New York (1984)

Unifying Zero-Knowledge Proofs of Knowledge

Ueli Maurer

Department of Computer Science
ETH Zurich
CH-8092 Zurich, Switzerland
maurer@inf.ethz.ch

Abstract. We present a simple zero-knowledge proof of knowledge protocol of which many protocols in the literature are instantiations. These include Schnorr's protocol for proving knowledge of a discrete logarithm, the Fiat-Shamir and Guillou-Quisquater protocols for proving knowledge of a modular root, protocols for proving knowledge of representations (like Okamoto's protocol), protocols for proving equality of secret values, a protocol for proving the correctness of a Diffie-Hellman key, protocols for proving the multiplicative relation of three commitments (as required in secure multi-party computation), and protocols used in credential systems.

This shows that a single simple treatment (and proof), at a high level of abstraction, can replace the individual previous treatments. Moreover, one can devise new instantiations of the protocol.

1 Introduction

1.1 Interactive Proofs

A conventional proof of a statement is a sequence of elementary, easily verifiable steps, which, starting from the axioms (or previously proven facts), in the last step yields the statement to be proved.

In contrast to a conventional proof, an *interactive proof* [7] is a protocol that is defined between a *prover*, usually called P or Peggy, and a *verifier*, usually called V or Vic. More formally, an interactive proof is a pair (P, V) of programs implementing the protocol steps Peggy and Vic are supposed to execute. An interactive proof must be complete and sound. Completeness means that an honest prover succeeds in convincing an honest verifier, and soundness means that a dishonest prover does not succeed in convincing Vic of a false statement.

There are several motivations, theoretical and practical, for considering interactive proofs as opposed to conventional proofs. One main motivation is that, in contrast to a conventional proof, an interactive proof can be performed in a way that transfers only the conviction that the claimed statement is true but does not leak any further information, in particular not a transferable proof. More precisely, an interactive proof is called *zero-knowledge* if the verifier could simulate the entire protocol transcript by himself, without interacting with the

B. Preneel (Ed.): AFRICACRYPT 2009, LNCS 5580, pp. 272–286, 2009.
© Springer-Verlag Berlin Heidelberg 2009

prover. In particular, this implies that the transcript is not convincing for any other party.

There are two types of interactive proofs: proofs of a mathematical statement and proofs of knowledge. A proof of knowledge proves that Peggy knows a value satisfying a certain predicate (a witness). Often a proof of a mathematical statement (e.g. that a number is a square modulo an RSA modulus) is carried out as a proof of knowledge of a witness for the statement (i.e., of a square root).

1.2 Contributions of This Paper

We introduce a new level of abstraction in a general type of proof of knowledge, namely of a preimage of a group homomorphism, and thereby unify and generalize a large number of protocols in the literature. We observe that actually the identical principle has been reused several times, where each result came with a separate proof. While the similarity of different protocols certainly did not go unnoticed, a viewpoint that shows them to be instantiations of the *same* protocol is new to our knowledge. We call this protocol, described in Section 5, the main protocol.

The relation of our results to Cramer's Σ-protocols [2] (see also [3]), also an abstraction of a general type of proofs of knowledge, will be discussed in Section 5.3. In short, we go a step further and not only abstract a protocol *type*, but actually show that many protocols are the *same* protocol when seen at the right level of abstraction.

The advantage of our abstract viewpoint is that one can provide a proof once and for all, and for each individual instantiation only needs to describe the group homomorphism underlying the particular example and check the conditions of Theorem 3, our main theorem. This requires just a few lines for a complete proof that a given protocol is a zero-knowledge proof of knowledge. Moreover, this approach leads to new protocols by using new instantiations of the group homomorphism.

1.3 Outline

The outline of the paper is as follows. In Section 2 we discuss two well-known examples of interactive proofs. In Section 3 we formalize the concept of a proof of knowledge. In Section 4 we define what it means for a protocol to be zero-knowledge. In Section 5 we present our new general protocol, referred to as the *main protocol* and prove that it is a zero-knowledge proof of knowledge. In Section 6 we show that many known and new protocols are instantiations of the main protocol. Sections 3 and 4 can be skipped if the reader is familiar with the topic and is just interested in the unified viewpoint.

We use the standard notions of efficient and negligible and point out that such definitions are asymptotic, i.e., for asymptotic families (of groups) depending on a security parameter (which we will not make explicit). Efficient is usually defined as polynomial-time and negligible as vanishing faster than the inverse of any polynomial. In general, we assume that the reader is familiar with the basic aspects of interactive proofs and cryptographic thinking.

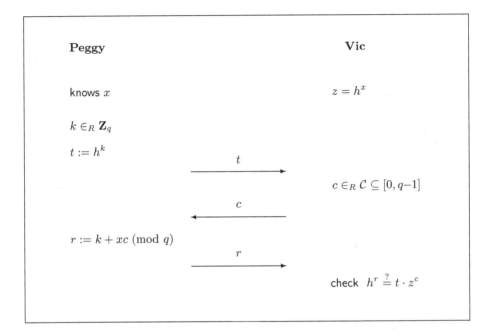

Fig. 1. The Schnorr protocol

2   Two Protocol Examples

In this section we discuss two classical examples of interactive proofs of knowledge.

2.1   The Schnorr Protocol

Consider a cyclic group H with prime order $|H| = q$ for which computing discrete logarithms (DL) is considered infeasible. Peggy wants to prove to Vic that she knows the DL x of an element z to the base h, i.e., that she knows x such that $z = h^x$. For example, z could be Peggy's public key and the protocol is then used as an identification protocol by which Peggy proves knowledge of the secret key.

The protocol, proposed by Schnorr [12], works as follows (see Figure 1). First, Peggy chooses $k \in \mathbf{Z}_q$ at random and sends the group element $t = h^k$ to Peggy. Then Vic chooses a challenge value $c \in \mathcal{C}$ at random from a challenge space $\mathcal{C}$ which is a subset of $[0, q-1]$. Then Peggy answers by sending the value $r = k + xc \pmod{q}$. Finally, Vic accepts the protocol execution if and only if $h^r = t \cdot z^c$.

Let us analyze the protocol. It is easy to see that if Peggy knows x and performs the protocol honestly, then Vic will accept (completeness). To argue about soundness, we observe that unless Peggy knows x, she cannot answer more than one challenge correctly. This can be seen as follows. If Peggy could answer, for a fixed t, two challenges c and c' by r and r', respectively, so that Vic accepts, then she could compute x (and hence knows it).[1] This can be shown as follows: We have $h^r = t \cdot z^c$ and $h^{r'} = t \cdot z^{c'}$ and thus

$$h^{r-r'} = z^{c-c'} = h^{x(c-c')}.$$

Therefore,

$$r - r' \equiv x(c - c') \pmod{q},$$

from which we obtain

$$x \equiv \frac{r_1 - r_2}{c_1 - c_2} \pmod{q}.$$

Note that it is important that q is prime since otherwise the inverse of $c_1 - c_2$ modulo q may not be defined, unless one restricts $\mathcal{C}$ in an artificial way.

One might be tempted to conclude from the above argument that any prover with success probability at least $2/|\mathcal{C}|$ can answer at least two challenges and therefore knows x. However, this argument is incorrect since it is not known how one could construct an *efficient* knowledge extractor (see Section 3).

Now we argue that the protocol is zero-knowledge. Without knowledge of x, one can, for any challenge c, generate a triple (t, c, r) with the distribution as it occurs in the protocol. One can prove (this is not entirely trivial) that even a dishonest verifier can simulate perfectly the entire transcript he would see in a protocol execution with Peggy, i.e., with the same probability distribution as it occurs in the real protocol (see Section 4). However, the simulation is efficient only if the size $|\mathcal{C}|$ of the challenge space is bounded to polynomial size. To obtain the zero-knowledge property, one may therefore choose $|\mathcal{C}|$ to be relatively small (e.g. on the order of 10^6), and repeat the protocol several (say s) times. Such a protocol is zero-knowledge but achieves the soundness guarantees corresponding to the size of the overall challenge space $\mathcal{C}^s$.

2.2   The Fiat-Shamir and Guillou-Quisquater Protocols

Consider a modulus m which is assumed to be difficult to factor. For concreteness, one can think of m as being an RSA-modulus [11]. For a given exponent e (with $\gcd(e, \varphi(m)) = 1$), breaking the RSA cryptosystem means to compute e-th roots modulo m. This is considered hard and, for a generic model of computation, has been proved to be equivalent to factoring m [1]. Unlike for RSA, in our context e is considered to be prime.

The Guillou-Quisquater (GQ) protocol [8] allows Peggy to prove to Vic that she knows the e-th root x modulo m of a given number $z \in \mathbf{Z}_m^*$, i.e., she knows x such that $x^e = z$ in $\mathbf{Z}_m^*$. (Again, z could be Peggy's public key for which she wants to prove knowledge of the corresponding private key.)

The protocol works as follows (see Figure 2). First, Peggy chooses $k \in \mathbf{Z}_m^*$ at random and sends the group element $t = k^e$ to Peggy. Then Vic chooses a

---

[1] A correct argument is more involved; one has to argue that there exists an efficient knowledge extractor (see Section 3).

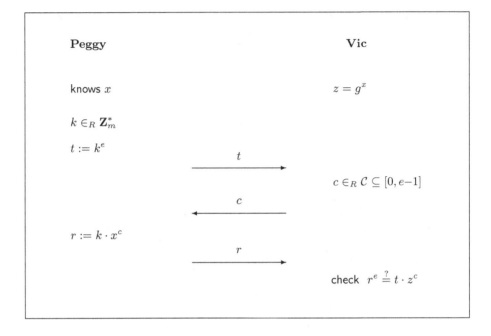

Fig. 2. The Guillou-Quisquater (GQ) protocol

challenge value $c \in \mathcal{C}$ at random from a challenge space $\mathcal{C} \subseteq [0, e-1]$. Then Peggy answers by sending the value $r = k \cdot x^c$ (in $\mathbf{Z}_m^*$, i.e., modulo m). Finally, Vic accepts the protocol execution if and only if $r^e = t \cdot z^c$.

This protocol is a generalization of the Fiat-Shamir protocol [6,5] which considers the special case $e = 2$. If $|\mathcal{C}|$ (and hence e) is sufficiently large, then a single execution of the protocol suffices. Otherwise, the protocol is repeated a sufficient number of times.

It is easy to see that if Peggy knows x and performs the protocol honestly, then Vic will accept (soundness). To argue about soundness, we observe again that unless Peggy knows x, she cannot answer more than one challenge correctly. This can be seen as follows. If Peggy could answer, for fixed t, both challenges c and c' by r and r', respectively (so that Vic accepts), then she could compute x (and hence knows it). This can be shown as follows: We have (in $\mathbf{Z}_m^*$) $r^e = t \cdot z^c$ and $r'^e = t \cdot z^{c'}$ and thus

$$\left(\frac{r}{r'}\right)^e \equiv z^{c-c'} \pmod{m}.$$

and hence

$$\frac{r}{r'} \equiv x^{c-c'} \pmod{m}.$$

In addition to $x^{c-c'}$, Peggy trivially knows another power of x, namely $z = x^e$. When e is prime, then $c-c'$ and e are relatively prime. From two different powers of x with relatively prime exponents one can compute x. Namely, application of Euclid's extended gcd-algorithm yields integers a and b such that

$$ea + (c - c')b = 1.$$

Therefore x can be computed as $x = x^{ea+(c-c')b}$, i.e., as

$$x \equiv z^a \cdot \left(\frac{r}{r'}\right)^b \pmod{m}.$$

2.3   Comparing the Two Protocols

The Schnorr protocol and the GQ protocol have a number of similarities, as already indicated by the fact that we chose to use the same names (k, t, c, and r) for the different quantities appearing in the protocol. But nevertheless the two protocols are quite different. The mathematical structure is different and so is the argument for proving that from the answers to two different challenges one can compute x. However, in Sections 5 and 6 we show that there is a level of abstraction at which the two protocols are identical, i.e., instantiations of the same protocol, as are many more protocols proposed in the literature.

3   Proofs of Knowledge

In this section we recall the definition of a proof of knowledge due to Feige, Fiat, and Shamir [5] and state a general theorem that can be used to easily prove that a protocol is a proof of knowledge.

The above soundness argument, namely that being able to answer two challenges implies knowledge of the secret value x, must be made more precise. Let us formalize the concept of a proof of knowledge. What constitutes knowledge, corresponding to a given value z, is defined by a (verification) predicate[2]

$$Q : \{0,1\}^* \times \{0,1\}^* \rightarrow \{\texttt{false}, \texttt{true}\}.$$

For a given value (a bit-string) z, Peggy claims to know a value (bit-string) x such that $Q(z, x) = \texttt{true}$.

The following classical definition'[5] captures the notion that being successful in the protocol implies knowledge of a witness x with $Q(z, x) = \texttt{true}$.

Definition 1. An interactive protocol (P, V) is a *proof of knowledge* for predicate Q if the following holds:

- (Completeness.) V accepts when P has as input an x with $Q(z, x) = \texttt{true}$.
- (Soundness.)   There is an efficient program K, called *knowledge extractor*, with the following property. For any (possibly dishonest) $\hat{P}$ with non-negligible probability of making V accept, K can interact with $\hat{P}$ and outputs (with overwhelming probability) an x such that $Q(z, x) = \texttt{true}$.[3]

---

[2] Equivalently, one can consider a relation on $\{0,1\}^*$.
[3] K must be able to choose the randomness of $\hat{P}$ and to reset $\hat{P}$.

We now capture the special property of a protocol which we proved for the Schnorr and the GQ protocols and which allowed us to argue about soundness.

Definition 2. Consider a predicate Q for a proof of knowledge. A three-move protocol round (Peggy sends t, Vic sends c, Peggy sends r) with challenge space $\mathcal{C}$ is *2-extractable*[4] if from any two triples (t, c, r) and (t, c', r') with distinct $c, c' \in \mathcal{C}$ accepted by Vic one can efficiently compute an x with $Q(z, x) = \mathtt{true}$.

The following theorem (see also [2,3]) states that for a protocol to be a proof of knowledge it suffices to check the 2-extractability condition for one (three-move) round of the protocol.

Theorem 1. *An interactive protocol consisting of s 2-extractable rounds with challenge space $\mathcal{C}$ is a proof of knowledge for predicate Q if $1/|\mathcal{C}|^s$ is negligible.*[5]

Proof. We need to exhibit a knowledge extractor K. It can be defined by the following simple procedure:

1. Choose the randomness for $\hat{P}$.
2. Generate two independent protocol executions between $\hat{P}$ and V (with the *same* chosen randomness for $\hat{P}$).
3. If V accepts in both executions and the challenge sequences were distinct, then identify the first round with different challenges c and c' (but, of course, the same t). Use 2-extractability to compute an x, and output it (and stop). Otherwise go back to step 1.

It is not very difficult to show that the expected running time of the knowledge extractor is polynomial if the success probability of $\hat{P}$ is non-negligible.

4   Zero-Knowledge Protocols

We now discuss the zero-knowledge property of a protocol. Informally, a protocol between P and V is zero-knowledge if even a dishonest V, which for this reason we call $\hat{V}$, does not learn anything from the protocol execution which he did not know before. This is captured by the notion of simulation [7]: $\hat{V}$ could simulate a protocol transcript by himself which is indistinguishable from a real transcript that would occur in an actual protocol execution between P and $\hat{V}$.

Definition 3. A protocol (P, V) is *zero-knowledge* if for every efficient program $\hat{V}$ there exists an efficient program S, the *simulator*, such that the output of S is indistinguishable from a transcript of the protocol execution between P and $\hat{V}$. If the indistinguishability is perfect,[6] i.e., the probability distribution of the simulated and the actual transcript are identical, then the protocol is called *perfect zero-knowledge.*

---

[4] It is also often called special soundness [2,3] when the challenge space is large.

[5] The last point implies that every particular challenge sequence $c_1, \ldots, c_s$ has negligible probability of being selected by an honest verifier.

[6] The indistinguishability could also be statistical or computational.

We now capture the special property of a protocol round, called c-simulatability, which is required to construct the zero-knowledge simulator.

Definition 4. *A three-move protocol round (Peggy sends t, Vic sends c, Peggy sends r) with challenge space $\mathcal{C}$ is c-simulatable[7] if for any value $c \in \mathcal{C}$ one can efficiently generate a triple (t, c, r) with the same distribution as occurring in the protocol (conditioned on the challenge being c).*

The following theorem states that for a protocol to be zero-knowledge it suffices to check the c-simulatability condition for one round of the protocol.

Theorem 2. *A protocol consisting of c-simulatable three-move rounds, with uniformly chosen challenge from a polynomially-bounded (per-round) challenge space $\mathcal{C}$, is perfect zero-knowledge.*

The proof of this theorem is not entirely trivial. We just describe the basic idea of the simulator. It simulates one round after the next. In each round (say the i-th), the simulator chooses a uniformly random challenge c_i, generates a triple (t_i, c_i, r_i) using the c-simulatability, and then checks whether $\hat{V}$ would actually issue challenge c_i if it were in the corresponding state in round i. If the check succeeds, then this round is appended to the simulated transcript as the i-th round, otherwise the simulation of the i-th round is restarted.[8]

5   Proving Knowledge of a Preimage of a Group Homomorphism

5.1   One-Way Group Homomorphisms

We consider two groups $(G, \star)$ and $(H, \otimes)$, where we intentionally use special symbols for the group operations, avoiding the addition and multiplication symbols "+" and "·". We assume that the group operations $\star$ and $\otimes$ are efficiently computable.

A function $f : G \to H$ is a homomorphism if

$$f(x \star y) = f(x) \otimes f(y).$$

We will consider the case where f is (believed to be) a one-way function, such that it is infeasible to compute x from $f(x)$ for a randomly chosen x.[9] In this case it is meaningful for a prover Peggy to prove that she knows an x such that for a given value z we have $z = f(x)$. To simplify the notation we write $[x]$ instead of $f(x)$.[10] We can consider $[x]$ to be an embedding of $x \in G$ in H. We point out

---

[7] This is sometimes also called *special honest-verifier zero-knowledge* [2,3].

[8] This requires access to the strategy of $\hat{V}$, or $\hat{V}$ must be rewindable.

[9] Note, however, that our treatment and claims do not depend on the one-way property. Should f not be one-way, then the protocols are perhaps less useful, but they still have the claimed properties.

[10] When we define a group homomorphism in terms of a given group homomorphism (denoted $[\cdot]$), then we write $[\![\cdot]\!]$ to avoid overloading the symbol $[\cdot]$.

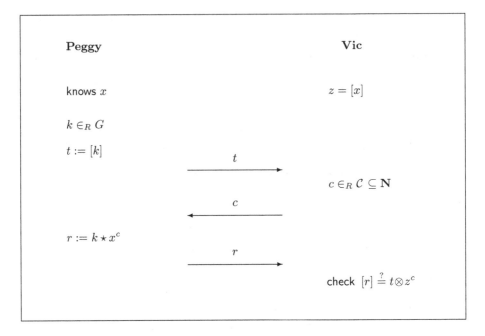

Fig. 3. Main protocol: Proof of knowledge, for a given value z, of a value x such that $z = [x]$, where $x \mapsto [x]$ is a (one-way) group homomorphism

that f need not be bijective and therefore a value $z = [x]$ does not necessarily determine x. But it is well-defined for which values x we have $z = [x]$.

Given embedded values $[x]$ and $[y]$ we can efficiently compute

$$[x \star y] = [x] \otimes [y],$$

without knowing x or y, due to the homomorphism.

5.2   The Main Protocol

The protocol in Figure 3, which we call the main protocol, is a proof of knowledge of a value x such that $z = [x]$, for a given value z, provided that the two conditions stated in the following theorem are satisfied. Note that G and H need not be commutative. The challenge space can be chosen as an arbitrary subset of $\mathbf{N}$. If it is chosen to be small,[11] then one needs several (three-move) rounds to reduce the soundness error to be negligible.

Theorem 3. *If values $\ell \in \mathbf{Z}$ and $u \in G$ are known such that*

---

[11] There can be at least two reasons for choosing a small challenge space. First, a larger space may not work, for example if e is small in the GQ protocol. Second, one may want the protocol to be zero-knowledge, which generally does not hold for large (per-round) challenge space.

(1) $\gcd(c_1 - c_2, \ell) = 1$ *for all* $c_1, c_2 \in \mathcal{C}$ *(with $c_1 \neq c_2$), and*
(2) $[u] = z^\ell$,

then the three-move protocol round described in Figure 3 is 2-extractable. Moreover, a protocol consisting of s rounds is a proof of knowledge if $1/|\mathcal{C}|^s$ is negligible, and it is zero-knowledge if $|\mathcal{C}|$ is polynomially bounded.

Proof. 2-extractability can be proved as follows: From r and r' such that $[r] = t \otimes z^c$ and $[r'] = t \otimes z^{c'}$ for two different challenges c and c' we can obtain $\tilde{x}$ satisfying $[\tilde{x}] = z$, as

$$\tilde{x} = u^a \star \left(r'^{-1} \star r\right)^b,$$

where a and b are computed using Euclid's extended gcd-algorithm such that

$$\ell a + (c - c')b = 1.$$

We make use of

$$[r'^{-1} \star r] = [r'^{-1}] \otimes [r] = z^{-c'} \otimes t^{-1} \otimes t \otimes z^c = z^{-c'} \otimes z^c = z^{c-c'}$$

to see that $[\tilde{x}] = z$:

$$\begin{aligned}
[\tilde{x}] &= [u^a \star (r'^{-1} \star r)^b] \\
&= [u]^a \otimes [r'^{-1} \star r]^b \\
&= (z^\ell)^a \otimes (z^{c-c'})^b \\
&= z^{\ell a + (c-c')b} = z.
\end{aligned}$$

Theorem 1 implies directly that the protocol is a proof of knowledge, and Theorem 2 implies that it is zero-knowledge if $|\mathcal{C}|$ is polynomially bounded since it is c-simulatable. This is easy to see: Given z and a challenge c, one chooses r at random and computes t as $t = [r] \otimes z^{-c}$.

5.3 Comparison with Other Work

Our result should be contrasted with another approach, due to Cramer [2] (see also [3]) to abstracting a general type of proofs of knowledge. Cramer introduced the notion of Σ-protocols which are basically three-move protocols, as discussed in this paper, which are both 2-extractable and c-simulatable. All the protocols we consider are Σ-protocols. However, we go further in that we show that a large class of protocols are not only of the same protocol type, but are actually the *same protocol*, thus requiring only one proof of the claimed properties. In order to apply our Theorem 3 one only needs to specify the groups G and H, the homomorphism, and check the two conditions of Theorem 3.

6   Special Cases of the Main Protocol

In this section we describe a number of protocols as instantiations of our main protocol.

6.1   Schnorr and GQ as Special Cases

The Schnorr protocol is the special case where $(G, \star) = (\mathbf{Z}_q, +)$ (with addition modulo q in $\mathbf{Z}_q$) and H is a group of order q with the group operation written as multiplication (i.e., ".", which can also be omitted). The (one-way) group homomorphism is defined by

$$G \to H : \quad x \mapsto [x] = h^x.$$

The challenge space $\mathcal{C}$ can be an arbitrary subset of $[0, q-1]$. The two conditions of Theorem 3 are satisfied for $\ell = q$ (if q is prime) and $u = 0$. Note that $\gcd(c_1 - c_2, \ell) = 1$ for all distinct $c_1, c_2 \in \mathcal{C}$, and $[u] = [0] = 1 = z^\ell$ since every element of H raised to the group order $|H| = q$ is the neutral element of H.

The GQ protocol is the special case where $(G, \star) = (\mathbf{Z}_m^*, \cdot) = (H, \otimes)$. The one-way homomorphism is defined by

$$G \to H : \quad x \mapsto [x] = x^e.$$

The challenge space $\mathcal{C}$ can be an arbitrary subset of $[0, e-1]$, provided e is prime. The conditions of Theorem 3 are satisfied for $\ell = e$ and $u = z$. Note that $\gcd(c_1 - c_2, \ell) = 1$ for all distinct $c_1, c_2 \in \mathcal{C}$, and $[u] = [z] = z^e = z^\ell$.

6.2   Proof of Knowledge of Several Values

Let

$$G_i \to H_i : \quad x \mapsto [x]^{(i)}$$

for $i = 1, \ldots, n$ be (possibly distinct) group homomorphisms for which, for the same ℓ, there exist $u_1, \ldots, u_n$ and $z_1, \ldots, z_n$ satisfying condition (2) in Theorem 3, i.e., $[u_i]^{(i)} = z_i^\ell$ for $i = 1, \ldots, n$. Then also

$$G_1 \times \cdots \times G_n \to H_1 \times \cdots \times H_n :$$

$$(x_1, \ldots, x_n) \mapsto [(x_1, \ldots, x_n)] = \left([x_1]^{(1)}, \ldots, [x_n]^{(n)} \right)$$

is a one-way group homomorphisms.[12] Therefore the main protocol proves in one stroke the knowledge of $x_1, \ldots, x_n$ such that for given $z_1 \in H_1, \ldots, z_k \in H_n$ we have $z_1 = [x_1]^{(1)}, \ldots, z_n = [x_n]^{(n)}$. This can be seen by setting $u = (u_1, \ldots, u_n)$ and $z = (z_1, \ldots, z_n)$ since

$$[u] = \left([u_1]^{(1)}, \ldots, [u_n]^{(n)} \right) = \left(z_1^\ell, \ldots, z_n^\ell \right) = z^\ell.$$

A typical application of this protocol is for proving knowledge of several discrete logarithms in (possibly distinct) groups of prime order q.

---

[12] The group operations in $G_1 \times \cdots \times G_n$ and $H_1 \times \cdots \times H_n$ are defined component-wise.

6.3 Proof of Equality of Embedded Values

Let again

$$G \to H_i : \quad x \mapsto [x]^{(i)}$$

for $i = 1, \ldots, n$ be one-way group homomorphisms as in the previous section, but with $u_1 = \cdots = u_n = u$. Then also

$$G \to H_1 \times \cdots \times H_n : \quad x \mapsto [x] = \left([x]^{(1)}, \ldots, [x]^{(n)}\right)$$

is a group homomorphisms (but not necessarily one-way). Therefore the main protocol proves the knowledge of x that is simultaneously a preimage of all n homomorphisms. More precisely, it proves knowledge of x such that for given $z_1 \in H_1, \ldots, z_k \in H_k$ we have $z_1 = [x]^{(1)}, \ldots, z_k = [x]^{(k)}$.[13] This can be seen by setting $z = (z_1, \ldots, z_n)$. A typical application of this protocol is for proving that several discrete logarithms in groups of prime order q are identical.

6.4 Proof of Knowledge of a Representation

Consider again a group H with prime order q, and let several generators $h_1, \ldots, h_m$ of H be given. A representation of an element $z \in H$ is a list $(x_1, \ldots, x_m)$ of exponents such that $z = h_1^{x_1} h_2^{x_2} \cdots h_m^{x_m}$. (Note that such a representation is not unique.) We want to prove knowledge of a representation of a given element z.

For the special case $m = 2$, a protocol for this purpose was proposed by Okamoto [9]. This is of interest, among other reasons, since Pedersen commitments [10] have this form.[14]

A protocol for proving knowledge of a representation can be obtained as another simple instantiation of our main protocol, using the homomorphism

$$\mathbf{Z}_q^m \to H : \quad (x_1, \ldots, x_m) \mapsto [(x_1, \ldots, x_m)] = h_1^{x_1} \cdots h_m^{x_m}.$$

The conditions of Theorem 3 are satisfied for the choice $\ell = q$ and $u := (0, \ldots, 0)$ since $[(0, \ldots, 0)] = h_1^0 \cdots h_m^0 = 1 = z^\ell$ for every $z \in H$.

6.5 Proof of Knowledge of a Set of Linear Representations

One can actually prove more general statements about the knowledge of representations, namely knowledge of values $x_1, \ldots, x_r$ that simultaneously satisfy several representation equations with respect to generators $h_1, \ldots, h_m$. Such protocols appear, for example, in the literature on credential systems.

For example, consider generators h_1, h_2, h_3 of H. For given values $z_1, z_2 \in H$ we can prove knowledge of values $x_1, x_2, x_3, x_4 \in \mathbf{Z}_q$ satisfying $z_1 = h_1^{x_3} h_2^{x_1}$ and

---

[13] Note that if the homomorphisms are bijective, then this protocol not only proves knowledge of x, but actually that all embedded values are identical.

[14] One commits to a value x by choosing a random r and sending $h_1^x h_2^r$ as the commitment (see also Section 6.7). This commitment scheme is information-theoretically hiding (but only computationally binding).

$z_2 = h_1^{x_2} h_2^{x_4} h_3^{x_1}$. The reader can figure out as an exercise how the homomorphism must be chosen such that our main protocol provides such a proof.

More generally, one can prove knowledge of $x_1, \ldots, x_r$ such that for given values $z_1, \ldots, z_s$ and, for sm linear (over $GF(q)$) functions $\phi_{11}, \ldots, \phi_{sm}$ from $GF(q)^r$ to $GF(q)$, we have

$$z_i = h_1^{\phi_{i1}(x_1,\ldots,x_r)} \cdot h_2^{\phi_{i2}(x_1,\ldots,x_r)} \cdots h_m^{\phi_{im}(x_1,\ldots,x_r)}$$

for $i = 1, \ldots, s$. The group homomorphism $\mathbf{Z}_q^r \to H^s$ is defined as

$$[(x_1, \ldots, x_r)] = \left(\prod_{j=1}^{m} h_j^{\phi_{1j}(x_1,\ldots,x_r)} \;,\; \ldots\;,\; \prod_{j=1}^{m} h_j^{\phi_{sj}(x_1,\ldots,x_r)} \right).$$

6.6   Proof of Correctness of Diffie-Hellman Keys

Let H be a group with prime order $|H| = q$ and generator h used in the Diffie-Hellman protocol [4]. As for the Schnorr protocol, we define a homomorphic embedding by

$$G \to H : \quad x \mapsto [x] = h^x.$$

Recall that in the Diffie-Hellman protocol, Alice chooses an $a \in \mathbf{Z}_q$ and sends $[a] = h^a$ to Bob and, symmetrically, Bob chooses a $b \in \mathbf{Z}_q$ and sends $[b] = h^b$ to Alice. The common secret key is $[ab] = h^{ab}$. It is believed that for general groups it is computationally hard to decide whether or not a given key $K \in H$ is the correct key, i.e., whether $K = h^{ab}$. This is known as the Decisional Diffie-Hellman (DDH) problem. For example, if a very powerful organization were willing to compute Diffie-Hellman keys as a commercial service (returning $[ab]$ when given $[a]$ and $[b]$), then the customer could not verify that the key is correct. In this context, as well as in other contexts, it is useful to be able to prove the correctness of a Diffie-Hellman key in zero-knowledge, in particular without leaking any information about a or b. This is again achieved by a simple instantiation of our main protocol.

Let values $A = [a]$, $B = [b]$ and $C = [c]$ be given. We wish to prove that $c = ab \pmod{q}$, i.e., that A, B, and C form a so-called Diffie-Hellman triple. For this purpose we define the following one-way group homomorphism which we denote by $\llbracket \cdot \rrbracket$ and which is defined in terms of the homomorphism $[\cdot]$ and of B:

$$\mathbf{Z}_q \to H \times H : \quad x \mapsto \llbracket x \rrbracket = ([x], [xb]) = (h^x, B^x).$$

Note that $[xb]$ can be computed efficiently from $B = [b]$ and x without knowing b. This yields, as a special case of the main protocol for the homomorphism $x \to \llbracket x \rrbracket$, the desired proof: One proves knowledge of a preimage x (namely $x = a$) such that

$$\llbracket x \rrbracket = (A, C).$$

Due to the particular choice of the homomorphism, this implies that $c = ab$.

While the protocol proves that the prover Peggy knows a, it does not prove that she knows b or c. (This does not contradict the fact that the claim $c = ab$ is indeed proved.) If desired, a proof of knowledge of b (and hence also of c) could be linked into the above proof using the technique of Section 6.3.

6.7   Multiplication Proof for Pedersen Commitments

An important step in secure multi-party computation (MPC) protocols is for a party to commit to the product of two values it is already committed to, and to prove that this product commitment is correct. We show how such a proof can be given for Pedersen commitments.

Recall that in the Pedersen commitment scheme one commits to a value $x \in \mathbf{Z}_q$ by choosing $\rho_x \in \mathbf{Z}_q$ at random and sending the value $g^x h^{\rho_x}$. (To avoid unnecessary indices we denote here the two generators as g and h instead of h_1 and h_2.) We consider the commitment one-way homomorphism

$$\mathbf{Z}_q \times \mathbf{Z}_q \to H : \quad (x, \rho_x) \mapsto [\![(x, \rho_x)]\!] = g^x h^{\rho_x}.$$

Since this commitment scheme is information-theoretically hiding, the value x is not determined by the commitment. What counts is *how* the committing party can open the commitment.

Let three commitments $A, B, C \in H$ by Peggy be given. In the following we assume that it is clear from the context that Peggy can open B as (b, ρ_b). If this were not the case, one could incorporate such a proof using the technique of Section 6.3. We describe a protocol that allows Peggy to prove that she can open A as (a, ρ_a) and C as (c, ρ_c) with $c = ab$.

For this purpose we define the following one-way group homomorphism:

$$\mathbf{Z}_q^3 \to H \times H : \quad (x, \rho_x, \sigma_x) \mapsto [\![(x, \rho_x, \sigma_x)]\!] = ([\![(x, \rho_x)]\!], [\![(xb, x\rho_b + \sigma_x)]\!]),$$

where the second component can be computed as

$$[\![(xb, x\rho_b + \sigma_x)]\!] = B^x h^{\sigma_x}$$

without knowledge of b and ρ_b (with $B = [\![(b, \rho_b)]\!]$).

The desired proof can now be obtained as a special case of the main protocol: Peggy proves that she knows a triple (x, ρ_x, σ_x) such that

$$[\![(x, \rho_x, \sigma_x)]\!] = (A, C).$$

As can easily be verified, this proof is successful for the choice

$$(x, \rho_x, \sigma_x) = (a, \rho_a, \rho_c - a\rho_b).$$

7   Conclusions

Our main protocol for proving knowledge of a preimage of a group homomorphism is the abstraction of a large class of protocols. The presented list of examples is by no means exhaustive. We encourage the reader to find other protocols in the literature which can be described as an instance of the main protocol.

Acknowledgments

It is a pleasure to thank Martin Hirt and Vassilis Zikas for their helpful comments.

References

1. Aggarwal, D., Maurer, U.: Breaking RSA generically is equivalent to factoring. In: Joux, A. (ed.) EUROCRYPT 2009. LNCS, vol. 5479, pp. 36–53. Springer, Heidelberg (2009)
2. Cramer, R.: Modular design of secure, yet practical cryptographic protocols, Ph.D. Thesis, University of Amsterdam (1996)
3. Damgård, I.: On Σ-protocols, Course notes. Århus University (2002)
4. Diffie, W., Hellman, M.E.: New directions in cryptography. IEEE Transactions on Information Theory 22(6), 644–654 (1976)
5. Feige, U., Fiat, A., Shamir, A.: Zero-knowledge proofs of identity. Journal of Cryptology 1, 77–94 (1988)
6. Fiat, A., Shamir, A.: How to prove yourself: practical solution to identification and signature problems. In: Odlyzko, A.M. (ed.) CRYPTO 1986. LNCS, vol. 263, pp. 186–194. Springer, Heidelberg (1987)
7. Goldwasser, S., Micali, S., Rackoff, C.: The knowledge complexity of interactive proof systems. SIAM Journal on Computing 18, 186–208 (1989)
8. Guillou, L.C., Quisquater, J.-J.: A practical zero-knowledge protocol fitted to security microprocessor minimizing both transmission and memory. In: Günther, C.G. (ed.) EUROCRYPT 1988. LNCS, vol. 330, pp. 123–128. Springer, Heidelberg (1988)
9. Okamoto, T.: Provably secure and practical identification schemes and corresponding signature schemes. In: Brickell, E.F. (ed.) CRYPTO 1992. LNCS, vol. 740, pp. 31–53. Springer, Heidelberg (1993)
10. Pedersen, T.: Non-interactive and information-theoretical secure verifiable secret sharing. In: Feigenbaum, J. (ed.) CRYPTO 1991. LNCS, vol. 576, pp. 129–140. Springer, Heidelberg (1992)
11. Rivest, R.L., Shamir, A., Adleman, L.: A method for obtaining digital signatures and public-key cryptosystems. Communications of the ACM 21(2), 120–126 (1978)
12. Schnorr, C.P.: Efficient identification and signatures for smart cards. In: Brassard, G. (ed.) CRYPTO 1989. LNCS, vol. 435, pp. 239–252. Springer, Heidelberg (1990)

Co-sound Zero-Knowledge with Public Keys⋆

Carmine Ventre[1] and Ivan Visconti[2]

[1] Department of Computer Science
University of Liverpool, UK
Carmine.Ventre@liverpool.ac.uk
[2] Dipartimento di Informatica ed Applicazioni
University of Salerno, Italy
visconti@dia.unisa.it

Abstract. In this paper we present two variations of the notion of *co-soundness* previously defined and used by [Groth et al. - EUROCRYPT 2006] in the common reference string model. The first variation holds in the Bare Public-Key (BPK, for short) model and closely follows the one of [Groth et al. - EUROCRYPT 2006]. The second variation (which we call weak co-soundness) is a weaker notion since it has a stronger requirement, and it holds in the Registered Public-Key model (RPK, for short).

We then show techniques to construct co-sound argument systems that can be proved secure under standard assumptions, more specifically:

1. in the main result of this paper we show a constant-round resettable zero-knowledge argument system in the BPK model using black-box techniques only (previously it was achieved in [Canetti et al. - STOC 2000, Di Crescenzo et al. - CRYPTO 2004] with complexity leveraging);
2. additionally, we show an efficient statistical non-interactive zero-knowledge argument system in the RPK model (previously it was achieved in [Damgård et al. - TCC 2006] with complexity leveraging).

We stress that no alternative solution preserving all properties enjoyed by ours is currently known using the classical notion of soundness.

Keywords: co-soundness, rZK, NIZK, public-key models.

1 Introduction

Standard complexity-theoretic assumptions concern well studied primitives that are believed to be unforgeable with respect to polynomial-time adversaries. Such primitives can then be used to prove that a protocol is secure by reducing a succeeding adversary for the protocol to a forgery for such primitives. Unfortunately we have in literature protocols that instead need the existence of primitives that

---

⋆ The work of the authors has been supported in part through the EPSRC grant EP/F069502/1, the EU ICT program under Contract ICT-2007-216646 ECRYPT II and the FP6 program under contract FP6-1596 AEOLUS.

B. Preneel (Ed.): AFRICACRYPT 2009, LNCS 5580, pp. 287–304, 2009.

are unforgeable also with respect to superpolynomial-time adversaries. Since such assumptions are less standard, the reduction of a succeeding adversary to a forgery for the primitive gives a proof of the *quasi* security of the protocol. The protocols that we will consider in this work are zero-knowledge argument systems and we will focus on models with public keys since they include two notable examples of quasi-secure protocols.

Quasi-security in zero-knowledge arguments. Concerning the design of zero knowledge argument systems, a major difficulty is the coexistence of two security properties, namely: soundness and zero knowledge. The need of designing proof systems with special properties (e.g., round efficiency, stronger notions of zero knowledge) motivated the use of various setup assumptions (e.g., common reference strings, trusted third parties, public key infrastructures) and of relaxed security notions (bounded concurrency, timing assumptions, quasi-security).

Here we focus on two setup assumptions where verifiers have identities represented by public keys. The first model is the Bare Public Key (BPK, for short) model proposed in [1], where first all verifiers announce their public keys (here no trusted entity is needed) and then proofs start. The second model is the Registered Public Key (RPK, for short) model where verifiers have to register their public keys to a trusted entity that will then give the public keys to the players [2]. Both models have been proposed as reasonable settings where zero-knowledge argument systems with special properties can be designed. Indeed the BPK model has been used to obtain constant-round resettable zero-knowledge [1] (rZK, for short) for $\mathcal{NP}$ using black-box techniques only, also enjoying concurrent soundness [3]. Instead, the RPK model has been proposed for the design of efficient statistical non-interactive zero knowledge (NIZK, for short) for many useful languages [2].

Unfortunately, in both cases the setup assumption does not seem to be sufficient. Indeed, for all above results the soundness is proved by using complexity leveraging, which in turn means that the soundness requirement holds if some primitives are secure against superpolynomial-time adversaries. The non-standard nature of this assumption has therefore naturally opened the problems of obtaining alternative argument systems that enjoy the wished special properties but under standard complexity-theoretic assumptions.

Co-soundness. We consider a variation of the classical notion of soundness, referred to as *co-soundness*, that has been proposed in [4] for obtaining perfect NIZK with respect to adaptive adversaries, which is then used for obtaining universally composable (UC, for short) NIZK. Given a language L in $\mathcal{NP}$, a co-sound proof system says that it is infeasible for an adversary to find an instance x along with a witness y for membership in $\bar{L}$, and at the same time to succeed in computing a proof that convinces the honest verifier that $x \in L$ (we are denoting by $\bar{L}$ the complement of L). In other words, the adversary cannot convince a verifier with a fake proof and at the same time know that the statement is false.

As already specified in [5] (we include here part of their discussion), in noninteractive arguments with perfect zero-knowledge, co-soundness makes sense comparably to standard soundness. Indeed, the problem with the standard definition appears when the adversary produces a statement x and a valid NIZK argument without knowing whether $x \in L$ or $x \notin L$. In these cases it may not be possible to reduce the adversary's output to a breach of some underlying (polynomial) cryptographic hardness assumption. Indeed in [6] it is shown that perfect NIZK arguments with direct black-box reductions to a cryptographic hardness assumption are possible only for languages in $\mathcal{P}/\text{poly}$. The observation that all known constructions of NIZK arguments rely on direct black-box reductions, indicates that the classical definition of soundness is not necessarily the only right definition of soundness for perfect NIZK arguments. We note that for NIZK proofs there is no such problem since they are not perfect zero-knowledge except for trivial languages; and in the case of interactive arguments with perfect zero-knowledge this problem does not appear either because the security proofs rely on rewinding techniques which make it possible to extract a witness for the statement being proven. This last argument given in [5] motivates also the use of co-soundness when rZK is considered, since it is known that rZK arguments of knowledge with a black-box extractor are possible only for trivial languages [1]. This is the reason that motivated the use of complexity leveraging for proving the soundness requirement with black-box techniques only.

The generalization to co-soundness makes it possible to get around the above problem as the adversary only breaks co-soundness when it knows a witness y for $x \notin L$. By choosing carefully the language, this witness can be used to reduce a successful co-soundness attack to a breach of a standard polynomial cryptographic complexity assumption.

Our results. In this paper we address the problem of removing complexity leveraging to obtain results comparable to those of [1,3,2] but under standard assumptions, and considering co-soundness instead of soundness. We define co-soundness in the BPK model following the definition given in [4], and weak co-soundness in the RPK model under a weaker definition where we also require that the adversary knows that the false theorem has been successfully proven.

We present computationally co-sound argument systems under standard assumptions for both problems. More precisely we show the following two results.

1. The main result of this work is the first constant-round concurrently co-sound rZK argument system in the BPK model under standard assumptions for all languages in $\mathcal{NP} \cap \text{co-}\mathcal{NP}$. This last result is obtained through a new primitive referred to as non-interactive trapdoor instance-dependent commitment scheme that we introduce and construct and that can be of independent interest.
2. We additionally present the first efficient transformation for obtaining efficient statistical weakly co-sound NIZK in the RPK model under standard assumptions for many useful languages, but under the above discussed weaker co-soundness notion.

The second result as in [2] requires that the adversarial prover runs at most a logarithmic number of proofs. We stress that for such problems, no alternative solution under classical soundness is currently known. Our techniques therefore show additional applications of the notion of co-soundness previously defined and used in [4] and can be of interest in other contexts where complexity leveraging is used. Our results can be seen as alternatives to the current state-of-the-art and can be preferred when co-soundness or weak co-soundness are considered useful and standard assumptions are required.

1.1   Tools

We now discuss the tools that will be used as building blocks for our results.

An *interactive proof (resp., argument) system* [7] for a language L is a pair of interactive Turing machines $\langle P, V \rangle$, satisfying the requirements of *completeness* and *soundness*. Informally, completeness requires that for any $x \in L$, at the end of the interaction between P and V, where P has on input a valid witness for $x \in L$, V rejects with negligible probability. Soundness requires that for any $x \notin L$, for any computationally unbounded (resp., probabilistic polynomial-time for arguments) $P^\star$, at the end of the interaction between $P^\star$ and V, V accepts with negligible probability. We denote by $\langle P, V \rangle(x)$ the output of the verifier V when interacting on common input x with prover P. Also, sometimes we will use the notation $\langle P(y), V \rangle(x)$ to stress that prover P receives as additional input witness y for $x \in L$.

Formally, we have the following definition.

Definition 1. *A pair of interactive Turing machines $\langle P, V \rangle$ is an* interactive proof system *for the language L, if V is probabilistic polynomial-time and*

1. *Completeness: There exists a negligible function $\nu(\cdot)$ such that for every $x \in L$ and for every $w \in W(x)$*

$$\mathrm{Prob}\left[\langle P(y), V \rangle(x) = 1\right] \geq 1 - \nu(|x|).$$

2. *Soundness: For every $x \notin L$ and for every interactive Turing machines $P^\star$ there exists a negligible function $\nu(\cdot)$ such that*

$$\mathrm{Prob}\left[\langle P^\star, V \rangle(x) = 1\right] < \nu(|x|).$$

If the soundness condition holds only with respect to probabilistic polynomial-time interactive Turing machines $P^\star$ then $\langle P, V \rangle$ is called an argument.

Σ-protocols. A Σ-protocol [8] is a 3-round interactive protocol between two probabilistic algorithms, an honest prover P and an honest verifier V. The algorithms receive as common input a statement "$x \in L$." P has as auxiliary input a witness y such that $(x, y) \in R_L$ where L is an $\mathcal{NP}$-language and x belongs to L. At the end of the protocol V decides whether the transcript is accepting with respect to the statement or not. The Σ-protocols we consider in this paper are public coin, enjoy special soundness (given two accepting transcripts with the same first message one can efficiently compute a witness) and special honest-verifier zero knowledge (an efficient simulator, on input a true statement

"$x \in L$" outputs for any c a pair (a, z) such that the triple (a, c, z) is statistically indistinguishable from the transcript of a conversation between P and V).

There are in the literature many Σ-protocols with these properties, most notably, the protocol of Blum [9] that is a Σ-protocol for the $\mathcal{NP}$-complete language Hamiltonicity (the statistical honest-verifier zero knowledge property requires a preprocessing message for implementing a two-round statistically hiding commitment scheme; without this message, Blum's protocol is only computational honest-verifier zero knowledge). Below, we will make use of Blum's Σ-protocol. This protocol can be instantiated on any 1-to-1 length-preserving one-way function since this suffices for non-interactive commitment schemes. We also stress that Blum's protocol is also a *special* 3-round witness-indistinguishable proof of knowledge (WIPoK, for short). Here, by special, we mean that the prover uses the witness only in the third round.

Σ-protocol with linear answer. A class of Σ-protocols we will consider in the following is that satisfying the following definition and already considered in [2].

Definition 2 ([2]). *A Σ-protocol with linear answer is a Σ-protocol where the prover's final message z is a sequence of integers, $z = (z_1, \ldots, z_m)$, where $z_j = u_j + v_j c$, and where u_j, v_j are integers that can be computed efficiently from x, P's random coins and its private input y.*

Public-Key Encryption Schemes. We will consider public-key encryption schemes with additional properties. A public-key encryption scheme is a triple of algorithms $(\mathcal{G}, \mathcal{E}, \mathcal{D})$ where $\mathcal{G}$ is the key generation algorithm, $\mathcal{E}$ is the encryption algorithm and $\mathcal{D}$ is the decryption algorithm. The key generation algorithm takes in input 1^k and outputs a pair (pk, sk). We will consider systems where plaintexts are integers from some interval $[0, n-1]$ where n can be computed from the public key. Given plaintext a and random coins r, the ciphertext is $\mathcal{E}_{pk}(a; r)$, and we require, of course, that $a = \mathcal{D}_{sk}(\mathcal{E}_{pk}(a; r))$.

We next give the standard definition of IND-CPA security [10].

Definition 3 (IND-CPA). *Let $\mathcal{PE} = (\mathcal{G}, \mathcal{E}, \mathcal{D})$ be a public-key encryption scheme, let $\mathcal{A} = (\mathcal{A}_1, \mathcal{A}_2)$ be an adversary. For $k \in \mathbb{N}$ let*

$$\mathbf{Adv}_{\mathcal{PE}, \mathcal{A}}^{ind-cpa}(k) = 2 \cdot \mathrm{Prob}\left[\mathrm{Expt}_{\mathcal{PE}, \mathcal{A}}^{ind-cpa}(k) = 1\right] - 1$$

where

$$\boxed{\begin{array}{l} \mathrm{Expt}_{\mathcal{PE}, \mathcal{A}}^{ind-cpa}(k): \\ (\mathrm{pk}, \mathrm{sk}) \leftarrow \mathcal{G}(1^k) \\ (x_0, x_1, s) \leftarrow \mathcal{A}_1(\mathrm{pk}) \\ b \leftarrow \{0, 1\} \\ c = \mathcal{E}_{\mathrm{pk}}(x_b) \\ g \leftarrow \mathcal{A}_2(x_0, x_1, s, c) \\ \textbf{return } 1 \textbf{ iff } g = b \end{array}}$$

Above it is mandatory that $|x_0| = |x_1|$. We say that $\mathcal{PE}$ is IND-CPA secure if $\mathcal{A}$ being polynomial-time implies $\mathbf{Adv}^{\text{ind}-\text{cpa}}_{\mathcal{PE},\,\mathcal{A}}(\cdot)$ is negligible.

We will be looking at systems that are *homomorphic*, in the following sense: the set of ciphertexts is an Abelian group, where the group operation is easy to compute given the public key. Furthermore, for any a, b, r_a, r_b it holds that $\mathcal{E}_{\text{pk}}(a; r_a) \cdot \mathcal{E}_{\text{pk}}(b; r_b) = \mathcal{E}_{\text{pk}}((a+b); s)$ for some s. We will assume throughout that n is a k-bit number. Note that by multiplying $\mathcal{E}_{\text{pk}}(a; r)$ by a random encryption of 0, one obtains a random and independently distributed encryption of a; we denote such operation with $\mathsf{randomize}(\mathcal{E}_{\text{pk}}(a; r))$.

A typical example of such homomorphic encryption is Paillier's cryptosystem [11], where pk is a k-bit RSA modulus n, and sk is the factorization of n. Here, $\mathcal{E}_{\text{pk}}(a; r) = (1+n)^a r^n \bmod n^2$, where r is uniformly chosen in Z_n^*.

Commitment schemes. We now give definitions for standard notions of commitment schemes. For readability we will use "for all m" to mean any possible message m of length polynomial in the security parameter. We start with the standard notion of commitment scheme with its two main variants (i.e., statistically binding and statistically hiding).

Definition 4. $(\mathsf{Com}, \mathsf{Ver})$ *is a* non-interactive commitment scheme *(NICS, for short) if:*

- **efficiency:** Com *and* Ver*, are polynomial-time algorithms;*
- **completeness:** *for all m it holds that*

$$\text{Prob}\left((\mathsf{com}, \mathsf{dec}) \leftarrow \mathsf{Com}(m) : \mathsf{Ver}(\mathsf{com}, \mathsf{dec}, m) = 1\right) = 1;$$

- **binding:** *for any probabilistic polynomial-time adversarial sender* **sen** *there is a negligible function ν such that for all sufficiently large k it holds that*

$$\text{Prob}\Big((\mathsf{com}, m_0, m_1, \mathsf{dec}_0, \mathsf{dec}_1) \leftarrow \mathbf{sen}(1^k) :$$

$$m_0 \neq m_1 \text{ and } \mathsf{Ver}(\mathsf{com}, \mathsf{dec}_0, m_0) = \mathsf{Ver}(\mathsf{com}, \mathsf{dec}_1, m_1) = 1\Big) \leq \nu(k);$$

- **hiding:** *for any probabilistic polynomial-time adversarial receiver* **rec** *there is a negligible function ν such that for all m_0, m_1 where $|m_0| = |m_1|$ and all sufficiently large k it holds that*

$$\text{Prob}\Big(b \leftarrow \{0,1\}; (\mathsf{com}, \mathsf{dec}) \leftarrow \mathsf{Com}(m_b) :: b \leftarrow \mathbf{rec}(\mathsf{com})\Big) < \frac{1}{2} + \nu(k).$$

If the binding property holds with respect to a computationally unbounded **sen***, the commitment scheme is said* statistically binding*; if instead, the hiding property holds with respect to a computationally unbounded* **rec***, the commitment scheme is said* statistically hiding*.*

2   Co-sound rZK in the BPK Model

Here we present a concurrently co-sound rZK argument in the BPK model with black-box techniques only. The construction that we give works for all languages $L \in \mathcal{NP}$ admitting a non-interactive trapdoor instance-dependent commitment scheme, a new primitive that we define and construct.

The Bare Public-Key model. In the BPK model: 1) there exists a public file F that is a collection of records, each containing a public information pk selected by each verifier according to a public-key generation procedure; 2) an (honest) prover is an interactive deterministic polynomial-time algorithm that takes as input a security parameter 1^k, F, a k-bit string x, such that $x \in L$ and L is an $\mathcal{NP}$-language, an auxiliary input y, a reference to an entry of F and a random tape; 3) an (honest) verifier V is an interactive deterministic polynomial-time algorithm that works in the following two stages: a) in a first stage on input a security parameter 1^k and a random tape, V runs the key generation procedure, stores pk in one entry of the file F and keeps the corresponding secret sk; b) in the second stage, V takes as input sk, a statement $x \in L$ and a random string, V performs an interactive protocol with a prover, and outputs 1 (i.e., "accept") or 0 (i.e., "reject"); 4) the first interaction of each prover starts after that all verifiers have completed their first stage.

Malicious provers in the BPK model. Let s be a positive polynomial and P^* be a probabilistic polynomial-time algorithm that takes as first input 1^k.

P^* is an *s-concurrent malicious* prover if on input a public key pk of V, can perform the following $s(k)$ interactive protocols with V: 1) if P^* is already running i protocols $0 \leq i < s(k)$ he can start a new protocol with V choosing the new statement to be proved; 2) he can output a message for any running protocol, receive immediately the response from V and continue.

Concurrent attacks in the BPK model. Given an s-concurrent malicious prover P^* and an honest verifier V, a *concurrent attack* is performed in the following way: 1) the first stage of V is run on input 1^k and a random string so that a pair (pk, sk) is obtained; 2) P^* is run on input 1^k and pk; 3) whenever P^* starts a new protocol choosing a statement, V is run on inputs the new statement, a new random string and sk.

Definition 5 (Co-soundness in the BPK Model). *Given a pair $\langle P, V \rangle$ enjoying the completeness requirement for an $\mathcal{NP}$-language L in the BPK model, we say that $\langle P, V \rangle$ is a concurrently co-sound interactive argument system for L if for all positive polynomial s, for all s-concurrent malicious prover P^*, for any false statement "$x \in L$" the probability that in an execution of a concurrent attack P^* outputs $\bar{y}$ such that $(x, \bar{y}) \in R_{\bar{L}}$ and V outputs 1 (i.e., "accept") for such a statement is negligible in $|x|$.*

Note that in [4], co-soundness is considered w.r.t. perfect NIZK arguments, therefore P^* has to output the witness for $x \in \bar{L}$ along with the computed proof π

for $x \in L$. Here instead we consider interactive zero-knowledge and thus it is not clear the instant in which $P^\star$ is required to output the witness for $x \in \bar{L}$. We will consider the worst case giving to $P^\star$ the chance of writing on the tape the witness at the end of the proof.

The strongest notion of zero knowledge, referred to as rZK, gives to a verifier the ability to rewind the prover to a previous state. This is significantly different from a scenario of multiple interactions between prover and verifier since after a rewind the prover uses the same random bits. We now give the formal definition of a black-box co-sound rZK argument system for $\mathcal{NP}$ in the BPK model.

Definition 6. *A co-sound interactive argument system $\langle P, V \rangle$ in the BPK model is black-box rZK if there exists a probabilistic polynomial-time algorithm S such that for any probabilistic polynomial time V^*, for any polynomials s, t, for any $x_i \in L$, $|x_i| = k$, $i = 1, \ldots, s(k)$, V^* runs in at most t steps and the following two distributions are indistinguishable:*

1. *the output of V^* that generates F with $s(k)$ entries and interacts (even concurrently) a polynomial number of times with each $P(x_i, y_i, j, r_k, F)$ where y_i is a witness for $x_i \in L$, $|x_i| = k$ and r_k is a random tape for $1 \leq i, j, k \leq s(k)$;*
2. *the output of S interacting with V^* on input $x_1, \ldots, x_{s(k)}$.*

We define such an adversarial verifier V^* as an (s, t)-resetting malicious verifier.

Related results in the BPK model. The original work by [1] showed how to achieve constant-round rZK with sequential soundness in the BPK model, using complexity leveraging and black-box techniques only. Complexity leveraging has been removed in [12] by using the non-black-box techniques of [13] and a larger (but still constant) number of rounds. Recently in [14], the result of [12] has been improved using the non-black-box techniques of [15] thus achieving concurrent soundness. In [3] the result of [1] was improved achieving concurrent soundness and rZK in 4 (optimal) rounds, using complexity leveraging and black-box techniques only. The current state-of-the-art therefore does not include any solution for the problem of constructing a constant-round rZK argument in the BPK model under standard assumptions and with black-box techniques only. Notice that the use of black-box techniques is so far crucial for obtaining efficient results. Indeed, the known[1] non-black-box techniques of [13,15] are not practical and require stronger complexity-theoretic assumptions (collision-resistant hash functions and families of claw-free permutations). Black-box techniques instead are efficiently instantiated in many cryptographic protocols.

2.1   Co-soundness in the BPK Model for All $\mathcal{NP} \cap$ Co-$\mathcal{NP}$

Here we show a completely new technique that exploits the co-soundness requirement to avoid the use of the complexity leveraging technique. We define and

---

[1] There are non-black-box assumptions (e.g., knowledge of exponent assumption) that allow one to keep efficiency, but they are not considered standard.

construct a non-interactive trapdoor instance-dependent commitment scheme (NITIDCS, for short). This will let us to obtain a constant-round co-sound resettable zero-knowledge argument system using standard assumptions and blackbox techniques only. This improves the quasi-security obtained in [3], where soundness is proved by means of complexity leveraging. We will also use some techniques previously used in [16,17,18].

Informally, a NITIDCS is a commitment scheme based on an instance x of an $\mathcal{NP}$ language L. If $x \in L$ then statistically binding holds. If instead $x \notin L$ then the commitment scheme is a trapdoor commitment scheme and the witness for $x \notin L$ is the corresponding trapdoor. This definition therefore makes sense for languages $L \in \mathcal{NP} \cap \text{co-}\mathcal{NP}$ (e.g., DDH). We now give the formal definition of a NITIDCS.

Definition 7 (Trapdoor Instance-Dependent Commitment). $(\text{Com}, \text{TCom}, \text{TDec}, \text{Ver})$ *is a (non-interactive) trapdoor instance-dependent commitment scheme ((NI)TIDCS, for short) with respect to a language L if given a statement $x \in L$, $\text{Com}, \text{Ver}, \text{TCom}$ and TDec, are polynomial-time algorithms and moreover:*

- *if $x \in L$ then $(\text{Com}', \text{Ver}')$ is a statistically binding non-interactive commitment scheme, where Com' and Ver' are defined as follows: $\text{Com}'(m) = \text{Com}(m, x)$ and $\text{Ver}'(\text{com}, \text{dec}, m) = \text{Ver}(\text{com}, \text{dec}, m, x)$.*
- **trapdoorness:** *if $x \notin L$ then $(\text{Com}, \text{TCom}, \text{TDec}, \text{Ver})$ for all m the probability distributions:* $\{(\text{com}, \text{dec}) \leftarrow \text{Com}(m, x) : (\text{com}, \text{dec}, m)\}$ *and* $\{(\text{com}', \text{aux}_{\text{com}'}) \leftarrow \text{TCom}(x, y); \text{dec}' \leftarrow \text{TDec}(\text{aux}_{\text{com}'}, m') : (\text{com}', \text{dec}', m')\}$ *are computationally indistinguishable, with y being a witness for $x \notin L$.[2]*

Ingredients. To design our protocol for a given language L we use the following tools: (i) A NITIDCS $(\text{Com}, \text{TCom}, \text{TDec}, \text{Ver})$ with respect to L; (ii) A 1-to-1 length-preserving one-way function which, in turn, implies the existence of a pseudo-random family of functions $\mathcal{R} = \{R_s\}$ and the existence of a non-interactive commitment scheme $(\text{Com}', \text{Ver}')$.

The public file. The public file F contains entries consisting in two images of a given one-way function f. That is, in the i-th entry of F there is $\langle f(a_i), f(b_i) \rangle$ for some values $a_i, b_i \in \{0, 1\}^*$.

Private inputs. The private input of the prover consists of a witness y for $x \in L$. The private input of the verifier consists of one of the preimages a_i and b_i corresponding to the verifier's entry in the public file F $\langle f(a_i), f(b_i) \rangle$.

Auxiliary languages. In the description of the protocol, we will consider the following languages: $L_i^1 = \{(A_i, B_i) | \exists\ a_i, b_i \text{ with } A_i = f(a_i) \vee B_i = f(b_i)\}$ and $L_i^2 = \{(A_i, B_i, x, \text{com}_0, \text{com}_1) | \exists\ y : ((A_i = f(y) \vee B_i = f(y)) \wedge \text{com}_0 \text{ is a commitment of } y) \vee ((x, y) \in R_L \wedge \text{com}_1 \text{ is a commitment of } y)\}^3$.

---

[2] With the notation $\{\beta_1, \ldots, \beta_k : \alpha\}$ we denote the probability distribution of α after the sequential executions of events $\beta_1, \ldots, \beta_k$.

[3] Language L_i^2 can be simplified including only one commitment, in this case the protocol would require also the existence of statistically hiding commitment schemes.

Observe that both languages are in $\mathcal{NP}$. Our protocol will use subprotocols proving membership in those languages. In particular, for the first language, the protocol will use a special 3-round WIPoK for an $\mathcal{NP}$-complete language (we can implement this by means of a reduction to the Hamiltonicity relation and then Blum's protocol is executed), and we will refer to these 3 messages as $(\mathtt{WI}_1, \mathtt{WI}_2, \mathtt{WI}_3)$. The special property that we require is that the prover uses the witness only for computing $\mathtt{WI}_3$ (and, as observed above, this special property is enjoyed by Blum's protocol). For the second language the protocol will use a Σ-protocol $(\sigma_1, \sigma_2, \sigma_3)$ for $\mathcal{NP}$ (again, we implement it by means of a reduction from this relation to the Hamiltonicity relation and then Blum's protocol is executed).

Protocol overview. Fix an $\mathcal{NP}$ language L admitting a NITIDCS and let x be the input statement. The protocol starts by letting the verifier commit, through the NITIDCS with respect to L, to all the challenges it will send in the execution of the forthcoming protocol's steps. Thereafter, the verifier starts the execution of a Blum's 3-round witness-indistinguishable proof of knowledge to prove that it knows the preimage of $f(a_i)$ or $f(b_i)$. This corresponds to the language L_i^1 defined above when i is the index of V in the public file F. (In order to use Blum's protocol the language L_i^1 is reduced to the Hamiltonicity relation.) The prover acts as verifier for this subprotocol, and in the second round of the (larger) protocol also sends a commitment, using a statistically binding commitment scheme, of a witness y for $x \in L$ and starts to prove through the Blum's Hamiltonicity Σ-protocol that the committed message is either a witness for $x \in L$ or one of the preimage of verifier's public file entry. This corresponds to the language L_i^2 defined above. (Again we stress that the language L_i^2 is reduced to the Hamiltonicity relation.) In the third round the verifier opens the challenge commitment and completes the first subprotocol. Finally, in the fourth round the second subprotocol is completed.

The protocol. The common input is a statement $x \in L$. We denote by i the index of the verifier in the public file so that the verifier knows either a_i or b_i associated with the F's i-th entry $\langle f(a_i), f(b_i) \rangle$.

In the first round the verifier V chooses a random challenge c for the Blum's Σ-protocol and then runs the commitment algorithm of the NITIDCS w.r.t. L on c, that is, $(\mathtt{com}_c, \mathtt{dec}_c) \leftarrow \mathtt{Com}(c, x)$. V also computes the the first message $\mathtt{WI}_1$ of the Blum's special 3-round witness-indistinguishable proof of knowledge for L_i^1 by executing the reduction from L_i^1 to Hamiltonicity language. We stress that V can run the steps of this subprotocol using as witness the secret chosen in $\{a_i, b_i\}$. Then, V sends $(\mathtt{com}_c, \mathtt{WI}_1)$ to the prover P.

In the second round P starts by picking a random seed s and computing $\rho = R_s(x \circ y \circ F \circ \mathtt{com}_c \circ \mathtt{WI}_1 \circ i)$ ("$\circ$" denotes concatenation) where $\{R_s\}$ is a family of pseudorandom functions. P uses ρ as its own randomness. Subsequently, P then computes the second message $\mathtt{WI}_2$ of the witness-indistinguishable subprotocol. Moreover, P runs $(\mathtt{com}_0, \mathtt{dec}_0) = \mathtt{Com}'(0^k)$, $(\mathtt{com}_1 = \mathtt{com}_y, \mathtt{dec}_y) = \mathtt{Com}'(y)$ using a statistically binding NICS. Finally, P computes the $\mathcal{NP}$-reduction from the language L_i^2 to Hamiltonicity language and the first message σ_1 of the Blum's

Σ-protocol on the obtained graph Hamiltonicity instance. We stress that P can run the steps of the Blum's protocol using as witness its private input y. P then sends $(\mathtt{WI}_2, \mathtt{com}_0, \mathtt{com}_1, \sigma_1)$ to V.

In the third round V computes the third message $\mathtt{WI}_3$ of the first subprotocol. V sends to P the following tuple $(\mathtt{WI}_3, \mathtt{dec}_c, c)$ where we stress that c is simply the second message of the second Σ-protocol (i.e., $\sigma_2 = c$).

Finally, in the fourth round, P checks that $\mathtt{dec}_c$ is actually the decommitment of c and verifies that $(\mathtt{WI}_1, \mathtt{WI}_2, \mathtt{WI}_3)$ is an accepting transcript for the first subprotocol, then computes the third message σ_3 of the second Σ-protocol and sends σ_3 to V. V accepts if and only if $(\sigma_1, \sigma_2, \sigma_3)$ is an accepting transcript for the second Σ-protocol.

We show a graphical illustration of the protocol in the BPK model in Fig. 1.

Theorem 1. *Under the assumption that 1-to-1 length-preserving one-way functions exist, all $\mathcal{NP}$ languages with a NITIDCS admit a 4-round concurrently co-sound rZK argument in the BPK model.*

For lack of space, the formal proof is omitted. Here however we give the intuitions to let the reader understand why our techniques solve this problem.

The proof of resettable zero-knowledge follows the standard approach used in previous works [1,3]. Completeness is easy by inspection. The co-soundness property instead is proved using our new techniques. Indeed, notice that when a false theorem is proved, the NITIDCS is a trapdoor commitment scheme. Moreover we know that a successful proof of a false theorem forces the adversarial prover $P^\star$ to give as output a witness y for $x \notin L$. Such a witness is also a trapdoor for the commitment scheme on the verifier's side. Therefore, the protocol is transformed in an argument of knowledge. Indeed, an extractor can use the trapdoor for opening the commitment in two different ways, therefore changing the challenge of the Σ-protocol proved by $P^\star$ and thus obtaining a witness for the proved statement. Since $x \notin L$, this will correspond to one of the two secret keys. We can now use the witness indistinguishability of the two subprotocols to obtain either the inversion of the one-way function (in case a secret extracted does not correspond to the ones used) or a contradiction of the witness indistinguishability property of the used subprotocols (in case the extracted secrets always correspond to the used ones).

Constructing NITIDCSs with respect to all $\mathcal{NP} \cap$ co-$\mathcal{NP}$. Let L be a language in $\mathcal{NP} \cap$ co-$\mathcal{NP}$ and consider the statement $x \in L$. We show how to construct a NITIDCS with respect to L. To this aim we need to define four efficient algorithms: $\mathtt{Com}, \mathtt{TCom}, \mathtt{TDec}$ and $\mathtt{Ver}$. The algorithm $\mathtt{Com}(m, x)$ is defined on top of a non-interactive statistically-binding commitment scheme $(\mathtt{SBCom}, \mathtt{SBVer})$ and goes as follows. $\mathtt{Com}$ uses the $\mathcal{NP}$ reduction from $\bar{L}$ (that is a language in $\mathcal{NP}$ being L in $\mathcal{NP} \cap$ co-$\mathcal{NP}$) to the Hamiltonicity relation to obtain a graph G (with q nodes) so that finding a Hamiltonian cycle in G is equivalent to finding a witness for $x \in \bar{L}$, or equivalently, for $x \notin L$. Then it goes as follows.

- To commit to 0, pick a random permutation π of the nodes of G, and commit to the entries of the adjacency matrix of the permuted graph one by

---

Common input: the public file F, n-bit string $x \in L$ and index i that specifies the i-th entry of F, $\langle f(a_i), f(b_i) \rangle$ the i-th entry of F.
P's private input: a witness y for $x \in L$.
V's private input: a randomly chosen value between a_i and b_i.

V-round-1:

1. randomly pick $c \leftarrow \{0,1\}^n$;
2. compute the $\mathcal{NP}$-reduction from L_i^1 to Hamiltonicity;
3. compute $(\mathtt{com}_c, \mathtt{dec}_c) \leftarrow \mathtt{Com}(c, x)$ and $\mathtt{WI}_1$;
4. send $(\mathtt{com}_c, \mathtt{WI}_1)$ to P;

P-round-2:

1. randomly pick $s \leftarrow \{0,1\}^n$ and compute $\rho = R_s(x \circ y \circ F \circ \mathtt{com}_c \circ \mathtt{WI}_1 \circ i)$;
2. compute $\mathtt{WI}_2$;
3. compute the $\mathcal{NP}$-reduction from L_i^2 to Hamiltonicity and compute σ_1;
4. compute $(\mathtt{com}_0, \mathtt{dec}_0) \leftarrow \mathtt{Com}'(0^k, x)$;
5. compute $(\mathtt{com}_1 = \mathtt{com}_y, \mathtt{dec}_y) \leftarrow \mathtt{Com}'(y, x)$;
6. send $(\mathtt{WI}_2, \mathtt{com}_0, \mathtt{com}_1, \sigma_1)$ to V

V-round-3:

1. compute $\mathtt{WI}_3$ by using either the $\mathcal{NP}$-reduction of a_i or b_i as witness;
2. send $(\mathtt{WI}_3, \mathtt{dec}_c, c)$ to P

P-round-4:

1. verify that $\mathtt{com}_c$ is the commitment of c using $\mathtt{dec}_c$;
2. verify that $(\mathtt{WI}_1, \mathtt{WI}_2, \mathtt{WI}_3)$ is the correct transcript of the 3-round witness indistinguishable proof on input instance $\langle f(a_i), f(b_i) \rangle$ reduced to Hamiltonicity;
3. compute σ_3 using the $\mathcal{NP}$-reduction of y as witness;
4. send σ_3 to V;

V-decision: verify that (σ_1, c, σ_3) is the correct transcript of Blum's Σ-protocol on input instance $(\mathtt{com}_0, \mathtt{com}_1, f(a_i), f(b_i))$ reduced to Hamiltonicity.

Fig. 1. The 4-round concurrently co-sound rZK argument system in the BPK model for languages with a NITIDCS $(\mathtt{Com}, \mathtt{Ver}, \mathtt{TCom}, \mathtt{TDec})$ and a statistically binding NICS $(\mathtt{Com}', \mathtt{Ver}')$

one, using $\mathtt{SBCom}$ (the underlying non-interactive statistically-binding commitment scheme). To decommit, send π and decommit to every entry of the adjacency matrix. The algorithm $\mathtt{Ver}$ verifies that the graph it received is $\pi(G)$ by means of $\mathtt{SBVer}$.

– To commit to 1, choose a randomly labeled q-cycle, and for all entries in the adjacency matrix that correspond to edges on the q-cycle, use $\mathtt{SBCom}$ to commit to 1 values. For all the other entries, use SBCom to commit to random values. (These will be indistinguishable from actual commitments due to the hiding property.) To decommit, open only the entries corresponding to the randomly chosen q-cycle in the adjacency matrix. The algorithm Ver verifies that the entries it received are indeed commitments of 1 by means of SBVer.

The algorithms TCom and TDec base on the following simple observation: having a witness for $x \notin L$ one can calculate via the $\mathcal{NP}$-reduction a Hamiltonian cycle and then it is possible to construct a commitment that can be opened either as 0 or as 1. It suffices to choose a random permutation of the graph G and commit using SBCom to all entries of the adjacency matrix. TCom then outputs the Hamiltonian cycle as auxiliary information for TDec and TDec can open the commitment in either ways.

We next argue that the above commitment scheme is indeed a trapdoor instance-dependent commitment scheme with respect to the language L. Efficiency and completeness of (Com, TCom, TDec, Ver) easily follow by inspection. Whenever $x \in L$ there is no witness for $x \notin L$. This in turn implies that the graph G obtained by the $\mathcal{NP}$ reduction does not have a Hamiltonian cycle and therefore, being the component commitment scheme (SBCom, SBVer) statistically binding, (Com, Ver) is statistically binding as well. Whenever $x \notin L$ then $x \in \bar{L}$ and as observed above a witness for $x \notin L$ is the trapdoor for TCom.

Theorem 2. *Under the assumption that 1-to-1 length-preserving one-way functions exist, for any language L in $\mathcal{NP} \cap$ co-$\mathcal{NP}$ there exists a non-interactive trapdoor instance-dependent commitment scheme with respect to L.*

Previous theorem follows by observing that under the assumption that 1-to-1 length-preserving one-way functions exist there exist non-interactive statistically-binding commitment schemes. Moreover, combining it with Theorem 1 we obtain the next theorem.

Theorem 3. *If there exist 1-to-1 length-preserving one-way functions then all languages in $\mathcal{NP} \cap$ co-$\mathcal{NP}$ have 4 (optimal) round concurrently co-sound rZK argument in the BPK model. Moreover, the security properties can be proved using black-box techniques only.*

3   Definitions and Constructions in the RPK Model

In this section, we show an efficient statistical NIZK argument system in the RPK model, but under a weaker co-soundness notion. We present a transformation that given a language L with a Σ-protocol Π with linear answer (see Definition 2), outputs an efficient weakly co-sound statistical NIZK argument for L in the RPK model. We start by giving the required definitions in the RPK model.

The RPK model. Next we briefly review the RPK model (introduced in [19]) following the definition of [2]. Let $KS(1^k)$ be a probabilistic polynomial-time algorithm which, on input a security parameter 1^k, outputs a private/public key pair. We write $KS(1^k; r)$ to denote the execution of KS using r as random coins. The RPK model [19] features a trusted authority, which the parties can invoke to register their key pairs and to retrieve other parties' public keys. Key-registration takes place by having the registrant privately sending the authority the random coins r that she used to create her key pair. The authority will then run $KS(1^k; r)$, store the resulting public key along with the identity of the registrant, and later give the public key to anyone who asks for it. Note that this in particular means that to register a public key one needs to know the corresponding private key. Note also that one does not need to have registered a public key of her own to ask the authority for somebody else's public key.

Variations of the RPK model and comparison with previous NIZK arguments. As already discussed in [2], the universally composable NIZK of [19] is only computational zero knowledge, is rather inefficient since it requires $\mathcal{NP}$ reductions and moreover requires a public key for both the prover and the verifier. However the nice property of the NIZK of [19] is that prover and verifier have only to trust their own trusted authority for the key-registration procedures. We focus on obtaining efficient statistical NIZK under standard assumptions (i.e., without complexity leveraging), and we will show how to achieve this result in the RPK model as defined in [2] but under a weak co-soundness notion. Under this formulation the key-registration authority is trusted by both the verifier for the key-registration procedure and the prover for obtaining a well-formed public key. For a list of possible implementations of the key-registration procedure see [19].

We finally stress that perfect NIZK has been achieved in the common reference string model in [4] and in [6]. The former result is based on a specific number-theoretic assumption on bilinear groups. The latter instead uses a non-standard non-black-box assumption.

NIZK with key setup. We present a definition of NIZK in the RPK model.[4] Let $KS(1^k)$ be the key setup for the key-registration authority, and let L be a language for which one can efficiently generate pairs $(x, y) \in R_L$ from a security parameter 1^k. A non-interactive system for L with key setup KS is a pair of efficient algorithms (P, V), where:

- $P(x, y, \mathrm{pk}_V)$ is a probabilistic algorithm run by the prover. It takes as input an instance x and y such that $(x, y) \in R_L$, along with the verifier registered public key pk_V, which the prover obtains from the authority. P outputs a string π as a non-interactive zero-knowledge proof that $x \in L$;
- $V(x, \pi, \mathrm{sk}_V)$ is a deterministic algorithm run by the verifier, satisfying the following correctness property: for all x and y such that $(x, y) \in R_L$, it holds that: $\mathrm{Prob}\left[V(x, \pi, \mathrm{sk}_V) = 1 | (\mathrm{pk}_V, \mathrm{sk}_V) \leftarrow KS(1^k); \pi \leftarrow P(x, y, \mathrm{pk}_V)\right] = 1$, where the probability is over the random coins of KS and P.

---

[4] In the definition we consider the case in which only the verifier has a registered public key. The definition simply extends to the case in which also the prover has a registered public key.

The system is zero-knowledge if there exists a probabilistic polynomial-time algorithm S, such that for all instances x and y such that $(x, y) \in R_L$, the following two ensembles are indistinguishable:

- V's Key Pair, Real Proof: $\{(\text{pk}_V, \text{sk}_V, \pi) | (\text{pk}_V, \text{sk}_V) \leftarrow KS(1^k); \pi \leftarrow P(x, y, \text{pk}_V)\}$,
- V's Key Pair, Simulated Proof: $\{(\text{pk}_V, \text{sk}_V, \pi) | (\text{pk}_V, \text{sk}_V) \leftarrow KS(1^k); \pi \leftarrow S(x, \text{pk}_V, \text{sk}_V)\}$.

As usual depending on the quality of above indistinguishability one obtains computational, statistical or perfect zero-knowledge.

To define weak co-soundness, we consider a probabilistic polynomial-time adversary P^* who plays the following game.

1. Execute $(\text{pk}_V, \text{sk}_V) \leftarrow KS(1^k)$ and give pk_V to P^*.
2. Repeat until P^* stops: P^* outputs[5] a quadruple (x, y, y', π) and receives $V(x, \pi, \text{sk}_V)$.

We say that P^* wins if it produces at least one quadruple $(x, \bar{y}, y', \pi)$ such that V accepts, where $(x, \bar{y}) \in R_{\bar{L}}$ and $((x, \pi, \text{pk}_V), y') \in R_\Lambda$ with $\Lambda = \{(x, \pi, \text{pk}_V) | V(x, \pi, \text{sk}_V) = 1\}$. In other words, P^* wins if it outputs an $x \notin L$ along with $\bar{y}$ a witness for $x \notin L$ *and* the following happens. A valid witness $\bar{y}$ for $x \in \bar{L}$ is given as output by P^* only along with a valid witness y' for membership in Λ meaning that P^* knows that the theorem is false (it outputs a witness $\bar{y}$ for membership in $\bar{L}$) and knows that the verifier will accept π (it outputs a witness y' for membership in Λ).

The NIZK argument is *weakly co-sound* if any P^* wins with probability negligible in k. We say that the NIZK argument is weakly co-sound for a particular number of proofs $m(k)$ if the game always stops after $m(k)$ proofs are generated.

Witnessing membership in Λ. The language Λ includes all the triples theorem, proofs, verifier's public key for which the verifier accepts the proof for the theorem in input the corresponding secret key. Our definition of weak co-soundness requires a malicious prover to output a witness y' of membership in Λ. We decided not to burden the definition of co-soundness giving details on y'. However, we next describe some natural ways to look at y'. First of all let us observe that Λ is in $\mathcal{NP}$ since sk_V is an $\mathcal{NP}$ witness. However, we cannot require a malicious prover to output the verifier secret key. A possible candidate as y' is then a 1-round zap [20] proving membership in Λ. Another possible candidate as y' would be the description of the algorithm used to compute the proof π along with the randomness used and a proof that such an algorithm outputs an accepting proof for a false theorem.

The stronger co-soundness requirement. We stress that the notion of weak co-soundness defined in the RPK model has a much stronger requirement with

---

[5] We stress that y, y' are given as output on another tape since they are not sent to the verifier.

respect to the co-soundness defined in the BPK model and the one defined in [4]. Indeed, here we also require that whenever the adversary outputs a valid witness for $x \notin L$, it also outputs a valid witness y' proving that π is an accepting proof. This models the intuition that the adversary also knows that the generated proof of a false statement is accepting and outputs the witness for $x \in \bar{L}$ together with a witness that certifies it. Summing up, we assume that the adversary never outputs a valid witness y for $x \in \bar{L}$ along with an invalid witness y' for membership in Λ. We will need this stronger requirement in the security proof of our construction. This makes the contribution of this second result as an additional example of the use of co-soundness, while the main the contribution of the paper remains the construction in the BPK model. We leave as an open problem the possibility of achieving (non-weak) co-soundness in the RPK model.

3.1   Efficient Co-sound Statistical NIZK in the RPK Model

We follow in part the approach of [2], and extend it with other techniques that result to be useful for proving the weak co-soundness property. We will make use of an IND-CPA homomorphic cryptosystem $\mathcal{PE} = (\mathcal{G}, \mathcal{E}, \mathcal{D})$. The protocol is a transformation from any Σ-protocol with linear answer Π. We now describe the key setup of the RPK model.

$KS(1^k)$ *(Key Setup for the Verifier).* Set $(\text{pk}) \leftarrow \mathcal{G}(1^k)$. Choose a challenge c as V would do in the given Σ-protocol, and set C_0, C_1 to be a random (homomorphic) encryptions of c under pk. The public key is now (pk, C_0, C_1) and the private key is (sk, c).

To show the idea behind the technique – which is similar to that used in [2] – we observe that because the Σ-protocol is with linear answer, it is possible to execute the prover's side of the protocol given the encryptions C_0 and C_1 of the challenge c. Namely, the prover starts by computing its first message a. Then, if the answer z is supposed to contain $z_j = u_j + v_j c$, the prover will be able (by definition of Σ-protocols with linear answers) to derive the values of u_j, v_j from x, its private input y and the random coins used to create a. At this point, the prover can compute $\mathcal{E}_{\text{pk}}(z_j)$ as $\text{randomize}(\mathcal{E}_{\text{pk}}(u_j) \cdot C_i^{v_j})$ for $i = 0, 1$. This can be decrypted and then checked as usual by V. We also stress here that even though the two different encryptions of the challenge c seem not to be necessary they will turn to be crucial in order to achieve co-soundness under standard assumptions, namely IND-CPA security of the encryption scheme, without relying on complexity leveraging as in [2].

We can show a protocol enjoying co-soundness by noticing that the adversarial prover will be able to prove false statements only when C_0 and C_1 correspond to the same plaintext. Indeed the special-soundness property of a Σ-protocol guarantees that there is no accepting pair of triples $((a, c, z), (a, c', z'))$ for a false statement when $c \neq c'$. The protocol goes as follows:

1. Given an instance x, a witness y, P on input V's public key (pk, C_0, C_1) (from the authority) computes the first message a in a proof according to Π. Let the

final message z be of the form $(u_1+v_1c, \ldots, u_m+v_mc)$; then, for $i = 1, \ldots, m$, P computes $s_i^0 \leftarrow \mathsf{randomize}(\mathcal{E}_{\mathrm{pk}}(u_j) \cdot C_0^{v_j})$ and $s_i^1 \leftarrow \mathsf{randomize}(\mathcal{E}_{\mathrm{pk}}(u_j) \cdot C_1^{v_j})$. P sends x, π to V , where $\pi = (a, (s_1^0, \ldots, s_m^0, s_1^1, \ldots, s_m^1))$.

2. On input x and a proof $\pi = (a, (s_1^0, \ldots, s_m^0, s_1^1, \ldots, s_m^1))$, V sets $u_i^j \leftarrow \mathcal{D}_{\mathrm{sk}}(s_i^j)$ for $j = 0, 1$, and then verifies that $x, (a, c, (u_1^0, \ldots, u_m^0))$ and $x, (a, c, (u_1^1, \ldots, u_m^1))$ would be accepted by the verifier of protocol $\varPi$, and accepts or rejects accordingly.

Theorem 4. *Above protocol is complete and statistical zero-knowledge in the RPK model. Moreover, it is a co-sound argument in the RPK model for provers generating $O(\log k)$ proofs.*

For lack of space, the proof is omitted from this extended abstract. The main idea behind the proof is that when playing with a fake public key where the two ciphertexts do not correspond to the same plaintext, the adversary can not prove a false theorem (by the special soundness property). This is then used to break the IND-CPA security of the encryption scheme.

The above theorem (and the observation that Paillier's encryption scheme [11] is an IND-CPA secure homomorphic public-key encryption scheme) leads to the following result.

Theorem 5. *Every language with a Σ-protocol with linear answer admits an efficient weak co-sound statistical NIZK argument in the RPK model under standard complexity-theoretic assumptions.*

References

1. Canetti, R., Goldreich, O., Goldwasser, S., Micali, S.: Resettable Zero-Knowledge. In: 32nd ACM Symposium on Theory of Computing (STOC 2000), pp. 235–244. ACM, New York (2000)
2. Damgård, I., Fazio, N., Nicolosi, A.: Non-interactive zero-knowledge from homomorphic encryption. In: Halevi, S., Rabin, T. (eds.) TCC 2006. LNCS, vol. 3876, pp. 41–59. Springer, Heidelberg (2006)
3. Di Crescenzo, G., Persiano, G., Visconti, I.: Constant-Round Resettable Zero Knowledge with Concurrent Soundness in the Bare Public-Key Model. In: Franklin, M. (ed.) CRYPTO 2004. LNCS, vol. 3152, pp. 237–253. Springer, Heidelberg (2004)
4. Groth, J., Ostrovsky, R., Sahai, A.: Perfect non-interactive zero knowledge for NP. In: Vaudenay, S. (ed.) EUROCRYPT 2006. LNCS, vol. 4004, pp. 339–358. Springer, Heidelberg (2006)
5. Groth, J., Lu, S.: A non-interactive shuffle with pairing based verifiability. In: Kurosawa, K. (ed.) ASIACRYPT 2007. LNCS, vol. 4833, pp. 51–67. Springer, Heidelberg (2007)
6. Abe, M., Fehr, S.: Perfect nizk with adaptive soundness. In: Vadhan, S.P. (ed.) TCC 2007. LNCS, vol. 4392, pp. 118–136. Springer, Heidelberg (2007)
7. Goldwasser, S., Micali, S., Rackoff, C.: The Knowledge Complexity of Interactive Proof-Systems. SIAM J. on Computing 18(6), 186–208 (1989)
8. Cramer, R., Damgård, I., Schoenmakers, B.: Proofs of Partial Knowledge and Simplified Design of Witness Hiding Protocols. In: Desmedt, Y.G. (ed.) CRYPTO 1994. LNCS, vol. 839, pp. 174–187. Springer, Heidelberg (1994)

9. Blum, M.: How to Prove a Theorem So No One Else Can Claim It. In: Proceedings of the International Congress of Mathematicians, pp. 1444–1451 (1986)

10. Bellare, M., Desai, A., Pointcheval, D., Rogaway, P.: Relations among notions of security for public-key encryption schemes. In: Krawczyk, H. (ed.) CRYPTO 1998. LNCS, vol. 1462, pp. 26–45. Springer, Heidelberg (1998)

11. Paillier, P.: Public-key cryptosystems based on composite degree residuosity classes. In: Stern, J. (ed.) EUROCRYPT 1999. LNCS, vol. 1592, pp. 223–238. Springer, Heidelberg (1999)

12. Barak, B., Goldreich, O., Goldwasser, S., Lindell, Y.: Resettably-Sound Zero-Znowledge and its Applications. In: Proceeding of the 42nd Symposium on Foundations of Computer Science (FOCS 2001), 1109 Spring Street, Suite 300, Silver Spring, MD 20910, USA, pp. 116–125. IEEE Computer Society Press, Los Alamitos (2001)

13. Barak, B.: How to Go Beyond the Black-Box Simulation Barrier. In: Proceeding of the 42nd Symposium on Foundations of Computer Science (FOCS 2001), 1109 Spring Street, Suite 300, Silver Spring, MD 20910, USA, pp. 106–115. IEEE Computer Society Press, Los Alamitos (2001)

14. Deng, Y., Lin, D.: Resettable zero knowledge arguments with concurrent soundness in the bare public-key model under standard assumptions. In: Pei, D., Yung, M., Lin, D., Wu, C. (eds.) INSCRYPT 2007. LNCS, vol. 4990, pp. 123–137. Springer, Heidelberg (2008)

15. Pass, R., Rosen, A.: Concurrent non-malleable commitments. In: Proc. of FOCS, pp. 563–572 (2005)

16. Di Crescenzo, G., Visconti, I.: Concurrent zero knowledge in the public-key model. In: Caires, L., Italiano, G.F., Monteiro, L., Palamidessi, C., Yung, M. (eds.) ICALP 2005. LNCS, vol. 3580, pp. 816–827. Springer, Heidelberg (2005)

17. Visconti, I.: Efficient zero knowledge on the internet. In: Bugliesi, M., Preneel, B., Sassone, V., Wegener, I. (eds.) ICALP 2006. LNCS, vol. 4052, pp. 22–33. Springer, Heidelberg (2006)

18. Ostrovsky, R., Persiano, G., Visconti, I.: Constant-round concurrent non-malleable zero knowledge in the bare public-key model. In: Aceto, L., Damgård, I., Goldberg, L.A., Halldórsson, M.M., Ingólfsdóttir, A., Walukiewicz, I. (eds.) ICALP 2008, Part II. LNCS, vol. 5126, pp. 548–559. Springer, Heidelberg (2008)

19. Barak, B., Canetti, R., Nielsen, J., Pass, R.: Universally Composable Protocols with Relaxed Set-up Assumptions. In: 44th IEEE Symposium on Foundations of Computer Science (FOCS 2004), pp. 394–403 (2004)

20. Groth, J., Ostrovsky, R., Sahai, A.: Non-interactive ZAPs and New Techniques for NIZK. In: Dwork, C. (ed.) CRYPTO 2006. LNCS, vol. 4117, pp. 97–111. Springer, Heidelberg (2006)

Another Look at Extended Private Information Retrieval Protocols*

Julien Bringer[1] and Hervé Chabanne[1,2]

[1] Sagem Sécurité
[2] TELECOM ParisTech

Abstract. Extended Private Information Retrieval (EPIR) has been introduced at CANS'07 by Bringer *et al.* as a generalization of the notion of Private Information Retrieval (PIR). The principle is to enable a user to privately evaluate a fixed and public function with two inputs, a chosen block from a database and an additional string.

The main contribution of our work is to extend this notion in order to add more flexibility during the system life. As an example, we introduce a general protocol enabling polynomial evaluations. We also revisit the protocol for Hamming distance computation which was described at CANS'07 to obtain a simpler construction. As to practical concern, we explain how amortizing database computations when dealing with several requests.

Keywords: Private Information Retrieval (PIR), Extended PIR (EPIR), Polynomial Evaluation, Batching.

1 Introduction

The notion of PIR models a protocol where a user is able to privately recover a block from a database. With application to privacy-preserving biometric authentication solutions as one motivation, [3] introduces an extension of this concept, named Extended Private Information Retrieval (EPIR). It is a generalization of PIR where instead of recovering a block, the user wants to retrieve the evaluation of a given function on a block. In this paper, we extend further this notion and design new protocols. In particular, we highlight the difference with the properties of PIR protocols by adding more flexibility in the choice of the function. Note that the work on this paper is also close to two-party private polynomial evaluations.

1.1 Related Works

PIR is introduced in Chor *et al.* [5,6], it enables a user to retrieve data (in the original definition, one bit) from a database without leaking which index is queried. Suppose a database is constituted with several bits, say N; to be secure, the protocol should satisfy two specific properties [12]: soundness and user

---

* Work partially supported by the French ANR RNRT project BACH.

B. Preneel (Ed.): AFRICACRYPT 2009, LNCS 5580, pp. 305–322, 2009.

privacy. Soundness (or correctness) states that when the user and the database follow the protocol, the result of the request is exactly the requested bit. User privacy is defined as the impossibility for the database to distinguish two requests made by the user. These properties are studied either under an information theoretic model for multi-database settings or under a computational model for single-database settings. Here we focus on single-database PIR only. Another additional property suggested by Gertner *et al.* in [12] is the notion of database privacy, measuring whether a user can learn more information than the requested data itself. If the user cannot determine whether he interacts with the real database or with a simulator knowing only the answer, the PIR achieves database privacy and is called a Symmetric PIR (SPIR) (also known as 1-out-of-N oblivious transfer [8]).

Among the known computational secure PIR protocols, block-based PIR – i.e. working directly on block of bits – allow to reduce efficiently the communication cost. The best performances are from Gentry and Ramzan [11] and Lipmaa [18] with a communication complexity polynomial in the logarithm of the number of blocks. See [10,21] for surveys of the subject.

[3] introduced the concept of EPIR, a kind of combination of PIR and oblivious function evaluation (a specific case of secure multiparty computation) based on the above mentioned works. Two EPIR protocols are described in [3], one for equality checking based on ElGamal [9] cryptosystem and the other one for Hamming distance computation based on the Boneh-Goh-Nissim (BGN) encryption scheme [2]. Our paper follows the same vein but aims to achieve more general constructions.

As two-party private evaluations, [4,20] are examples of oblivious function evaluation protocols related to our problem. [4] studies the problem of evaluating a public function on private inputs stored on one or more databases and which inputs indexes are chosen by the user. [20] describes protocols for two-party oblivious polynomial evaluation where the input is owned by one party and the polynomial by the other party. Our revisited notion of EPIR can thus be seen also as a generalization of these problems. Note that a similar goal is explored by Lipmaa in [19] by means of branching programs and pre-processing techniques for Selective Private Function Evaluation (SPFE). EPIR is in fact a specific case of SPFE, and we notably explain here a solution which does not need such pre-processing.

1.2   Our Contributions

In a single-database context, we generalize the concept of EPIR introduced in [3]. Instead of setting a fixed and public function to be evaluated on two inputs – one user's input and one database's block, we introduce more flexibility in the choice of the function. We then adapt the security model, namely the user privacy and database privacy requirements, accordingly. In particular we want to ensure privacy against the database for the function chosen by the user.

By imposing non-disclosure of the function, this additional property of flexibility is not straightforward, as we need to encrypt the chosen function in a way which let the capability to evaluate it directly in the encrypted domain. One of our solutions investigated here is to embed an encrypted description of the function directly in the query which is sent to the database. Note that the EPIR examples of [3] do not have this capability.

Our motivation with this generalization is also to underline the distinction between PIR and EPIR. We indeed design two new EPIR protocols. The first one is a new EPIR protocol for Hamming distance computation, based on additive homomorphic encryption, and lighter than the previous one from [3]. Our second construction is a generic protocol enabling evaluation of any polynomial freely chosen by the user at the request time. Then, we discuss batching solutions to amortize database workload when several requests are made.

Note that when letting the user choose freely the function to be evaluated, we face a new problem which is to avoid the use of trivial functions – e.g. the identity – which would enable the user to learn the value of the database blocks. We discuss this problem in Section 2.3 but we do not describe a practical solution to solve it. It is left as an open problem.

2 EPIR Revisited

2.1 Definitions

Following the definition of [3], a single-database EPIR protocol is a protocol between a database $\mathcal{DB}$ which holds a set of N blocks $(R_1, R_2, \cdots, R_N)$ where $R_j \in \{0,1\}^{\ell_1}$ and a user $\mathcal{U}$ which wants to retrieve the value of a function $f(R_i, X)$ for a value $X \in \{0,1\}^{k_1}$ and an index i chosen by the user. Here N is public and [3] assumes that the description of f is set for once and also public.

Let $\mathcal{F}$ be a subset of the set of functions from $\{0,1\}^{\ell_1}$ to $\{0,1\}^*$. From a dual point of view, we now define an EPIR protocol as a protocol – with a support of evaluation $\mathcal{F}$ – which enables the user $\mathcal{U}$ to retrieve the value $F(R_i)$ for a chosen index i and a chosen function $F \in \mathcal{F}$. $\mathcal{F}$ is assumed to be public. The query made by the user is denoted below as a retrieve(F, i) query.

In [3], the first construction corresponds to $\mathcal{F} = \{\text{IsEqual}(., X), \forall X \in \{0,1\}^{\ell_1}\}$ where IsEqual checks the equality of its two inputs and the second construction is given for $\mathcal{F} = \{d_w(., X), \forall X \in \{0,1\}^{\ell_1}, \forall w \in \mathbb{N}^{\ell_1}\}$ where d_w denotes a Hamming distance weighted via w a vector of weights. Our goal is to continue the study of EPIR looking for more constructions and investigating the general situation where functions may vary over time. Due to this extension, we slightly adapt the security model. In particular, we want to ensure that the function F chosen by the user remains hidden from the database. Nevertheless, as this adaptation is quite natural, we give only a quick overview of the definitions and refer to [3] for further details.

2.2 Security Model

We define correctness of an EPIR protocol as usual by saying that the protocol is correct if any query retrieve(F, i) returns the correct value $F(R_i)$ with an overwhelming probability when $\mathcal{U}$ and $\mathcal{DB}$ follow the specification.

In the sequel, an EPIR protocol is said to be secure if any adversary has only negligible advantages against user privacy and database privacy following the experiments described below.

Notation. The security is evaluated by an experiment between an attacker and a challenger, where the challenger simulates the protocol executions and answers the attacker's oracle queries. Without further notice, algorithms are assumed to be polynomial-time. For $\mathcal{A}$ a probabilistic algorithm, we denote $\mathcal{A}(\mathcal{O}, \mathsf{retrieve})$ as the action to run $\mathcal{A}$ with access to any polynomial number of retrieve queries generated or answered (depending on the position of the adversary) by the oracle $\mathcal{O}$. We recall the definitions for negligible and overwhelming probabilities.

Definition 1. *The function $P(\ell) : \mathbb{Z} \to \mathbb{R}$ is said negligible if, for all polynomials $f(\ell)$, there exists an integer N_f such that $P(\ell) \le \frac{1}{f(\ell)}$ for all $\ell \ge N_f$. If $P(\ell)$ is negligible, then the probability $1 - P(\ell)$ is said to be overwhelming.*

User Privacy. This requirement is inspired by the notion of user privacy in the case of PIR. It captures the notion that for any $\mathsf{retrieve}(F, i)$ query, $\mathcal{DB}$ learns nothing about block index queried nor the function F chosen by $\mathcal{U}$. Particularly, this implies in general that the result of the evaluation remains private.

Formally, an EPIR protocol is said to respect user privacy if any attacker $\mathcal{A} = (\mathcal{A}_1, \mathcal{A}_2, \mathcal{A}_3, \mathcal{A}_4)$, acting as a malicious database, has only a negligible advantage in the following game, where the attacker's advantage is $|\Pr[b' = b] - \frac{1}{2}|$.

$$
\begin{aligned}
&\mathbf{Exp}_{\mathcal{A}}^{\text{user-privacy}} \\
&\left|
\begin{aligned}
(R_1, R_2, \cdots, R_N) &\leftarrow \mathcal{A}_1(1^\ell) \\
1 \le i_0, i_1 \le N; F_0, F_1 \in \mathcal{F} &\leftarrow \mathcal{A}_2(Challenger; \mathsf{retrieve}) \\
b &\xleftarrow{R} \{0, 1\} \\
\emptyset &\leftarrow \mathcal{A}_3(Challenger; \mathsf{retrieve}(F_b, i_b)) \\
b' &\leftarrow \mathcal{A}_4(Challenger; \mathsf{retrieve})
\end{aligned}
\right.
\end{aligned}
$$

Database Privacy. This property is also inspired by PIR security model. It is the analog of symmetric property of SPIR [12]. This models the requirement that a malicious user $\mathcal{U}$ learns nothing else than $F'(R_{i'})$ for some $1 \le i' \le N$ and $F' \in \mathcal{F}$ via a query retrieve. This is proved via a simulation principle, for which $\mathcal{U}$ cannot determine if he interacts with a simulator which takes only $(i', F'(R_{i'}))$ as input, or with $\mathcal{DB}$.

Let $\mathcal{S}_0$ be the database $\mathcal{DB}$. An EPIR protocol achieves database privacy, if there exists a simulator $\mathcal{S}_1$ such that any adversary $\mathcal{A} = (\mathcal{A}_1, \mathcal{A}_2)$, acting as a malicious user, has only a negligible advantage in the following game, where the advantage is defined as $|\Pr[b' = b] - \frac{1}{2}|$. For every retrieve query, $\mathcal{S}_1$ has an auxiliary input, one couple $(i', F'(R_{i'}))$ for some $1 \le i' \le N$ and $F' \in \mathcal{F}$, coming from a hypothetical oracle $\mathcal{O}$. This way, $\mathcal{S}_1$ simulates a database which possesses only the information of the requested value.

$$
\begin{aligned}
&\mathbf{Exp}_{\mathcal{A}}^{\text{database-privacy}} \\
&\left|
\begin{aligned}
b &\xleftarrow{R} \{0, 1\} \\
(R_1, R_2, \cdots, R_N) &\leftarrow \mathcal{A}_1(1^\ell) \\
b' &\leftarrow \mathcal{A}_2(\mathcal{S}_b; \mathsf{retrieve})
\end{aligned}
\right.
\end{aligned}
$$

Similarly as for SPIR, we consider that the hypothetical oracle $\mathcal{O}$ may have unlimited computing resources, so that $\mathcal{S}_1$ always learns exactly the input related to the request made by the adversary.

Remark 1. As discussed in next subsection, $(i', F'(R_{i'}))$ may give some information on $R_{i'}$ to the adversary (for instance if F' is chosen invertible): in practice additional means should be envisaged to reduce this effect.

2.3 Controlling the Set $\mathcal{F}$ of Authorized Functions

If we do not need to respect database privacy, evaluation of a function on a block of the database is feasible by running a PIR to recover R_i directly, and then to compute at the user side the value $F(R_i)$.

However, if we need database privacy and in particular if we want to disallow the option for someone to recover directly the value of a block, we cannot let the user choose among all the possible functions as the result will leak information on the block itself (which is clear if F is a polynomial of degree 1). Such awkward situation will not happen if the function F is fixed by another entity. Another possibility is to authorize only a family of functions to be evaluated on the database blocks. A trivial solution to this problem of restricting the functions is to use, when available, a kind of certification authority, to validate and sign the function. A more interesting idea to investigate is to rely on techniques on Conditional Disclosure of Secret (CDS) transformations [17,1] which are exactly proper for that purpose. With CDS the database would be able to output the result only if F is a member of the set of authorized functions. More works are needed to integrate these techniques into our proposal in order to obtain the description of a scheme combining the ideas of these papers and ours.

3 A New EPIR for Hamming Distance

We now describe an EPIR construction with the same functionality as the second one described in [3] but – as an improvement – without the expensive utilization of the BGN cryptosystem [2] which needs pairing computations in that case. Despite this difference, the remaining of the protocol is quite close to this one from [3] with a similar pre-processing step.

Note that this first new EPIR has to be read as an introductive example, following [3], where the evaluation is fixed (refer to Section 4 for a construction with more flexibility).

3.1 Description

For simplicity, we describe the protocol for classical Hamming distance. It is straightforward to adapt it to weighted distance as in [3]. With a chosen vector X, the function $F = F_X$ is defined as $R \mapsto d(X, R)$, where $d(X, R)$ is the Hamming distance between X and R.

For an ℓ_1-bit vector X, we denote the l-th bit of X as $X^{(l)}$. We assume that every block in $\mathcal{DB}$ has a bit length ℓ_1. Let $(\mathsf{Gen}, \mathsf{Enc}, \mathsf{Dec})$ be a semantically secure encryption scheme homomorphic with respect to addition. For instance, one can think at the additive version [7] of ElGamal encryption scheme [9]. As another example, one can have the Paillier encryption scheme [22].

Remark 2. In the following, the second parameter – i.e. the public key – of Enc is omitted. For instance, in the description of our EPIR protocol just below, $\mathsf{Enc}(m)$ stands for the encryption of message m under a given public key.

Moreover, one should note there is an abuse of notation regarding the equality between two ciphertexts: the underlying messages are equal but not the ciphertexts (the randomness is different each time a semantic encryption is made).

Finally, note that $(\mathsf{Gen}, \mathsf{Enc}, \mathsf{Dec})$ has not to be necessarily efficiently decryptable.

The EPIR protocol is as follows.

1. $\mathcal{U}$ generates via Gen key material and system parameters and sends corresponding public key to the database $\mathcal{DB}$.
2. To retrieve the distance $F_X(R_i)$ between X and R_i, for any $1 \le i \le N$ and $X \in \{0,1\}^{\ell_1}$, $\mathcal{U}$ first computes $C = \mathsf{Enc}(i)$ and for all $1 \le l \le \ell_1$, $D_l = \mathsf{Enc}(X^{(l)})$. He then sends C and the D_l to $\mathcal{DB}$ (with a zero knowledge proof that the $X^{(l)}$ are binary valued).
3. $\mathcal{DB}$ computes a private database of scalar products of X and every R_j (a similar step is done in [13]): the temporary version R'_j of the blocks for every $1 \le j \le N$ is obtained as follows.
   (a) For every $1 \le l \le \ell_1$, compute $D_l \times D_l^{-2R_j^{(l)}} \times \mathsf{Enc}\left(R_j^{(l)}\right)$, which leads to $\mathsf{Enc}\left(X^{(l)} \oplus R_j^{(l)}\right)$
   (b) then compute R'_j, where r'_j is randomly chosen and

$$R'_j = C^{r'_j} \times \mathsf{Enc}\left(-j\right)^{r'_j} \prod_{l=1}^{\ell_1} \mathsf{Enc}\left(X^{(l)} \oplus R_j^{(l)}\right) = \mathsf{Enc}\left(r'_j\left(i-j\right)\right) \mathsf{Enc}\left(d\left(X, R_j\right)\right)$$

   (Note that Step 3.(b) is done in order to randomize all elements of temporary database with an index different than i, see [1,17])
4. $\mathcal{U}$ and $\mathcal{DB}$ apply an arbitrary CPIR protocol to transfer the ith element of this temporary database to $\mathcal{U}$.
5. $\mathcal{U}$ decrypts the result to recover $d\left(X, R_i\right)$. (Note that if Paillier cryptosystem is in use, Step 5 is immediate. For the additive version of ElGamal, as long as ℓ_1 is not too big, $\mathcal{U}$ is able to retrieve the value of the Hamming distance $d(X, R_i)$, either by trying exponentiations g^t from $t = 1$ up to ℓ_1, or by comparing the value $g^{d(X,R_i)}$ with a pre-computed lookup table, where g is a generator of a cyclic subgroup $\mathbb{G}$, see Appendix A).

Thanks to the randomization in Step 3.(b), the PIR used in Step 4 does not need to achieve database privacy (i.e. to be a Symmetric PIR). Indeed thanks to the introduction of $r'_j \times (i - j)$, the decryption values of the R'_j for $j \neq i$ are uniformly and independently distributed, which enables a perfect simulation of the temporary database containing the R'_j only with the knowledge of R'_i.

With respect to the performances, the difference with the protocol using BGN encryption [2] is to avoid the use of pairing and to decrease slightly the number of exponentiations. The communication complexity is still dependent on that of the PIR protocol with the additional cost of the transmission of C and D_l $(1 \leq l \leq \ell_1)$.

Remark 3. Similar ideas have been presented in literature before as computing Hamming distance can be seen as a special case of the protocol of [19].

3.2   Security

The security analysis is only briefly overviewed here as it is a direct adaptation of the analysis done in [3] for the protocol using BGN encryption. If the PIR protocol is correct then the EPIR protocol for Hamming distance is correct. As to the user privacy, the data received and computed by $\mathcal{DB}$ are coming from or are ciphertexts, they do not bring any information on their content due to semantic security, thus the result below.

Lemma 1. *If the PIR protocol respects user privacy and if* (Gen, Enc, Dec) *is semantically secure, then the EPIR protocol for Hamming distance achieves user privacy.*

The database privacy is ensured by construction without needing to introduce a symmetric PIR protocol as all the R'_j are randomly distributed for $j \neq i$.

Lemma 2. *The EPIR protocol for Hamming distance achieves database privacy.*

4   EPIR Construction for Polynomial Evaluation

The previous protocol shows an improvement of an existing protocol [3] but does not bring new functionalities. Now, our motivation is twofold: 1. to construct new EPIR allowing the use of more generic function evaluation, 2. to introduce flexibility during the life of the database by letting the possibility for a user or an outsider to choose the function.

In this section we present a generic construction of an EPIR protocol which enables $\mathcal{U}$ to use any polynomial function of his choice and to recover its evaluation on a block from $\mathcal{DB}$.

4.1   Preamble

The two protocols of [3], either for equality testing or for weighted Hamming distance computation, are based in fact on a pre-processing stage. This is the same for our first protocol in Section 3.1 as well. The idea is to send a first message computed by the user thanks to the data he wants to compared the requested block with. Then the database uses this message to generate a tempo-rary database which contains encrypted results of comparisons of the data with each block. Thereafter, the user is able to run a PIR to get back only the result corresponding to the queried block.

A first idea would be to use the same trick again: Given a function F to be evaluated on the block R_i, we send F to the database which generates the temporary database $(F(R_1), \ldots, F(R_N))$ and the user executes a PIR query to recover $F(R_i)$. Although in [3] it is possible to do it via the transmission of a kind of encrypted version of F (for F corresponding either to the equality check or the distance computation) to avoid the database to learn the function F. It seems quite difficult to achieve the same for any functions. Moreover, the pre-processing step of computing the temporary database is one of the most costly steps of these constructions.

An improvement would be to encrypt F in a way which let the capability to evaluate it directly in the encrypted domain. One can think to send the encrypted version of the coefficients of F and to use homomorphic properties to obtain an encrypted evaluation of F but here we go further by embedding directly an encrypted description of the function in the query which is sent to $\mathcal{DB}$. This enables us to avoid disclosure of F and pre-computation of a temporary database.

4.2   Parameters

The construction is based on the ElGamal public key cryptographic scheme [9] (cf. Appendix A) over the multiplicative group of an extension field. Let p be a prime number and $K = \mathrm{GF}(p)$ be the Galois field with p elements, $L = \mathrm{GF}(p^n)$ be an extension of degree n of K and $\mathbb{G} = L^\times$ its multiplicative group with a generator g of order $q = p^n - 1$. The particular instance below of ElGamal cryptosystem is related to L. Let $(\mathsf{Gen}, \mathsf{Enc}, \mathsf{Dec})$ be as follows.

The key generation algorithm Gen takes a security parameter 1^k as input and generates the private key x which is randomly chosen from $\mathbb{Z}_q$, and the public key is $y = g^x$. The encryption algorithm Enc and the decryption algorithm Dec are defined as usual (cf. Appendix A).

Now, we assume that all blocks of $\mathcal{DB}$ are in $L^\times$ and that F is a polynomial function over $L = \mathrm{GF}(p^n)$. Let α be a primitive element of L/K, i.e. such that $L = K(\alpha)$. We know that there exist polynomials $G, Y \in \mathrm{GF}(p)[X]$ of degree at most $n-1$ such that $g = G(\alpha)$ and $y = Y(\alpha)$ as $(1, \alpha, \alpha^2, \ldots, \alpha^{n-1})$ is a basis of L over K. So doing, an encryption of a message m is seen as a couple of polynomials in α. We exploit this polynomial representation in our protocol. In the sequel, we assume that the database contains blocks $R_1, \ldots, R_N$, such that for all $1 \leq j \leq N$, $Y(R_j) = G(R_j)^x$ and $G(R_j)$ is invertible (and consequently $Y(R_j)$ as well).

Remark 4. Instead of the ElGamal encryption scheme, other homomorphic en-cryption schemes are imaginable in our protocol as soon they are defined on a cyclic subgroup of multiplicative group of a finite field.

4.3   Description of the Protocol

Our new EPIR protocol is described below. We first explain a communication linear scheme and discuss after the way to improve the communication complexity.

1. $\mathcal{U}$ generates an ElGamal key pair (pk, sk), where $pk = (q, g, y)$, $y = g^x$, and $sk = x$ is randomly chosen from $\mathbb{Z}_q$. $\mathcal{U}$ also sends pk to let $\mathcal{DB}$ the possibility to verify the validity of pk as an ElGamal public key. In practice, the validity of pk can be certified by a TTP, and the same pk can be used by the user for all his queries.

2. For any polynomial function $F : \mathrm{GF}(p^n) \to \mathrm{GF}(p^n)$ and any index $1 \le i \le N$, $\mathcal{U}$ computes $C_1, \ldots, C_N$ and sends them to $\mathcal{DB}$ where
   - $C_i = \mathsf{Enc}(F(\alpha) + r) = (G(\alpha)^{r_i}, Y(\alpha)^{r_i}(F(\alpha) + r))$
   - and $C_j = \mathsf{Enc}(1) = (G(\alpha)^{r_j}, Y(\alpha)^{r_j})$ for all $j \ne i$,

   with randomly chosen $r \in \mathrm{GF}(p)$, $r_j \in \mathbb{Z}_q$ $(1 \le j \le N)$. Each C_j can be written as $C_j = (V_j(\alpha), W_j(\alpha))$ where V_j and W_j are polynomial over $\mathrm{GF}(p)$ of degree at most $n - 1$.

3. After reception of the C_j, the database $\mathcal{DB}$ checks that they are nontrivial ElGamal ciphertexts and computes $C_j(R_j) = (V_j(R_j), W_j(R_j))$ by replacing each occurrence of α (resp. α^l for all power $l < n$) with R_j (resp. with R_j^l).

4. Then $\mathcal{DB}$ performs the product of all the C_j together with a random encryption of 1, $\mathsf{Enc}(1) = (g^{r'}, y^{r'})$, to obtain

$$\mathsf{Enc}(1) \times \prod_{j=1}^{N} C_j(R_j) = \left(g^{r'} \prod_{j=1}^{N} G(R_j)^{r_j}, y^{r'} \left(\prod_{j=1}^{N} Y(R_j)^{r_j}\right)(F(R_i) + r)\right).$$

5. $\mathcal{DB}$ finally sends the result to $\mathcal{U}$ which is able to decrypt via $\mathsf{Dec}(., sk)$ in order to recover the value $F(R_i)$.

The reason why the last step works as expected is that $\prod Y(R_j)^{r_j} = Y'$ is equal to $(\prod G(R_j)^{r_j})^x = G'^x$, i.e. x is the private key associated to G', Y'. In our case, as we do not use any encoding algorithm for ElGamal encryption, it is clear that the semantic security will remain only if $F + r$ does not vanish to 0 either on α or R_i, which has a negligible probability to happen. We see that the goal of r is to mask F to avoid any such deterministic behavior (for instance against a function F chosen by a malicious database ; for an honest database, this mask can be skipped when F is appropriately designed). Note that with our solution, here, even the degree of F remains hidden from $\mathcal{DB}$.

It is straightforward to see that our scheme is based on an adaptation of the generic construction of a PIR (via group-homomorphic encryption cf. Appendix B) where we embed the value of F. Hence concerning the performances, the communication complexity is still linear. And like any computational PIR, the number of computations is mostly dependent of the database operations which is linear in N.

To decrease the communication overhead, we could follow the same technique as the one explained in [21, sec. 2.2]. It considers the database as a d-dimensional cube and roughly converts the linear query into a $O\left(N^{\frac{1}{d}}\right)$ query by applying an iterative method as in Lipmaa protocol [18]. The query is made of small queries allowing the database to isolate iteratively rows of dimension smaller and smaller.

Remark 5. If communication complexity needs to be reduced further, we can choose a more efficient PIR algorithm if we apply the pre-processing technique: $\mathcal{U}$ computes and sends $C = \mathsf{Enc}(g^i)$ and $D = \mathsf{Enc}\left(F\left(\alpha\right)\right)$, then $\mathcal{DB}$ computes temporary blocks R'_j for $1 \leq j \leq N$ as

$$R'_j = \left(C/\mathsf{Enc}\left(g^j\right)\right)^{r_j} \times \mathsf{Enc}\left(F\left(R_j\right)+r\right) = \mathsf{Enc}\left(g^{r_j(i-j)}\right) \times \mathsf{Enc}\left(F\left(R_j\right)+r\right)$$

Finally $\mathcal{U}$ executes a PIR query to recover R'_i and then decrypts it to learn $F\left(R_i\right)$.

In that case, one can also envisage a solution with the Paillier cryptosystem where $\mathcal{U}$ sends the encrypted coefficients of F instead of F directly. However in both situations, this increases the computation cost due to the generation of a temporary database with blocks $R'_1, \ldots, R'_N$ on which the PIR protocol is applied.

4.4   Security

A query gives the expected result as soon as there is no index j for which one of the values $G(R_j)$ or $Y(R_j)$ is zero, which may occur only with a negligible probability in practice, leading to the correctness of the EPIR protocol.

The construction is based on ElGamal encryption, which is semantically secure. In addition a random mask is used to avoid that a zero value of F helps to distinguish a challenge. Except with a negligible probability, all the values $G(R_j), Y(R_j)$ for all j and $F(R_i) + r$ remains in $\mathrm{GF}(p^n)^\times$. This enables to achieve user privacy against a malicious database.

Lemma 3. *If the ElGamal scheme is semantically secure, i.e. when the DDH assumption holds, the EPIR protocol for polynomial evaluation achieves user privacy.*

The generic design of PIR on which is based our EPIR protocol is symmetric thanks to the ElGamal encryption of the single returned value. The decryption will give information only on the result and as the $G(R_j)$ are all invertible, it is not possible to distinguish from the ciphertext between the database and the simulator which knows only the answer of the query.

Lemma 4. *The EPIR protocol for polynomial evaluation achieves database privacy.*

The proofs are reported in Appendix C.

4.5   Applications and Extensions

A great advantage of this technique is the freedom given to $\mathcal{U}$ to choose the function F while $\mathcal{DB}$ will not learn information on this choice. This shows clearly an improvement upon the notion and the related constructions introduced in [3]. Still, the way to restrict the choice of the function in the set of authorized ones $\mathcal{F}$ is an open problem, as mentioned in Section 2.3.

Several interesting applications can be handled with our protocol. For instance, when $\mathcal{U}$ is willing to check if one R_i is member of a set of elements $S = \{d_1, \ldots, d_l\} \subset \mathrm{GF}(p^n)^\times$ only known by $\mathcal{U}$, then he takes $F(X) = r + r' \prod_{k=1}^{l}(X - d_k)$ with r, r' randomly chosen into $\mathrm{GF}(p^n)^\times$. Thereafter he executes the EPIR query toward the database to recover $F(R_i)$ and if the value is r, then he learns that R_i is a member of S. If S contains only one element, this yields directly to an equality check thus it can be seen as a generalization of the construction from [3].

Closer to general SPFE, another interesting application is the evaluation of multivariate polynomials with separated variables. To evaluate

$$F(X_1, \ldots, X_k) = F_1(X_1) \cdots F_k(X_k)$$

on $(R_{i_1}, \ldots, R_{i_k})$, $\mathcal{U}$ will send the following C_j: $C_i = \mathsf{Enc}(F_i(\alpha))$ for all $i \in \{i_1, \ldots, i_k\}$ and $C_j = \mathsf{Enc}(1)$ otherwise. For instance, this can be exploited for AND checks: to verify that for all i_t, R_{i_t} is in a given set S_{i_t}, we will construct F_{i_t} as $r_t + r'_t \prod_{d \in S_{i_t}}(X - d)$. After reception of the answer to his query, then the user will check if the decrypted result is equal to $\prod_{t=1}^{k} r_t$ to determine whether the assertion is true.

Remark 6. Due to the lack of space, we cannot include more practical applications such as the use of EPIR for Hamming distance in biometric system [3].

5   Batching

Another interesting question regarding EPIR protocols is whether we can amortize database workload. In our generic construction for polynomial evaluation of Section 4, in the construction for Hammming distance Section 3 and in the examples from [3], the retrieval algorithm is based on or is an adaption of PIR algorithm, so we can try to apply methods which has been studied for PIR.

There are two different situations, either the same user is making different queries to the database, or multiple users are involved. The first setting is analyzed in [15] for PIR where the user wants to retrieve k several blocks at the same time. A solution in the second setting can be found in [16].

We assume here that there is only one user accessing the database at a time. The goal is to decrease the database computations below the linear complexity $O(k \times N)$ (which is not possible for $k = 1$ with the user privacy constraint) corresponding to the answers to k separate PIR queries. The communication complexity is also a factor taken in account. The idea appearing in [15] is the use of batch codes which are coming from a generalization of the following: pick a random hash function from $\{1, \ldots, N\}$ to $\{1, \ldots, k\}$, send h to the database to split the database in k buckets of about N/k blocks and run a number of PIR queries toward each bucket. It is shown that for an appropriate choice of h, the user will get all k blocks except for a probability $2^{-\Omega(\sigma)}$ with at most $k \times \sigma \log k$ PIR queries. By counting the length of the buckets, this leads to at most $n \times \sigma \log k$ operations for the database to compute the different answers. Batch codes are in fact a way to find an optimal splitting of a dataset into m buckets such that you need to read at most t elements of each buckets to recover any subset of k elements of your dataset. The general idea is thus to split efficiently the database by a specific encoding procedure in small buckets in order to run on each bucket a small number of PIR queries. When k is large, [15] achieves a computational complexity about $n^{1+o(1)}$ for the database to answer to the k queries.

The first technique with hash-splitting is directly compatible with EPIR. EPIR queries can be made accordingly on the small buckets. However, when in particular a specific pre-processing is needed to answer the EPIR query as our construction in Section 3, we can wonder whether it is possible to increase further the efficiency on each small bucket.

The second solution, due to the encoding with the batch code is applicable to EPIR only when the k queries are made with the same function F, which is not always the case, thus introducing a specific difficulty for batching EPIR. For instance into [3] for the applications of EPIR to biometric authentication, new query corresponds to a new biometric template captured so far. Hence, the function F evolves with each capture.

In our generic protocol, as no pre-processing is needed, the batching technique via hash-splitting is already a fair solution by itself. So we consider below only the case of algorithms with pre-processing.

A generic batch procedure for EPIR which pre-processes the database would be to generate k parallel temporary databases, one for each request, and then using an adapted version of a batch PIR request to recover the k expected values. However, the adaption seems not trivial and one has to take in account that the pre-processing step is linear in k, as the generation of the k temporary databases costs $k \times N$ operations.

We now introduce an improvement on two examples by managing the pre-processing step in order to reduce further the number of PIR requests.

5.1   Batching an EPIR for Hamming Distance Computation

We place ourselves in the case where (Gen, Enc, Dec) correspond to the additive version of the ElGamal cryptosystem. At the end of our protocol for Hamming distance computation (see Section 3.1), the user $\mathcal{U}$ retrieves an exponentiation of the expected distance. For common settings, ℓ_1 could be relatively small, hence we can try to encode several distance values in the same exponent.

Let $\ell_2 = \lceil \log_2(\ell_1 + 1) \rceil$ and assume that we want to encode k different distances $d(X_1, R_{i_1}), \ldots, d(X_k, R_{i_k})$ as an exponent, then we consider

$$g' = g^{d(X_1, R_{i_1}) + 2^{\ell_2} d(X_2, R_{i_2}) + \cdots + 2^{(k-1)\ell_2} d(X_k, R_{i_k})}.$$

If $2^{k\ell_2}$ stays below a reasonable size T (i.e. $2^k \leq T^{1/\ell_2}$), with T lower than the computational capacity of $\mathcal{U}$ and lower than the order of g (i.e. q), then $\mathcal{U}$ is still able to retrieve the k distance values from g' by performing a search in basis g up to the exponent T (which takes expected time $O(\sqrt{T})$ with the Pollard's lambda method).

To this end, we repeat the Steps 2 and 3 for each request with X_t and index i_t, so that we obtain k temporary databases with blocks $(R(1)_1, \ldots, R(1)_N), \ldots,$ $(R(k)_1, \ldots, R(k)_N)$ where $R(t)_j = \mathsf{Enc}\,(r_{t,j}\,(i_t - j))\,\mathsf{Enc}\,(d\,(X_t, R_j))$. $\mathcal{U}$ transmits also k randomly chosen permutations $\sigma_1, \ldots, \sigma_k$ of $\{1, \ldots, N\}$ such that there exists a constant i with $\sigma_t(i_t) = i$ for all $1 \leq t \leq k$. When received, $\mathcal{DB}$ computes $R'_j = \prod_{t=1}^k R(t)_{\sigma_t(j)}^{2^{(t-1)\ell_2}}$. This leads to the encryption of a random element from $\mathbb{G}$ when $j \neq i$ and to the value $R'_i = \prod_{t=1}^k \mathsf{Enc}\,\left(2^{(t-1)\ell_2} d\,(X_t, R_{i_t})\right)$ for $j = i$, which is the expected result. Finally, $\mathcal{U}$ performs one single PIR query to recover the block of index i before proceeding to its decryption and to the recovery of each distance $d(X_t, R_{i_t})$. Note that the knowledge of the permutations $\sigma_1, \ldots, \sigma_N$ does not bring information to the database, under the assumption that the k indexes of the requests are independent.

As to the performance, this technique is only available for small k as we impose $k \leq (\log_2 T)/\ell_2$ (e.g. $k \leq 10$ for $\ell_1 = 15$ and $T = 2^{40}$). For a larger number $k' = \alpha.k$ of EPIR requests, this can be seen as a first step: first compute α intermediate databases with k distances each and then execute a batched PIR query for α queries.

Note that the limitation with respect to T does not hold with Paillier encryption (which is efficiently decryptable) so that it becomes more efficient than with ElGamal: $k \leq (\log_2 n)/\ell_2$ with n the order of the message space (e.g. $k \leq 512$ for $\ell_1 = 15$ and $\log_2 n = 2048$).

5.2   Batching an EPIR for Equality Check

The same technique is applicable to the first protocol of [3] but with a lower computation cost as it is possible to generate more directly the temporary database on which the PIR query is performed. That protocol, which is conceived for equality check, has a similar structure as the protocol of Section 3.1. The Step 1 remains unchanged. Steps 2 to 5 become:

Step 2. $\mathcal{U}$ sends an ElGamal ciphertext $C = \left(g^r, y^r g^{i\|X}\right) = \mathsf{Enc}(g^{i\|X})$ to $\mathcal{DB}$.

Step 3. After reception of C, $\mathcal{DB}$ computes the temporary blocks R'_j as

$$R'_j = C^{r_j} \times \left(g^{r'_j}, y^{r'_j}\left(g^{j\|R_j}\right)^{-r_j}\right) = \left(g^{r'_j}(g^r)^{r_j}, y^{r'_j}\left(y^r g^{i\|X}\left(g^{j\|R_j}\right)^{-1}\right)^{r_j}\right).$$

Steps 4.&5. $\mathcal{U}$ runs a PIR query to retrieve R'_i, decrypts it and determines the equality between X and R_i by comparing the result with the value $g^0 = 1$.

To batch these steps for k queries, we suggest to introduce a hash function h with output of length $2^{\ell_2} - 1$ and to check the equalities all on the hashed domain in order to reduce the length of the data involved. Then we applied the same technique as in Section 5.1.

$\mathcal{U}$ sends $C = \left(g^r, y^r g^{\sum_{t=1}^{k} 2^{(t-1)\ell_2} h(X_t)} \right)$ and $D = \mathsf{Enc}(g^i)$ together with the permutations $\sigma_1, \ldots, \sigma_k$ as defined previously to $\mathcal{DB}$ and $\mathcal{DB}$ computes the temporary blocks R'_j as $R'_j = \left(D/\mathsf{Enc}\left(g^j \right) \right)^{r_j} \times C/\mathsf{Enc}\left(g^{\sum_{t=1}^{k} 2^{(t-1)\ell_2} h\left(R_{\sigma_t(j)} \right)} \right)$. At the end, when receiving the answer to his PIR query, $\mathcal{U}$ determines equalities or not within each couple (X_t, R_{i_t}) by retrieving the discrete logarithm of the plaintext. Here again, the number k of simultaneous queries is quite small when we want to check all the inequalities separately. Otherwise, if we want to check whether all equalities hold at the same time, then we can increase k as we are only interested to determine whether the plaintext equals to $g^0 = 1$. Adaptation to other homomorphic encryption schemes is also possible, e.g. replacing additive ElGamal with Paillier enables a more efficient decryption (as in Section 5.1).

6   Conclusion

Our work extends the notion and constructions of EPIR to new perspectives. The way to restrict the choice of functions to a set of authorized functions is left as an open problem. As discussed in Section 2.3, the link with Conditional Disclosure of Secret to check the validity of the request is an interesting direction for further improvements and toward clear distinction with notion of PIR.

Acknowledgments. The authors thank David Pointcheval, Qiang Tang and the reviewers for their thorough comments.

References

1. Aiello, W., Ishai, Y., Reingold, O.: Priced oblivious transfer: How to sell digital goods. In: Pfitzmann, B. (ed.) EUROCRYPT 2001. LNCS, vol. 2045, pp. 119–135. Springer, Heidelberg (2001)
2. Boneh, D., Goh, E.-J., Nissim, K.: Evaluating 2-DNF formulas on ciphertexts. In: Kilian, J. (ed.) TCC 2005. LNCS, vol. 3378, pp. 325–341. Springer, Heidelberg (2005)
3. Bringer, J., Chabanne, H., Pointcheval, D., Tang, Q.: Extended private information retrieval and its application in biometrics authentications. In: Bao, F., Ling, S., Okamoto, T., Wang, H., Xing, C. (eds.) CANS 2007. LNCS, vol. 4856, pp. 175–193. Springer, Heidelberg (2007)
4. Canetti, R., Ishai, Y., Kumar, R., Reiter, M.K., Rubinfeld, R., Wright, R.N.: Selective private function evaluation with applications to private statistics. In: PODC 2001: Proceedings of the twentieth annual ACM symposium on Principles of distributed computing, pp. 293–304. ACM Press, New York (2001)
5. Chor, B., Goldreich, O., Kushilevitz, E., Sudan, M.: Private information retrieval. In: FOCS, pp. 41–50 (1995)

6. Chor, B., Kushilevitz, E., Goldreich, O., Sudan, M.: Private information retrieval. J. ACM 45(6), 965–981 (1998)
7. Cramer, R., Gennaro, R., Schoenmakers, B.: A secure and optimally efficient multi-authority election scheme. In: Fumy, W. (ed.) EUROCRYPT 1997. LNCS, vol. 1233, pp. 103–118. Springer, Heidelberg (1997)
8. Crescenzo, G.D., Malkin, T., Ostrovsky, R.: Single database private information retrieval implies oblivious transfer. In: Preneel, B. (ed.) EUROCRYPT 2000. LNCS, vol. 1807, pp. 122–138. Springer, Heidelberg (2000)
9. Gamal, T.E.: A public key cryptosystem and a signature scheme based on discrete logarithms. In: Blakley, G.R., Chaum, D. (eds.) CRYPTO 1984. LNCS, vol. 196, pp. 10–18. Springer, Heidelberg (1985)
10. Gasarch, W.: A survey on private information retrieval, http://www.cs.umd.edu/~gasarch/pir/pir.html
11. Gentry, C., Ramzan, Z.: Single-database private information retrieval with constant communication rate. In: Caires, L., Italiano, G.F., Monteiro, L., Palamidessi, C., Yung, M. (eds.) ICALP 2005. LNCS, vol. 3580, pp. 803–815. Springer, Heidelberg (2005)
12. Gertner, Y., Ishai, Y., Kushilevitz, E., Malkin, T.: Protecting data privacy in private information retrieval schemes. In: STOC, pp. 151–160 (1998)
13. Goethals, B., Laur, S., Lipmaa, H., Mielikäinen, T.: On private scalar product computation for privacy-preserving data mining. In: Park, C., Chee, S. (eds.) ICISC 2004. LNCS, vol. 3506, pp. 104–120. Springer, Heidelberg (2005)
14. Goldwasser, S., Micali, S.: Probabilistic encryption and how to play mental poker keeping secret all partial information. In: STOC, pp. 365–377. ACM, New York (1982)
15. Ishai, Y., Kushilevitz, E., Ostrovsky, R., Sahai, A.: Batch codes and their applications. In: Babai, L. (ed.) STOC, pp. 262–271. ACM, New York (2004)
16. Ishai, Y., Kushilevitz, E., Ostrovsky, R., Sahai, A.: Cryptography from anonymity. In: FOCS, pp. 239–248. IEEE Computer Society, Los Alamitos (2006)
17. Laur, S., Lipmaa, H.: A new protocol for conditional disclosure of secrets and its applications. In: Katz, J., Yung, M. (eds.) ACNS 2007. LNCS, vol. 4521, pp. 207–225. Springer, Heidelberg (2007)
18. Lipmaa, H.: An oblivious transfer protocol with log-squared communication. In: Zhou, J., López, J., Deng, R.H., Bao, F. (eds.) ISC 2005. LNCS, vol. 3650, pp. 314–328. Springer, Heidelberg (2005)
19. Lipmaa, H.: Private branching programs: On communication-efficient cryptocomputing. Cryptology ePrint Archive, Report 2008/107 (2008), http://eprint.iacr.org/
20. Naor, M., Pinkas, B.: Oblivious polynomial evaluation. SIAM J. Comput. 35(5), 1254–1281 (2006)
21. Ostrovsky, R., Skeith III, W.E.: A survey of single-database private information retrieval: Techniques and applications. In: Okamoto, T., Wang, X. (eds.) PKC 2007. LNCS, vol. 4450, pp. 393–411. Springer, Heidelberg (2007)
22. Paillier, P.: Public-key cryptosystems based on composite degree residuosity classes. In: Stern, J. (ed.) EUROCRYPT 1999. LNCS, vol. 1592, pp. 223–238. Springer, Heidelberg (1999)
23. Stern, J.P.: A new efficient all-or-nothing disclosure of secrets protocol. In: Ohta, K., Pei, D. (eds.) ASIACRYPT 1998. LNCS, vol. 1514, pp. 357–371. Springer, Heidelberg (1998)

24. Wu, Q., Qin, B., Wang, C., Chen, X., Wang, Y.: -out-of- string/bit oblivious transfers revisited. In: Deng, R.H., Bao, F., Pang, H., Zhou, J. (eds.) ISPEC 2005. LNCS,
vol. 3439, pp. 410–421. Springer, Heidelberg (2005)

A   ElGamal Public Key Cryptosystem

Let $(\mathsf{Gen}, \mathsf{Enc}, \mathsf{Dec})$ be an ElGamal scheme [9]. It is defined as follows.

1. The key generation algorithm Gen takes a security parameter 1^k as input and
   generates p, q two primes with $q | p - 1$, g a generator of the cyclic subgroup
   $\mathbb{G}$ of order q in $\mathbb{Z}_p^*$, a public key $y = g^x$ with the private key x randomly
   chosen from $\mathbb{Z}_q$.
2. The encryption algorithm Enc takes a message m in $\mathbb{G}$ and the public key
   y as input, and outputs the ciphertext $c = (c_1, c_2) = (g^r, y^r m)$ where r is
   randomly chosen from $\mathbb{Z}_q^*$.
3. The decryption algorithm Dec takes a ciphertext $c = (c_1, c_2)$ and the private
   key x as input, and outputs the message $m = (c_1^{-x} c_2)$.

The ElGamal scheme is semantically secure based on the Decisional Diffie-
Hellman (DDH) assumption.

B   Generic PIR Construction from Homomorphic Encryption

As described in [21], with the exploitation of any semantically secure (i.e. IND-
CPA, cf. [14]) group-homomorphic public key encryption scheme, the following
design yields to a secure PIR protocol. The use of homomorphic encryption for
PIR was first used by [23]. We can also note that there is no essential difference
with the generic construction (by removing the knowledge proof for database
privacy) of [24].

Let $(\mathsf{Gen}, \mathsf{Enc}, \mathsf{Dec})$ be an asymmetric cryptosystem with a group G as plaintexts space and a group G' as ciphertexts space. Assume that this cryptosystem
is homomorphic from G to G', i.e. that for any $a, b \in G$,

$$\mathsf{Dec}\left(\mathsf{Enc}\left(a, pk\right) \times_{G'} \mathsf{Enc}\left(b, pk\right), sk\right) = a \times_G b$$

where $\times_G$ (resp. $\times_{G'}$) represents the group law in G (resp. G') and (pk, sk) is a
couple of public and private keys generated thanks to Gen.

If a user $\mathcal{U}$ wants to retrieve a block of index i from a database $\mathcal{DB}$ containing
N blocks $R_1, \dots, R_N$ (with $R_1, \dots, R_N$ integers).

1. He generates a key pair (pk, sk) and sends to $\mathcal{DB}$ a query containing pk and
   encrypted values $C_1 = \mathsf{Enc}\left(\delta_{1,i}, pk\right), \dots, C_N = \mathsf{Enc}\left(\delta_{N,i}, pk\right)$ where for all
   $1 \le j \le N$,
   - if $j \ne i$, then $\delta_{j,i}$ is 1_G the identity element of G, i.e. $C_j = \mathsf{Enc}(1_G, pk)$;
   - if $j = i$, then $\delta_{i,i}$ is an element $g \ne 1_G$.

2. After reception, the database computes

$$R = R_1.C_1 \times_{G'} \cdots \times_{G'} R_N.C_N$$

where . denotes the $\mathbb{Z}$-module action (e.g. $R_1.C_1 = R_1 \times C_1$, if $\times_{G'}$ is an additive law). Then $\mathcal{DB}$ sends the result to $\mathcal{U}$.

3. By homomorphism,

$$R = \mathsf{Enc}\,(R_1.\delta_{1,i} \times_G \cdots \times_G R_N.\delta_{N,i}) = \mathsf{Enc}(R_i.g).$$

Note that the simplification above comes from the relation $x.1_G = 1_G$. Thus, $\mathcal{U}$ obtains by decryption the value $R_i.g$

If the law is additive, than by dividing by g it gives the expected result (R_i). If the law is multiplicative, one can obtain R_i from g^{R_i} either if the discrete logarithm problem is easy or if R_i is small.

This protocol is clearly correct and the user privacy comes from the semantically secure encryption of the C_j which do not give any information to $\mathcal{DB}$ which may help to distinguish queries.

In [21], different ways to improve this basic construction are discussed. This has be applied to various known homomorphic cryptosystems, like Goldwasser-Micali scheme [14] (homomorphic from $(\mathbb{Z}_2, \oplus)$ to $(\mathbb{Z}_n^\times, \times)$), Paillier cryptosystem [22] (from $(\mathbb{Z}_n, +)$ to $(\mathbb{Z}_{n^2}^\times, \times)$), ElGamal encryption scheme [9] (from $(G, \times)$ to $(G^2, \times)$ for a cyclic group G), ...

C Security Proofs

This section contains the proofs of Lemma 3 and Lemma 4 of Section 4.4.

Lemma 3. *If the ElGamal scheme is semantically secure, the EPIR protocol for polynomial evaluation achieves user privacy.*

Proof. Assume that the scheme does not achieve user privacy, i.e. that an adversary $\mathcal{A} = (\mathcal{A}_1, \mathcal{A}_2, \mathcal{A}_3, \mathcal{A}_4)$ has a non-negligible advantage in the game described Section 2. We construct an adversary $\mathcal{A}'$ against the semantic security of ElGamal.

1. After generation of (pk, sk) by Gen, $\mathcal{A}'$ receives pk and uses $\mathcal{A}_1$ to generate a database.
2. Then $\mathcal{A}'$ runs $\mathcal{A}_2$ to receive $\{i_0, i_1, F_0, F_1\}$ after a number of responses to retrieve queries.
3. $\mathcal{A}'$ forwards the messages (m_0, m_1) to the ElGamal IND-CPA challenger with $m_0 = F_0(\alpha)$ (which is non-zero with an overwhelming probability) and $m_1 = 1$ and obtains a challenge $c = \mathsf{Enc}(m_b)$ for a coin toss b.
4. $\mathcal{A}'$ defines $C_{i_e} = c$ together with $C_j = \mathsf{Enc}(1)$ for $j \neq i_e$ for the choice $e = 0$. He sends $C_1, \ldots, C_N$ to the adversary $\mathcal{A}_3$.
5. Finally, $\mathcal{A}_4$, after several responses to queries, outputs his guess e' and $\mathcal{A}$ outputs $b' = e'$.

Let E be the event that $e = b$ in the game; obviously $\Pr[E] = \frac{1}{2}$. If $e \neq b$, then C_{i_e} is an encryption of 1 as all the C_j, so the advantage of $\mathcal{A}_4$ is negligible (this implies $|\Pr[e' \neq e|\neg E] - \frac{1}{2}| = \epsilon$). Otherwise, if $e = b$, then $\mathcal{A}_4$ will output e with a non-negligible advantage $Adv = |\Pr[e' = e|E] - \frac{1}{2}|$. We have $\Pr[b = b'] = \Pr[E]\Pr[e' = e|E] + \Pr[\neg E]\Pr[e' \neq e|\neg E]$, leading to

$$\left|\Pr[b = b'] - \frac{1}{2}\right| = \frac{1}{2}\left|\left(\Pr[e' = e|E] - \frac{1}{2}\right) + \left(\Pr[e' \neq e|\neg E] - \frac{1}{2}\right)\right|,$$

so that $|\Pr[b = b'] - \frac{1}{2}| \geq \frac{1}{2}(Adv - \epsilon)$, which contradicts semantic security of El-Gamal (as the time complexity of the Steps 1 to 5 remains polynomial, including the running of $\mathcal{A}$ and the time for the challenger to answer the requests). $\square$

Lemma 4. *The EPIR protocol for polynomial evaluation achieves database privacy.*

Proof. We briefly explain here how works the simulator $\mathcal{S}_1$ based on the auxiliary input $(i', F'(R_{i'}))$ from the hypothetical oracle $\mathcal{O}$ (recall that via $\mathcal{O}$, $\mathcal{S}_1$ is assumed to known the good input, i.e. that corresponding exactly to the query of the adversary). After reception of pk and a query $C_1, \ldots, C_N$, the simulator merely computes $\mathsf{Enc}(F'(R_{i'}))$ and sends the result to the adversary which cannot make the difference with the encryption sent by the legitimate database, thanks to the use of $\mathsf{Enc}(1)$ in Step 5 to randomize the response. $\square$

Constructing Universally Composable Oblivious Transfers from Double Trap-Door Encryptions

Huafei Zhu and Feng Bao

C&S, I^{2}R, A-star, Singapore

Abstract. In this paper, a new implementation of universally composable, 1-out-of-2 oblivious transfer in the presence of static adversaries is presented and analyzed. Our scheme is constructed from the state-of-the-art Bresson-Catalano-Pointcheval's double trap-door public-key encryption scheme, where a trapdoor string comprises a master key and local keys. The idea behind our implementation is that the master key is used to extract input messages of a corrupted sender (as a result, a simulator designated for the corrupted sender is constructed) while the local keys are used to extract input messages of a corrupted receiver (as a result, a simulator designated for the corrupted receiver is defined). We show that the proposed oblivious transfer protocol realizes universally composable security in the presence of static adversaries in the common reference model assuming that the decisional Diffie-Hellman problem over a squared composite modulus of the form $N = pq$ is hard.

Keywords: Oblivious transfer, simulator, universally composable.

1 Introduction

Oblivious transfer (OT) introduced by Rabin [20], and generalized by Even, Goldreich and Lempel [10] and Brassard, Crépeau and Robert [1] is one of the most basic and widely used protocol primitives in cryptography. OT stands at the center of the fundamental results on secure two-party and multi-party computation showing that any efficient functionality can be securely computed ([22] and [11]). Due to its general importance, the task of constructing efficient oblivious transfer protocols has attracted much interest, see e.g., [9], [13] and [15].

OT constructions in the semi-simulation model: Naor and Pinkas ([15], [16] and [17]) have constructed the first efficient oblivious transfer protocols based on the decisional Diffie-Hellman assumption. Tauman [21] generalized Naor and Pinkas' results based on a variety of concrete assumptions building on the top of projective hash framework of Cramer and Shoup [8]. The primary drawback of these constructions is that their security is only proven according to a semi-simulation definition of security. Namely, a receiver security is defined by requiring that a sender's view of the protocol when the receiver chooses an index σ_0 is indistinguishable from a view of the protocol when the receiver chooses an

B. Preneel (Ed.): AFRICACRYPT 2009, LNCS 5580, pp. 323–333, 2009.

index σ_1. A sender security follows the real/ideal world paradigm and guarantees that any malicious receiver in the real world can be mapped to a receiver in an idealized game in which the oblivious transfer protocol is implemented by a trusted third party.

OT constructions in the full-simulation model: Very recently, Camenisch, Neven and Shelat [3] proposed oblivious transfer protocols that are provable secure according to a full-simulation definition where the security employs the real/ideal world paradigm for both sender and receiver. The difficulty in obtaining secure oblivious transfer protocols in this model is the strict security requirement of the simulation-based definition. Subsequently, Green and Hohenberger [12] proposed fully simulatable oblivious transfer protocols based on a weaker set of static assumptions on bilinear groups.

Lindell [14] presented an interesting implementation of fully-simulatable oblivious bit transfer protocols under the decisional Diffie-Hellman problem, the Nth residuosity and the quadratic residuosity assumptions as well as the assumption that the homomorphic encryptions exist. All protocols are nice since they are provably secure in the presence of malicious adversaries under the real/ideal model simulation paradigm without using the general zero-knowledge proofs under the standard complexity assumptions. Lindell's construction makes use of the cut-and-choose technique avoiding the need for zero-knowledge proofs, and allowing a simulator to rewind the malicious party so that an expected polynomial time simulator under the standard cryptographic primitives can be defined. All these schemes are not known to be secure when composed in parallel and concurrently.

OT constructions in the universally composable model: At Crypto'08, Peikert, Vaikuntanathan and Waters [19] proposed a practical implementation of universally composable secure oblivious transfer protocols. Their protocols are based on a new notion called dual-mode cryptosystem. Such a system starts with a setup phase that produces a common reference string, which is made available to all parties. The cryptosystem is set up in one of two modes: extraction mode or decryption mode. A crucial of the dual-mode cryptosystem is that no adversary can distinguish the common reference string between two modes. To prove the security against a malicious sender, a simulator must run a trapdoor extractable algorithm that given a trapdoor t, outputs (pk, sk_0, sk_1), where pk is a public encryption key and sk_0 and sk_1 are corresponding secret keys for index 0 and 1 respectively. To prove the security against a malicious receiver, a simulator must run a find-lossy algorithm that given a trapdoor t and pk, outputs an index corresponding to the message-lossy index of pk.

1.1   This Work

We present a new framework for constructing universally composable 1-out-of-2 oblivious transfer protocols based on Bresson-Catalano-Pointcheval's double trapdoor public-key encryption scheme, where a trapdoor string inherently consists of two parts: a master key and local keys. This master key allows us to define a simulator so that it can efficiently extract two input messages from a malicious sender (for simulating malicious behaviors of this corrupted sender, the simulator needs not to know any local key). The local keys generated by a receiver allow it to decrypt the received ciphertexts and benefit the ideal world adversary to simulate malicious behaviors of the corrupted receiver. We will show that our protocol is universally composable secure in the presence of static adversaries in the common reference string model assuming that the decisional Diffie-Hellman problem over a squared composite modulus of the form $N = pq$ is hard.

There are significant differences between our idea and the idea presented in [19] that is mostly related to this work. The trapdoor information generated in the extraction mode including the find-lossy algorithm defined in [19] plays the role of the master key generation algorithm in our scheme; The trapdoor information generated in the decryption mode including the trapdoor extraction of keys decryptable on both branches c_0 and c_1 in [19] plays the role of the local key generation algorithm in our scheme. The importance of our methodology is that the local key generation algorithm introduced in this paper allows us to generate polynomial pairs of local keys while the double mode only allows to generate two-pairs of local keys (*as a result, our methodology may be applied to construct universally composable, adaptive and k-out-of-n oblivious transfer protocols. We however, restrict our research on the constructions of universally composable, non-adaptive and 1-out-of-2 oblivious transfer protocols throughout the paper*). We thus provide a different and more natural implementation for oblivious transfers in the universally composable paradigm.

We provide the following comparisons with the cryptosystem based on the Decisional Diffie-Hellman assumption over Z_p^* in [19] that is the most related to our implementations.

	security assumption	crs generation mode	extension to $OT_{k \times 1}^n$
[19]	DDH over Z_p^*	double-mode	no
this paper	DDH over $Z_{N^2}^*$	uniform-mode	yes

Road map: The rest of this paper is organized as follows: in Section 2, universally composable framework is sketched. A very useful tool called double decryption cryptosystem is sketched in Section 3; In Section 4, a round optimal universally composable 1-out-of-2 oblivious transfer is presented and analyzed. We conclude our work in Section 5.

2   Definitions

This paper works in the standard universally composable framework of Canetti [4]. The universally composable framework defines a probabilistic polynomial time (PPT) environment machine $\mathcal{Z}$. $\mathcal{Z}$ oversees the execution of a protocol π in the real world involving PPT parties and a real world adversary $\mathcal{A}$ and the execution of a protocol in the ideal world involving dummy parties and an ideal world adversary $\mathcal{S}$ (a simulator). In the real world, parties (some of them are corrupted) interact with each other according to a specified protocol π. In the ideal world, dummy parties (some of them are corrupted) interact with an ideal functionality $\mathcal{F}$. The task of $\mathcal{Z}$ is to distinguish between two executions. We refer to [4] for a detailed description of the executions, and definitions of $\text{IDEAL}_{\mathcal{F},\mathcal{S},\mathcal{Z}}$ and $\text{REAL}_{\pi,\mathcal{A},\mathcal{Z}}$.

Common reference string model: Canetti and Fischlin have shown that OT cannot be UC-realized without a trusted setup assumption [6]. We thus assume the existence of an honestly-generated Common Reference String (crs) and work in the so called $\mathcal{F}_{\text{crs}}^{\mathcal{D}}$-hybrid model. The functionality of common reference string model assumes that all participants have access to a common string that is drawn from some specified distribution $\mathcal{D}$ (see Fig. 1 for more details).

Functionality $\mathcal{F}_{\text{crs}}^{\mathcal{D}}$

$\mathcal{F}_{\text{crs}}^{\mathcal{D}}$ proceeds as follows, when parameterized by a distribution $\mathcal{D}$.

- when receiving a message (sid, P_i, P_j) from P_i, let crs $\leftarrow \mathcal{D}(1^n)$ and send (sid, crs) to P_i, and send (crs, sid, P_i, P_j) to the adversary, where sid is a session identity. Next when receiving (sid, P_i, P_j) from P_j (and only from P_j), send (sid, crs) to P_j and to the adversary, and halt.

Fig. 1. Functionality $\mathcal{F}_{\text{crs}}^{\mathcal{D}}$ (due to [5])

The functionality of oblivious transfer: Oblivious transfer is a two-party functionality, involving a sender S with input (x_0, x_1) ($|x_0| = |x_1|$) and a receiver R with input $\sigma \in \{0, 1\}$. The receiver learns x_σ (and nothing else) and the sender learns nothing at all. These requirements are captured by the specification of the oblivious transfer functionality $\mathcal{F}_{\text{OT}}$ from [7], given in Figure 2.

Functionality $\mathcal{F}_{\text{OT}}$

$\mathcal{F}_{\text{OT}}$ interacts with a sender S and a receiver R

- Upon receiving a message (sid, sender, x_0, x_1) from S, where $x_i \in \{0, 1\}^l$, store (x_0, x_1) (the length of the string l is fixed and known to all parties)
- Upon receiving a message (sid, receiver, σ) from R, check if a (sid, sender, $\cdots$) message was previously sent. If yes, send (sid, x_σ) to R and sid to the adversary S and halt. If not, send nothing to R.

Fig. 2. The oblivious transfer functionality $\mathcal{F}_{\text{OT}}$ (due to [7])

Definition 1. *Let $\mathcal{F}_{OT}$ be the oblivious transfer functionality described in* Fig. 2. *A protocol π is said to universally composably realize $\mathcal{F}_{OT}$ if for any adversary $\mathcal{A}$, there exists a simulator $\mathcal{S}$ such that for all environments $\mathcal{Z}$, the ensemble* $\text{IDEAL}_{\mathcal{F}_{OT},\mathcal{S},\mathcal{Z}}$ *is computationally indistinguishable with the ensemble* $\text{REAL}_{\pi,\mathcal{A},\mathcal{Z}}$.

3   Double Trapdoor Public-Key Encryption Scheme

Decisional Diffie-Hellman assumption over $G \subseteq Z_{N^2}^*$: Let N be a product of two large safe primes p and q and $N' = p'q'$, where $p = 2p'+1$ and $q = 2q'+1$. Let $G \subseteq Z_{N^2}^*$ be a cyclic group of order N'. The decisional Diffie-Hellman assumption over G assumes that for every probabilistic polynomial time algorithm A, there exists a negligible function $\nu()$ such that for sufficiently large κ,

$\Pr[A(N, X, Y, Z_b) = b \mid p, q \leftarrow \text{Prime-Generator}(\kappa); N = pq; G \leftarrow \text{Group-Generator}(p, q), g \leftarrow G; x, y, z \leftarrow_R [1, \text{ord}(G)]; X = g^x \mod N^2; Y = g^y \mod N^2, Z_0 = g^z \mod N^2; Z_1 = g^{xy} \mod N^2; b \in \{0,1\}] = 1/2 + \nu(\kappa)$.

Bresson, Catalano and Pointcheval [2] proposed a "lite" Cramer-Shoup variant sketched in Fig. 3 based on Paillier's encryption scheme [18]:

Lemma 1. *(due to [2]) Assuming the decisional Diffie-Hellman assumption over $G \subseteq Z_{N^2}^*$ is hard, Bresson-Catalano-Pointcheval public key encryption scheme is semantically secure.*

Bresson-Catalano-Pointcheval's Double Trap-door Public-key Encryption

- Key generation algorithm KeyGen: on input a security parameter κ, it outputs two large safe primes p and q such that $|p| = |q| = \kappa$;
  KeyGen outputs a cyclic group $G \subseteq Z_{N^2}^*$ of order N' and a random generator g of G.
  KeyGen then randomly chooses $z \in [0, N^2/2)$ and sets $h = g^z \mod N^2$. The public key is (N, g, h), the secret key is z.
- Encryption algorithm E_{pk}: given a message $m \in Z_N$, E_{pk} chooses $r \in_R [0, N/4]$, and computes $u = g^r \mod N^2$, $v = h^r(1+mN) \mod N^2$. The ciphertext is (u, v).
- Decryption algorithm D_{sk}: given a ciphertext c $(=(u,v))$, D_{sk} decrypts the ciphertext c below:
  - given the auxiliary string z, D_{sk} computes m by the procedure $m = \frac{(v/u^z-1)\mod N^2}{N}$;
  - if the factorization of the modulus is provided, then there is alternative decryption procedure. That is, given $sk = (p, q)$, D_{sk} computes m by the following procedure

  $$m = (\frac{v^{\lambda(N)} - 1)\mod N^2}{N} \times (\lambda(N)^{-1}\mod N))\mod N$$

Fig. 3. Bresson-Catalano-Pointcheval encryption scheme

Let G be a cyclic group of $Z_{N^2}^*$ of order N'. For each $x \in [1, N']$, define $\mathrm{DLog}_G(x)$ $= \{(g, g^x) | g \in G\}$. Let g, g', h and h' be generators of G. Let $u = g^s h^t \bmod N^2$ and $v = g'^s h'^t \bmod N^2$, where $s, t \in_R [1, N']$. The following statement is true.

Lemma 2. *(due to [15] and [19]) If $(g, g') \in \mathrm{DLog}_G(x)$ and $(h, h') \in \mathrm{DLog}_G(x)$, then (u, v) is uniformly random in $\mathrm{DLog}_G(x)$; If $(g, g') \in \mathrm{DLog}_G(x)$ and $(h, h') \in \mathrm{DLog}_G(y)$ for some $x \neq y$, then (u, v) is uniformly random in G^2.*

4   UC-Secure OT_1^2 Protocols

Our new construction of universally composable 1-out-of-2 oblivious transfer protocols, is described in Fig. 4.

Universally composable OT_1^2 protocol π

- The input of a sender S is two messages (m_0, m_1) ($m_0 \in Z_N$, $m_1 \in Z_N$, where N is specified by a common reference string generation algorithm described below); The output of S is $\perp$. The input of a receiver R is a bit $\sigma \in \{0, 1\}$; the output of R is the message m_σ;
- Common reference string generation algorithm $\mathsf{OTcrsGen}(1^\kappa)$ for oblivious transfer protocols: on input a security parameter κ, $\mathsf{OTcrsGen}(1^\kappa)$ generates a composite modulus of the form $N = pq$ that is a product of two safe primes p and q such that $p = 2p' + 1$ and $q = 2q' + 1$; $\mathsf{OTcrsGen}(1^\kappa)$ then outputs a cyclic group $G \subseteq Z_{N^2}^*$ of order N' $(= p'q')$ and two random generators g_0 and g_1 of G. $\mathsf{OTcrsGen}(1^\kappa)$ randomly chooses $x_0, x_1 \in_R [0, N^2/2)$ and sets $h_i = g_i^{x_i} \bmod N^2$ $(i = 0, 1)$. The common reference string crs is $< (g_0, h_0, g_1, h_1), N >$. The auxiliary trapdoor information is $< (p, q), (x_0, x_1) >$. The auxiliary string (p, q) is called the master key and the auxiliary strings x_0 and x_1 are called the local keys.
- Upon receiving a common reference string crs, R runs a key generation algorithm KeyGen described below:
  - On input crs and $\sigma \in \{0, 1\}$, KeyGen chooses $r \in [0, N^2/2)$ uniformly at random, and computes the cipher-text (g, h) where $g = g_\sigma^r \bmod N^2$ and $h = h_\sigma^r \bmod N^2$.
  - Let pk be (g, h) and sk be r; The public key (g, h) is sent to S.
- On input crs and pk, S checks the following three conditions: 1) $g \in Z_{N^2}^*$ and $h \in Z_{N^2}^*$; 2) $g^{2N} \neq 1$ and $h^{2N} \neq 1$. If any of two conditions are violated, then outputs $\perp$; otherwise, S runs $\mathsf{E}_{pk}^{\mathsf{crs}}$ to perform the following computations:
  - randomly choosing $s_i, t_i \in [0, N/4]$ and computing $u_i = g_i^{s_i} h_i^{t_i}$, $v_i = g^{s_i} h^{t_i} (1 + N)^{m_i}$ $(i = 0, 1)$;
  - the output of $\mathsf{E}_{pk}^{\mathsf{crs}}$ is two ciphertexts (c_0, c_1), where $c_i = (u_i, v_i)$ $(i = 0, 1)$.
- Upon receiving ciphertexts (c_0, c_1), R first checks that $u_i \in [0, N^2 - 1]$ and $v_i \in [0, N^2 - 1]$ $(i = 0, 1)$. If any of conditions are violated, R outputs $\perp$; Otherwise, R runs the decryption algorithm $\mathsf{D}_{sk}^{\mathsf{crs}}$ to recover the message $m_\sigma \in Z_N$ from the equation $v_\sigma / u_\sigma^r \bmod N^2 = 1 + m_\sigma N$. The output of R is the message m_σ.

Fig. 4. The description of OT_1^2 protocol π

We claim that

Theorem 1. *The oblivious transfer protocol π is universally composable secure in the $\mathcal{F}_{crs}^{\mathcal{D}}$-hybrid model in the presence of static adversaries assuming that the decisional Diffie-Hellman problem over G is hard.*

Proof. Let $\mathcal{A}$ be a static adversary that interacts with the parties S and R running the protocol π described in Fig. 4. We will construct an ideal world adversary $\mathcal{S}$ interacting with the ideal functionality $\mathcal{F}_{OT}$ such that no environment $\mathcal{Z}$ can distinguish an interaction with $\mathcal{A}$ in the protocol π from an interaction with the simulator $\mathcal{S}$ in the ideal world.

Simulating the case where only the sender S is corrupted: When S is corrupted and R is honest, the adversary $\mathcal{A}$ gets S's input from the environment $\mathcal{Z}$, and generates all the messages from S. The goal of an simulator $\mathcal{S}$, then, is to generate the remaining messages (namely, messages from R) so that the entire transcript is indistinguishable from the real interaction between S and R. To do so, we will define a notion of implicit input of the corrupted sender S. In other words, for any message that a corrupted sender S sends out, there is a unique implicit input of S that is consistent with the message.

Game 1: The description of an ideal-world adversary $\mathcal{S}$ when S is corrupted

- when the environment $\mathcal{Z}$ queries to the ideal functionality $\mathcal{F}_{crs}^{\mathcal{D}}$ a common reference string crs, the simulator $\mathcal{S}$ runs the common reference string generation algorithm $\mathsf{OTcrsGen}(1^\kappa)$ described in the real-world protocol π to obtain crs and the corresponding trapdoor information t, where crs $=< (g_0, h_0)$, $(g_1, h_1), N >$ together with a description of a cyclic group G with order N', and $t =< (p,q), (x_0, x_1) >$ such that $N = pq$ and $h_i = g_i^{x_i}$ $(i = 0, 1)$. $\mathcal{F}_{crs}^{\mathcal{D}}$ returns (sid, crs) to the environment $\mathcal{Z}$;
- when the honest receiver R is activated on input (sid, crs), $\mathcal{S}$ chooses two random generators g and h of G (both are of order N') and sets $pk = (g, h)$. Since $\mathcal{S}$ holds the master key (p, q), it is a trivial task for $\mathcal{S}$ to generate such a public key pk. $\mathcal{S}$ sends pk to the adversary $\mathcal{A}$;
- when $\mathcal{S}$ obtains (sid, c_0, c_1) from $\mathcal{A}$, $\mathcal{S}$ performs the following computations
  - if any of conditions $u_i \in [0, N^2 - 1]$ and $v_i \in [0, N^2 - 1]$ $(i = 0, 1)$ are violated, $\mathcal{S}$ sends an empty string ϵ to the functionality $\mathcal{F}_{OT}$ and outputs what $\mathcal{A}$ outputs;
  - if all conditions $u_i \in [0, N^2 - 1]$ and $v_i \in [0, N^2 - 1]$ $(i = 0, 1)$ are satisfied, $\mathcal{S}$ sends the extracted messages (m_0', m_1') to the ideal functionality $\mathcal{F}_{OT}$ and outputs what $\mathcal{A}$ outputs.

Let IDEAL$_{\mathcal{F}_{OT}, \mathcal{S}, \mathcal{Z}}$ be the view of ideal world adversary $\mathcal{S}$ according to the description of **Game 1** and REAL$_{\pi, \mathcal{A}, \mathcal{Z}}$ be the view of real world adversary $\mathcal{A}$ according to the description of protocol π. The only difference between REAL$_{\pi, \mathcal{A}, \mathcal{Z}}$

and IDEAL$_{\mathcal{F},\mathcal{S},\mathcal{Z}}$ is the different strategies to generate the public key (g, h) in π and the public key (g, h) in **Game 1**.

Let $M = N^2/2$, and let Z denote the random variable $[y \leftarrow_R [0, M-1], z \leftarrow y \mod N' : z]$ and $Z' = [z \leftarrow_R [0, N'-1] : z]$. Define s, t by $M = sN' + t$, where $0 \le t < N' - 1$.

- if $z \in [0, t-1]$, then $\Pr[Z = z] = \frac{s+1}{M}$;
- if $z \in [t, N'-1]$, then $\Pr[Z = z] = \frac{s}{M}$;

$$\delta(k) = 1/2 \sum_z |\Pr[Z = z] - \Pr[Z' = z]|$$
$$= 1/2(t|\frac{s+1}{M} - \frac{1}{N'}| + (N' - t)|\frac{s}{M} - \frac{1}{N'}|)$$
$$= \frac{t(N-t)}{MN'}$$
$$\le \frac{N'}{N^2}$$

This means that the random variables Z and Z' are statistically $\delta(k)$-close. It follows that $\mathrm{REAL}_{\pi,\mathcal{A},\mathcal{Z}} \approx \mathrm{IDEAL}_{\mathcal{F}_{\mathrm{OT}},\mathcal{S},\mathcal{Z}}$.

Simulating the case where only the receiver R is corrupted: When S is honest and R is corrupted, the adversary gets R's input from the environment $\mathcal{Z}$, and generates all the messages from R. The goal of the simulator, then, is to generate the remaining messages (namely, all messages from S) so that the entire transcript is indistinguishable from the real interaction between S and R.

Game 2: The description of an ideal-world adversary $\mathcal{S}$ when R is corrupted

- when the environment $\mathcal{Z}$ queries to the ideal functionality $\mathcal{F}_{\mathrm{crs}}^{\mathcal{D}}$ a common reference string crs, the simulator $\mathcal{S}$ runs the common reference string generation algorithm $\mathsf{OTcrsGen}(1^\kappa)$ described in the real-world protocol to obtain crs and the corresponding trapdoor information t, where crs $=< (g_0, h_0), (g_1, h_1), N >$ together with a description of a cyclic group G with order N', and $t =< (x_0, x_1) >$ such that $h_i = g_i^{x_i}$ ($i = 0, 1$). $\mathcal{F}_{\mathrm{crs}}^{\mathcal{D}}$ returns (sid, crs) to the environment $\mathcal{Z}$;
- on upon receiving (g, h), the sender S checks the following three conditions: 1) $g \in Z_{N^2}^*$ and $h \in Z_{N^2}^*$; 2) $g^{2N} \mod N^2 \ne 1$ and $h^{2N} \mod N^2 \ne 1$. If any of the conditions are violated, then the simulator $\mathcal{S}$ sends an empty string ϵ to the oblivious transfer functionality $\mathcal{F}_{\mathrm{OT}}$ and outputs what $\mathcal{A}$ outputs; otherwise, the simulator $\mathcal{S}$ extracts an index $i \in \{0, 1\}$ by testing the equation $h \overset{?}{=} g^{x_i}$ ($i = 0, 1$).
  - if no such an index $i \in \{0, 1\}$ exists, the simulator $\mathcal{S}$ sends an empty string ϵ to the oblivious transfer functionality $\mathcal{F}_{\mathrm{OT}}$ and outputs what $\mathcal{A}$ outputs;

- otherwise, the simulator S sends the bit $i \in \{0,1\}$ to the ideal functionality $\mathcal{F}_{OT}$ and obtains m_i.
  - Finally, the simulator S sends (c_0, c_1) to the adversary $\mathcal{A}$, where c_i is an encryption of m_i, c_{1-i} is a dummy ciphertext and outputs what $\mathcal{A}$ outputs.

Let $\text{IDEAL}_{\mathcal{F}_{OT}, S, Z}$ be the view of ideal world adversary S according to the description of **Game 2** and $\text{REAL}_{\pi, \mathcal{A}, Z}$ be the view of real world adversary $\mathcal{A}$ according to the description of protocol π. The only difference between $\text{REAL}_{\pi, \mathcal{A}, Z}$ and $\text{IDEAL}_{\mathcal{F}_{OT}, S, Z}$ is the strategies to generate ciphertexts c_0 and c_1. Namely, $c_{1-\sigma}$ is a dummy ciphertext in **Game 2** while $c_{1-\sigma}$ is a ciphertext of the $m_{1-\sigma}$ in the protocol π. Since Bresson-Catalano-Pointcheval's double trapdoor encryption scheme is semantically secure assuming that the decisional Diffie-Hellman problem over a squared composite modulus of the form $N = pq$ is hard, it follows that $\text{REAL}_{\pi, \mathcal{A}, Z} \approx \text{IDEAL}_{\mathcal{F}_{OT}, S, Z}$.

Simulating the case where both S and R are honest: If both sender and receiver are honest, we define a simulator S below:

Game 3: The description of an ideal-world adversary S when S and R are honest

- S internally runs the honest S on input (sid, $m_0 = 0 \in Z_N$, $m_1 = 0 \in Z_N$);
- S internally runs the honest R on input (sid, $\sigma = i$), where $i \in_R \{0,1\}$;
- S activates the protocol π when the corresponding dummy party is activated in the ideal execution, and delivering all messages between its internal R and S to $\mathcal{A}$.

Let $\text{IDEAL}_{\mathcal{F}_{OT}, S, Z}$ be the view of ideal world adversary S according to the description of **Game 3** and $\text{REAL}_{\pi, \mathcal{A}, Z}$ be the view of real world adversary $\mathcal{A}$ according to the description of protocol π. Since Bresson-Catalano-Pointcheval's double trapdoor encryption scheme is semantically secure assuming that the decisional Diffie-Hellman problem over a squared composite modulus of the form $N = pq$ is hard, it follows that $\text{REAL}_{\pi, \mathcal{A}, Z} \approx \text{IDEAL}_{\mathcal{F}_{OT}, S, Z}$.

Simulating the case where both S and R are corrupted: when both S and R are corrupted, the adversary $\mathcal{A}$ gets both S's and R's inputs from the environment Z and generates both the sender's messages and the receiver's messages which put together, forms the entire transcript. Thus, the simulator's task is trivial in this case.   □

5   Conclusion

This paper is a further investigation of universally composable oblivious transfer protocols. We have proposed a new implementation of universally composable oblivious transfer based on Bresson-Catalano-Pointcheval's double trapdoor encryption scheme. We have shown that the proposed non-adaptive oblivious transfer protocol is universally composable secure in the common reference string model in the presence of static adversaries assuming that the decisional Diffie-Hellman problem over a squared composite modulus of the form $N = pq$ is hard.

Acknowledgment

The authors are grateful to Helger Lipmaa and anonymous AfricaCrypt'99 program committee members for their invaluable comments.

References

1. Brassard, G., Crépeau, C., Robert, J.-M.: All-or-Nothing Disclosure of Secrets. In: Odlyzko, A.M. (ed.) CRYPTO 1986. LNCS, vol. 263, pp. 234–238. Springer, Heidelberg (1987)
2. Bresson, E., Catalano, D., Pointcheval, D.: A Simple Public-Key Cryptosystem with a Double Trapdoor Decryption Mechanism and Its Applications. In: Laih, C.-S. (ed.) ASIACRYPT 2003. LNCS, vol. 2894, pp. 37–54. Springer, Heidelberg (2003)
3. Camenisch, J., Neven, G., Shelat, A.: Simulatable Adaptive Oblivious Transfer. In: Naor, M. (ed.) EUROCRYPT 2007. LNCS, vol. 4515, pp. 573–590. Springer, Heidelberg (2007)
4. Canetti, R.: A new paradigm for cryptographic protocols. In: FOCS 2001, pp. 136–145 (2001)
5. Canetti, R.: Obtaining Universally Composable Security: Towards the Bare Bones of Trust. In: Kurosawa, K. (ed.) ASIACRYPT 2007. LNCS, vol. 4833, pp. 88–112. Springer, Heidelberg (2007)
6. Canetti, R., Fischlin, M.: Universally Composable Commitments. In: Kilian, J. (ed.) CRYPTO 2001. LNCS, vol. 2139, pp. 19–40. Springer, Heidelberg (2001)
7. Canetti, R., Lindell, Y., Ostrovsky, R., Sahai, A.: Universally composable two-party and multi-party secure computation. In: STOC 2002, pp. 494–503 (2002)
8. Cramer, R., Shoup, V.: Universal Hash Proofs and a Paradigm for Adaptive Chosen Ciphertext Secure Public-Key Encryption. In: Knudsen, L.R. (ed.) EUROCRYPT 2002. LNCS, vol. 2332, pp. 45–64. Springer, Heidelberg (2002)
9. Crépeau, C.: Equivalence Between Two Flavours of Oblivious Transfers. In: Pomerance, C. (ed.) CRYPTO 1987. LNCS, vol. 293, pp. 350–354. Springer, Heidelberg (1988)
10. Even, S., Goldreich, O., Lempel, A.: A Randomized Protocol for Signing Contracts. Commun. ACM 28(6), 637–647 (1985)
11. Goldreich, O., Micali, S., Wigderson, A.: How to Play any Mental Game or A Completeness Theorem for Protocols with Honest Majority. In: STOC 1987, pp. 218–229 (1987)
12. Green, M., Hohenberger, S.: Blind Identity-Based Encryption and Simulatable Oblivious Transfer. In: Kurosawa, K. (ed.) ASIACRYPT 2007. LNCS, vol. 4833, pp. 265–282. Springer, Heidelberg (2007)
13. Kilian, J.: Founding Cryptography on Oblivious Transfer STOC, pp. 20–31 (1988)
14. Lindell, Y.: Efficient Fully-Simulatable Oblivious Transfer. In: Malkin, T.G. (ed.) CT-RSA 2008. LNCS, vol. 4964, pp. 52–70. Springer, Heidelberg (2008)
15. Naor, M., Pinkas, B.: Efficient oblivious transfer protocols. In: SODA 2001, pp. 448–457 (2001)
16. Naor, M., Pinkas, B.: Oblivious Transfer with Adaptive Queries. In: Wiener, M. (ed.) CRYPTO 1999. LNCS, vol. 1666, pp. 573–590. Springer, Heidelberg (1999)
17. Naor, M., Pinkas, B.: Computationally Secure Oblivious Transfer. J. Cryptology 18(1), 1–35 (2005)

18. Paillier, P.: Public-Key Cryptosystems Based on Composite Degree Residuosity Classes. In: Stern, J. (ed.) EUROCRYPT 1999. LNCS, vol. 1592, pp. 223–238. Springer, Heidelberg (1999)
19. Peikert, C., Vaikuntanathan, V., Waters, B.: A Framework for Efficient and Composable Oblivious Transfer. In: Wagner, D. (ed.) CRYPTO 2008. LNCS, vol. 5157, pp. 554–571. Springer, Heidelberg (2008)
20. Rabin, M.O.: How to exchange secrets by oblivious transfer. Technical Report TR-81, Aiken Computation Laboratory, Harvard University (1981)
21. Kalai, Y.T.: Smooth Projective Hashing and Two-Message Oblivious Transfer. In: Cramer, R. (ed.) EUROCRYPT 2005. LNCS, vol. 3494, pp. 78–95. Springer, Heidelberg (2005)
22. Yao, A.C.-C.: Protocols for Secure Computations (Extended Abstract). In: FOCS 1982, pp. 160–164 (1982)

Exponent Recoding and Regular Exponentiation Algorithms

Marc Joye[1] and Michael Tunstall[2]

[1] Thomson R&D France
Technology Group, Corporate Research, Security Laboratory
1 avenue de Belle Fontaine, 35576 Cesson-Sévigné Cedex, France
marc.joye@thomson.net
http://www.geocities.com/MarcJoye/
[2] Department of Computer Science, University of Bristol
Merchant Venturers Building, Woodland Road
Bristol BS8 1UB, United Kingdom
tunstall@cs.bris.ac.uk

Abstract. This paper describes methods of recoding exponents to allow for regular implementations of m-ary exponentiation algorithms. Recoding algorithms previously proposed in the literature do not lend themselves to being implemented in a regular manner, which is required if the implementation needs to resist side-channel attacks based on simple power analysis. The advantage of the algorithms proposed in this paper over previous work is that the recoding can be readily implemented in a regular manner. Recoding algorithms are proposed for exponentiation algorithms that use both signed and unsigned exponent digits.

Keywords: Exponent recoding, exponentiation algorithms, side-channel analysis.

1   Introduction

Exponentiation algorithms have been shown to be vulnerable to side-channel analysis, where an attacker observes the power consumption [17] or electromagnetic emanations [9,24]. These attacks are referred to as Simple Power Analysis (SPA) and Simple Electromagnetic Analysis (SEMA). A naïvely implemented exponentiation algorithm will reveal the exponent used, as the operations required are dependent on the (bitwise) representation of the exponent.

Recoding algorithms have been proposed to decrease the number of operations required to compute an exponentiation [4,15]. However, these recoding algorithms are designed to produce efficient exponentiation algorithms, and not to produce side-channel resistant exponentiation algorithms.

There are several recoding algorithms that have been proposed in the literature [19,20,21,22,23,26,27] in order to thwart SPA or SEMA. However, as noted in [25], to achieve a regular exponentiation algorithm any recoding that is used also needs to be regular. We define "regular" to mean that an algorithm executes

B. Preneel (Ed.): AFRICACRYPT 2009, LNCS 5580, pp. 334–349, 2009.

the same instructions in the same order for any input values. There is, therefore, no leakage through simply inspecting a side-channel. It could be argued that this recoding could be done when an exponent is generated. However, if an exponent is combined with a random value to prevent differential side-channel analysis, as detailed in [7,16], the recoding will have to be conducted just prior to the exponentiation algorithm.

Algorithms are proposed for signed and unsigned exponentiation algorithms. They are described in general terms and can be readily adapted for use in $(\mathbb{Z}/N\mathbb{Z})^*$, $\mathbb{F}_p^*$, etc. It should be noted that differential side channel analysis and the required countermeasures are not considered in this paper (the interested reader is referred to [18] for a discussion of this topic). Only exponentiation algorithms that lend themselves to being implemented with a regular structure are considered in this paper. For a more generic survey of (fast) exponentiation algorithms, the interested reader is referred to [11].

The rest of this paper is organised as follows. In the next section, we review basic methods for evaluating an exponentiation. In Section 3, we explain how exponent recoding may prevent certain side-channel attacks and present new recoding methods. For further efficiency, we consider exponents recoded to both signed and unsigned digits. In Section 4, we describe (higher-radix) right-to-left exponentiation methods and point out why they are more suited for secure implementations. Finally, we conclude in Section 5. Specific applications of our recoding methods can be found in Appendix A.

2 Basic Methods

2.1 Square-and-Multiply Method

The simplest algorithm for computing an exponentiation is the square-and-multiply algorithm. This is where an exponent is read left-to-right bit-by-bit, a zero results in a squaring operation being performed, and a one results in a squaring operation followed by a multiplication. This algorithm is detailed in Algorithm 1, where we define the function $\mathrm{bit}(n, i)$ as a function returning the ith bit of n.[1] The input is an element x in a (multiplicatively written) group $\mathbb{G}$ and a positive ℓ'-bit integer n; the output is the element $z = x^n$ in $\mathbb{G}$.

It has been shown that bit values of exponent n can be distinguished by observing a suitable side channel, such as the power consumption [17] or electromagnetic emanations [9,24]. One countermeasure to prevent an attacker from being able to recover the bit values of an exponent is to execute the same code (i.e., without conditional branchings) whatever the value of exponent n, referred to as side-channel atomicity [5]. Specific implementations described in [5] assume that the multiplications and squaring operations behave similarly. However, it has been shown that squaring operations and multiplications may be distinguished by differences in the power consumption of hardware implementations [1], or the distribution of the Hamming weight of the result of the single precision operations required to compute multi-precision operations [2].

---

[1] By convention, the first bit is bit number 0.

Algorithm 1. Square-and-Multiply Algorithm

Input: $x \in \mathbb{G}$, $n \geq 1$, ℓ' the binary length of n (i.e., $2^{\ell'-1} \leq n < 2^{\ell'}$)
Output: $z = x^n$

$A \leftarrow x; R \leftarrow x$
for $i = \ell' - 2$ **down to** 0 **do**
$\quad A \leftarrow A^2$
$\quad$ **if** $(\text{bit}(n, i) \neq 0)$ **then** $A \leftarrow A \cdot R$
end

---

return A

2.2 Square-and-Multiply-Always Method

The first regular exponentiation algorithm was proposed in [7], where a multiplication is performed for each bit of an exponent. The value of the bit in question determines whether the result is used or discarded.

Algorithm 2. Square-and-Multiply-*Always* Algorithm

Input: $x \in \mathbb{G}$, $n \geq 1$, ℓ' the binary length of n
Output: $z = x^n$

$R[0] \leftarrow x; R[1] \leftarrow x; R[2] \leftarrow x$
for $i = \ell' - 2$ **down to** 0 **do**
$\quad R[1] \leftarrow R[1]^2$
$\quad b \leftarrow \text{bit}(n, i); R[b] \leftarrow R[b] \cdot R[2]$
end

---

return $R[1]$

This algorithm is less efficient than many other exponentiation algorithms. More importantly, it has been shown that this algorithm can be attacked by a safe-error fault attack [29,30]. If a fault is injected into a multiplication, it will change the output only if the result of that multiplication is used. An attacker could potentially use this to determine bits of the exponent by targeting chosen multiplications. A description of fault attacks that could potentially be applied to an exponentiation is beyond the scope of this paper (see [3,10]). However, it would be prudent to avoid algorithms that are vulnerable to such attacks.

2.3 m-Ary Exponentiation

In order to speed up the evaluation of $z = x^n$ it is possible to precompute some values that are small multiples of x by breaking up the exponent into ℓ words in radix m [15]. Typically, m is chosen to be equal to 2^k, for some convenient value of k, to enable the relevant digits to simply be read from the exponent. The m-ary algorithm is shown in Algorithm 3, where function $\text{digit}(n, i)$ returns the ith digit of n (in radix m).

Algorithm 3. m-ary Exponentiation Algorithm

Input: $x \in \mathbb{G}$, $n \geq 1$, ℓ the m-ary length of n
Output: $z = x^n$
Uses: A, $R[i]$ for $i \in \{1, 2, \ldots, m-1\}$

$R[1] \leftarrow x$
for $i = 2$ **to** $m - 1$ **do** $R[i] \leftarrow R[i-1] \cdot x$

$d \leftarrow \mathrm{digit}(n, \ell - 1); \; A \leftarrow R[d]$

for $i = \ell - 2$ **down to** 0 **do**
    $A \leftarrow A^m$
    $d \leftarrow \mathrm{digit}(n, i)$
    if $(d \neq 0)$ **then** $A \leftarrow A \cdot R[d]$
end

return A

This algorithm is more efficient than the square-and-multiply algorithm and has the advantage that it is more regular and will, therefore, leak less information. The algorithm is not entirely regular since no multiplication is conducted when a digit is equal to zero. It would, therefore, be expected that an attacker could determine the digits of n that are equal to zero by observing a suitable side channel.

3 Exponent Recoding

Several methods for recoding exponents have been proposed in the literature. The most commonly known example is Non-Adjacent Form (NAF) recoding [4] that recodes the bits of an exponent with values in $\{-1, 0, 1\}$. This reduces the number of multiplications that are required in the subsequent exponentiation algorithm, and can be generalised to a m-ary recoding [15]. Some methods of NAF recoding to produce a regular exponentiation algorithm have been proposed as a countermeasure to side-channel analysis [22,23]. However, these recoding algorithms have been shown to be vulnerable to attack, since the recoding algorithm is not itself regular [25].

Other recoding algorithms have been proposed that make the subsequent exponentiation algorithm regular. In [19], Möller describes a recoding algorithm for m-ary exponentiation where each digit that is equal to zero is replaced with $-m$, and the next most significant digit is incremented by one. This leads to an exponent recoded with digits in the set $\{1, \ldots, m-1\} \cup \{-m\}$. Combined with the m-ary exponentiation algorithm, this implies that x^{-m} should be precomputed. While this computation is "easy" on elliptic curves, this is not the case for the multiplicative group of integers modulo N. An unsigned version of Möller's algorithm is described in [27] where the digits are recoded in the set $\{1, \ldots, m\}$: each zero digit is replaced with m and the next digit is decremented by one.

The problem with the recoding algorithms proposed in [19,27] is that they cannot easily be implemented in a regular manner. In this section we present some recoding methods for regular exponentiation, where the exponent can be simply recoded in a regular fashion.

3.1 Unsigned-Digit Recoding

The goal is to compute $z = x^n$ for some integer n. Let $n = \sum_{i=0}^{\ell-1} d_i\, m^i$ denote the expansion of n in radix m (typically, as above, $m = 2^k$). Consider positive integer $s < n$ and define $n' := n - s$. If $n' = \sum_{i=0}^{\ell-1} d'_i\, m^i$ and $s = \sum_{i=0}^{\ell-1} s_i\, m^i$ respectively denote the m-expansion of n' and s, it follows that

$$x^n = x^{n'+s}$$
$$= x^{\sum_{i=0}^{\ell-1} \kappa_i\, m^i} \quad \text{with } \kappa_i := d'_i + s_i \ .$$

We define the most significant digit of s in radix m to be zero. This means that the significant digit of n' in radix m (i.e., $\kappa_{\ell-1}$) will remain greater than, or equal to, zero. Otherwise, the recoding would no longer be unsigned and would not be suitable for groups where computing inversions is expensive.

There are several possible realisations of this idea. We detail below two such implementations.

First implementation. Choose $s = \sum_{i=0}^{\ell-2} m^i$. This can be seen as setting all digits of s to 1, namely $s = (1, \dots, 1)_m$. Since $n'_i \in \{0, \dots, m-1\}$, it follows that $\kappa_i \in \{1, \dots, m\}$. We, therefore, obtain the following algorithm.

Algorithm 4. Unsigned-Digit Recoding Algorithm (I)

Input: $n \geq 1$, $m = 2^k$, ℓ the m-ary length of n
Output: $n = (\kappa_{\ell-1}, \dots, \kappa_0)_m$ with $\kappa_i \in \{1, \dots, m\}$, $0 \leq i \leq \ell - 2$

$s \leftarrow (1, \dots, 1)_m$; $n \leftarrow n - s$
for $i = 0$ **to** $\ell - 2$ **do**
    $d \leftarrow n \bmod m$; $n \leftarrow \lfloor n/m \rfloor$
    $\kappa_i \leftarrow d + 1$
end
$\kappa_{\ell-1} \leftarrow n$

It is interesting to note that recoding with digits in the set $\{1, \dots, m\}$ is unique. Indeed, suppose that $n = \sum_i \kappa_i\, m^i = \sum_i \kappa_i^*\, m^i \geq m + 1$. This implies $\kappa_0 \equiv \kappa_0^* \pmod{m}$, which in turn yields $\kappa_0 = \kappa_0^*$ since, given their definition range, $|\kappa_0 - \kappa_0^*| \leq m - 1$. The same argument applies to $n \leftarrow (n - \kappa_0)/m$ which implies $\kappa_1 = \kappa_1^*$ and so on. Therefore, the recoded digits obtained by the algorithm in [27] correspond to those obtained by Algorithm 4. But, contrarily to [27], the proposed algorithm is *itself* regular.

Example 1. Consider, for example, an exponent $n = 31415$ whose binary representation is given by $(1,1,1,1,0,1,0,1,0,1,1,0,1,1,1)_2$. For $m = 2$, Algorithm 4 produces the equivalent representation $(2,2,2,1,2,1,2,1,2,2,2,1,1,1)_2$

in radix 2 with digits in the set $\{1, 2\}$. In radix $m = 4$, it produces $(1, 3, 2, 2, 2, 2, 3, 1, 3)_4$ with digits in the set $\{1, 2, 3\}$.

A possible drawback of Algorithm 4 is that it requires the knowledge of the m-ary length of n (i.e., of ℓ) ahead of time. We describe hereafter an alternative implementation overcoming this limitation.

Second implementation. Looking in more detail into the subtraction step, we have

$$d_i' = (d_i - s_i + \gamma_i) \bmod m \quad \text{and} \quad \gamma_{i+1} = \left\lfloor \frac{d_i - s_i + \gamma_i}{m} \right\rfloor \in \{-1, 0\}$$

where the "borrow" is initialised to 0 (i.e., $\gamma_0 = 0$). This is the classical schoolboy algorithm. Since $d_i, s_i \in \{0, \dots, m-1\}$, we get

$$\kappa_i = d_i' + s_i = \begin{cases} d_i + \gamma_i & \text{if } d_i + \gamma_i \geq s_i, \\ d_i + \gamma_i + m & \text{otherwise}. \end{cases}$$

Hence, we see that any choice for s_i with $s_i \neq 0$ when $d_i \in \{0, 1\}$ leads to a non-zero value for κ_i. As in our first implementation, we choose $s = \sum_{i=0}^{\ell-2} m^i$. Further, to only resort to unsigned arithmetic, we define $\gamma_i' = \gamma_i + 1$ where $\gamma_i \in \{0, 1\}$.

Algorithm 5. Unsigned-Digit Recoding Algorithm (II)

Input: $n \geq 1$, $m = 2^k$
Output: $n = (\kappa_{\ell-1}, \dots, \kappa_0)_m$ with $\kappa_i \in \{1, \dots, m\}$, $0 \leq i \leq \ell - 2$

$i \leftarrow 0$; $\gamma' \leftarrow 1$
while $(n \geq m + 1)$ **do**
    $d \leftarrow n \bmod m$; $d' \leftarrow d + \gamma' + m - 2$
    $\kappa_i \leftarrow (d' \bmod m) + 1$; $\gamma' \leftarrow \lfloor d'/m \rfloor$
    $n \leftarrow \lfloor n/m \rfloor$
    $i \leftarrow i + 1$
end
$\kappa_i \leftarrow n + \gamma' - 1$

3.2 · Signed-Digit Recoding

Again, we let $m = 2^k$. The goal is to rewrite the exponent into digits that take odd values in $\{-(m-1), \dots, -1, 1, \dots, m-1\}$,

$$n = \sum_{i=0}^{\ell-1} \kappa_i m^i \quad \text{with } \kappa_i \in \{\pm 1, \dots, \pm(m-1)\} \text{ and } \kappa_i \text{ odd}.$$

An m-ary exponentiation algorithm using these signed digits would require the same amount of precomputed values as the unsigned version. When computing inverses is easy (for example on elliptic curves), it has been suggested that the exponent digits κ_i could be broken up into an unsigned integer value and the sign, i.e. $\kappa_i = s_i \tau_i$ where τ_i is unsigned and $s_i \in \{1, -1\}$ [12,20,21,26]. In this case, only the values multiplied by the values in $\{1, \ldots, m-1\}$ need to be computed, and the inversion applied as required. The advantage of using this method is that the number of precomputed values is halved.

We rely on the observation that any odd integer in the range $[0, 2m)$ can be written as

$$1 = m + (-(m-1))$$
$$3 = m + (-(m-3))$$

$$\vdots$$

$$m - 1 = m + (-1)$$
$$m + 1 = m + (1)$$

$$\vdots$$

$$2m - 3 = m + (m-3)$$
$$2m - 1 = m + (m-1)$$

which yields the following algorithm.

Algorithm 6. (Odd) Signed-Digit Recoding Algorithm

Input: n odd, $m = 2^k$
Output: $n = (\kappa_{\ell-1}, \ldots, \kappa_0)_{\pm m}$ with $\kappa_i \in \{\pm 1, \ldots, \pm(m-1)\}$ and κ_i odd

$i \leftarrow 0$
while $(n > m)$ **do**
    $\kappa_i \leftarrow (n \bmod 2m) - m$
    $n \leftarrow (n - \kappa_i)/m$
    $i \leftarrow i + 1$
end
$\kappa_i \leftarrow n$

The correctness of the recoding follows by inspecting that for an odd integer $n > m$, we have:

1. $|\kappa_i| \in [1, m-1]$: $1 - m \leq (n \bmod 2m) - m \leq m - 1$;
2. $(n \bmod 2m) - m$ is odd: $(n \bmod 2m) - m \equiv n \equiv 1 \pmod{2}$;
3. $(n - \kappa_i)/m < n$: for $n > m > 1$, $n + m \leq nm \implies (n + m - 1)/m < n$.

Moreover, it is easy to see that the updating step, $n \leftarrow (n - \kappa_i)/m$, does not change the parity:

$$(n - \kappa_i)/m \bmod 2 = \frac{(n - \kappa_i) \bmod 2m}{m} = \frac{m \bmod 2m}{m} = 1 \ .$$

Example 2. Again, with the example of exponent $n = 31415$, Algorithm 6 produces the equivalent representation $(1,1,1,1,1,\bar{1},1,\bar{1},1,\bar{1},1,1,1,\bar{1},1,1)_{\pm 2}$ for $m = 2$. For $m = 4$, it produces $(1,3,3,\bar{1},\bar{1},\bar{1},1,3)_{\pm 4}$.

Algorithm 6 requires that n is odd. It is noted in [13] that when n is even it can be replaced with $n' = n+1$ and that the result is multiplied by the inverse of the input, i.e. x^{-1}. To make this regular for any exponent an equivalent treatment can be conducted when n is odd. In this case, n is replaced with $n' = n+2$ and the result is multiplied by x^{-2}.

It is also possible to modify Algorithm 6 so that the length of the recoded integer is fixed. This allows one to hide the Hamming weight, by setting consecutive digits more significant than the most significant word in base m to $-m$ and one, as described in [19]. This, assuming an even number of digits are available, will not change the output of the exponentiation algorithm.

4   Right-to-Left Exponentiation Algorithms

4.1   Right-to-Left m-Ary Exponentiation

While it is well-known that the square-and-multiply algorithm can be improved by using a higher radix, it is sometimes believed that such improvements only apply to left-to-right versions (e.g., in [6, p. 10]). In [15, §4.6.3], Knuth suggests as an exercise to design an exponentiation algorithm that is analogous to the right-to-left binary method, but based on a general radix m.

For $m = 2$, the right-to-left binary method works as follows. Let $n = \sum_{i=0}^{\ell'-1} b_i\, 2^i$ with $b_i \in \{0,1\}$. The algorithm evaluates $z = x^n$ in $\mathbb{G}$ as

$$z := x^{\sum_{i=0}^{\ell'-1} b_i\, 2^i} = \prod_{\substack{0 \le i \le \ell'-1 \\ b_i \ne 0}} x^{2^i} = \prod_{\substack{0 \le i \le \ell'-1 \\ b_i \ne 0}} Y_i \quad \text{with} \begin{cases} Y_0 = x \\ Y_i = Y_{i-1}^2, & i \ge 1 \end{cases}.$$

Likewise, for a general radix m, we can consider the m-ary expansion of n, $n = \sum_{i=0}^{\ell-1} d_i\, m^i$ where $0 \le d_i < m$. We can then write $z = x^n$ as

$$z = \prod_{\substack{0 \le i \le \ell-1 \\ d_i = 1}} x^{m^i} \cdot \prod_{\substack{0 \le i \le \ell-1 \\ d_i = 2}} x^{2 \cdot m^i} \cdots \prod_{\substack{0 \le i \le \ell-1 \\ d_i = j}} x^{j \cdot m^i} \cdots \prod_{\substack{0 \le i \le \ell-1 \\ d_i = m-1}} x^{(m-1) \cdot m^i}$$

$$= \prod_{j=1}^{m-1} (T_j)^j \quad \text{where } T_j = \prod_{\substack{0 \le i \le \ell-1 \\ d_i = j}} x^{m^i}.$$

Therefore, the evaluation of $z = x^n$ in $\mathbb{G}$ amounts to evaluating the T_j's followed by $\prod_j (T_j)^j$. It is worth noting that for a given index i, only one T_j is affected (i.e. T_j is only affected by $j = d_i$).

In Algorithm 7, we use $(m-1)$ temporary variables, $R[1], \ldots, R[m-1]$, each of them initialised to $1_{\mathbb{G}}$, the identity element of $\mathbb{G}$. At the end of the loop, $R[j]$

Algorithm 7. Right-to-left m-ary Exponentiation

Input: $x \in \mathbb{G}$, $n \geq 1$
Output: $z = x^n$
Uses: A, $R[j]$ for $j \in \{1, \ldots, m-1\}$

`// Step 1: Evaluation of` T_j `for` $1 \leq j \leq m-1$
for $j = 1$ **to** $m-1$ **do** $R[j] \leftarrow 1_{\mathbb{G}}$
$A \leftarrow x$
while $(n \geq m)$ **do**
    $d \leftarrow n \bmod m$
    if $(d \neq 0)$ **then** $R[d] \leftarrow R[d] \cdot A$
    $A \leftarrow A^m$
    $n \leftarrow \lfloor n/m \rfloor$
end
$R[n] \leftarrow R[n] \cdot A$

`// Step 2: Evaluation of` $z = \prod_{j=1}^{m-1} (T_j)^j$
$A \leftarrow R[m-1]$
for $j = m-2$ **down to** 1 **do**
    $R[j] \leftarrow R[j] \cdot R[j+1]$
    $A \leftarrow A \cdot R[j]$
end

return A

---

will contain the value of T_j. We also make use of an accumulator A that keeps track of the successive values of x^{m^i}. For $i = 0, \ldots, \ell-1$, if at iteration i, digit d_i of n is equal to d then, provided that $d \neq 0$, $R[d]$ is updated as $R[d] \leftarrow R[d] \cdot A$. Accumulator A is initialised to x and, at iteration i, it is updated by computing $A \leftarrow A^m$ so that it contains the value of $x^{m^{i+1}}$ for the next iteration. To avoid the useless computation of x^{m^ℓ} (i.e., when $i = \ell-1$), we stop the loop at iteration $\ell-2$, and only update $R[d_{\ell-1}]$ as $R[d_{\ell-1}] \leftarrow R[d_{\ell-1}] \cdot A$ for the last iteration.

It now remains to compute $\prod_{j=1}^{m-1} (R[j])^j$ to get the value of $z = x^n$, as expected. This can be done with only $(2m-4)$ multiplications in $\mathbb{G}$. Indeed, letting $U_{\bar{j}} := \prod_{j=\bar{j}}^{m-1} T_j$ and $V_{\bar{j}} := U_{\bar{j}} \cdot \prod_{j=\bar{j}}^{m-1} (T_j)^{j-\bar{j}}$, we observe that $V_1 = \prod_{j=1}^{m-1} (T_j)^j = z$ and it is easy to check that

$$U_{\bar{j}} = T_{\bar{j}} \cdot U_{\bar{j}+1} \quad \text{and} \quad V_{\bar{j}} = V_{\bar{j}+1} \cdot U_{\bar{j}} .$$

Consequently, $\prod_{j=1}^{m-1} (R[j])^j$ can be evaluated by using accumulator A initialised to $R[m-1]$ and then containing the successive values of V_j for $j = m-2, \ldots, 1$. Further, as the content of $R[j]$ (i.e., T_j) is only needed in iteration j for the evaluation of U_j, $R[j]$ can be used to store the value of U_j. Accumulator A is so updated as $A \leftarrow A \cdot R[j]$.

4.2 Right-to-Left vs. Left-to-Right Exponentiation

As with the left-to-right m-ary algorithm, it would be expected that an attacker would be able to determine the digits of n that are equal to zero in Algorithm 7.

An easy way to prevent this is to treat the digit 0 as the other digits. Namely, we can remove the "**if** $(d \neq 0)$ **then**" by using an additional temporary variable $R[0]$. However, as for the left-to-right version, the resulting implementation would become vulnerable to a safe-error attack. Alternatively, the recent attack of [14] can be extended to recover the zero digits of n. All these attacks can be prevented by making use of the recoding methods described in Section 3. This will be exemplified by the algorithms detailed in Appendix A.

Overall, right-to-left exponentiation methods are superior than their left-to-right counterparts, from a security viewpoint. Known left-to-right exponentiations repeatedly multiply the accumulator by the *same* precomputed value for a non-zero exponent digit. While this can be advantageous from an efficiency viewpoint (for example when the precomputed values feature some properties allowing for a faster multiplication), this can have dramatic consequences from a security viewpoint. In [28], Walter exploits the fact the same values are used to mount what he calls "Big Mac" attacks. Surprisingly, Big Mac attacks are more powerful when there are more precomputed values (i.e. when the algorithms are faster). Right-to-left exponentiation methods are not subject to these attacks.

Another class of attacks exploiting specific properties of left-to-right methods relies on collisions for carefully chosen inputs. These attacks were introduced by Fouque and Valette [8] and subsequently extended by Yen et al. [31]. The most general presentation of these attacks with several extensions applying to all known left-to-right methods (including the Montgomery ladder) is given by Homma et al. [14]. Although not described in [14], as aforementioned, it is an easy exercise to show that a similar attack can be extended to recover the *zero* digits of the exponent — but not all the digits as in the left-to-right algorithms — in the *basic* right-to-left m-ary exponentiation algorithm (Algorithm 7).

5   Conclusion

In this paper, we describe regular recoding algorithms to aid in the implementation of regular m-ary exponentiation algorithms. If a recoding algorithm is not regular then it may itself become subject to a side channel attack, as noted in [25]. In this paper we detail secure implementations for recoding exponents into signed and unsigned digits. The side-channel resistance of the left-to-right m-ary exponentiation algorithm is compared to that of the right-to-left m-ary exponentiation algorithm. This provides a base for a side channel resistant implementation of an exponentiation algorithm, but further countermeasures will need to be included to prevent differential side channel analysis. (Some examples of how the recoded exponents can be used to provide regular exponentiation algorithms are presented in the appendix.)

Acknowledgements

The work described in this paper has been supported in part by the European Commission IST Programme under Contract IST-2002-507932 ECRYPT and EPSRC grant EP/F039638/1 "Investigation of Power Analysis Attacks".

References

1. Akishita, T., Takagi, T.: Power analysis to ECC using differential power between multiplication and squaring. In: Domingo-Ferrer, J., Posegga, J., Schreckling, D. (eds.) CARDIS 2006. LNCS, vol. 3928, pp. 151–164. Springer, Heidelberg (2006)
2. Amiel, F., Feix, B., Tunstall, M., Whelan, C., Marnane, W.P.: Distinguishing multiplications from squaring operations. In: Selected Areas in Cryptography — SAC 2008. LNCS, Springer, Heidelberg (2008) (to appear)
3. Bar-El, H., Choukri, H., Naccache, D., Tunstall, M., Whelan, C.: The sorcerer's apprentice guide to fault attacks. Proceedings the IEEE 94(2), 370–382 (2006); Earlier version in Proc. of FDTC 2004
4. Blake, I., Seroussi, G., Smart, N.: Elliptic Curves in Cryptography. London Mathematical Society Lecture Note Series, vol. 265. Cambridge University Press, Cambridge (1999)
5. Chevallier-Mames, B., Ciet, M., Joye, M.: Low-cost solutions for preventing simple side-channel analysis: Side-channel atomicity. IEEE Transactions on Computers 53(6), 760–768 (2004)
6. Cohen, H.: A Course in Computational Algebraic Number Theory. Graduate Texts in Mathematics, vol. 138. Springer, Heidelberg (1993)
7. Coron, J.-S.: Resistance against differential power analysis for elliptic curve cryptosystems. In: Koç, Ç.K., Paar, C. (eds.) CHES 1999. LNCS, vol. 1717, pp. 292–302. Springer, Heidelberg (1999)
8. Fouque, P.-A., Valette, F.: The doubling attack — Why upwards is better than downwards. In: Walter, C.D., Koç, Ç.K., Paar, C. (eds.) CHES 2003. LNCS, vol. 2779, pp. 269–280. Springer, Heidelberg (2003)
9. Gandolfi, K., Mourtel, C., Olivier, F.: Electromagnetic analysis: Concrete results. In: Koç, Ç.K., Naccache, D., Paar, C. (eds.) CHES 2001. LNCS, vol. 2162, pp. 251–261. Springer, Heidelberg (2001)
10. Giraud, C., Thiebeauld, H.: A survey on fault attacks. In: Quisquater, J.-J., et al. (eds.) Smart Card Research and Advanced Applications VI (CARDIS 2004), pp. 159–176. Kluwer, Dordrecht (2004)
11. Gordon, D.M.: A survey of fast exponentiation methods. Journal of Algorithms 27(1), 129–146 (1998)
12. Hédabou, M., Pinel, P., Bénéteau, L.: A comb method to render ECC resistant against side channel attacks. Report 2004/342, Cryptology ePrint Archive (2004), http://eprint.iacr.org/
13. Hédabou, M., Pinel, P., Bénéteau, L.: Countermeasures for preventing comb method against SCA attacks. In: Deng, R.H., Bao, F., Pang, H., Zhou, J. (eds.) ISPEC 2005. LNCS, vol. 3439, pp. 85–96. Springer, Heidelberg (2005)
14. Homma, N., Miyamoto, A., Aoki, T., Satoh, A., Shamir, A.: Collision-based power analysis of modular exponentiation using chosen-message pairs. In: Oswald, E., Rohatgi, P. (eds.) CHES 2008. LNCS, vol. 5154, pp. 15–29. Springer, Heidelberg (2008)
15. Knuth, D.E.: The Art of Computer Programming, 2nd edn. Seminumerical Algorithms, vol. 2. Addison-Wesley, Reading (1981)
16. Kocher, P.: Timing attacks on implementations of Diffie-Hellman, RSA, DSS, and other systems. In: Koblitz, N. (ed.) CRYPTO 1996. LNCS, vol. 1109, pp. 104–113. Springer, Heidelberg (1996)
17. Kocher, P., Jaffe, J., Jun, B.: Differential power analysis. In: Wiener, M. (ed.) CRYPTO 1999. LNCS, vol. 1666, pp. 388–397. Springer, Heidelberg (1999)

18. Mangard, S., Oswald, E., Popp, T.: Power Analysis Attacks: Revealing the Secrets of Smart Cards. Springer, Heidelberg (2007)
19. Möller, B.: Securing elliptic curve point multiplication against side-channel attacks. In: Davida, G.I., Frankel, Y. (eds.) ISC 2001. LNCS, vol. 2200, pp. 324–334. Springer, Heidelberg (2001)
20. Möller, B.: Parallelizable elliptic curve point multiplication method with resistance against side-channel attacks. In: Chan, A.H., Gligor, V.D. (eds.) ISC 2002. LNCS, vol. 2433, pp. 402–413. Springer, Heidelberg (2002)
21. Möller, B.: Fractional windows revisited: Improved signed-digit representation for effcient exponentiation. In: Park, C.-s., Chee, S. (eds.) ICISC 2004. LNCS, vol. 3506, pp. 137–153. Springer, Heidelberg (2005)
22. Okeya, K., Takagi, T.: A more flexible countermeasure against side channel attacks using window method. In: Walter, C.D., Koç, Ç.K., Paar, C. (eds.) CHES 2003. LNCS, vol. 2779, pp. 397–410. Springer, Heidelberg (2003)
23. Okeya, K., Takagi, T.: The width-w NAF method provides small memory and fast elliptic scalar multiplications secure against side channel attacks. In: Joye, M. (ed.) CT-RSA 2003. LNCS, vol. 2612, pp. 328–342. Springer, Heidelberg (2003)
24. Quisquater, J.-J., Samyde, D.: Electromagnetic analysis (EMA): Measures and counter-measures for smart cards. In: Attali, S., Jensen, T. (eds.) E-smart 2001. LNCS, vol. 2140, pp. 200–210. Springer, Heidelberg (2001)
25. Sakai, Y., Sakurai, K.: A new attack with side channel leakage during exponent recoding computations. In: Joye, M., Quisquater, J.-J. (eds.) CHES 2004. LNCS, vol. 3156, pp. 298–311. Springer, Heidelberg (2004)
26. Thériault, N.: SPA resistant left-to-right integer recodings. In: Preneel, B., Tavares, S.E. (eds.) Selected Areas in Cryptograhy (SAC 2005). LNCS, vol. 3156, pp. 345–358. Springer, Heidelberg (2004)
27. Vuillaume, C., Okeya, K.: Flexible exponentiation with resistance to side channel attacks. In: Zhou, J., Yung, M., Bao, F. (eds.) ACNS 2006. LNCS, vol. 3989, pp. 268–283. Springer, Heidelberg (2006)
28. Walter, C.D.: Sliding windows succumbs to Big Mac attack. In: Koç, Ç.K., Naccache, D., Paar, C. (eds.) CHES 2001. LNCS, vol. 2162, pp. 286–299. Springer, Heidelberg (2001)
29. Yen, S.-M., Joye, M.: Checking before output may not be enough against fault based cryptanalysis. IEEE Transactions on Computers 49(9), 967–970 (2000)
30. Yen, S.-M., Kim, S.-J., Lim, S.-G., Moon, S.-J.: A countermeasure against one physical cryptanalysis may benefit another attack. In: Kim, K.-c. (ed.) ICISC 2001. LNCS, vol. 2288, pp. 414–427. Springer, Heidelberg (2002)
31. Yen, S.-M., Lien, W.-C., Moon, S.-J., Ha, J.: Power analysis by exploiting chosen message and internal collisions — Vulnerability of checking mechanism for RSA-decryption. In: Dawson, E., Vaudenay, S. (eds.) Mycrypt 2005. LNCS, vol. 3715, pp. 183–195. Springer, Heidelberg (2005)

A   Regular Exponentiation Algorithms

A.1   Unsigned-Digit Recoding

A regular equivalent of Algorithm 7 can designed using the exponent recoding as described in Section 3.1.

The resulting algorithm requires m temporary variables for storing the respective accumulated values for recoded digits $d' + 1$ with $d' \in \{0, 1, \ldots, m - 1\}$. In Algorithm 8 these values are stored in locations $R[i]$ for $i \in \{0, 1, \ldots, m - 1\}$ which avoids having to add 1 to the index in the main loop.

Algorithm 8. Regular Right-to-left m-ary Exponentiation

Input: $x \in \mathbb{G}$, $n \geq 1$, $k \geq 1$, $m = 2^k$
Output: $z = x^n$
Uses: A, $R[i]$ for $i \in \{0, 1, \ldots, m - 1\}$

for $j = 0$ to $m - 1$ do $R[j] \leftarrow 1_\mathbb{G}$
$A \leftarrow x$; $\gamma' \leftarrow 1$; $n' \leftarrow \lfloor n/m \rfloor$
while $(n' \geq m + 1)$ do
$\quad$ $d \leftarrow n \bmod m$; $d' \leftarrow d + \gamma' + m - 2$; $\gamma' \leftarrow \lfloor d'/m \rfloor$
$\quad$ $d' \leftarrow d' \bmod m$; $n \leftarrow n'$; $n' \leftarrow \lfloor n/m \rfloor$
$\quad$ $R[d'] \leftarrow R[d'] \cdot A$
$\quad$ $A \leftarrow A^m$
end
$d \leftarrow n \bmod m$; $d' \leftarrow d + \gamma' + m - 2$
$\gamma' \leftarrow \lfloor d'/m \rfloor$; $d' \leftarrow d' \bmod m$
$R[d'] \leftarrow R[d'] \cdot A$
$d' \leftarrow n' + \gamma' - 1$
if $(d' \neq 0)$ then
$\quad$ while $(d' < 1)$ do
$\quad\quad$ $d' \leftarrow 2 \cdot d'$; $k \leftarrow k - 1$
$\quad$ end
$\quad$ $A \leftarrow A^{2^k}$; $R[d' - 1] \leftarrow R[d' - 1] \cdot A$
end
$A \leftarrow R[m - 1]$
for $j = m - 2$ down to 0 do
$\quad$ $R[j] \leftarrow R[j] \cdot R[j + 1]$; $A \leftarrow A \cdot R[j]$
end
$A \leftarrow A \cdot R[0]$

return A

The final digit is in $\{0, 1, \ldots, m\}$ and can, therefore, require special treatment. If the final digit is in $\{1, \ldots, m\}$ the digit can be treated like the rest of the digits of the exponent. If the final digit is equal to zero the final multiplication can be avoided or replaced with a multiplication with $1_\mathbb{G}$ (where such a multiplication would not be apparent by observing a side channel).

Another potential problem is the first multiplication for each $R[i]$, for $i \in \{0, 1, \ldots, m - 1\}$, which will be with $1_\mathbb{G}$. Depending on the arithmetic involved with this multiplication, it may be visible in a side channel. In a side channel resistant implementation one would expect the operands to be blinded such that there are numerous values that represent $1_\mathbb{G}$ [7,16]. In this case, multiplications with $1_\mathbb{G}$ would no longer be visible.

A.2   Signed-Digit Recoding

The same exponent recoding presented in Algorithm 6 in Section 3.2 can be used in a right-to-left algorithm. The algorithm given in Algorithm 9 describes how the recoded exponent can be used to compute an exponentiation.

In Algorithm 9 the variable $R[i]$ for $i \in \{0, \ldots, m/2 - 1\}$ is used to store the product of the negative digits of the recoded exponent without inverting them. The product of the positive digits is stored in $R[i]$ for $i \in \{m/2, \ldots, m-1\}$. When the main loop is terminated the result of combining $R[i]$ for $i \in \{0, \ldots, m/2 - 1\}$ can be inverted to produce the value of x raised to the power of the negative digits. This can then be combined with the values corresponding to the positive digits to produce the result of the exponentiation.

Algorithm 9. Right-to-left Signed Recoding m-ary Exponentiation (I)

Input: $x \in \mathbb{G}$, $n \geq 1$ and odd, $m = 2^k$
Output: $z = x^n$
Uses: A, $R[i]$ for $i \in \{0, 1, \ldots, m - 1\}$

for $j = 0$ **to** $m - 1$ **do** $R[j] \leftarrow 1_{\mathbb{G}}$

$A \leftarrow x$
while $(n > m)$ **do**
$\quad \kappa \leftarrow (n \bmod 2m) - m;\ d' \leftarrow \lfloor (\kappa + m)/2 \rfloor$
$\quad R[d'] \leftarrow R[d'] \cdot A$
$\quad A \leftarrow A^m$
$\quad n \leftarrow (n - \kappa)/m$
end
$d' \leftarrow \lfloor (n + m)/2 \rfloor;\ R[d'] \leftarrow R[d'] \cdot A$

$A \leftarrow R[0]$
for $j = 1$ **to** $m/2 - 1$ **do**
$\quad A \leftarrow A \cdot R[j - 1];\ R[j] \leftarrow R[j] \cdot R[j - 1];\ A \leftarrow A \cdot R[j]$
end
$R[0] \leftarrow A;\ A \leftarrow R[m - 1]$
for $i = m - 2$ **down to** $m/2$ **do**
$\quad A \leftarrow A \cdot R[j + 1];\ R[j] \leftarrow R[j] \cdot R[j + 1];\ A \leftarrow A \cdot R[j]$
end
$A \leftarrow A \cdot R[0]^{-1}$

return A

As with Algorithm 8, the initial multiplications with $1_{\mathbb{G}}$ may be visible in a side channel in Algorithm 9. This problem is easier to deal with in this algorithm since an inversion is computed at the end of the algorithm. For example, the line "$R[j] \leftarrow 1_{\mathbb{G}}$" can be replaced with "$R[j] \leftarrow x$". This will, effectively, blind the exponentiation and be removed when "$A \leftarrow A \cdot R[0]^{-1}$" is computed at the end of the algorithm.

Another option, when a random element in $\mathbb{G}$ can be computed efficiently, is to replace the initialisation loop, " **for** $j = 0$ **to** $m - 1$ **do** $R[j] \leftarrow 1_{\mathbb{G}}$", with

$$\textbf{for } j = 0 \textbf{ to } m/2 - 1 \textbf{ do}$$
$$R[j] \leftarrow \texttt{RandomElement}()$$
$$R[m - j - 1] \leftarrow R[j]$$
$$\textbf{end}$$

where the function $\texttt{RandomElement}()$ returns a random element in $\mathbb{G}$. Given that the random elements are balanced, the computation of a group inversion in the last line will remove the effect of these group elements. This provides more security than if x is used, since an attacker may attempt to manipulate x to produce specific effects in a side channel.

Algorithm 9 can be rewritten so that it uses $m/2$ values stored in memory. This means that inversions will need to be computed as required, rather than deferred to the end of the algorithm.

Algorithm 10. Right-to-left Signed Recoding m-ary Exponentiation (II)

Input: $x \in \mathbb{G}$, $n \geq 1$ and odd, $m = 2^k$ with $k > 1$
Output: $z = x^n$
Uses: $R[i]$ for $i \in \{0, 1, \ldots, m/2\}$

$\textbf{for } j = 1 \textbf{ to } m/2 \textbf{ do } R[j] \leftarrow 1_{\mathbb{G}}$

$R[0] \leftarrow x$
$\textbf{while } (n > m) \textbf{ do}$
$\quad \kappa \leftarrow (n \bmod 2m) - m; \; d' \leftarrow \lceil |\kappa|/2 \rceil$
$\quad s \leftarrow \text{sign}(\kappa); \; d \leftarrow [(1 + s)((d' \bmod (m/2)) + 1)]/2$
$\quad R[d] \leftarrow R[d]^{-1}; \; R[d'] \leftarrow R[d'] \cdot R[0]; \; R[d] \leftarrow R[d]^{-1}$
$\quad R[0] \leftarrow R[0]^m$
$\quad n \leftarrow (n - \kappa)/m$
$\textbf{end}$
$d' \leftarrow \lceil n/2 \rceil; \; R[d'] \leftarrow R[d'] \cdot R[0]$

$R[0] \leftarrow R[m/2]$
$\textbf{for } j = m/2 - 1 \textbf{ down to } 1 \textbf{ do}$
$\quad R[0] \leftarrow R[0] \cdot R[j + 1]; \; R[j] \leftarrow R[j] \cdot R[j + 1]; \; R[0] \leftarrow R[0] \cdot R[j]$
$\textbf{end}$

$\textbf{return } R[0]$

Depending on the group $\mathbb{G}$ and the target device, this may be problematic to compute in a regular manner, as discussed in Section 3.2. Some methods of achieving a regular solution when using elliptic curve arithmetic can be envisaged where κ_i is broken down into s_i and τ_i, e.g.:

- In $\mathbb{F}_{2^m}$ the unsigned representation of s_i can be used as a look up to either zero or the binary representation of the irreducible polynomial used to represent $\mathbb{F}_q$ over $\mathbb{F}_2$ which will be XOR-ed with the value of τ_i to produce the required value for the κ_i.
- An inversion over $\mathbb{F}_p$ can be achieved if we add p to the y coordinate and subtract zero or two times the value of the y coordinate, depending on the value of s_i. This will produce the required value corresponding to the value

of κ_i in a regular manner. However, this will be vulnerable to a safe-error attack on the function that multiplies the y coordinate by two.

Another approach for elliptic curve arithmetic would be to compute the triplet $\{x, y, -y\}$ for each point produced by multiplying the initial point by $\{1, 3, \ldots, m-1\}$. The unsigned representation of s_i can then be used to select y or $-y$ as required. The memory requirements are reduced when compared to a standard m-ary algorithm, but not as much as would be possible if the computation of an inversion were completely free.

Yet another approach, provided that inversion can be computed for free, with fewer memory requirements, is to apply the previous trick but "on-the-fly". This requires an additional temporary variable, say B. In the **while**-loop, at each step, the inverse of A is computed and stored in B. The sign of digit d' is then used to compute $R[d'] \leftarrow R[d'] \cdot A$ or $[d'] \leftarrow R[d'] \cdot B$ as required. Provided that $k > 1$, the same effect can be obtained but without additional temporary variable with two inversions. This is detailed in Algorithm 10, where, to make the presentation easier, we rename accumulator A to $R[0]$ and reorganise the other variables: $R[i]$ for $i \in \{1, 2, \ldots, m/2\}$ is used to accumulated the values for recoded digits $\{1, 3, \ldots, m-1\}$ (in absolute value).

Efficient Acceleration of Asymmetric Cryptography on Graphics Hardware

Owen Harrison and John Waldron

Computer Architecture Group, Trinity College Dublin, Dublin 2, Ireland
harrisoo@cs.tcd.ie, john.waldron@cs.tcd.ie

Abstract. Graphics processing units (GPU) are increasingly being used for general purpose computing. We present implementations of large integer modular exponentiation, the core of public-key cryptosystems such as RSA, on a DirectX 10 compliant GPU. DirectX 10 compliant graphics processors are the latest generation of GPU architecture, which provide increased programming flexibility and support for integer operations. We present high performance modular exponentiation implementations based on integers represented in both standard radix form and residue number system form. We show how a GPU implementation of a 1024-bit RSA decrypt primitive can outperform a comparable CPU implementation by up to 4 times and also improve the performance of previous GPU implementations by decreasing latency by up to 7 times and doubling throughput. We present how an adaptive approach to modular exponentiation involving implementations based on both a radix and a residue number system gives the best all-around performance on the GPU both in terms of latency and throughput. We also highlight the usage criteria necessary to allow the GPU to reach peak performance on public key cryptographic operations.

Keywords: Graphics Processor, Public-Key Cryptography, RSA, Residue Number System.

1 Introduction

The graphics processing unit (GPU) has enjoyed a large increase in floating point performance compared with the CPU in the last number of years. The traditional CPU has leveled off in terms of clock frequency as power and heat concerns increasingly become dominant restrictions. The latest GPU from Nvidia's GT200 series reports a peak throughput of almost 1 TeraFlop, whereas the latest Intel CPUs reported throughput is in the order of 100 GigaFlops [1]. This competitive advantage of the GPU comes at the price of a decreased applicability to general purpose computing. The latest generation of graphics processors, which are DirectX 10 [2] compliant, support integer processing and give more control over the processor's threading and memory model compared to previous GPU generations. We use this new generation of GPU to accelerate public key cryptography. In particular we use an Nvidia 8800GTX GPU with CUDA [3] to investigate the possibility of high speed 1024-bit RSA decryption. We focus

B. Preneel (Ed.): AFRICACRYPT 2009, LNCS 5580, pp. 350–367, 2009.

on 1024-bit RSA decryption as it shows a high arithmetic intensity, ratio of arithmetic to IO operations, and also allows easy comparison with CPU implementations. We exploit the new GPU's flexibility to support a GPU sliding window [4] exponentiation implementation, based on Montgomery exponentiation [5] using both radix and residue number system (RNS) representations. We investigate both types of number representation showing how GPU occupancy and inter thread communication plays a central role to performance. Regarding the RNS implementations, we exploit the GPU's flexibility to use a more optimised base extension approach than was previously possible. We also explore various GPU implementations of single precision modular multiplication for use within the exponentiation approaches based on RNS.

2   GPU Background

The GPU that we have used in our implementations is Nvidia's 8800GTX (a 2006 release), part of the G80 series, which was the first DirectX 10 [2] compliant GPU released by Nvidia. It is Nvidia's first processor that supports the CUDA API [3] and as such all implementations using this API are forward compatible with newer CUDA compliant devices. All CUDA compatible devices support 32-bit integer processing. The 8800GTX consists of 16 SIMD processors, called Streaming Multiprocessors (SM). Each SM contains 8 ALUs which operate in lockstep controlled by a single instruction unit. The simplest instructions are executed by the SMs in 4 clock cycles, which creates an effective SIMD width of 32. Each SM contains a small amount of fast local storage which consists of a register file, a block of shared memory, a constant memory cache and a texture memory cache. The main bulk of the GPU's storage is off-chip global memory, and is thus considerably slower than the on-chip storage. A lot of the programming effort concerning a GPU is the careful usage of the different types of available storage. This is due to the scarcity of fast storage and so performance can degrade dramatically with naive implementations.

The code which runs on the GPU is referred to as a kernel. A kernel call is a single invocation of the code which runs until completion. The GPU follows a Single Process, Multiple Data (SPMD, or recently SIMT) threading model [1]. All threads must run from the same static code and must all finish before a kernel can finish. Via the CUDA API, the programmer can specify the number of threads that are required for execution during a kernel call. Threads are grouped into a programmer defined number of CUDA blocks, where each block of threads is guaranteed to run on a single SM. Threads within a block can communicate using a synchronisation barrier which ensures that all threads within the block fully commit to the same instruction before proceeding. The number of threads per block is also programmer defined. They should be allocated in groups of 32, called a CUDA warp, to match the effective SIMD width mentioned above. If the thread execution path within a warp diverge, all paths must be executed serially on the SM. An important consideration for GPU performance is its level of occupancy. Occupancy refers to the number of threads available for execution at any one time. It is normally desirable to have a high level of occupancy as it facilitates the hiding of memory latency.

3   Related Work

The earliest attempts at accelerating cryptography using GPUs involved symmetric primitives [6,7,8,9]. Concerning public key acceleration, Moss et al. [10] presented an implementation of 1024-bit exponentiation using an RNS approach. This paper demonstrated the feasibility of a public key implementation on a GPU with 1024-bit exponentiation throughput rates of 5.7 ms/op (a theoretical rate of 1.42 ms/op (5.7/4) for 1024-bit RSA decryption). Their approach was restricted due to the use of the Nvidia 7800GTX GPU, which is not DirectX 10 compliant. Fleissner et al. [11] presented an acceleration of 192-bit exponentiation also on the 7800GTX and as such has similar restrictions as the above approach. The paper unfortunately only presents its results in terms of ratio comparisons with its own CPU implementation so objective comparisons cannot be made.

Most recently, Szerwinski et al. [15] presented implementations of 1024 and 2048-bit modular exponentiations based on both radix and RNS approaches using an Nvidia 8800GTS (slightly slower GPU than the one used in this paper). The maximum throughput achieved was via a radix based approach, resulting in 1.2 ms/op for 1024 bit modular exponentiation, though with a latency of 6.9 seconds (~0.3 ms/op with 1.7 seconds latency for a 1024-bit RSA decryption). Note, we multiply by 4 the throughput and correspondingly divide by 4 the latency to give an approximation of the RSA-1024 performance, i.e. 2 512-bit modular exponentiations [4]; this approximation approach is used in the results sections later in this paper. Due to this high latency the paper concludes that the GPU's maximum throughput can be achieved only in contexts where latency is irrelevant and thus its usefulness is quite restricted. The RNS approach presented displayed better latency characteristics, though with a throughput rate of 2.3 ms/op (0.57 ms/op for RSA-1024), which is only marginally better than the Crypto++ CPU implementation [12]. We show that with careful memory management, Section 4.1, one can reduce the high latency of a radix approach by over 7 times. We also show that through a detailed analysis of the underlying modular operations used, Section 5.2, and the optimisation of the employed RNS algorithm, Section 5.3, we can double the throughput of the RNS approach. The paper, [15], also restricts the use of the GPU to a single exponent per batch of processing, we outline how different keys can be effectively used in both radix and RNS approaches.

4   Standard Montgomery Exponentiation on the GPU

We present two different GPU implementations with varying degrees of parallelism incorporating the Montgomery reduction method in radix representation and pencil-and-paper multiplication. One observation that applies to all implementations of exponentiation on a CUDA compatible device is that it is only suitable to use a single exponent per CUDA warp, and in some scenarios per CUDA block. The reason for this is that the exponent largely determines the flow of control through the code. These conditional code paths dependant on the exponent cause thread divergence. When threads within a CUDA warp diverge on a single processor, all code paths are executed serially, thus a large performance overhead is incurred for threads that diverge for large portions of code. If inter thread communication is required, a synchronisation barrier must be used to prevent race conditions occurring. All threads within a CUDA block that perform a synchronisation barrier must not be divergent at the point of synchronisation. Thus all threads within a single CUDA block are required to execute the same path at points of synchronisation and so it follows that for exponentiation that uses inter thread communication, only one exponent can be used per CUDA block.

4.1 Serial Approach

Each thread within this implementation performs a full exponentiation without any inter thread communication or cooperation. This is a standard optimised implementation of an exponentiation using the Quisquater and Couvreur CRT approach [21], operating on two independent pairs of 16 limb numbers. The approach also uses the sliding window technique to reduce the number of Montgomery multiplies and squares required. As a single thread computes an exponentiation independently, a single exponent must be used across groups of 32 threads. In terms of RSA, assuming peak performance, this implementation is restricted to using a maximum of 1 key per 32 primitives (or messages). As we are using the CRT based approach to split the input messages in two, we also use two different exponents for a single message. Thus a message must be split into different groups of 32 threads to avoid guaranteed thread divergence. We have adopted a simple strategy to avoid divergence, whereby CUDA blocks are used in pairs. The first block handles all 16 limb numbers relating to the modulus p and the second block handles all numbers relating to the modulus q, where $n = pq$ and n is the original modulus. Note that there is an underlying assumption that the input data ensures the maximum number of keys per primitives is upheld (as mentioned above). The threading model employed is illustrated in Figure 1. This separation of p and q related data is also used in the implementations in Section 4.2 and 5.

The added support for integers, bitwise operations and increased memory flexibility such as scatter operations, in the 8800GTX, allows this implementation to execute largely in a single kernel call. The byte and bit manipulation operations required for the efficient implementation of sliding window are now straightforward. The macro level details of this algorithm are largely standard, consisting of: CRT and sliding window; Montgomery multiplication across normal radix representation; input data is first converted to Montgomery representation; the data is then multiplied or squared according to the exponent; and finally a Montgomery multiplication by 1 to undo the initial Montgomery representation. As such, we do not list the high level steps of the algorithm, however we draw

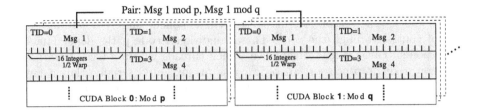

Fig. 1. Serial Thread Model

attention to the following optimisations that were applied within the implementation: all $N \times N$ limb multiplies used cumulative addition to reduce memory operations [13]; all squaring requirements were optimised to reduce the number of required multiplies [4]; $N \times N$ limb multiplies mod R were truncated, again to remove redundant multiplies; and the final two steps within Montgomery multiplication were combined into a single $N \times N$ multiply and accumulate. These optimisations are listed here as they are relevant to the implementation in Section 4.2.

Memory usage: The concept of a uniform, hierarchical memory structure such as a CPU's L1/L2 cache etc does not exist on the GPU and performance cliffs can be encountered without careful memory planning. The following are the highlights of the various memory interactions of this implementation. Note that the implementations in Section 4.2 and Section 5 use similar adaptive memory approaches as described below.

Adaptive memory approaches: The sliding window technique requires the pre-calculation of various powers of the input data. This data is used during the exponentiation process to act as one of the N limb inputs into an $N \times N$ multi-precision multiplication. There are two options on how to handle the storage and retrieval of this pre-calculated data. **1.** The pre-calculation is done on the GPU and is written to global memory. The data is stored in a single array with a stride width equal to the number messages being processed in a single kernel call multiplied by the message size. Reads are then made subsequently from this array direct from global memory. In this scenario only a single kernel call is required for the exponentiation. **2.** Unfortunately the data reads cannot be coalesced as each thread reads a single limb which is separated by 16 integers from the next message. Coalesced global reads require the data to start at a 128-bit boundary for a warp and require each thread of the warp to read consecutively from memory with a stride of up to 4 32-bit integers wide. Non-coalesced reads generate separate memory transactions thus significantly reducing load/store throughput. To ameliorate this the sliding window pre-calculation data is first generated in an initialisation kernel writing its results to global memory. A texture can then be bound to this memory and the subsequent exponentiation kernel can use the pre-calculation data via texture references. Note that texture access uses the texture cache, which is a local on chip cache, however textures cannot be written to directly hence the need for a separate initialisation kernel. The first approach described above is better for smaller amounts of data. The second approach is beneficial for larger amounts of data when the advantage of texture use outweighs the fixed overhead of the extra kernel call.

Another adaptive memory approach concerns the exponent. As mentioned, the exponent must be the same across a warp number of threads, thus all threads within a warp, when reading the exponent, access the same memory location at any one time. Constant memory has by far the best performance under this scenario [9], however is limited to 64KB on the G80. As each exponent requires 32 integers worth of storage, in an RSA 1024-bit context we can use constant memory for up to 512 different keys. If the amount of exponents exceed this threshold (in practice lower than 512 different keys as a small amount of constant memory is used for other purposes and a new key is used for at least each new block whether needed or not for lookup efficiency) then texture memory is used.

Other memory considerations: In an aim to increase the $N \times N$ multiplication performance we have allocated all of the on chip fast shared memory for storing and retrieving the most commonly used N limb multiplicand of the $N \times N$ operation. The less frequently accessed multiplier is retrieved from textures when possible. The input and output data is non exceptional in this implementation save that it cannot be coalesced due to the message stride within memory. A convolution of multiple messages could be an option to offset the lack of coalescing though this has not been explored here and would seem to be just adding extra steps to the CPU processing side. The other per key variables, $-n^{-1}(\mathrm{mod}\ R)$ and $R^2(\mathrm{mod}\ n)$ (for use in generating the Montgomery representation of the input) for both moduli p and q, where $n = pq$, are stored and loaded via texture references. In the context of RSA decryption these variables are assumed to be pre-calculated and it should be noted that performance will degrade slightly if these have to be calculated with a high frequency. The results for this implementation are presented in Section 4.3 in conjunction with the parallel approach described below. Note that two parts of the exponentiation are not included in these implementations, the initial $x(\mathrm{mod}\ p)$, $x(\mathrm{mod}\ q)$ and the final CRT to recombine, these are done on the CPU. This is also the case for all implementations reported in this paper. These steps contribute little to the overall exponentiation runtime and so the performance impact is expected to minor.

4.2   Parallel Approach

This approach uses the same macro structure as the algorithm used above, however it executes the various stages within the algorithm in parallel. Each thread is responsible for loading a single limb of the input data, with 16 threads combining to calculate the exponentiation. Each thread undergoes the same high level code flow, following the sliding window main loop, however the Montgomery multiplication stages are implemented in parallel. This approach relies heavily on inter thread communication, which has a performance overhead as well as an implication that only one exponent is supported per CUDA block. As the number of threads per block in this implementation is limited to 256, due to shared resource constraints, the number of 1024-bit RSA primitives per key in effect is limited to a minimum of 16. This is a hard limit in that the code is defined to have undefined behaviour if threads are divergent at points of synchronisation. The threading model uses the same separation of message pairs, for p and q, as in Figure 1, however, a single thread reads only a single integer.

The intensive $N \times N$ multiplies within the Montgomery multiplication are parallelised by their straight forward separation into individual $1 \times N$ limb multiplications. Each thread is independently responsible for a single $1 \times N$ limb multiply. This is followed by a co-operative reduction across all threads to calculate the partial product additions. This parallel reduction carries with it an overhead where more and more threads are left idle. Figure 2 shows the distribution of the $N \times N$ operation across the 16 threads and its subsequent additive reduction. It also shows the use of shared memory to store the entire operations output and input of each stage. As previously noted, the number of threads per block is limited to 256, so each RSA primitive can use up to 1KB (16KB / (256/16)) worth of shared memory. This allows for the entire $N \times N$ calculation to fit within shared memory. Also shown in the Figure 2 are the synchronisation points used to ensure all shared memory writes are committed before subsequent reads are performed. As this code is the most intensive part of the exponentiation, these synchronisation calls add a significant burden to the performance.

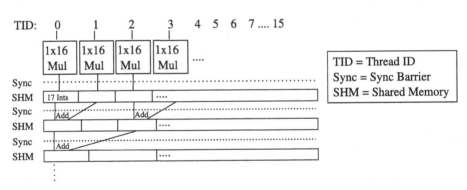

Fig. 2. $N \times N$ limb multiplication in parallel on a CUDA device

The optimisations applied to the different $N \times N$ multiplies, listed in the serial approach, are not possible in the parallel approach. The squaring optimisation, and also the modulo multiplication step, in general only execute half the limb multiplies that are required compared to a full $N \times N$ multiply. However, the longest limb within the $N \times N$ multiply dictates its overall execution time as all threads within a warp must execute in lock step. Thus, although one thread only executes a single multiply, it must wait until the largest $1 \times N$ multiply finishes.

Also, as each thread executes its own $1 \times N$ multiply separately, the cumulative approach to addition must also be separated from the multiplication process. The results for this approach are presented below.

4.3   Results

Figure 3 illustrates the performance of both the parallel and serial approaches presented above. All measurements presented represent the number of 1024-bit RSA decrypt primitives executed per second. The GPU implementations show their dependence on an increasing number of messages per batch of work to approach their peak performance. This is due to having more threads available to hide memory read/write latency and also an increased ratio of kernel work compared to the fixed overheads associated with data transfer and kernel calls. We can see the advantage of the parallel approach over the serial approach at lower primitives per kernel call due to an higher level of occupancy. However the performance bottlenecks of excessive synchronisations and lack of optimisations limit the parallel approach.

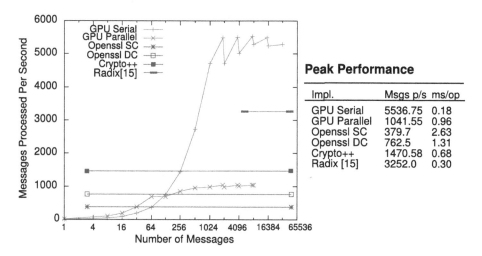

Peak Performance		
Impl.	Msgs p/s	ms/op
GPU Serial	5536.75	0.18
GPU Parallel	1041.55	0.96
Openssl SC	379.7	2.63
Openssl DC	762.5	1.31
Crypto++	1470.58	0.68
Radix [15]	3252.0	0.30

Fig. 3. GPU Radix based Montgomery Exponentiation: 1024-bit RSA Decryption

Also included in Figure 3, is the fastest implementation reported on the Crypto++ [12] website for a 1024-bit RSA decrypt, which is running on an AMD Opteron 2.4 GHz processor (note that this figure does not include memory bandwidth so will likely be slower). Both the CPU and the GPU used in these tests are slightly old, however their results are expected to scale in a straightforward manner. Also included are the performance measurements for Openssl's [14] speed test for 1024-bit RSA decryption running in both single (SC) and dual core (DC) modes on an AMD Athlon 64 X2 Dual Core 3800+. As can be seen at peak performance, the serial approach on the GPU is almost 4 times the speed of the fastest CPU implementation at 5536.75 primitives per second. We can see that the serial approach becomes competitive with the fastest CPU implementation at batches of 256 primitives. We also include the results for the radix approach presented in Szerwinski et al. [15], where the maximum throughput was 3252 primitives per second, though with a minimum latency of 1.7 seconds for all throughput rates. As a point of comparison we achieve a throughput rate of 4707 primitives per second at a latency of 218ms or 1024 primitives per batch of processing, i.e. a latency reduction of over 7 times. The increase in performance is largely due to the lack of reliance on global memory and the increased use of the faster on chip memory stores. We have only plotted the graph for [15] from the lowest number of message per batch corresponding to the reported minimum latency; the time taken to process smaller batches of messages remains the same.

5   Montgomery Exponentiation in RNS on the GPU

The motivation for using RNS is to improve on the parallel approach in Section 4, with an aim to reducing the number of primitives required before the GPU becomes competitive with the CPU. In RNS representation a number x, is denoted as $< x >_a$, where $< x >_a = (|x|_{a1}, |x|_{a2}...|x|_{an})$, $|x|_{ai} = x(\text{mod } a_i)$ and $a = \{a_1, a_2...a_n\}$, called the RNS base, is a set whose members are co-prime. $< x >_a$ can be converted into radix form, x, by the use of the Chinese Remainder Theorem (CRT). The CRT conversion of $< x >_a$ into radix form produces $x(\text{mod } A)$, where $A = \prod_{i=1}^{n} a_i$, called the RNS's dynamic range. Numbers in RNS form have an advantage whereby multiplication, addition and subtraction can be performed as: $< x >_a$ op $< y >_a = (|(|x|_{a1} \; op \; |y|_{a1})|_{a1}, ...|(|x|_{an} \; op \; |y|_{an})|_{an})$, where op is $+, -$ or $*$. As each operation is executed independently, this form of modular arithmetic is suited to parallel processing. However, divide within an RNS is difficult to perform, and as such presents an issue with regard to exponentiation as we will see.

5.1   Montgomery in RNS

As Montgomery multiplication consists primarily of multiplication and addition, there is a temptation to perform this via RNS, thus simply parallelising the entire exponentiation process. However, two parts of the algorithm cause problems within RNS - the mod R operation and the final divide by R, referring to notation used in Section 4. Unfortunately both of these operations cannot be performed in a single RNS base. As has been presented in papers [16] and [17], a way around this issue is to use 2 bases, one for the execution of the mod R operation, and the other for execution of the divide. Thus one of the bases in effect acts as Montgomery's R, the other acts as a facilitator to be able to represent R^{-1} and perform a division by inverse multiplication. Table 1 contains an outline of Montgomery reduction in RNS as presented by Kawamura et al. [17], which we largely base our implementations upon. Note the RNS bases are denoted as a and b, and B $(\prod_{i=1}^{n} b_i)$ is equivalent to R in the standard algorithm.

Table 1. Kawamura et al. [17] Montgomery RNS

Input: $< x >_{a \cup b}, < y >_{a \cup b}$, (where $x, y < 2N$)	
Output: $< w >_{a \cup b}$ (where $w \equiv xyB^{-1}(\mathrm{mod}\ N), w < 2N$)	
Base a Operation	Base b Operation
1: $< s >_a \leftarrow < x >_a \cdot < y >_a$	$< s >_b \leftarrow < x >_b \cdot < y >_b$
2a: —	$< t >_b \leftarrow < s >_b \cdot < -N^{-1} >_b$
2b: $< t >_{a \cup b} \Leftarrow < t >_b$	
3: $< u >_a \leftarrow < t >_a \cdot < N >_a$	—
4: $< v >_a \leftarrow < s >_a + < u >_a$	—
5a: $< w >_a \leftarrow < v >_a \cdot < B^{-1} >_a$	—
5b: $< w >_a \Rightarrow < w >_{a \cup b}$	

The missing steps are mostly due to the observation that as B represents the Montgomery R, v is a multiple of B and thus all its residues will be zero. As can be seen, stage 2b and 5b consist of the conversion between bases, this is called base extension. The most computationally expensive part of this algorithm are these two base extensions. A base extension in its essence is the conversion from $< x >_a$ into its weighted radix form x and then into $< x >_b$. However, the conversion from $< x >_a$ into x via a full CRT is inefficient and is impractical to form part of the inner loop of an exponentiation. Szabo and Tanaka [18] detailed an early base extension algorithm which is based on preliminary conversion to a mixed radix representation (MRS) before extension into the new base. Although more efficient than a full CRT round trip, the conversion into MRS is inherently serial and thus difficult to accelerate on a parallel device. Our approach below is based on a more efficient base conversion technique using CRT as presented by Posch et al. [19] and Kawamura et al. [17]. This base extension technique does not undergo a full CRT round trip and allows for a greater degree of parallelism than the MRS technique. The details of this base extension are presented in detail below in Section 5.3.

5.2   Single Precision Modular Multiplication on the GPU

The most executed primitive operation within Montgomery RNS is a single precision modular multiplication. On the Nvidia CUDA hardware series the integer operations are limited to 32-bit input and output. Integer multiplies are reported to take 16 cycles, where divides are not quoted in cycles but rather a recommendation to avoid if possible [1]. Here we present an investigation into 6 different techniques for achieving single precision modular multiplication suitable for RNS based exponentiation implementations. The techniques are limited to those applicable within the context of RNS and 1024-bit RSA.

1. 32-bit Simple Long Division: given two 32-bit unsigned integers we use the native multiply operation and the __umulhi(x,y) CUDA intrinsic to generate the low and high 32-bit parts of the product. We then use the product as a

4 16-bit limb dividend and divide by the 2 16-bit limb divisor using standard multi-precision division [13] to generate the remainder.

2. 32-bit Division by Invariant Integers using Multiplication: we make the observation that the divisors within an RNS Montgomery implementation, i.e. the base's moduli, are static. Also, as we select the moduli, they can be chosen to be close to the word size of the GPU. Thus we can assume that all invariant divisors, within the context of our implementation, are normalised (i.e. they have their most significant bit set). These two observations allow us to use an optimised variant of Granlund and Montgomery's approach for division by invariants using multiplication [20]. The basic concept used by [20] to calculate n/d is to find a sufficiently accurate approximation of $1/d$ in the form $m/2^x$. Thus the division can be performed by the multiplication of $n * m$ and cheap byte manipulation for division. We pre-calculate m for each of the base residues used and transfer them to the GPU for use via texture lookups. The algorithm in Table 2 removes all normalisation requirements from the original algorithm. It also rearranges some of the calculations to suit the efficient predication available on the GPU. Inputs: N is the word bit length on the GPU; single word multiplier and multiplicand x and y; m is a pre-calculated value dependent on d alone; d is the divisor. Output: r, the remainder. Some of these operations require 2 word precision and thus require extra instructions on the GPU. hiword() indicates the most significant word of a two word integer, where loword() indicates the least significant word. For a thorough explanation of the concepts involved, refer to [20].

Table 2. Granlund and Montgomery's division by invariants optimised for GPU and RNS

$$n = x * y, n1 = \text{hiword}(n), n0 = \text{loword}(n)$$
$$ns = n0 >> (N - 1)$$
$$if(ns > 0)\ n0+ = d$$
$$t = \text{hiword}((m * (n1 + ns)) + n0)$$
$$q1 = n1 + t$$
$$dr = (n - (d << N)) + ((2^N - 1 - q1) * d)$$
$$r = \text{loword}(dr) + (d\ \&\ \text{hiword}(dr))$$

3. 32-bit Reduction by Residue Multiplication: in this approach we use the observation that the moduli comprising the RNS bases can be selected close the GPU's maximum single word value. For 1024-bit RSA we can determine that for all moduli, d, the following holds $|2^N|_d < 2^{11}$, where N is the word bit length of the GPU, i.e. 32. As such, given a single precision multiplication $n = xy$, and using the convention that $n1$ is the most signification word of n, and $n0$ the least significant word, we can rewrite n as $|n1 * 2^N + n0|_d$. By repeatedly applying this representation to the most signification part of the equation, and using the pre-calculated value $r = |2^N|_d$, we can derive an algorithm for executing modular multiplication with multiplies and additions only. This observation is

Table 3. Algorithm for 32-bit Reduction by Residue Multiplication

Observation:	Pseudocode:
$\|x * y\|_d = \|n\|_d$	$n = x * y$
$\qquad = \|\|n1\|_d * \|2^N\|_d + \|n0\|_d\|_d$	$n0 = \text{loword}(n), n1 = \text{hiword}(n)$
Let $r = \|2^N\|_d$ /* $r < 2^{11}$ */,	$n1r = n1 * r$
$\|n\|_d = \|\|n1\|_d * r + \|n0\|_d\|_d$	$n1r0 = \text{loword}(n1r)$
$\qquad = \|\|n1r\|_d + \|n0\|_d\|_d$ /* $n1r < 2^{43}$ */	$n1r1 = \text{hiword}(n1r)$
$\|n1r\|_d = \|\|n1r1\|_d * r + \|n1r0\|_d\|_d$	$n1r1r = \text{loword}(n1r1 * r)$
$\qquad = \|\|n1r1r\|_d + \|n1r0\|_d\|_d$ /* $n1r1r < 2^{22}$ */	$r = n1r1r + n1r0 + n0$
Thus:	if$(r < d)r- = d$
$\|n\|_d \equiv \|n1r1r\|_d + \|n1r0\|_d + \|n0\|_d$, which is $< 3d$	if$(r < d)r- = d$.

more formally stated in Table 3 (left), and the resultant pseudocode is also listed (right). Note that this approach benefits from being able to use CUDA's _umul24 limited precision fast multiply operation in the calculation of $n1r1r$.

4. 32-bit Native Reduction using CRT: for the RNS bases we can use moduli that are the product of two co-prime factors, and also co-prime to each other. Using a modulus with two co-prime factors p and q, we can represent the modular multiplication input values, x and y, as $\|x\|_p, \|x\|_q, \|y\|_p, \|y\|_q$. Thus we have a mini RNS representation and as such can multiply these independently. We use CRT to recombine to give the final product. As p and q can be 16-bit, we are able to use the GPU's native integer modulus operator while maintaining 32-bit operands for our modular multiplication. This approach is described in more detail in the Moss et al. paper [10].

5. 16-bit Native Reduction: we can use 16-bit integers as the basic operand size of our modular multiplication, both input and output. We can then simply use the GPU's native multiply and modulus operators without any concern of overflow. However, we need to maintain the original dynamic range of the RNS bases when using 32-bit moduli. We can achieve this by doubling the number of moduli used in each base (note there is plenty of extra dynamic range when using 17 32-bit integers to accommodate this simple doubling).

6. 12-bit Native Reduction: this is the same concept as the 16-bit native approach above, however using 12-bit input and outputs we can use the much faster floating point multiplies and modulus operators without overflow concerns. Again we need to maintain the dynamic range by approximately tripling the original 32-bit moduli. Also there is an issue where the Kawamura approximations require the base moduli to be within a certain range of the next power of 2. This is not discussed further here, though note that a full 12-bit implementation would require the use of a different base extension method than the one described below.

Results: All tests of the above approaches processed the same amount of data, 2^{32} bytes, executing modular multiplication operations, reading and accumulating from and to shared memory. Thus the 16-bit implementation performed twice as many operations as the 32-bit approaches, with similar logic applied to the 12-bit implementation. The results can be seen in Table 4. We can see that the 12-bit and 16-bit approaches show the best performance, however a correction step is required for these figures. As we will see, the base extension executes in $O(n)$ time across n processors, where n is the number of moduli in the RNS base. In the context of 1024-bit RSA, the 12-bit approach requires a minimum of 43 moduli (512 bits / 12 bits) compared to 17 32-bit moduli for each RNS base. Also, the base extension step in Montgomery RNS is the most intensive part of our implementations consuming 80% of the execution time. A minimum approximation correction for the 12-bit result presented here is a division of 2, and for 16-bit 1.5. In effect the most efficient approach for use in Montgomery RNS is **Reduction by Residue Multiplication**.

Table 4. GPU Modular Multiplication throughput using a variety of techniques

	Modular Multiplication Approach	Modular multiplications per second
1.	32-bit LongDiv	$2.89 * 10^9$
2.	32-bit Inverse Mul	$3.63 * 10^9$
3.	32-bit Residue Mul	$4.64 * 10^9$
4.	32-bit Native+CRT	$1.12 * 10^9$
5.	16-bit Native	$4.71 * 10^9$
6.	12-bit Native	$7.99 * 10^9$

5.3   Exponentation Using Kawamura on the GPU

Our Montgomery RNS implementation is based on Kawamura et al.'s [17] approach. Its base extension algorithm relies on the following representation of CRT.

$x = \sum_{i=1}^{n}(|x|_{mi}|M_i^{-1}|_{mi} \bmod m_i)M_i - kM$, where m is a set of moduli, $M = \prod_{i=1}^{n} m_i$, $M_i = M/m_i$ and $|M_i^{-1}|_{mi}$ is the multiplicative inverse of $M_i(\bmod m_i)$. To base extend from $< x >_m$ to a single moduli m_1' and letting $E_i = |x|_{mi}|M_i^{-1}|_{mi} \bmod m_i$ we can write $|x|_{m'1} = |\sum_{i=1}^{n}(|E_iM_i|_{m'1}) - k|M|_{m'1}|_{m'1}$.

Here E_i, for each i, can be calculated independently and thus in parallel. $|M_i|_{m'1}$ and $|M|_{m'1}$ are based on invariant moduli and as such can be precalculated. To calculate the base extension into multiple new moduli, m', each residue can be calculated independently. Also, if k can be calculated in parallel, we can use a parallel reduction which executes a base extension in $O(\log n)$ time, as reported in papers based on this technique [19]. However, in practice this requires a growth of processing units in the order of $O(n^2)$, where n is the number of moduli in the new base. On an SPMD machine, and with regards to throughput and not latency, it is more efficient to apply the effect of k serially for each new modulus.

The generation of k above can be written as $\lfloor \sum_{i=1}^{n} E_i/m_i \rfloor$. Kawamura et al. calculate this divide using an approximation based on the observation that

m_i can be close to a power of 2. Also E_i is approximated, using a certain number of its most signification bits (the emphasis for Kawamura's approach is on VLSI design). In Table 5 we present a modified version of Kawamura's base extension algorithm which does not use the approximation of E_i and is suitable for a 32-bit processor. Inputs: $< x >_m$, m, m', α, Output: $< z >_{m \cup m'} = (< x >_m, < x >'_m)$, where m is the source base, m' is the new destination base and α is used to compensate for the approximations introduced, typically set to $2^{N/2}$.

Table 5. Optimised Kawamura base extension for 32-bit processor

$E_i = \|\|x\|_{mi} M_i^{-1}\|_{mi} (\forall i)$
for $j = 1$ to n
$\quad \alpha 0 = \alpha,\ \alpha + = E_i\ /*$ note α wraps at 2^{32} on GPU*/
$\quad if(\alpha < \alpha 0)\ r_i = \|r_i + (\| - M\|_{m'i})\|_{m'i} (\forall i)$
$\quad r_i = \|r_i + E_j\|M_j\|_{m'i}\|_{m'i} (\forall i)$

α must be higher than the maximum error caused by the approximations above, however lower than 2^N [17], where N is the word bit length of the GPU. As we have removed the error due to approximation of E_i the only determinant of the size of the error is the distance between the moduli used and their next power of 2. This puts a restriction on the number of moduli that can be used with this technique. This base extension can be used in the context of 1024-bit RSA with 32 and 16-bit moduli, while 12-bit moduli require the use of a different method.

Our most performance competitive implementation is based on the Reduction by Residue Multiplication approach for modular multiplication as described above, and as such uses 32-bit moduli. For 1024-bit RSA, we require two sets of 17 32-bit moduli for use as the two bases, assuming an implementation based on Quisquater et al.'s CRT approach [21]. Groups of 17 consecutive threads within a CUDA block operate co-operatively to calculate an exponentiation. Each thread reads in two residues, $|x|_{ai}$ and $|x|_{bi}$, thus a single thread executes both the left and right sides of the Montgomery RNS algorithm, see Table 1, for a pair of residues, ensuring each thread is continuously busy.

Memory Usage: As the residues are in groups of 17, we employ a padding scheme for the ciphertext/plaintext data whereby the start address used by the first thread of a CUDA block is aligned to a 128-bit boundary. We also pad the number of threads per block to match this padding of input data, which allows a simple address mapping scheme while allowing for fully coalesced reads. The CUDA thread allocation scheme for ensuring even distribution across available SMs and also correct padding is show in Table 6, where RNS_SIZE is the number of moduli per base, MAX_THREADS_PER_BLOCK is a predefined constant dependant on shared resource pressure of the kernel and BLOCK_GROUP is the number of SMs on the GPU.

Each thread is responsible for calculating a single E_i, after which point a synchronisation barrier is used to ensure all new E_i values can be safely used

Table 6. CUDA thread allocation scheme for RNS based modular exponentiation

$total\_threads = noMsgs * RNS\_SIZE * 2$
$blocks\_required = \lceil total\_threads / MAX\_THREADS\_PER\_BLOCK \rceil$
$blocks\_required = \lceil blocks\_required \rceil^{BLOCK\_GROUP}$
$threads\_per\_block = \lceil total\_threads / blocks\_required \rceil$
$threads\_per\_block = \lceil \lceil threads\_per\_block \rceil^{RNS\_SIZE} \rceil^{WARP\_SIZE}$

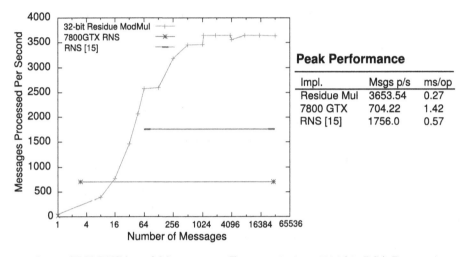

Peak Performance

Impl.	Msgs p/s	ms/op
Residue Mul	3653.54	0.27
7800 GTX	704.22	1.42
RNS [15]	1756.0	0.57

Fig. 4. GPU RNS based Montgomery Exponentiation: 1024-bit RSA Decryption

by all threads. As discussed in Section 4, this synchronisation barrier, along with the general performance issues with thread divergence dictates that only a single exponent can be used for each CUDA block of threads. In practice, for peak performance with 1024-bit RSA, a single exponent should be used a maximum of once for every 15 primitives (256 threads per block / 17 threads per primitive). With regards to the shared memory use, we use two different locations for storing the values of E_i. The storage locations are alternated for each subsequent base extension. This permits a single synchronisation point to be used, rather than two - one before and one after the generation of E_i, which is necessary when only one location is used for storing the values of E_i. We also use shared memory to accelerate the most intensive table lookup corresponding to $|M_j|_{m'i}$ in the base extension. At the start of each kernel call, all threads within a block cooperate in loading into shared memory the entirety of the two arrays, $|A_j|_{bi}$ and $|B_j|_{ai}$ (one for each base), via texture lookups.

5.4   Results

Figure 4 shows the throughput of our RNS implementation using CRT, sliding window and 32-bit modular reduction using the Reduction by Residue Multiplication as described previously. Also included is the peak RNS performance from

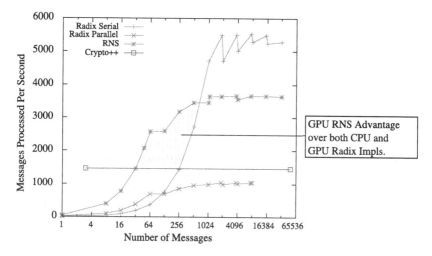

Fig. 5. RNS vs Radix: 1024-bit RSA Decryption

Szerwinski et al. [15] at 1756 primitives per second, we can see that our peak performance has 2 times higher throughput. The gains in performance is deemed mainly to come from the improved single precision modular multiplication and optimised base extension algorithm. Like in the radix results, we have only plotted the graph for [15] from the point of the reported minimum latency. We have also included as a historical reference, the previous RNS implementation on an Nvidia 7800GTX by Moss et al. [10] to show the improvements possible due to the advances in hardware and software libraries available. [10] reports a performance of 175.5 true 1024-bit exponentiation operations per second, which we have multiplied by 4 [4] and use as a best case scenario estimate of the runtime for a 1024-bit RSA decrypt, i.e. split into 2 512-bit exponentiations. Comparing the peak performance of the Crypto++ CPU implementations listed in Figure 3 we can see that our peak performance for our RNS implementation has up to 2.5 times higher throughput.

5.5   Radix vs. RNS on the GPU

We use Figure 5 to illustrate the effectiveness of our RNS implementation at accelerating RSA in comparison to our radix based implementations. As can be seen the RNS implementation gives superior throughput with much smaller number of messages per kernel call. The point at which the serial radix approach becomes faster than the CPU is at 256 messages, where the RNS approach has better performance at 32 messages per kernel. The greater performance at smaller message numbers is due to a higher GPU occupancy for the RNS approach over the serial radix approach. The RNS approach also does not suffer from the extreme levels of synchronisation during a Montgomery multiplication as the parallel radix approach. Using RNS can greatly improve the GPU's ability to provide feasible acceleration for RSA decryption, or any public key cryptographic scheme where the modulus and exponent change with a low frequency. It is clear from Figure 5 that an adaptive approach to exponentiation would provide the best overall performance, switching from an RNS based implementation at low message requirements, to a radix based approach at high message requirements.

6   Conclusions

In this paper we have presented implementations of modular exponentiation suitable for public key cryptography. We have focused on 1024-bit RSA decryption running on an Nvidia 8800GTX and demonstrated a peak throughput of 0.18 ms/op giving a 4 times improvement over a comparable CPU implementation. We have also shown that the significant problems with latency in previous GPU implementations can be overcome while also improving on throughput rates. We have shown that an adaptive approach to modular exponentiation on the GPU provides the best performance across a range of usage scenarios. A standard serial implementation of Montgomery exponentiation gives the best performance in the context of a high number of parallel messages, while an RNS based Montgomery exponentiation gives better performance with fewer messages. We show that an optimised RNS approach is more performant than a CPU implementation at 32 messages per kernel call and that the pencil-and-paper approach proves better than the RNS approach at 256 messages.

Also covered in the paper is the applicability of the GPU to general public key cryptography, where the observation is made that peak performance is only achievable in the context of substantial key reuse. In the case of 1024-bit RSA using RNS, peak performance requires the key to change at a maximum rate of once per 15 messages, and once per 32 messages when using a serial pencil-and-paper approach. Thus the GPU can be effectively used in an RSA decryption capacity where it is common for a server to use a limited number of keys.

RNS based approaches are highly dependent on efficient support of single precision modular multiplication, which the paper illustrates is non trivial on the GPU. In this context we have explored a variety of techniques for achieving efficient modular multiplication and show that a novel approach suited to RNS and the GPU gives the best performance. The GPU could offer improved performance with RNS based approaches if future architectures provide efficient native modular operations. Future work consists of extending the implementations presented to cater for larger public key exponents and messages such those used in 2048 and 4096-bit RSA. This may lead to the feasibility of other forms of big integer multiplication such as Karatsuba or Toom-Cook techniques.

References

1. Nvidia CUDA Programming Guide, Version 2.0 (2008)
2. Microsoft, Direct X Technology, http://msdn.microsoft.com/directx/

3. Nvidia Corporation, "CUDA",
   http://developer.nvidia.com/object/cuda.html
4. Menezes, A., van Oorschot, P., Vanstone, S.: Handbook of Applied Cryptography. CRC Press, Boca Raton (1996) ISBN 0-8493-8523-7
5. Montgomery, P.L.: Modular Multiplication Without Trial Division. Mathematics of Computation 44, 519–521 (1985)
6. Cook, D., Ioannidis, J., Keromytis, A., Luck, J.: CryptoGraphics: Secret Key Cryptography Using Graphics Cards. In: Menezes, A. (ed.) CT-RSA 2005. LNCS, vol. 3376, pp. 334–350. Springer, Heidelberg (2005)
7. Harrison, O., Waldron, J.: AES encryption implementation and analysis on commodity graphics processing units. In: Paillier, P., Verbauwhede, I. (eds.) CHES 2007. LNCS, vol. 4727, pp. 209–226. Springer, Heidelberg (2007)
8. Yang, J., Goodman, J.: Symmetric Key Cryptography on Modern Graphics Hardware. In: Kurosawa, K. (ed.) ASIACRYPT 2007. LNCS, vol. 4833, pp. 249–264. Springer, Heidelberg (2007)
9. Harrison, O., Waldron, J.: Practical Symmetric Key Cryptography on Modern Graphics Hardware. In: 17th USENIX Security Symposium, San Jose, CA, July 28 - August 1 (2008)
10. Moss, A., Page, D., Smart, N.P.: Toward Acceleration of RSA Using 3D Graphics Hardware. In: 11th IMA International Conference on Cryptography and Coding, Cirencester, UK, December 18-20 (2007)
11. Fleissner, S.: GPU-Accelerated Montgomery Exponentiation. In: Shi, Y., van Albada, G.D., Dongarra, J., Sloot, P.M.A. (eds.) ICCS 2007. LNCS, vol. 4487, pp. 213–220. Springer, Heidelberg (2007)
12. AMD 64 RSA Benchmarks, http://www.cryptopp.com/benchmarks-amd64.html
13. Knuth, D.E.: The Art of Computer Programming, 3rd edn., vol. 2. Addison-Wesley, Reading (1997)
14. OpenSSL Open Source Project, http://www.openssl.org/
15. Szerwinski, R., Güneysu, T.: Exploiting the Power of GPUs for Asymmetric Cryptography. In: Oswald, E., Rohatgi, P. (eds.) CHES 2008. LNCS, vol. 5154, pp. 79–99. Springer, Heidelberg (2008)
16. Posch, K.C., Posch, R.: Modulo Reduction in Residues Numbers Systems. IEEE Trans. on Parallel and Distributed Systems 6(5), 449–454 (1995)
17. Kawamura, S., Koike, M., Sano, F., Shimbo, A.: Cox-Rower Architecture for Fast Parallel Montgomery Multiplication. In: Preneel, B. (ed.) EUROCRYPT 2000. LNCS, vol. 1807, pp. 523–538. Springer, Heidelberg (2000)
18. Szabo, N.S., Tanaka, R.I.: Residue Arithmetic and its Applications to Computer Technology. McGraw-Hill, New York (1967)
19. Posch, K.C., Posch, R.: Base Extension Using a Convolution Sum in Residue Number Systems. Computing 50, 93–104 (1993)
20. Granlund, T., Montgomery, P.: Division by Invariant Integers using Multiplication. In: SIGPLAN 1994 Conference on Programming Language Design and Implementation, Orlando, Florida (June 1994)
21. Quisquater, J.-J., Couvreur, C.: Fast Decipherment Algorithm for RSA Public-Key Cryptosystem. Electronics Letters 18(21), 905–907 (1982)

Fast Elliptic-Curve Cryptography on the Cell Broadband Engine

Neil Costigan[1] and Peter Schwabe[2,*]

[1] School of Computing
Dublin City University, Glasnevin, Dublin 9, Ireland
neil.costigan@computing.dcu.ie
[2] Department of Mathematics and Computer Science
Technische Universiteit Eindhoven, P.O. Box 513, 5600 MB Eindhoven, Netherlands
peter@cryptojedi.org

Abstract. This paper is the first to investigate the power of the Cell Broadband Engine for state-of-the-art public-key cryptography. We present a high-speed implementation of elliptic-curve Diffie-Hellman (ECDH) key exchange for this processor, which needs 697080 cycles on one Synergistic Processor Unit for a scalar multiplication on a 255-bit elliptic curve, including the costs for key verification and key compression. This cycle count is independent of inputs therefore protecting against timing attacks.

This speed relies on a new representation of elements of the underlying finite field suited for the unconventional instruction set of this architecture.

Furthermore we demonstrate that an implementation based on the multi-precision integer arithmetic functions provided by IBM's multi-precision math (MPM) library would take at least 2227040 cycles.

Comparison with implementations of the same function for other architectures shows that the Cell Broadband Engine is competitive in terms of cost-performance ratio to other recent processors such as the Intel Core 2 for public-key cryptography.

Specifically, the state-of-the-art Galbraith-Lin-Scott ECDH software performs 27370 scalar multiplications per second using all four cores of a 2.5GHz Intel Core 2 Quad Q9300 inside a $296 computer, while the new software reported in this paper performs 27474 scalar multiplications per second on a Playstation 3 that costs just $221. Both of these speed reports are for high-security 256-bit elliptic-curve cryptography.

Keywords: Cell Broadband Engine, elliptic-curve cryptography (ECC), efficient implementation.

* The first author was supported by the Irish Research Council for Science, Engineering and Technology (IRCSET). The second author was supported by the European Commission through the ICT Programme under Contract ICT–2007–216499 CACE, and through the ICT Programme under Contract ICT-2007-216646 ECRYPT II. Permanent ID of this document: a33572712a64958c0bf522e608f25f0d. Date: Mar 30, 2009.

1   Introduction

This paper describes a high-speed implementation of state-of-the-art public-key cryptography for the Cell Broadband Engine (CBE). More specifically we describe an implementation of the `curve25519` function, an elliptic-curve Diffie-Hellman key exchange (ECDH) function introduced in [3].

Implementations of this function have been achieving speed records for high-security ECDH software on different platforms [3], [9]. Benchmarks of our implementation show that the CBE is competitive (in terms of cost-performance ratio) to other recent processors as the Intel Core 2 for public-key cryptography.

Our implementation needs 697080 cycles on one Synergistic Processor Unit (SPU) of the CBE. This number includes not only scalar multiplication on the underlying 255-bit elliptic curve, but furthermore costs for key compression, key validation and protection against timing attacks. We put our implementation into the public domain to maximize reusability of our results. It is available as part of the SUPERCOP benchmarking suite [4] and at `http://cryptojedi.org/crypto/index.shtml#celldh`.

We wish to thank Dan Bernstein, Tanja Lange, Ruben Niederhagen, and Michael Scott for their invaluable feedback and encouragement. Neil Costigan would also like to thank Luleå University of Technology, Sweden.

1.1   How These Speeds Were Achieved

Elliptic-curve cryptography (ECC) is usually implemented as a sequence of arithmetic operations in a finite field. IBM provides a multi-precision math (MPM) library developed especially for the Cell Broadband Engine as part of the standard Cell Software Development Kit (SDK). The obvious approach for the implementation of ECC on the CBE is thus to use this library for the underlying finite field arithmetic.

However, we will show that the targeted performance cannot be achieved following this approach, not even with optimizing some functions of the MPM library for arithmetic in fields of the desired size.

Instead, the speed of our implementation is achieved by

- Parting with the traditional way of implementing elliptic-curve cryptography which uses arithmetic operations in the underlying field as smallest building blocks,
- Representing finite field elements in a way that takes into account the special structure of the finite field and the unconventional SPU instruction set, and
- Careful optimization of the code on assembly level.

Related work. Implementations of public-key cryptography for the Cell Broadband Engine have not yet been extensively studied. In particular we don't know of any previous implementation of ECC for the Cell Broadband Engine.

Costigan and Scott investigate in [5] the use of IBM's MPM library to accelerate OpenSSL on a Sony Playstation 3. The paper reports benchmarks for

RSA with different key lengths; RSA signature generation with a 2048 bit key is reported to take 0.015636s corresponding to 50035200 cycles on one SPU.

An implementation of the Digital Signature Algorithm (DSA) supporting key lengths up to 1024 bits is included in the SPE Cryptographic Library [16].

In [18] Shimizu et al. report 4074000 cycles for 1024-bit-RSA encryption or decryption and 1331000 cycles for 1024-bit-DSA key generation. Furthermore they report 2250000 cycles for 1024-bit-DSA signature generation and 4375000 cycles for 1024-bit-DSA signature verification.

The Cell Broadband Engine has recently demonstrated its power for cryptanalysis of symmetric cryptographic primitives [20], [19].

Organization of the paper. In Section 2 we will briefly review the features of the CBE which are relevant for our implementations. Section 3 describes the `curve25519` function including some necessary background on elliptic-curve arithmetic. Section 4 describes IBM's MPM library including optimizations we applied to accelerate arithmetic in finite fields of the desired size. We show that an implementation based on this library cannot achieve the targeted performance. In Section 5 we detail our implementation of `curve25519`. We conclude the paper with a discussion of benchmarking results and a comparison to ECDH implementations for other architectures in Section 6.

2   The Cell Broadband Engine

When it became apparent that multi-core chip design rather than increased frequency was the gateway to more efficient CPUs, IBM, Sony and Toshiba took a novel approach: Instead of developing a chip with every core being the same, they came up with the Cell Broadband Engine Architecture (CBEA). Currently two different CBEA processors are available: the Cell Broadband Engine (CBE) and the PowerXCell 8i processor. Both are multi-core processors consisting of a traditional central processor and 8 specialized high performance processors called Synergistic Processor Units (SPUs).

These units are combined across a high bandwidth (204 GB/s) [12] bus to offer a multi-core environment with two instruction sets and enormous processing power. Compared with the CBE the PowerXCell 8i processor has highly increased double precision floating point performance. The implementation described in this paper is optimized for the CBE, we will therefore in the following focus on the description of this processor.

The Cell Broadband Engine can be found in the Sony Playstation 3 and the IBM QS20 and QS21 blade server series. Note that the CBE in the Playstation 3 makes just 6 out of 8 SPUs available for general purpose computations. Toshiba equips several laptops of the Qosmio series with the SpursEngine consisting of 4 SPUs intended for media processing. This SpursEngine can also be found in a PCI Express card called WinFast pxVC1100 manufactured by Leadtek which is currently available only in Japan.

The primary processor of the Cell Broadband Engine is a 64-bit Power Processor Unit (PPU). This PPU works as a supervisor for the other cores. Currently operating at 3.2GHz, the PPU is a variant of the G5/PowerPC product line, a RISC driven processor found in IBM's servers and Apple's PowerMac range.

2.1   The Cell's SPU

The real power of the CBE is in the additional SPUs. Each SPU is a specialist processor with a RISC-like SIMD instruction set and a 128-element array of 128-bit registers. It has two pipelines (pipeline 0 and pipeline 1); each cycle it can dispatch one instruction per pipeline. Whether or not the SPU really dispatches two instructions in a given cycle is highly dependent on instruction scheduling and alignment. This is subject to the following conditions:

- Execution of instructions is purely in-order.
- The two pipelines execute disjoint sets of instructions (i.e. each instruction is either a pipeline-0 or a pipeline-1 instruction).
- The SPU has a fetch queue that can contain at most two instructions.
- Instructions are fetched into the fetch queue only if the fetch queue is empty.
- Instructions are fetched in pairs; the first instruction in such a pair is from an even word address, the second from an odd word address.
- The SPU executes two instructions in one cycle only if two instructions are in the fetch queue, the first being a pipeline-0 instruction and the second being a pipeline-1 instruction and all inputs to these instructions being available and not pending due to latencies of previously executed instructions.

Hence, instruction *scheduling* has to ensure that pipeline-0 and pipeline-1 instructions are interleaved and that latencies are hidden; instruction *alignment* has to ensure that pipeline-0 instructions are at even word addresses and pipeline-1 instructions are at odd word addresses.

Both our implementation and the MPM library build the finite field arithmetic on the integer arithmetic instructions of the SPU. This is due to the fact that single-precision floating-point arithmetic offers a too small mantissa and that double-precision floating-point arithmetic causes excessive pipeline stalls on the SPU and is therefore very inefficient.

All integer arithmetic instructions (except shift and rotate instructions) are SIMD instructions operating either on 4 32-bit word elements or on 8 16-bit halfword elements or on 16 8-bit byte elements of a 128-bit register.

Integer multiplication constitutes an exception to this scheme: The integer multiplication instructions multiply 4 16-bit halfwords in parallel and store the 32-bit results in the 4 word elements of the result register.

The following instructions are the most relevant for our implementation; for a detailed description of the SPU instruction set see [14], for a list of instruction latencies and associated pipelines see [13, Appendix B].

a: Adds each 32-bit word element of a register a to the corresponding word element of a register b and stores the results in a register r.

mpy: Multiplies the 16 least significant bits of each 32-bit word element of a register a with the corresponding 16 bits of each word element of a register

b and stores the resulting four 32-bit results in the four word elements of a register r.

mpya: Multiplies 16-bit halfwords as the mpy instruction but adds the resulting four 32-bit word elements to the corresponding word elements of a register c and stores the resulting sum in a register r.

shl: Shifts each word element of a register a to the left by the number of bits given by the corresponding word element of a register b and stores the result in a register r.

rotmi: Shifts of each word element of a register a to the right by the number of bits given in an immediate value and stores the result in a register r.

shufb: Allows to set each byte of the result register r to either the value of an arbitrary byte of one of two input registers a and b or to a constant value of 0, 0x80 or 0xff.

2.2 Computation Micro-kernels for the SPU

Alvaro, Kurzak and Dongarra [1] introduce the idea of a *computation micro-kernel* for the SPU where the restricted code and data size of the SPU become important design criteria but issues such as inter-chip communication and synchronization are not considered. The kernel focuses on utilization of the wide registers and the instruction level parallelism. Furthermore, for security aware applications such as those using ECC, there is an interesting security architecture where an SPU can run in *isolation mode*, where inter-chip communications, loading and unloading program code incur significant overhead. In this paper we describe such a computation micro-kernel implementation running on one SPU of the CBE.

3   ECDH and the curve25519 Function

3.1   Elliptic-Curve Diffie-Hellman Key Exchange (ECDH)

Let $\mathbb{F}$ be a finite field and $E/\mathbb{F}$ an elliptic curve defined over $\mathbb{F}$. Let $E(\mathbb{F})$ denote the group of $\mathbb{F}$-rational points on E. For any $P \in E(\mathbb{F})$ and $k \in \mathbb{Z}$ we will denote the k-th scalar multiple of P as $[k]P$.

The Diffie-Hellman key exchange protocol [6] can now be carried out in the group $\langle P \rangle \subseteq E(\mathbb{F})$ as follows: User A chooses a random $a \in \{2, \ldots, |\langle P \rangle| - 1\}$, computes $[a]P$ and sends this to user B. User B chooses a random $b \in \{2, \ldots, |\langle P \rangle| - 1\}$, computes $[b]P$ and sends this to user A. Now both users can compute $Q = [a]([b]P) = [b]([a]P) = [(a \cdot b)]P$. The joint key for secret key cryptography is then extracted from Q; a common way to do this is to compute a hash value of the x-coordinate of Q.

3.2   Montgomery Arithmetic

For elliptic curves defined by an equation of the form $By^2 = x^3 + Ax^2 + x$, Montgomery introduced in [17] a fast method to compute the x-coordinate of a point $R = P + Q$, given the x-coordinates of two points P and Q and the x-coordinate of their difference $P - Q$.

These formulas lead to an efficient algorithm to compute the x-coordinate of $Q = [k]P$ for any point P. This algorithm is often referred to as the Montgomery ladder. In this algorithm the x-coordinate x_P of a point P is represented as (X_P, Z_P), where $x_P = X_P/Z_P$; for the representation of the point at infinity see the discussion in Appendix B of [3]. See Algorithms 1 and 2 for a pseudocode description of the Montgomery ladder.

Algorithm 1. The Montgomery ladder for x-coordinate-based scalar multiplication on the elliptic curve $E : By^2 = x^3 + Ax^2 + x$

Input: A scalar $0 \leq k \in \mathbb{Z}$ and the x-coordinate x_P of some point P
Output: $(X_{[k]P}, Z_{[k]P})$ fulfilling $x_{[k]P} = X_{[k]P}/Z_{[k]P}$
$\quad t = \lceil \log_2 k + 1 \rceil$
$\quad X_1 = x_P;\ X_2 = 1;\ Z_2 = 0;\ X_3 = x_P;\ Z_3 = 1$
$\quad$ **for** $i \leftarrow t - 1$ downto 0 **do**
$\quad\quad$ **if** bit i of k is 1 **then**
$\quad\quad\quad (X3, Z3, X2, Z2) \leftarrow \text{LADDERSTEP}(X1, X3, Z3, X2, Z2)$
$\quad\quad$ **else**
$\quad\quad\quad (X2, Z2, X3, Z3) \leftarrow \text{LADDERSTEP}(X1, X2, Z2, X3, Z3)$
$\quad\quad$ **end if**
$\quad$ **end for**
$\quad$ **return** (X_2, Z_2)

Each "ladder step" as described in Algorithm 2 requires 5 multiplications, 4 squarings, 8 additions and one multiplication with the constant $a24 = (A+2)/4$ in the underlying finite field.

3.3 The curve25519 Function

Bernstein proposed in [3] the curve25519 function for elliptic-curve Diffie-Hellman key exchange. This function uses arithmetic on the elliptic curve defined by the equation $E : y^2 = x^3 + Ax^2 + x$ over the field $\mathbb{F}_p$, where $p = 2^{255} - 19$ and $A = 486662$; observe that this elliptic curve allows for the x-coordinate-based scalar multiplication described above.

The elliptic curve and underlying finite field are carefully chosen to meet high security requirements and to allow for fast implementation, for a detailed discussion of the security properties of curve25519 see [3].

The curve25519 function takes as input two 32-byte strings, one representing the x-coordinate of a point P and the other representing a 256-bit scalar k. It gives as output a 32-byte string representing the x-coordinate x_Q of $Q = [k]P$. For each of these values curve25519 is assuming little-endian representation.

For our implementation we decided to follow [3] and compute x_Q by first using Algorithm 1 to compute (X_Q, Z_Q) and then computing $x_Q = Z_Q^{-1} \cdot X_Q$.

Algorithm 2. One "ladder step" of the Montgomery ladder

---

const $a24 = (A+2)/4$ (A from the curve equation)
function LADDERSTEP($X_{Q-P}, X_P, Z_P, X_Q, Z_Q$)
$\quad t_1 \leftarrow X_P + Z_P$
$\quad t_6 \leftarrow t_1^2$
$\quad t_2 \leftarrow X_P - Z_P$
$\quad t_7 \leftarrow t_2^2$
$\quad t_5 \leftarrow t_6 - t_7$
$\quad t_3 \leftarrow X_Q + Z_Q$
$\quad t_4 \leftarrow X_Q - Z_Q$
$\quad t_8 \leftarrow t_4 \cdot t_1$
$\quad t_9 \leftarrow t_3 \cdot t_2$
$\quad X_{P+Q} \leftarrow (t_8 + t_9)^2$
$\quad Z_{P+Q} \leftarrow X_{Q-P} \cdot (t_8 - t_9)^2$
$\quad X_{[2]P} \leftarrow t_6 \cdot t_7$
$\quad Z_{[2]P} \leftarrow t_5 \cdot (t_7 + a24 \cdot t_5)$
$\quad$**return** $(X_{[2]P}, Z_{[2]P}, X_{P+Q}, Z_{P+Q})$
end function

---

4   The MPM Library and ECC

4.1   Implementation Hierarchy

Implementations of elliptic-curve cryptography are often described as a hierarchy of operations. In [11, Section 5.2.1] Hankerson, Menezes and Vanstone outline a hierarchy of operations in ECC as protocols, point multiplication, elliptic-curve addition and doubling, finite-field arithmetic. Fan, Sakiyama & Verbauwhede expand this in [7] to describe a 5-layer pyramid of

1. Integrity, confidentially, authentication,
2. Elliptic-curve scalar multiplication $[k]P$,
3. Point addition and doubling,
4. Modular operations in $\mathbb{F}_p$,
5. Instructions of a w-bit core.

4.2 $\mathbb{F}_p$ Arithmetic Using the MPM Library

In Section 3 we described how the upper 3 layers of this hierarchy are handled. Hence, the obvious next step is to look at efficient modular operations in $\mathbb{F}_p$ and how these operations can be mapped to the SIMD instructions on 128-bit registers of the SPU.

For this task of mapping operations on large integers to the SPU instruction set, IBM provides a vector-optimized multi-precision math (MPM) library [15] as part of the software development kit (SDK) offered for the Cell Broadband Engine.

This MPM library is provided in source code and its algorithms are generic for arbitrary sized numbers. They operate on 16-bit halfwords as smallest units, elements of our 255-bit field are therefore actually handled as 256-bit values.

As our computation is mostly bottlenecked by costs for multiplications and squarings in the finite field we decided to optimize these functions for 256-bit input values.

At a high level the original MPM multiplication functions are structured by an array declaration section, then array initialization via loop dictated by the input sizes, a main body consisting of calculating partial products inside nested loops (again determined by the input sizes), and finally a gather section where the partial products feed into the result.

We optimized this pattern for fixed 256-bit inputs by using the large register array to remove the implicit array calls and then fully unrolled the loops. The Montgomery routine also lets us use a __builtin_expect compiler directive at the final overflow test to direct the hardware to which branch will be taken and avoid an expensive stall. While these manual unroll and branch hint techniques help both the GCC and IBM XLC compilers it should be noted that the GCC-derived compiler achieves a 10% improvement over the XLC compiler[1].

The MPM library supplies a specialized function for squaring where significant optimizations should be made over a general multiply by reusing partial products. However our timings indicate that such savings are not achieved until the size of the multi-precision inputs exceeds 512-bits. We therefore take the timings of a multiplication for a squaring.

4.3 What Speed Can We Achieve Using MPM?

The Montgomery ladder in the curve25519 computation consists of 255 ladder steps, hence, computation takes 1276 multiplications, 1020 squarings, 255 multiplications with a constant, 2040 additions and one inversion in the finite field $\mathbb{F}_{2^{255}-19}$. Table 1 gives the number of CPU cycles required for each of these operations (except inversion).

For finite field multiplication and squaring we benchmarked two possibilities: a call to mpm_mul followed by a call to mpm_mod and the Montgomery multiplication function mpm_mont_mod_mul. Addition is implemented as a call to mpm_add and a conditional call (mpm_cmpge) to mpm_sub. For multiplication we include timings of the original MPM functions and of our optimized versions. The original MPM library offers a number of options for each operation. We select the inlined option with equal input sizes for fair comparison.

From these numbers we can compute a lower bound of 2227040 cycles (1276M + 1020S + 2040A, where M, S and A stand for the costs of multiplication, squaring and addition respectively) required for the curve25519 computation when using MPM. Observe that this lower bound still ignores costs for the inversion and for multiplication with the constant.

---

[1] IBM XL C/C++ for Multicore Acceleration for Linux, V10.1. CBE SDK 3.1.

Table 1. MPM performance for arithmetic operations in a 256-bit finite field

Operation	Number of cycles
Addition/Subtraction	86
Multiplication (original MPM)	4334
Multiplication (optimized)	4124
Montgomery Multiplication (original MPM)	1197
Montgomery Multiplication (optimized)	892

The high cost for modular reduction in these algorithms results from the fact, that the MPM library cannot make use of the special form of the modulus $2^{255} - 19$; an improved, specialized reduction routine would probably yield a smaller lower bound. We therefore investigate what lower bound we get when entirely ignoring costs for modular reduction. Table 2 gives numbers of cycles for multiplication and addition of 256-bit integers without modular reduction. This yields a lower bound of 934080 cycles. Any real implementation would, of course, take significantly more time as it would need to account for operations not considered in this estimation.

Table 2. MPM performance for arithmetic operations on 256-bit integers

Operation	Number of cycles
Addition/Subtraction	52
Multiplication (original MPM)	594
Multiplication (optimized)	360

5   Implementation of curve25519

As described in Section 3 the computation of the curve25519 function consists of two parts, the Montgomery ladder computing (X_Q, Z_Q) and the inversion of Z_Q.

We decided to implement the inversion as an exponentiation with $p - 2 = 2^{255} - 21$ using the the same sequence of 254 squarings and 11 multiplications as [3]. This might not be the most efficient algorithm for inversion, but it is the easiest way to implement an inversion algorithm which takes constant time independent of the input.

The addition chain is specialized for the particular exponent and cannot be implemented as a simple square-and-multiply loop; completely inlining all multiplications and squarings would result in an excessive increase of the overall code size. We therefore implement multiplication and squaring functions and use calls to these functions.

However for the first part—the Montgomery ladder—we do not use calls to these functions but take one ladder step as smallest building block and implement the complete Montgomery ladder in one function. This allows for a higher degree of data-level parallelism, especially in the modular reductions, and thus yields a significantly increased performance.

For the speed-critical parts of our implementation we use the qhasm programming language [2], which offers us all flexibility for code optimization on the assembly level, while still supporting a more convenient programming model than plain assembly. We extended this language to also support the SPU of the Cell Broadband Engine as target architecture.

In the description of our implementation we will use the term "register variable". Note that for qhasm (unlike C) the term register variable refers to variables that are forced to be kept in registers.

5.1   Fast Arithmetic

In the following we will first describe how we represent elements of the finite field $\mathbb{F}_{2^{255}-19}$ and then detail the three algorithms that influence execution speed of curve25519 most, namely finite field multiplications, finite field squaring and a Montgomery ladder step.

5.2   Representing Elements of $\mathbb{F}_{2^{255}-19}$

We represent an element a of $\mathbb{F}_{2^{255}-19}$ as a tuple $(a_0, \ldots, a_{19})$ where

$$a = \sum_{i=0}^{19} a_i 2^{\lceil 12.75i \rceil}. \tag{1}$$

We call a coefficient a_i reduced if $a_i \in [0, 2^{13} - 1]$. Analogously we call the representation of an element $a \in \mathbb{F}_{2^{255}-19}$ reduced if all its coefficients $a_0, \ldots, a_{19}$ are reduced.

As described in section 2 the Cell Broadband Engine can only perform 16-bit integer multiplication, where one instruction performs 4 such multiplications in parallel. In order to achieve high performance of finite field arithmetic it is crucial to properly arrange the values $a_0, \ldots a_{19}$ in registers and to adapt algorithms for field arithmetic to make use of this SIMD capability.

Multiplication and Squaring in $\mathbb{F}_{2^{255}-19}$. As input to field multiplication we get two finite field elements $(a_0, \ldots, a_{19})$ and $(b_0, \ldots, b_{19})$. We assume that these field elements are in reduced representation. This input is arranged in 10 register variables a03, a47, a811, a1215, a1619, b03, b47, b811, b1215 and b1619 as follows: Register variable a03 contains in its word elements the coefficients a_0, a_1, a_2, a_3, register variable a47 contains in its word elements the coefficients a_4, a_5, a_6, a_7, and so on.

The idea of multiplication is to compute coefficients $r_0, \ldots, r_{38}$ of $r = ab$ where:

$$r_0 = a_0 b_0$$
$$r_1 = a_1 b_0 + a_0 b_1$$
$$r_2 = a_2 b_0 + a_1 b_1 + a_0 b_2$$
$$r_3 = a_3 b_0 + a_2 b_1 + a_1 b_2 + a_0 b_3$$
$$r_4 = a_4 b_0 + 2a_3 b_1 + 2a_2 b_2 + 2a_1 b_3 + a_0 b_4$$
$$r_5 = a_5 b_0 + a_4 b_1 + 2a_3 b_2 + 2a_2 b_3 + a_1 b_4 + a_0 b_5$$
$$r_6 = a_6 b_0 + a_5 b_1 + a_4 b_2 + 2a_3 b_3 + a_2 b_4 + a_1 b_5 + a_0 b_6$$
$$r_7 = a_7 b_0 + a_6 b_1 + a_5 b_2 + a_4 b_3 + a_3 b_4 + a_2 b_5 + a_1 b_6 + a_0 b_7$$
$$r_8 = a_8 b_0 + 2a_7 b_1 + 2a_6 b_2 + 2a_5 b_3 + a_4 b_4 + 2a_3 b_5 + 2a_2 b_6 + 2a_1 b_7 + a_0 b_8$$

$$\vdots$$

This computation requires 400 multiplications and 361 additions. Making use of the SIMD instructions, at best 4 of these multiplications can be done in parallel, adding the result of a multiplication is at best for free using the mpya instruction, so we need at least 100 instructions to compute the coefficients $r_0, \ldots, r_{38}$. Furthermore we need to multiply some intermediate products by 2, an effect resulting from the non-integer radix 12.75 used for the representation of finite field elements. As we assume the inputs to have reduced coefficients, all result coefficients r_i fit into 32-bit word elements.

We will now describe how the coefficients $r0, \ldots, r38$ can be computed using 145 pipeline-0 instructions (arithmetic instructions). This computation requires some rearrangement of coefficients in registers using the shufb instruction but with careful instruction scheduling and alignment these pipeline-1 instructions do not increase the number of cycles needed for multiplication. From the description of the arithmetic instructions it should be clear which rearrangement of inputs is necessary.

First use 15 shl instructions to have register variables
b03s1 containing $b_0, b_1, b_2, 2b_3$,
b03s2 containing $b_0, b_1, 2b_2, 2b_3$,
b03s3 containing $b_0, 2b_1, 2b_2, 2b_3$,
b47s1 containing $b_4, b_5, b_6, 2b_7$ and so on.

Now we can proceed producing intermediate result variables
r03 containing $a_0 b_0, a_0 b_1, a_0 b_2, a_0 b_3$ (one mpy instruction),
r14 containing $a_1 b_0, a_1 b_1, a_1 b_2, 2a_1 b_3$ (one mpy instruction),
r25 containing $a_2 b_0, a_2 b_1, 2a_2 b_2, 2a_2 b_3$ (one mpy instruction),
r36 containing $a_3 b_0, 2a_3 b_1, 2a_3 b_2, 2a_3 b_3$ (one mpy instruction),
r47 containing $a_4 b_0 + a_0 b_4, a_4 b_1 + a_0 b_5, a_4 b_2 + a_0 b_6, a_4 b_3 + a_0 b_7$ (one mpy and one mpya instruction),
r58 containing $a_5 b_0 + a_1 b_4, a_5 b_1 + a_1 b_5, a_5 b_2 + a_1 b_6, 2a_5 b_3 + 2a_1 b_7$ (one mpy and one mpya instruction) and so on. In total these computations need 36 mpy and 64 mpya instructions.

As a final step these intermediate results have to be joined to produce the coefficients $r_0, \ldots r_{38}$ in the register variables r03, r47,... r3639. We can do this using 30 additions if we first combine intermediate results using the shufb instruction. For example we join in one register variable the highest word of r14 and the three lowest words of r58 before adding this register variable to r47.

The basic idea for squaring is the same as for multiplication. We can make squaring slightly more efficient by exploiting the fact that some intermediate results are equal.

For a squaring of a value a given in reduced representation $(a_0, \ldots, a_{19})$, formulas for the result coefficients $r_0, \ldots, r_{38}$ are the following:

$$r_0 = a_0 a_0$$
$$r_1 = 2a_1 a_0$$
$$r_2 = 2a_2 a_0 + a_1 a_1$$
$$r_3 = 2a_3 a_0 + 2a_2 a_1$$
$$r_4 = 2a_4 a_0 + 4a_3 a_1 + 2a_2 a_2$$
$$r_5 = 2a_5 a_0 + 2a_4 a_1 + 4a_3 a_2$$
$$r_6 = 2a_6 a_0 + 2a_5 a_1 + 2a_4 a_2 + 2a_3 a_3$$
$$r_7 = 2a_7 a_0 + 2a_6 a_1 + 2a_5 a_2 + 2a_4 a_3$$
$$r_8 = 2a_8 a_0 + 4a_7 a_1 + 4a_6 a_2 + 4a_5 a_3 + a_4 a_4$$

$$\vdots$$

The main part of the computation only requires 60 multiplications (24 mpya and 36 mpy instructions). However, some partial results have to be multiplied by 4; this requires more preprocessing of the inputs, we end up using 35 instead of 15 shl instructions before entering the main block of multiplications. Squaring is therefore only 20 cycles faster than multiplication.

During both multiplication and squaring, we can overcome latencies by interleaving independent instructions.

5.3   Reduction

The task of the reduction step is to compute from the coefficients $r_0, \ldots r_{38}$ a reduced representation $(r_0, \ldots, r_{19})$. Implementing this computation efficiently is challenging in two ways: In a typical reduction chain every instruction is dependent on the result of the preceding instruction. This makes it very hard to vectorize operations in SIMD instructions and to hide latencies.

We will now describe a way to handle reduction hiding most instruction latencies but without data level parallelism through SIMD instructions.

The basic idea of reduction is to first reduce the coefficients r_{20} to r_{38} (producing a coefficient r_{39}), then add $19r_{20}$ to r_0, $19r_{21}$ to r_1 and so on until adding $19r_{39}$ to r_{19} and then reduce the coefficients r_0 to r_{19}.

Multiplications by 19 result from the fact, that the coefficient a_{20} stands for $a_{20} \cdot 2^{255}$ (see equation (1)). By the definition of the finite field $\mathbb{F}_{2^{255}-19}$, $2^{255}a_{20}$ is the same as $19a_{20}$. Equivalent statements hold for the coefficients $a_{21}, \ldots, a_{39}$.

The most speed critical parts of this reduction are the two carry chains from r_{20} to r_{39} and from r_0 to r_{19}. In order to overcome latencies in these chains we break each of them into four parallel carry chains, Algorithm 3 describes this structure of our modular reduction algorithm.

Algorithm 3. Structure of the modular reduction

Carry from r_{20} to r_{21}, from r_{24} to r_{25}, from r_{28} to r_{29} and from r_{32} to r_{33}
Carry from r_{21} to r_{22}, from r_{25} to r_{26}, from r_{29} to r_{30} and from r_{33} to r_{34}
Carry from r_{22} to r_{23}, from r_{26} to r_{27}, from r_{30} to r_{31} and from r_{34} to r_{35}
Carry from r_{23} to r_{24}, from r_{27} to r_{28}, from r_{31} to r_{32} and from r_{35} to r_{36}

Carry from r_{24} to r_{25}, from r_{28} to r_{29}, from r_{32} to r_{33} and from r_{36} to r_{37}
Carry from r_{25} to r_{26}, from r_{29} to r_{30}, from r_{33} to r_{34} and from r_{37} to r_{38}
Carry from r_{26} to r_{27}, from r_{30} to r_{31}, from r_{34} to r_{35} and from r_{38} to r_{39}
Carry from r_{27} to r_{28}, from r_{31} to r_{32} and from r_{35} to r_{36}

Add $19r_{20}$ to r_0, add $19r_{21}$ to r_1, add $19r_{22}$ to r_2 and add $19r_{23}$ to r_3
Add $19r_{24}$ to r_4, add $19r_{25}$ to r_5, add $19r_{26}$ to r_6 and add $19r_{27}$ to r_7
Add $19r_{28}$ to r_8, add $19r_{29}$ to r_9, add $19r_{30}$ to r_{10} and add $19r_{31}$ to r_{11}
Add $19r_{32}$ to r_{12}, add $19r_{33}$ to r_{13}, add $19r_{34}$ to r_{14} and add $19r_{35}$ to r_{15}
Add $19r_{36}$ to r_{16}, add $19r_{37}$ to r_{17}, add $19r_{38}$ to r_{18} and add $19r_{39}$ to r_{19}

Carry from r_{16} to r_{17}, from r_{17} to r_{18}, from r_{18} to r_{19} and from r_{19} to r_{20}
Add $19r_{20}$ to r_0

Carry from r_0 to r_1, from r_4 to r_5, from r_8 to r_9 and from r_{12} to r_{13}
Carry from r_1 to r_2, from r_5 to r_6, from r_9 to r_{10} and from r_{13} to r_{14}
Carry from r_2 to r_3, from r_6 to r_7, from r_{10} to r_{11} and from r_{14} to r_{15}
Carry from r_3 to r_4, from r_7 to r_8, from r_{11} to r_{12} and from r_{15} to r_{16}

Carry from r_4 to r_5, from r_8 to r_9, from r_{12} to r_{13} and from r_{16} to r_{17}
Carry from r_5 to r_6, from r_9 to r_{10}, from r_{13} to r_{14} and from r_{17} to r_{18}
Carry from r_6 to r_7, from r_{10} to r_{11}, from r_{14} to r_{15} and from r_{18} to r_{19}
Carry from r_7 to r_8, from r_{11} to r_{12} and from r_{15} to r_{16}

Each of the carry operations in Algorithm 3 can be done using one **shufb**, one **rotmi** and one **a** instruction. Furthermore we need 8 masking instructions (bitwise **and**) for each of the two carry chains.

In total, a call to the multiplication function (including reduction) takes 444 cycles, a call to the squaring function takes 424 cycles. This includes 144 cycles for multiplication (124 cycles for squaring), 244 cycles for reduction and some more cycles to load input and store output. Furthermore the cost of a function call is included in these numbers.

Montgomery ladder step. For the implementation of a Montgomery ladder step we exploit the fact that we can optimize a fixed sequence of arithmetic instructions in $\mathbb{F}_{2^{255}-19}$ instead of single instructions. This makes it much easier to make efficient use of the SIMD instruction set, in particular, for modular reduction.

The idea is to arrange the operations in $\mathbb{F}_{2^{255}-19}$ into blocks of 4 equal or similar instructions, similar meaning that multiplications and squarings can be grouped together and additions and subtractions can be grouped together as well. Then these operations can be carried out using the 4-way parallel SIMD instructions in the obvious way; for example for 4 multiplications $r = a \cdot b$, $s = c \cdot d$, $t = e \cdot f$ and $u = g \cdot h$ we first produce register variables $\texttt{aceg0}$ containing in its word elements a_0, c_0, e_0, g_0 and $\texttt{bdgh0}$ containing b_0, d_0, e_0, g_0 and so on. Then the first coefficient of r, s, t and u can be computed by applying the $\texttt{mpy}$ instruction on $\texttt{aceg0}$ and $\texttt{bdfh0}$. All other result coefficients of r, s, t and u can be computed in a similar way using $\texttt{mpy}$ and $\texttt{mpya}$ instructions.

This way of using the SIMD capabilities of CPUs was introduced in [10] as "digit-slicing". In our case it not only makes multiplication slightly faster (420 arithmetic instructions instead of 576 for 4 multiplications), it also allows for much faster reduction: The reduction algorithm described above can now be applied to 4 results in parallel, reducing the cost of a reduction by a factor of 4.

In Algorithm 4 we describe how we divide a Montgomery ladder step into blocks of 4 similar operations. In this algorithm the computation of Z_{P+Q} in the last step requires one multiplication and reduction which we carry out as described in the previous section. The computation of a ladder step again requires rearrangement of data in registers using the $\texttt{shufb}$ instruction. Again we can hide these pipeline-1 instructions almost entirely by interleaving with arithmetic pipeline-0 instructions.

One remark regarding subtractions occurring in this computation: As reduction expects all coefficients to be larger than zero, we cannot just compute the difference of each coefficient. Instead, for the subtraction $a - b$ we first add $2p$ to a and then subtract b. For blocks containing additions and subtractions in Algorithm 4 we compute the additions together with additions of $2p$ and perform the subtraction in a separate step.

In total one call to the ladder-step function takes 2433 cycles.

6   Results and Comparison

6.1   Benchmarking Methodology

In order to make our benchmarking results comparable and verifiable we use the SUPERCOP toolkit, a benchmarking framework developed within eBACS, the benchmarking project of ECRYPT II [4]. The software presented in this paper passes the extensive tests of this toolkit showing compatibility to other $\texttt{curve25519}$ implementations, in particular the reference implementation included in the toolkit.

Algorithm 4. Structure of a Montgomery ladder step (see Algorithm 2) optimized for 4-way parallel computation

$t_1 \leftarrow X_P + Z_P$
$t_2 \leftarrow X_P - Z_P$
$t_3 \leftarrow X_Q + Z_Q$
$t_4 \leftarrow X_Q - Z_Q$
Reduce t_1, t_2, t_2, t_3

$t_6 \leftarrow t_1^2$
$t_7 \leftarrow t_2^2$
$t_8 \leftarrow t_4 \cdot t_1$
$t_9 \leftarrow t_3 \cdot t_2$
Reduce t_6, t_7, t_8, t_9

$t_{10} = a24 \cdot t_6$
$t_{11} = (a24 - 1) \cdot t_7$

$t_5 \leftarrow t_6 - t_7$
$t_4 \leftarrow t_{10} - t_{11}$
$t_1 \leftarrow t_8 - t_9$
$t_0 \leftarrow t_8 + t_9$
Reduce t_5, t_4, t_1, t_0

$Z_{[2]P} \leftarrow t_5 \cdot t_4$
$X_{P+Q} \leftarrow t_0^2$
$X_{[2]P} \leftarrow t_6 \cdot t_7$
$t_2 \leftarrow t_1^2$
Reduce $Z_{[2]P}, X_{P+Q}, X_{[2]P}, t_2$

$Z_{P+Q} \leftarrow X_{Q-P} \cdot t_2$
Reduce Z_{P+Q}

For scalar multiplication software, SUPERCOP measures two different cycle counts: The `crypto_scalarmult` benchmark measures cycles for a scalar multiplication of an arbitrary point; the `crypto_scalarmult_base` benchmark measures cycles needed for a scalar multiplication of a fixed base point.

We currently implement `crypto_scalarmult_base` as `crypto_scalarmult`; faster implementations would be useful in applications that frequently call `crypto_scalarmult_base`.

Two further benchmarks regard our `curve25519` software in the context of Diffie-Hellman key exchange: The `crypto_dh_keypair` benchmark measures the number of cycles to generate a key pair consisting of a secret and a public key. The `crypto_dh` benchmark measures cycles to compute a joint key, given a secret and a public key.

6.2   Results

We benchmarked our software on `hex01`, a QS21 blade containing two 3200 MHz Cell Broadband Engine processors (revision 5.1) at the Chair for Operating Systems at RWTH Aachen University. We also benchmarked the software on `node001`, a QS22 blade at the Research Center Jülich containing two 3200 MHz PowerXCell 8i processors (Cell Broadband Engine (revision 48.0)). Furthermore we benchmarked the software on `cosmovoid`, a Sony Playstation 3 containing a 3192 MHz Cell Broadband Engine processor (revision 5.1) located at the Chair for Operating Systems at RWTH Aachen University. All measurements used one SPU of one CBE.

Table 3. Cycle counts of our software on different machines

SUPERCOP benchmark	hex01	node001	cosmovoid
crypto_scalarmult	697080	697080	697040
crypto_scalarmult_base	697080	697080	697080
crypto_dh_keypair	720120	720120	720200
crypto_dh	697080	697080	697040

6.3   Comparison

As the software described in this paper is the first implementation of ECC for the Cell Broadband Engine, we cannot compare to previous results. To give an impression of the power of the Cell Broadband Engine for asymmetric cryptography we compare our results on a cost-performance basis with ECDH software for Intel processors.

For this comparison we consider the cheapest hardware configuration containing a Cell Broadband Engine, namely the Sony Playstation 3, and compare the results to an Intel-Core-2-based configuration running the ECDH software presented in [8]. This is the currently fastest implementation of ECDH for the Core 2 processor providing a similar security as `curve25519`. Note that this software is not protected against timing attacks.

We benchmarked this software on a machine called `archer` at the National Taiwan University. This machine has a 2500MHz Intel Core 2 Q9300 processor with 4 cores; measurements used one core.

SUPERCOP reports 365363 cycles for the `crypto_dh` benchmark (this software is not benchmarked as scalar-multiplication software). Key-pair generation specializes the scalar multiplication algorithm for the known basepoint; the `crypto_dh_keypair` benchmark reports 151215 cycles.

To estimate a price for a complete workstation including an Intel Core 2 Quad Q9300 processor we determined the lowest prices for processor, case, motherboard, memory, hard disk and power supply from different online retailers using Google Product Search yielding $296 (Mar 30, 2009).

To determine the best price for the Sony Playstation 3 we also used Google Product Search. The currently (Mar 30, 2009) cheapest offer is \$221 for the Playstation 3 with a 40 GB hard disk.

The Sony Playstation 3 makes 6 SPUs available for general purpose computations. Using our implementation running at 697080 cycles (crypto_dh on cosmovoid) on 6 SPUs operating at 3192MHz yields 27474 curve25519 computations per second. Taking the \$221 market price for the Playstation as a basis, the cheapest CBE-based hardware can thus perform 124 computations of curve25519 per second per dollar.

The Q9300-based workstation has 4 cores operating at 2.5GHz, using the above-mentioned implementation which takes 365363 cycles, we can thus perform 27368 joint-key computations per second. Taking \$296 market price for a Q9300-based workstation as a basis, the cheapest Core-2-based hardware can thus perform 92 joint-key computations per second per dollar.

Note, that this comparison is not fair in several ways: The cheapest Q9300-based workstation has for example more memory than the Playstation 3 (1GB instead of 256MB).

On the other hand we only use the 6 SPUs of the CBE for the curve25519 computation, the PPU is still available for other tasks, whereas the performance estimation for the Core-2-based system assumes 100% workload on all CPU cores.

Furthermore hardware prices are subject to frequent changes and different price-performance ratios are achieved for other Intel or AMD processors.

In any case the above figures demonstrate that the Cell Broadband Engine, when used properly, is one of the best available CPUs for public-key cryptography.

References

1. Alvaro, W., Kurzak, J., Dongarra, J.: Fast and small short vector SIMD matrix multiplication kernels for the synergistic processing element of the CELL processor. In: Bubak, M., van Albada, G.D., Dongarra, J., Sloot, P.M.A. (eds.) ICCS 2008, Part I. LNCS, vol. 5101, pp. 935–944. Springer, Heidelberg (2008)
2. Bernstein, D.J.: qhasm: tools to help write high-speed software, http://cr.yp.to/qhasm.html (accessed January 1, 2009)
3. Bernstein, D.J.: Curve25519: new Diffie-Hellman speed records. In: Yung, M., Dodis, Y., Kiayias, A., Malkin, T.G. (eds.) PKC 2006. LNCS, vol. 3958, pp. 207–228. Springer, Heidelberg (2006)
4. Bernstein, D.J., Lange, T. (eds.): eBACS: ECRYPT benchmarking of cryptographic systems (November 2008), http://bench.cr.yp.to/ (accessed January 1, 2009)
5. Costigan, N., Scott, M.: Accelerating SSL using the vector processors in IBM's Cell Broadband Engine for Sony's Playstation 3. In: Proceedings of SPEED workshop (2007), http://www.hyperelliptic.org/SPEED/
6. Diffie, W., Hellman, M.E.: New directions in cryptography. IEEE Transactions on Information Theory IT-22(6), 644–654 (1976), http://citeseer.ist.psu.edu/diffie76new.html

7. Fan, J., Sakiyama, K., Verbauwhede, I.: Elliptic curve cryptography on embedded multicore systems. In: Workshop on Embedded Systems Security - WESS 2007, Salzburg, Austria, pp. 17–22 (2007)
8. Galbraith, S.D., Lin, X., Scott, M.: Endomorphisms for faster elliptic curve cryptography on general curves (2008), http://eprint.iacr.org/2008/194
9. Gaudry, P., Thomé, E.: The mp$\mathbb{F}_q$ library and implementing curve-based key exchanges. In: Proceedings of SPEED workshop (2007), http://www.loria.fr/~gaudry/publis/mpfq.pdf
10. Grabher, P., Großchädl, J., Page, D.: On software parallel implementation of cryptographic pairings. In: Selected Areas in Cryptography – SAC 2008. LNCS, vol. 5381, pp. 34–49. Springer, Heidelberg (to appear)
11. Hankerson, D., Menezes, A.J., Vanstone, S.A.: Guide to Elliptic Curve Cryptography. Springer, Berlin (2003)
12. IBM DeveloperWorks. Cell broadband engine architecture and its first implementation (November 2005), http://www-128.ibm.com/developerworks/power/library/pa-cellperf/
13. IBM DeveloperWorks. Cell broadband engine programming handbook (version 1.1) (April 2007), http://www-01.ibm.com/chips/techlib/techlib.nsf/techdocs/9F820A5FFA3ECE8C8725716A0062585F
14. IBM DeveloperWorks. SPU assembly language specification (version 1.6) (September 2007), http://www-01.ibm.com/chips/techlib/techlib.nsf/techdocs/EFA2B196893B550787257060006FC9FB
15. IBM DeveloperWorks. Example library API reference (version 3.1) (September 2008), http://www.ibm.com/developerworks/power/cell/documents.html.
16. IBM DeveloperWorks. SPE cryptographic library user documentation 1.0 (September 2008), http://www.ibm.com/developerworks/power/cell/documents.html.
17. Montgomery, P.L.: Speeding the Pollard and elliptic curve methods of factorization. Mathematics of Computation 48(177), 243–264 (1987)
18. Shimizu, K., Brokenshire, D., Peyravian, M.: Cell Broadband Engine support for privacy, security, and digital rights management applications. White paper, IBM (October 2005), http://www-01.ibm.com/chips/techlib/techlib.nsf/techdocs/3F88DA69A1C0AC40872570AB00570985
19. Sotirov, A., Stevens, M., Appelbaum, J., Lenstra, A., Molnar, D., Osvik, D.A., de Weger, B.: MD5 considered harmful today (December 2008), http://www.win.tue.nl/hashclash/rogue-ca/ (accessed January 4, 2009)
20. Stevens, M., Lenstra, A., de Weger, B.: Nostradamus – predicting the winner of the 2008 US presidential elections using a Sony PlayStation 3 (November 2007), http://www.win.tue.nl/hashclash/Nostradamus/ (accessed January 4, 2009)

On Modular Decomposition of Integers

Billy Bob Brumley[1] and Kaisa Nyberg[1,2]

[1] Department of Information and Computer Science,
Helsinki University of Technology,
P.O. Box 5400, FI-02015 TKK, Finland
{billy.brumley,kaisa.nyberg}@tkk.fi
[2] Nokia Research Center, Finland
kaisa.nyberg@nokia.com

Abstract. At Crypto 2001, Gallant et al. showed how to exploit fast endomorphisms on some specific classes of elliptic curves to obtain fast scalar multiplication. The GLV method works by decomposing scalars into two small portions using multiplications, divisions, and rounding operations in the rationals. We present a new simple method based on the extended Euclidean algorithm that uses notably different operations than that of traditional decomposition. We obtain strict bounds on each component. Additionally, we examine the use of random decompositions, useful for key generation or cryptosystems requiring ephemeral keys. Specifically, we provide a complete description of the probability distribution of random decompositions and give bounds for each component in such a way that ensures a concrete level of entropy. This is the first analysis on distribution of random decompositions in GLV allowing the derivation of the entropy and thus an answer to the question first posed by Gallant in 1999.

Keywords: elliptic curve cryptography, GLV method, integer decompositions.

1 Introduction

Elliptic curve cryptography is a field rich with methods to obtain fast implementations. A reasonable choice when high speed ECC is needed is a curve which admits a fast endomorphism. While methods using the Frobenius endomorphism exist, for example Koblitz curves, the classes of curves for which this is useful is somewhat limited.

Gallant, Lambert, and Vanstone [1] showed a novel method of using fast endomorphisms on larger classes of curves. Given a point P of prime order n on a curve that admits a fast endomorphism ϕ, then ϕ acts as on P the multiplication map $[\lambda]$. Given a scalar $k \in \mathbb{Z}_n$, the GLV method works by computing a decomposition $k_1, k_2 \approx \sqrt{n}$ such that $k \equiv k_1 + k_2\lambda \pmod{n}$. The scalar multiplication calculation kP is then carried out as $kP = (k_1 + k_2\lambda)P = k_1P + k_2\phi(P)$ using any number of variations of the Straus-Shamir method [2,3]; the immediate effect is that half of the point doublings are eliminated.

B. Preneel (Ed.): AFRICACRYPT 2009, LNCS 5580, pp. 386–402, 2009.

For further motivation in this area, Galbraith et al. [4] recently showed how to apply the GLV method to larger classes of curves by working in quadratic extension fields to induce a similar type of endomorphism. The work of [4] has produced one of the fastest, if not the fastest known methods for scalar multiplication in software.

A method of obtaining such a scalar decomposition is outlined in [1]. It uses a number of multiplications, divisions, rounding operations, and computations in the rationals. The first contribution we present is a new method for scalar decomposition. The method is based on the extended Euclidean algorithm, and uses very different operations than those carried out in traditional decomposition. The new method is more flexible in the respect that it allows a fairly arbitrary ratio of balance between the size of the components.

Some cryptosystems require only an ephemeral key, such as Diffie-Hellman key agreement or ECDSA signature generation. In these environments as well as key generation, the natural approach is to simply start with a random decomposition. This raises concerns about the distribution of such random decompositions, as first observed by Gallant at ECC'99 [5]. This question remained unanswered.

We answer this question in our second contribution. We first state and prove a theorem which provides a complete description of the distribution of the integers $\kappa\lambda \mod n$ on the interval $[0, n]$. We then present a method for deriving bounds on each portion of the decomposition in such a way that the distribution of the resulting random decompositions can be explicitly calculated, and hence we obtain an exact formula for the achieved entropy. As far as we are aware, this is the first work on to examine this problem for GLV.

We begin in Sec. 2 with a brief overview of elliptic curves and the GLV method using curves with fast endomorphisms. In Sec. 3 we present the new method for scalar decomposition. We study the distribution of random decompositions in Sec. 4 and provide the method for deriving bounds to ensure a certain level of entropy. We conclude in Sec. 5.

2   Preliminaries

As far as speed is concerned, elliptic curves that admit fast endomorphisms allow for significantly faster scalar multiplication compared to random curves. In [1], Gallant et al. showed a novel approach to speeding up scalar multiplication on such curves; while previous methods exploited the Frobenius endomorphism, the novelty of the GLV method was the use of other fast endomorphisms to speed up the operation. We now provide a brief background on the topic.

2.1   Curves with Fast Endomorphisms

We denote E an elliptic curve and $\phi : E \to E$ an endomorphism. For some $P \in E$ we consider the main subgroup $\langle P \rangle$ of prime order n. It follows $\phi(P) \in \langle P \rangle$ and ϕ acts as the multiplication map $[\lambda]$; that is, there exists $\lambda \in \mathbb{Z}_n$ such that $\phi(Q) = \lambda Q$ for all $Q \in \langle P \rangle$. Such a λ is a root of the characteristic polynomial of ϕ modulo n. The following examples are given in [1]; such curves are standardized in [6,7].

Example 1. Let p prime satisfy $p \equiv 1 \pmod 4$. There exists an element $\alpha \in \mathbb{Z}_p^*$ of multiplicative order 4 and the curve $E_1(\mathbb{F}_p) : y^2 = x^3 + ax$ admits an efficient endomorphism $\phi : E_1(\mathbb{F}_p) \to E_1(\mathbb{F}_p)$ defined by $\phi : (x, y) \mapsto (-x, \alpha y)$, $\mathcal{O} \mapsto \mathcal{O}$ and ϕ satisfies the characteristic equation $X^2 + 1 = 0$.

Example 2. Let p prime satisfy $p \equiv 1 \pmod 3$. There exists an element $\beta \in \mathbb{Z}_p^*$ of multiplicative order 3 and the curve $E_2(\mathbb{F}_p) : y^2 = x^3 + b$ admits an efficient endomorphism $\phi : E_2(\mathbb{F}_p) \to E_2(\mathbb{F}_p)$ defined by $\phi : (x, y) \mapsto (\beta x, y)$, $\mathcal{O} \mapsto \mathcal{O}$ and ϕ satisfies the characteristic equation $X^2 + X + 1 = 0$.

More recently, the work of Galbraith et al. [4] showed how to induce a similar fast endomorphism on much larger classes of elliptic curves. In short, working with curves defined over a quadratic extension field can induce a similar endomorphism. The result is a significantly larger class of curves for which the GLV method can be applied.

To take advantage of these specific endomorphisms, given a scalar $k \in \mathbb{Z}_n$ and $P \in E(\mathbb{F}_p)$ the GLV method finds a decomposition of k such that $k \equiv k_1 + k_2\lambda$ $\pmod n$ with k_i of size $O(\sqrt{n})$. Then the computation $kP = (k_1 + k_2\lambda)P = k_1P + k_2\phi(P)$ is carried out using some form of the Straus-Shamir method [2,3], effectively cutting the number of point doublings in half. Such a decomposition is sometimes called a *balanced length-two representation* [8]; a description follows.

2.2   Integer Decompositions

A method for obtaining such an integer representation given [1]; it uses Babai's rounding method [9].

Let $f : \mathbb{Z}^2 \to \mathbb{Z}_n$ denote the homomorphism defined by $f : (a, b) \mapsto a + b\lambda$ mod n. First find two linearly independent vectors $v_1 = (a_1, b_1)$ and $v_2 = (a_2, b_2)$ in $\mathbb{Z}^2$ of size $\approx \sqrt{n}$ such that $f(v_1) = f(v_2) = 0$. This is done offline using the extended Euclidean algorithm with input n and λ, obtaining a sequence of equations satisfying $u_i n + t_i \lambda = r_i$ with remainders r_i strictly decreasing, see Section 3.1. We recall the following lemma from [1], instrumental in finding the two vectors v_1 and v_2:

Lemma 1 (iv). $r_{i-1}|t_i| + r_i|t_{i-1}| = n$ *for all* $i \geq 1$.

Denoting m as the largest index such that $r_m \geq \sqrt{n}$ holds, the vectors $v_1 = (r_{m+1}, -t_{m+1})$ and $v_2 = (r_m, -t_m)$ with components of size $\approx \sqrt{n}$ satisfy $f(v_1) = f(v_2) = 0$.

Given a scalar k, express the vector $(k, 0) \in \mathbb{Q}^2$ as $(k, 0) = \gamma_1 v_1 + \gamma_2 v_2$ with $\gamma_i \in \mathbb{Q}$. Denoting $c_i = \lceil \gamma_i \rfloor$ as the integer closest to γ_i, the vector $v = c_1 v_1 + c_2 v_2$ satisfies $f(v) = 0$ and is close to $(k, 0)$, and the vector $u = (k_1, k_2) = (k, 0) - v$ satisfies $f(u) = k$ with k_1, k_2 of size $O(\sqrt{n})$. While these bounds are more heuristic [1], tighter bounds are given in [10,11], as well as a similar decomposition method in [12] with tight bounds.

2.3   Implementing GLV

A computation of the form $kP + jQ$ needed in the GLV method is quite common in cryptography and has been the subject of much research. Most methods are a variation of the Straus-Shamir method [2,3], in which one accumulator is used for the result and the point doublings are shared, moving left-to-right scanning the coefficients of k and j. Two popular general strategies exist to reduce the number of point additions: either opting for a low-weight joint representation of k and j, or interleaving and focusing on individual low-weight representations of k and j. Generally, the former is used when little storage is available, and the latter if more storage is available.

One can see there are several components involved in carrying out scalar multiplication using the GLV method. The choice for these components is highly platform dependent, depending on how much storage and what instructions are available. A somewhat typical implementation is show in Fig. 1. When a scalar multiplication is to be performed, the balanced length-two representation of the scalar is first obtained, then a low-weight joint representation is calculated (shown as SJSF [13]), and finally the curve operations begin from left-to-right.

{Scalar multiplication, Straus-Shamir/SJSF}
Input: Integer k, point $P \in E$
Output: kP
$(k_1, k_2) \leftarrow$ DECOMPOSE(k)
$\begin{pmatrix} x_\ell \cdots x_0 \\ y_\ell \cdots y_0 \end{pmatrix} \leftarrow$ SJSF(k_1, k_2)
Precompute $\phi(P)$, $P + \phi(P)$, $P - \phi(P)$
$Q \leftarrow (x_\ell P + y_\ell \phi(P))$
for $i = \ell - 1$ down to 0 **do**
  $Q \leftarrow 2Q$
  $Q \leftarrow Q + (x_i P + y_i \phi(P))$
end for
return Q

{DECOMPOSE, balanced length-two rep. of k}
Input: Integers λ, k; vectors $v_1 = (a_1, b_1)$, $v_2 = (a_2, b_2)$
Output: Integers $k_1, k_2 \approx \sqrt{n} : k \equiv k_1 + k_2 \lambda$ (mod n)
$c_1 \leftarrow \lfloor b_2 k/n \rceil$
$c_2 \leftarrow \lfloor -b_1 k/n \rceil$
$k_1 \leftarrow k - c_1 a_1 - c_2 a_2$
$k_2 \leftarrow -c_1 b_1 - c_2 b_2$
return (k_1, k_2)

{Simple Joint Sparse Form SJSF}
Input: Integers x, y
Output: Simple Joint Sparse Form of x, y
$j \leftarrow 0$
while $x \neq 0$ or $y \neq 0$ **do**
  $x_j \leftarrow x \mod 2$, $y_j \leftarrow y \mod 2$
  if $x_j = y_j = 1$ **then**
    if $(x - x_j)/2 \equiv 1$ (mod 2) **then**
      $x_j \leftarrow -x_j$
    end if
    if $(y - y_j)/2 \equiv 1$ (mod 2) **then**
      $y_j \leftarrow -y_j$
    end if
  else if $x_j \neq y_j$ **then**
    if $(x - x_j)/2 \not\equiv (y - y_j)/2$ (mod 2) **then**
      $x_j \leftarrow -x_j, y_j \leftarrow -y_j$
    end if
  end if
  $x \leftarrow (x - x_j)/2$, $y \leftarrow (y - y_j)/2$
  $j \leftarrow j + 1$
end while
return $\begin{pmatrix} x_{j-1} \cdots x_0 \\ y_{j-1} \cdots y_0 \end{pmatrix}$

Fig. 1. Components of a typical implementation of GLV scalar multiplication

3   A Modular Decomposition

As show in Fig. 1, computing the traditional balanced length-two representation involves a number of multiplications, divisions, rounding operations, and computations in the rationals. In this section we present a new, modular method for decomposition of integers with a desired ratio of balance between the size of the components. Naturally when using the GLV method one prefers a balanced representation with each component of size $O(\sqrt{n})$. The new method also allows an unbalanced representation (for example, components of size $n^{1/3}$ and $n^{2/3}$), although we have yet to find a reasonable application of this feature without fixed parameters, such as the generator. Subsequently balancing exponent lengths with fixed parameters is examined in [14], but the appeal of GLV is the use of a fast endomorphism—in the case of fixed P this is no longer necessary as the computation can be done offline and stored.

We begin with the derivation of the method in Sec. 3.2, then proceed to prove bounds on each component of the decomposition. Implementation considerations are given in Sec. 3.3.

3.1   Extended Euclidean Algorithm

In this section we give the notation used for the extended Euclidean algorithm in this paper. Running it on n and $0 < \lambda < n$, we obtain a sequence of positive quotients $q_1, \ldots, q_m$, decreasing sequence of positive remainders $r_1, \ldots, r_m$, sequence of integers $t_1, \ldots, t_{m+1}$, and sequence of integers $u_1, \ldots, u_{m+1}$. With $r_0 = n$, $r_1 = \lambda$, $t_0 = 0$, $t_1 = 1$, $u_0 = 1$, $u_1 = 0$, these satisfy

$$r_{i-1} = q_i r_i + r_{i+1}$$
$$t_{i+1} = t_{i-1} - q_i t_i \text{ and} \tag{1}$$
$$r_i \equiv t_i \lambda + u_i n$$

for all $i = 1, \ldots, m$. The integers t_i satisfy $t_i < 0$ and $u_i > 0$ for $i > 0$ even and $t_i > 0$ for $i > 0$ odd, and $u_i < 0$ for $i \geq 3$ odd.

3.2   Description

Let the index $m > 0$ be fixed but arbitrary. Given $k \in \mathbb{Z}_n$, we iteratively divide k by $r_1, \ldots, r_m$:

$$k = a_1 r_1 + j_1 \tag{2}$$
$$j_1 = a_2 r_2 + j_2$$
$$\cdots$$

$$j_{m-1} = a_m r_m + j_m, \text{ and denoting } j_m = k_1 \text{ it follows}$$

$$k = k_1 + \sum_{i=1}^{m} a_i r_i.$$

constructed in a greedy fashion. Note $0 \leq j_i < r_i$ and in particular k_1 satisfies $0 \leq k_1 < r_m$. We denote $k_2 = \sum_{i=1}^{m} a_i t_i$ where, notably, the construction above and the inequalities $0 \leq k < n$ and $0 \leq j_i < r_i$ guarantee $0 \leq a_i \leq q_i$ for all $i = 1, \ldots, m$.

Lemma 2. $k_1 + k_2 \lambda \equiv k \pmod{n}$.

Proof. The extended Euclidean algorithm guarantees $r_i = t_i\lambda + u_i n$, see Section 3.1. Then

$$k = k_1 + \sum_{i=1}^{m} a_i(t_i\lambda + u_i n) \equiv k_1 + \lambda \sum_{i=1}^{m} a_i t_i \equiv k_1 + k_2\lambda \pmod{n} \qquad \square$$

We now give bounds on k_1 and k_2. Denoting $A_m = r_m$ and $B_m = |t_m| + |t_{m+1}|$, the bound $0 \leq k_1 < A_m$ on k_1 is immediate.

Theorem 1. $0 \leq |k_2| < B_m$.

Proof. With $0 \leq k < n$ and $0 < j_i < r_i$ it immediately follows $0 \leq a_i \leq q_i$ for all $i = 1, \ldots, m$. We have

$$\left| \sum_{i=1}^{m} a_i t_i \right| \leq \sum_{\substack{i=1 \\ i \text{ odd}}}^{m} a_i t_i + \sum_{\substack{i=1 \\ i \text{ even}}}^{m} a_i |t_i| \leq \sum_{\substack{i=1 \\ i \text{ odd}}}^{m} q_i t_i + \sum_{\substack{i=1 \\ i \text{ even}}}^{m} q_i |t_i|.$$

Recalling (1), we have $q_i t_i = t_{i-1} - t_{i+1}$ for i odd, and $q_i|t_i| = t_{i+1} - t_{i-1}$ for i even. Summing up the telescopic sums we obtain

$$|k_2| \leq \begin{cases} t_0 - t_{m+1} + t_m - t_1 = |t_m| + |t_{m+1}| - 1 & \text{, for } m \text{ odd} \\ t_0 - t_m + t_{m+1} - t_1 = |t_m| + |t_{m+1}| - 1 & \text{, for } m \text{ even} \end{cases}$$

and the claim holds. $\qquad \square$

3.3   Implementation

As previously mentioned, the new method above allows one to select a desired level of balance between the size of the components. For the GLV method, to maximize efficiency we would thus be interested in obtaining a decomposition similar in size to that of the balanced length-two representation. To achieve this, the index m is selected in such a way that $r_m, r_{m+1}, |t_m|, |t_{m+1}|$ are all as close as possible in size to $\sqrt{n}$ and $|t_m| + |t_{m+1}| \approx r_m/2$.

There are many possible implementations obtaining the decomposition inline with the extended Euclidean algorithm, using no precomputation; we show one in Fig. 2. There is inherently a need to divide subsequent remainders of k as shown in (2). We choose to perform this division first, then use the result to lessen the burden of the second required division; we thus ensure that $r_{i-1} = q_i r_i + r_{i+1}$ in (1) holds by obtaining q as the sum of the two quotients a, b. The **if** statement ensures the remainder is indeed smaller than the divisor.

The trade-off is clear. Traditional decomposition uses 2 divisions, 2 rounds, and 6 multiplications. The algorithm in Fig. 2 would be implemented with a large[1] number of shifts and adds since the quotients are small on average. Although this is unlikely to be faster, it can be easier and even smaller to implement. This effect is documented in [16] in the similar context of reduction

---

[1] cf. [15, Chapter 4.5] for lengthy discussions on the behavior of the Euclidean algorithm.

Input: Integer k, modulus n, λ, bound index m
Output: $(k_1, k_2) : k_1 + k_2\lambda \equiv k \pmod{n}$ and $0 \leq k_1 < A_m$, $|k_2| < B_m$

$k_0 \leftarrow k, k_1 \leftarrow 0, k_2 \leftarrow 0, r_0 \leftarrow n, r_1 \leftarrow \lambda, t_0 \leftarrow 0, t_1 \leftarrow 1$
for $i = 1$ to m **do**
    $(a, k_1) \leftarrow \text{DIVREM}(k_0, r_1)$
    $(b, c) \leftarrow \text{DIVREM}(r_0 - k_0, r_1)$
    $r_0 \leftarrow r_1, r_1 \leftarrow k_1 + c, q \leftarrow a + b, k_0 \leftarrow k_1$
    if $r_1 \geq r_0$ **then**
        $r_1 \leftarrow r_1 - r_0, q \leftarrow q + 1$
    end if
    $k_2 \leftarrow k_2 + qt_1, (t_0, t_1) \leftarrow (t_1, t_0 - t_1 q)$
end for
return (k_1, k_2)

Fig. 2. Integer decomposition using the extended Euclidean algorithm

for Koblitz curves, where simpler algorithms using shifts and adds are shown to be advantageous for FPGA implementations over those using large divisions and multiplications. We conclude that the intrigue of the new decomposition method is unlikely to lie in a faster implementation, but rather in the fact that the derivation of the algorithm and the types of operations performed differ significantly from those of the traditional GLV balanced length-two representation. It remains an open question whether the new method may offer computational advantages for some types of platforms.

4   Random Decompositions

As stated in Sec. 2.3, a typical implementation of GLV involves several components: Scalar multiplication logic, integer decomposition routines, as well as scalar recoding mechanisms. To obtain the desired efficient implementations, all of these components are inherently necessary when presented with a specific scalar, say with signature verification.

However, in some cryptographic settings this may not be the case; it might be possible to do away with one or more of these components. There are many applications where the scalar is chosen at random; the specific value of this key/scalar is not particularly important, only that is is taken in a suitably random fashion. Such applications include key generation and any cryptosystems using ephemeral keys such as Diffie-Hellman key agreement and signature generation with ElGamal based schemes such as ECDSA.

In the context of GLV, to minimize implementation area, memory requirements, and maximize efficiency, for these applications the better solution would be to start with a random decomposition $k_1 + k_2\lambda \mod n$. The natural concern is the distribution of such decompositions; that is, specifically what bounds should be chosen for k_i to ensure proper distribution or a certain level of entropy? In fact, this concern was first mentioned by Gallant in his ECC'99 presentation [5]:

If we get to *choose* k as $k_0 + k_1 \cdot \lambda$, this works, but we might worry about the distribution of $k_0 + k_1 \cdot \lambda \mod n$ as $0 \leq k_0, k_1 \leq \sqrt{n} \ldots$

Gallant used this concern in part to motivate an algorithmic method for integer decomposition. Lange and Shparlinski [17] gave an analysis of the distribution of random base-ϕ expansions in a higher norm setting. When working with trace zero varieties and the Frobenius endomorphism, results on obtaining such injective sets are given in [18, pp. 386]. We now examine the distribution of random GLV decompositions in detail.

4.1   Distribution of Random Decompositions

To facilitate the analysis, we introduce the following notation. Let $\mathcal{R}, \mathcal{T}$ be sets (in practice, intervals) of integers; we denote the function $f : \mathcal{R} \times \mathcal{T} \rightarrow \mathbb{Z}_n$ defined as

$$f : v = (k_1, k_2) \mapsto k_1 + k_2 \lambda \mod n.$$

We denote the function $F : \mathbb{Z}_n \rightarrow \mathbb{Z}$ defined as

$$F : k \mapsto \#\{v \in \mathcal{R} \times \mathcal{T} : f(v) = k\}.$$

Note that calculation of F for all $k \in \mathbb{Z}_n$ leads directly to a formula for achieved entropy. For efficiency, it would be advantageous to have $\mathcal{R}, \mathcal{T}$ roughly the same cardinality and with elements of size roughly less than or equal to $\sqrt{n}$; this is, naturally, very heuristic. Hence with n prime it is not practical that f be bijective. Two seemingly interesting use cases exist depending on the application, surjective and injective generation, which will be examined as examples of applications of our random generation method in more detail in Sec. 4.3.

Surjective Random Generation. If the entire keyspace is not being covered, then the search space is reduced. If this is a concern, one would like to ensure $\min_{k \in \mathbb{Z}_n} F(k) \geq 1$ and f is surjective. Some attacks (for example, the attack by Bleichenbacher on DSA mentioned in [19]) exploit a large bias in ephemeral key generation. If certain key values occur significantly more often than others, on a range or interval of keys, this can be exploited. Hence one would like to ensure that $\max_{k \in \mathbb{Z}_n} F(k)$ is as small as possible. Using our method one can ensure that $\max_{k \in \mathbb{Z}_n} F(k) = 2$ for surjective generation.

Injective Random Generation. If reduction in the keyspace is not a concern, collisions can be avoided. In this case one would like to ensure $\max_{k \in \mathbb{Z}_n} F(k) = 1$ and f is injective. Even in this case, it is usually desirable to cover as much of the keyspace as possible. We provide the following example for illustration.

Example 3. Consider the curve $E_2(\mathbb{F}_{61987}) : y^2 = x^3 + 22364$. This curve is of the type given in Sec. 2, Example 2 ($p = 1 \mod 3$) and has prime order $\#E = n = 61543$. We obtain $\lambda = 50568$ as a solution to $\lambda^2 + \lambda + 1 \equiv 0 \pmod{n}$. Gallant questioned [5] defining $\mathcal{R} = \mathcal{T} = \{x \in \mathbb{Z} : 0 \leq x \leq \lfloor \sqrt{n} \rfloor\}$ and we have

$\#\{\mathcal{R} \times \mathcal{T}\} > n$; we might expect fairly good coverage of $\mathbb{Z}_n$. By exhaustive search (given no other recourse) we obtain the following values for the distribution:

$$\#\{k \in \mathbb{Z}_n : F(k) = 0\} = 10582$$
$$\#\{k \in \mathbb{Z}_n : F(k) = 1\} = 39921$$
$$\#\{k \in \mathbb{Z}_n : F(k) = 2\} = 11040.$$

Surprisingly, we fail to cover $\approx 17.2\%$ of $\mathbb{Z}_n$ and are notably lacking uniformity. Given these frequency counts, the achieved entropy is ≈ 15.56 bits while maximum entropy is $\lg n \approx 15.91$.

This example illustrates the numerous concerns with this approach:

- How should the sets $\mathcal{R}, \mathcal{T}$ be defined?
- How can values of F—and thus the achieved entropy—be obtained efficiently?
- For large n, what can we say about $\max_{k \in \mathbb{Z}_n} F(k)$?
- How could we make f injective—ensure (in the best case) $\max_{k \in \mathbb{Z}_n} F(k) = 1$ holds, but at the same time maximize entropy?
- How could we make f surjective—ensure $\min_{k \in \mathbb{Z}_n} F(k) \geq 1$ holds, minimize $\max_{k \in \mathbb{Z}_n} F(k)$, and maximize entropy?

We answer these questions in the following sections. We start by giving a theorem on the distribution of integer multiples of λ modulo n.

4.2   Interval Splitting

The main result is given in the following theorem where briefly the strategy is to find points $\{\kappa\lambda \mod n : \kappa \in \mathbb{Z}\}$ such that we obtain the distance between these neighboring points when placed on the interval $[0, n]$. We turn again to the extended Euclidean algorithm, see Sec. 3.1, to analyze this problem.

Theorem 2. *Suppose that sequences r_i, q_{i+1}, t_i and u_i, $i \geq 0$, have been obtained by running the extended Euclidean algorithm for integers n and λ, $n > \lambda$. Then, for any fixed $i \geq 1$ and j such that $0 \leq j < q_{i+1}$, the points $\kappa\lambda \mod n$, $0 < \kappa < |t_i| + (j+1)|t_{i+1}|$, split the interval $[0, n]$ into $|t_i| + (j+1)|t_{i+1}|$ subintervals, out of which $|t_i| + j|t_{i+1}|$ are of length r_{i+1} and $|t_{i+1}|$ are of length $r_i - jr_{i+1}$.*

The proof of this theorem is based on the following lemma:

Lemma 3. *Suppose that sequences r_i, q_{i+1}, t_i and u_i, $i \geq 0$ have been obtained by running the extended Euclidean algorithm for integers n and λ. Then, for any $i \geq 1$, j such that $0 \leq j < q_{i+1}$, and κ such that $0 < |\kappa| < |t_i| + (j+1)|t_{i+1}|$, there exist integers a and b such that*

$$\kappa\lambda \mod n = a(r_i - jr_{i+1}) + br_{i+1},$$

where $0 \leq a \leq |t_{i+1}|$ and $0 \leq b \leq |t_i| + j|t_{i+1}|$. Moreover, $a = 0$ and $b = 0$ do not hold simultaneously, and similarly, $a = |t_{i+1}|$ and $b = |t_i| + j|t_{i+1}|$ do not hold simultaneously.

Proof. For clarity, we give the proof in the case where $j = 0$. At the end of the proof we explain how to modify the proof for the case $j > 0$. By the extended Euclidean algorithm

$$r_i = u_i n + t_i \lambda$$

where $u_0 = t_1 = 1$ and $t_0 = u_1 = 0$ and for all $i \geq 2$ $u_i u_{i+1} < 0$, $t_i t_{i+1} < 0$ and $u_i t_i < 0$. By Lemma 1 and the analogous result for λ we have

$$\lambda = |u_i| r_{i+1} + |u_{i+1}| r_i \tag{3}$$
$$n = |t_i| r_{i+1} + |t_{i+1}| r_i, \text{ and} \tag{4}$$
$$1 = ||u_i||t_{i+1}| - |u_{i+1}||t_i|| = |u_i t_{i+1} - u_{i+1} t_i|, \tag{5}$$

for all $i \geq 0$.

Let us now fix κ such that $0 < \kappa < |t_i| + |t_{i+1}|$ arbitrarily. To derive the required representation $\kappa \lambda \mod n = a r_i + b r_{i+1}$, we use (3) to write

$$\kappa \lambda = \kappa |u_i| r_{i+1} + \kappa |u_{i+1}| r_i.$$

To do reduction modulo n, we aim at using (4). With this goal in mind we investigate the coefficients of r_i and r_{i+1} in the equation above and write

$$\kappa |u_i| = \alpha_i |t_i| + \beta_i$$
$$\kappa |u_{i+1}| = \alpha_{i+1} |t_{i+1}| + \beta_{i+1}$$

with quotients $0 \leq \alpha_i, 0 \leq \alpha_{i+1}$ and remainders $0 \leq \beta_i < |t_i|, 0 \leq \beta_{i+1} < |t_{i+1}|$. Then,

$$\left| \frac{\kappa |u_i|}{|t_i|} - \frac{\kappa |u_{i+1}|}{|t_{i+1}|} \right| = \left| \alpha_i - \alpha_{i+1} + \frac{\beta_i}{|t_i|} - \frac{\beta_{i+1}}{|t_{i+1}|} \right| \tag{6}$$
$$\geq |\alpha_i - \alpha_{i+1}| - \left| \frac{\beta_i}{|t_i|} - \frac{\beta_{i+1}}{|t_{i+1}|} \right|$$
$$> |\alpha_i - \alpha_{i+1}| - 1, \tag{7}$$

since $0 \leq \beta_i / |t_i| < 1$ and $0 \leq \beta_{i+1}/|t_{i+1}| < 1$. On the other hand, it follows using (5) that

$$\left| \frac{\kappa |u_i|}{|t_i|} - \frac{\kappa |u_{i+1}|}{|t_{i+1}|} \right| = \kappa \left| \frac{|u_i||t_{i+1}| - |t_i||u_{i+1}|}{|t_i||t_{i+1}|} \right| = \frac{\kappa}{|t_i||t_{i+1}|}$$
$$< \frac{|t_i| + |t_{i+1}|}{|t_i||t_{i+1}|} = \frac{1}{|t_{i+1}|} + \frac{1}{|t_i|} < 1, \tag{8}$$

for $i \geq 2$. Combining this with (7) we have $|\alpha_i - \alpha_{i+1}| \leq 1$. If $\alpha_i = \alpha_{i+1}$ we denote their common value by α and obtain

$$\kappa \lambda \mod n = \beta_i r_{i+1} + \beta_{i+1} r_i + \alpha(|t_i| r_{i+1} + |t_{i+1}| r_i) \mod n$$
$$= \beta_i r_{i+1} + \beta_{i+1} r_i + \alpha n \mod n$$
$$= \beta_i r_{i+1} + \beta_{i+1} r_i,$$

as desired.

Assume now that $\alpha_i = \alpha_{i+1} + 1$. By (6) and (8) we have

$$1 + \frac{\beta_i}{|t_i|} - \frac{\beta_{i+1}}{|t_{i+1}|} = \left| 1 + \frac{\beta_i}{|t_i|} - \frac{\beta_{i+1}}{|t_{i+1}|} \right| < \frac{1}{|t_{i+1}|} + \frac{1}{|t_i|},$$

from where it follows that

$$1 < \frac{\beta_{i+1} + 1}{|t_{i+1}|} - \frac{\beta_i - 1}{|t_i|},$$

which can hold if and only if $\beta_i = 0$ as $\beta_{i+1} < |t_{i+1}|$. It follows that we obtain a representation

$$\begin{aligned}
\kappa\lambda \bmod n &= |t_i|r_{i+1} + \beta_{i+1}r_i + \alpha_{i+1}(|t_i|r_{i+1} + |t_{i+1}|r_i) \bmod n \\
&= |t_i|r_{i+1} + \beta_{i+1}r_i + \alpha_{i+1}n \bmod n \\
&= |t_i|r_{i+1} + \beta_{i+1}r_i,
\end{aligned}$$

where $|\beta_{i+1}| < |t_{i+1}|$ as desired.

To handle the case $\alpha_{i+1} = \alpha_i + 1$ we obtain by (6) and (8) that

$$1 - \frac{\beta_i}{|t_i|} + \frac{\beta_{i+1}}{|t_{i+1}|} = \left| -1 + \frac{\beta_i}{|t_i|} - \frac{\beta_{i+1}}{|t_{i+1}|} \right| < \frac{1}{|t_{i+1}|} + \frac{1}{|t_i|}.$$

Hence it suffices to interchange the indices i and $i+1$ in the derivations above to get

$$\kappa\lambda \bmod n = \beta_i r_{i+1} + |t_{i+1}|r_i,$$

as desired. This completes the proof for the case $j = 0$.

The proof of the case where $j > 0$ is identical to the proof shown above after we make the following replacements:

$$\begin{aligned}
|t_i| \text{ is replaced by } & |t_i| + j|t_{i+1}|, \\
|u_i| \text{ is replaced by } & |u_i| + j|u_{i+1}|, \\
r_i \text{ is replaced by } & r_i - jr_{i+1}, \\
\lambda = & (|u_i| + j|u_{i+1}|)r_{i+1} + |u_{i+1}|(r_i - jr_{i+1}), \text{ and} \\
n = & (|t_i| + j|t_{i+1}|)r_{i+1} + |t_{i+1}|(r_i - jr_{i+1}).
\end{aligned}$$

To complete the proof, let us consider the case where $-|t_i| - (j+1)|t_{i+1}| < \kappa < 0$. By the proof for a positive κ, there exist integers a and b such that

$$\begin{aligned}
\kappa\lambda \bmod n &= n - (-\kappa)\lambda \bmod n \\
&= n - a(r_i - jr_{i+1}) - br_{i+1} \\
&= (|t_{i+1}| - a)(r_i - jr_{i+1}) + (|t_i| + j|t_{i+1}| - b)r_{i+1} \\
&= a'(r_i - jr_{i+1}) + b'r_{i+1},
\end{aligned}$$

where $0 \le a' \le |t_{i+1}|$ and $0 \le b' \le |t_i| + j|t_{i+1}|$. Moreover, $a' = 0$ and $b' = 0$ do not hold simultaneously, and similarly, $a' = |t_{i+1}|$ and $b' = |t_i| + j|t_{i+1}|$ do not hold simultaneously. $\qquad \square$

Based on this lemma we can give the proof of Theorem 2.

Proof. The proof is given by induction on $i \geq 0$. Suppose $i = 0$. If $q_1 = 1$, then $j = 0$, in which case the claim is empty. If $q_1 > 1$, then $j > 0$, and $\kappa\lambda \bmod n = \kappa\lambda$, for all $0 < \kappa < |t_0| + (j+1)|t_1| = j+1$. Then the interval $[0, n]$ is split into $j + 1$ subintervals, out of which j is of length $r_1 = \lambda$ and $|t_1| = 1$ is of length $r_0 - jr_i = n - j\lambda$, for all j, $0 \leq j < q_1$, as desired.

Now we make the induction hypothesis that the claim holds for i and for all j, $0 \leq j < q_{i+1}$. Then, in particular, the points

$$\kappa\lambda \bmod n, \quad 0 < \kappa < |t_i| + q_{i+1}|t_{i+1}| = |t_{i+2}| \tag{9}$$

split the interval $[0, n]$ into

$$|t_i| + (q_{i+1} - 1)|t_{i+1}| = |t_{i+2}| - |t_{i+1}| \text{ subintervals of length } r_{i+1}$$
$$|t_{i+1}| \text{ subintervals of length } r_i - (q_{i+1} - 1)r_{i+1} = r_{i+1} + r_{i+2}.$$

Let us now prove that under the induction hypothesis the claim holds also for $i + 1$ and $j = 0$ and the splitting given by the points

$$\kappa\lambda \bmod n, \quad 0 < \kappa < |t_{i+1}| + |t_{i+2}|.$$

By arranging these points in increasing order we get $|t_{i+1}| + |t_{i+2}| + 1$ points $y_\ell = \kappa_\ell\lambda \bmod n$ such that

$$0 = y_0 < y_1 < \dots < y_{|t_{i+1}|+|t_{i+2}|-1} < n = y_{|t_{i+1}|+|t_{i+2}|}.$$

By Lemma 3 there exist nonnegative numbers a_ℓ and b_ℓ such that

$$y_{\ell+1} - y_\ell = (\kappa_{\ell+1} - \kappa_\ell)\lambda \bmod n = a_\ell r_{i+1} + b_\ell r_{i+2}, \tag{10}$$

where $a_\ell > 0$ or $b_\ell > 0$, for all $\ell = 0, 1, \dots, |t_{i+1}| + |t_{i+2}| - 1$.

Note that the $|t_{i+2}| - 1$ points given by (9), which split the interval into $|t_{i+2}|$ subintervals of length either r_{i+1} or $r_{i+1} + r_{i+2}$ are included in the set $\{y_\ell \mid \ell = 1, 2, \dots, |t_{i+1}| + |t_{i+2}| - 1\}$. Then it is possible to satisfy (10) only if the new $|t_{i+1}|$ points y_ℓ are placed on the interval in such a way that they split the $|t_{i+1}|$ intervals of length $r_{i+1} + r_{i+2}$ into two intervals of length r_{i+1} and r_{i+2}. This means that $y_{\ell+1} - y_\ell = r_{i+1}$ or $y_{\ell+1} - y_\ell = r_{i+2}$ for all $\ell = 0, 1, \dots, |t_{i+1}| + |t_{i+2}| - 1$, as desired.

Assume now that we have achieved the splitting of the desired form for $i + 1$ and $j < q_{i+2} - 1$. Then repeating the reasoning given above, we get that the claim holds also for $j + 1$. In this manner we get that the claim holds for $i + 1$ and all j such that $0 \leq j < q_{i+2}$. □

4.3   Coverings of the Interval and Establishing Entropy

Given a positive integer B, points $\{\kappa\lambda : 0 \leq \kappa < B\} = \{y_\ell : 0 \leq \ell < B, 0 = y_0 < y_1 < \dots < y_{B-1} < y_B = n\}$ and a positive integer A, a (partial) covering of the interval $[0, n]$ is comprised of the subintervals

$$\{x \in \mathbb{Z} : y_\ell \leq x < y_\ell + A\}, \quad 0 \leq \ell < B. \tag{11}$$

We restrict our analysis to special types of coverings resulting from random decompositions. Specifically, we assume that for all ℓ, $0 \le \ell < B$, we have either $y_{\ell+1} - y_\ell = \Delta_1$ or $y_{\ell+1} - y_\ell = \Delta_2$, where $0 < \Delta_1 < \Delta_2$. It is immediate to verify that such a covering of the set $\{ x \in \mathbb{Z} : 0 \le x < n \}$ is surjective if and only if $A \ge \Delta_2$ and the resulting overlap of the surjective covering is minimal with the smallest value of A, $A = \Delta_2$.

Next we make an additional assumption that the lengths of the subintervals are close to each other, more precisely, $\Delta_2 < 2\Delta_1$. Under this assumption, no point on $[0, n]$ is covered by more than two adjacent subintervals of the covering (11). To prove this, let us assume the contrary, i.e., there is $x \in \mathbb{Z}$ and ℓ, $0 \le \ell < B - 2$ such that

$$y_\ell \le x < y_\ell + A$$
$$y_{\ell+1} \le x < y_{\ell+1} + A$$
$$y_{\ell+2} \le x < y_{\ell+2} + A.$$

Then $2\Delta_1 \le (y_{\ell+2}-y_{\ell+1})+(y_{\ell+1}-y_\ell) = y_{\ell+2}-y_\ell < x-(x-A) = A = \Delta_2 < 2\Delta_1$, where we have a contradiction. To obtain an injective covering, it is necessary that $A \le \Delta_1$, in which case the coverage is maximized for $A = \Delta_1$.

We summarize the results in the following theorem. For the definitions of f and F we refer to Sec. 4.1. Given a covering (11) the sets $\mathcal{R}$ and $\mathcal{T}$ are defined as

$$\mathcal{R} = \{x \in \mathbb{Z} : 0 \le x < A\}$$
$$\mathcal{T} = \{\kappa \in \mathbb{Z} : 0 \le \kappa < B\}.$$

Theorem 3. *Suppose that in the setting described above a covering (11) of the set $\{x \in \mathbb{Z} : 0 \le x < n\}$ satisfies $y_{\ell+1} - y_\ell = \Delta_1$ or $y_{\ell+1} - y_\ell = \Delta_2$, where $0 < \Delta_1 < \Delta_2$. If $A = \Delta_2$, then $F(k) = 2$ for $AB - n$ values of k and $F(k) = 1$ for $2n - AB$ values. In this case the generation of random decomposition is surjective and the entropy is equal to $\log AB + \frac{2n}{AB} - 2$. If $A = \Delta_1$ then $F(k) = 1$ for AB values of k and $F(k) = 0$ for $n - AB$ values. In this case the generation is injective and the entropy is equal to $\log AB$.*

Theorem 2 gives many options to generate random points $\kappa\lambda \bmod n$, $0 \le \kappa < B$, for splitting the interval $[0, n]$ to subintervals of two lengths, Δ_1 and Δ_2, $0 < \Delta_1 < \Delta_2$. Moreover, splittings that satisfy the additional assumption $\Delta_2 < 2\Delta_1$ can be achieved. We get the following corollary.

Corollary 1. *Let m be a positive integer. Then in the setting of Theorem 2, we have $0 < r_m < r_{m-1} - jr_m$, for all $0 \le j < q_m$. Moreover, $r_{m-1} - jr_m < 2r_m$ if and only if $j = q_m - 1$. If this is the case then the length of the longer interval is equal to $r_{m-1} - jr_m = r_m + r_{m+1}$, the length of the shorter interval is equal to r_m and the number of subintervals is equal to $|t_{m+1}|$.*

Proof. By Theorem 2, $(j + 2)r_m - r_{m-1} \le -r_{m+1}$ for all $0 \le j \le q_m - 2$, and $(q_m + 1)r_m - r_{m-1} = r_m - r_{m+1} > 0$. □

Example 4. Continuing with the curve given in Example 3, the extended Euclidean algorithm yields the following values (u_i omitted for brevity):

m	0	1	2	3	4	5	6	7	8	9	10	11	12	13	14	15
r_m	61543	50568	10975	6668	4307	2361	1946	415	286	129	28	17	11	6	5	1
t_m	0	1	-1	5	-6	11	-17	28	-129	157	-443	1929	-2372	4301	-6673	10974

The splitting is illustrated in Fig. 3 where we have chosen $m = 8$ and $j = 0 = q_8 - 1$ in Corollary 1. The top figure depicts points $\{\kappa\lambda \bmod n : 0 \le \kappa < |t_7| + |t_8| = 157\}$; the multiples κ are represented by the gradient color value, from bottom to top. Multiples of n make up the y-axis. The middle figure is conceptually the top figure collapsed modulo n. We can see that, as Corollary 1 shows, the distance between points is two-valued, in this case $|t_7| = 28$ points with distance $r_8 = 286$ and $|t_8| = 129$ with distance $r_7 = 415 = r_8 + r_9$. The distance given is from the nearest point to the right. The bottom figure shows a smaller range in more detail.

We now investigate the distributions obtained using this splitting according to Corollary 1. We set $m = 8$ and $B = |t_9| = 157$ and obtain the following values for the distribution by exhaustive search. To obtain a surjective distribution, we set $A = r_8 + r_9 = 415$. Then $AB = 65155$ and

$$\#\{k \in \mathbb{Z}_n : F(k) = 0\} = 0$$
$$\#\{k \in \mathbb{Z}_n : F(k) = 1\} = 57931$$
$$\#\{k \in \mathbb{Z}_n : F(k) = 2\} = 3612.$$

This yields entropy of ≈ 15.88 as expected according to Corollary 1.

By setting $A = r_8 = 286$, and $B = |t_9| = 157$ as above, the distribution of the random decomposition is injective and

$$\#\{k \in \mathbb{Z}_n : F(k) = 0\} = 16484$$
$$\#\{k \in \mathbb{Z}_n : F(k) = 1\} = 45059$$
$$\#\{k \in \mathbb{Z}_n : F(k) = 2\} = 0.$$

This yields entropy of ≈ 15.46. In the injective case the entropy is the larger the larger is the coverage and the value $AB = |t_{m+1}|r_m$. A significantly better entropy for the injective generation is achieved for $m = 9$. In this case $|t_{m+1}|r_m = 443 \cdot 129 = 57147$ and the distribution is:

$$\#\{k \in \mathbb{Z}_n : F(k) = 0\} = 4396$$
$$\#\{k \in \mathbb{Z}_n : F(k) = 1\} = 57147$$
$$\#\{k \in \mathbb{Z}_n : F(k) = 2\} = 0.$$

This yields entropy of ≈ 15.80 of the probability distribution of the injective random generation.

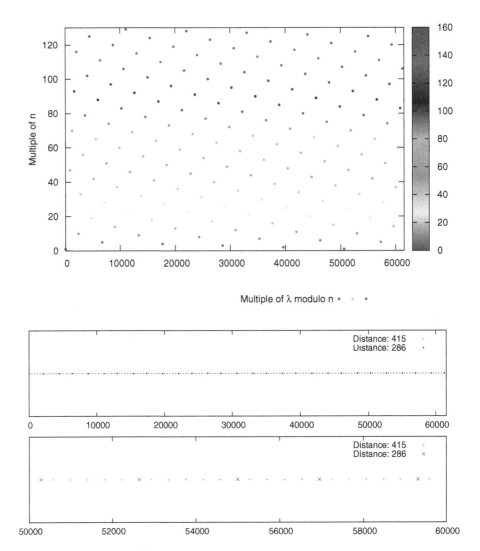

Fig. 3. Splitting for the curve in Example 3. Top: Calculating multiples (the gradient) of λ; multiples of n are given on the y-axis. Middle: The top figure collapsed modulo n; distances are given to the nearest point to the right. Bottom: A detailed view of a smaller range.

Remark 1. Typically one would select m such that the sets $\mathcal{R}$ and $\mathcal{T}$ are about equal size, but we stress that our analysis allows unbalanced generation of the components of k to achieve a more uniform distribution if needed. This is illustrated in the example above. A slightly larger m was chosen to improve the entropy in the injective case. Thus given the values from the extended Euclidean algorithm, one is free to choose the most appropriate m.

5   Conclusion

The GLV method by Gallant, Lambert, and Vanstone [1] has become a well-established method for high speed elliptic curve cryptography. The recent work of Galbraith et al. [4] further encourages its use. The seminal method for balanced length-two representations given in [1] uses a number of multiplications, divisions, round operations, and computations with rationals.

We described a new, conceptually simple, method for modular decomposition of integers based on the extended Euclidean algorithm. We derived strict bounds for each component of the decomposition, leading to a representation with a desired level of balance. The construction of the decomposition is fundamentally different than that of traditional GLV decomposition.

Furthermore, we examined the strategy of starting with random decompositions, useful in key generation and cryptosystems requiring ephemeral keys. The distribution of such random decompositions was first questioned by Gallant in 1999 [5] and remained unanswered in the context of GLV until now. We gave a method of deriving proper bounds for each component of the decomposition in such a way that ensures a tractable distribution and hence an explicit formula for the achieved entropy. Our method also allows enforcement of special properties such as injectivity or surjectivity of the generation of random decompositions.

Extension to higher dimensions remains an open problem. The next interesting case for [4] is dimension 4. Algebraic number theoretic methods could prove very useful in such a generalization, as the proofs given here become tedious as the dimension increases.

Acknowledgements

We thank Lars Lone Rasmussen for running some initial experiments on generation of random GLV decompositions. We are also grateful to the anonymous referees for thorough reading of our submission and proposing several improvements to the presentation of the paper.

References

1. Gallant, R.P., Lambert, R.J., Vanstone, S.A.: Faster point multiplication on elliptic curves with efficient endomorphisms. In: Kilian, J. (ed.) CRYPTO 2001. LNCS, vol. 2139, pp. 190–200. Springer, Heidelberg (2001)
2. Bellman, R., Straus, E.G.: Problems and Solutions: Solutions of Advanced Problems: 5125. Amer. Math. Monthly 71(7), 806–808 (1964)
3. ElGamal, T.: A public key cryptosystem and a signature scheme based on discrete logarithms. IEEE Trans. Inform. Theory 31(4), 469–472 (1985)
4. Galbraith, S.D., Lin, X., Scott, M.: Endomorphisms for faster elliptic curve cryptography on a large class of curves. In: Advances in cryptology—EUROCRYPT 2009. LNCS, Springer, Heidelberg (2009) (to appear)
5. Gallant, R.: Faster elliptic curve cryptography using efficient endomorphisms. In: 3rd workshop on Elliptic Curve Cryptography—ECC 1999 (1999) (presentation slides)

6. SECG: Recommended elliptic curve domain parameters. Standards for Efficient Cryptography SEC 2 (September 20, 2000)

7. ANSI: Public key cryptography for the financial services industry: Key agreement and key transport using elliptical curve cryptography (2001) ANSI X9.63

8. Hankerson, D., Menezes, A.J., Vanstone, S.A.: Guide to Elliptic Curve Cryptography. Springer, New York (2004)

9. Babai, L.: On Lovász' lattice reduction and the nearest lattice point problem. Combinatorica 6(1), 1–13 (1986)

10. Kim, D., Lim, S.: Integer decomposition for fast scalar multiplication on elliptic curves. In: Nyberg, K., Heys, H.M. (eds.) SAC 2002. LNCS, vol. 2595, pp. 13–20. Springer, Heidelberg (2003)

11. Sica, F., Ciet, M., Quisquater, J.J.: Analysis of the Gallant-Lambert-Vanstone method based on efficient endomorphisms: elliptic and hyperelliptic curves. In: Nyberg, K., Heys, H.M. (eds.) SAC 2002. LNCS, vol. 2595, pp. 21–36. Springer, Heidelberg (2003)

12. Park, Y.H., Jeong, S., Kim, C.H., Lim, J.: An alternate decomposition of an integer for faster point multiplication on certain elliptic curves. In: Naccache, D., Paillier, P. (eds.) PKC 2002. LNCS, vol. 2274, pp. 323–334. Springer, Heidelberg (2002)

13. Grabner, P.J., Heuberger, C., Prodinger, H.: Distribution results for low-weight binary representations for pairs of integers. Theoret. Comput. Sci. 319(1-3), 307–331 (2004)

14. Antipa, A., Brown, D., Gallant, R., Lambert, R., Struik, R., Vanstone, S.: Accelerated verification of ECDSA signatures. In: Prenel, B., Tavares, S. (eds.) SAC 2005. LNCS, vol. 3897, pp. 307–318. Springer, Heidelberg (2006)

15. Knuth, D.E.: The Art of Computer Programming, Volume II: Seminumerical Algorithms, 2nd edn. Addison-Wesley, Reading (1981)

16. Järvinen, K., Forsten, J., Skyttä, J.: Efficient circuitry for computing τ-adic non-adjacent form. In: Proceedings of the 13th IEEE International Conference on Electronics, Circuits and Systems, ICECS 2006, pp. 232–235. IEEE, Los Alamitos (2006)

17. Lange, T., Shparlinski, I.E.: Distribution of some sequences of points on elliptic curves. J. Math. Cryptol. 1(1), 1–11 (2007)

18. Cohen, H., Frey, G. (eds.): Handbook of elliptic and hyperelliptic curve cryptography. CRC Press, Boca Raton (2005)

19. IEEE: IEEE P1363 working group for public-key cryptography standards. meeting minutes (November 15, 2000),
http://grouper.ieee.org/groups/1363/WorkingGroup/minutes/Nov00.txt

Breaking KEELOQ in a Flash: On Extracting Keys at Lightning Speed*

Markus Kasper, Timo Kasper, Amir Moradi, and Christof Paar

Horst Görtz Institute for IT Security, Ruhr University Bochum, Germany
{mkasper,tkasper,moradi,cpaar}@crypto.rub.de

Abstract. We present the first simple power analysis (SPA) of software implementations of KEELOQ. Our attack drastically reduces the efforts required for a complete break of remote keyless entry (RKE) systems based on KEELOQ. We analyze implementations of KEELOQ on microcontrollers and exploit timing vulnerabilities to develop an attack that allows for a practical key recovery within seconds of computation time, thereby significantly outperforming all existing attacks: Only one single measurement of a section of a KEELOQ decryption is sufficient to extract the 64 bit master key of commercial products, without the prior knowledge of neither plaintext nor ciphertext. We further introduce techniques for effectively realizing an automatic SPA and a method for circumventing a simple countermeasure, that can also be applied for analyzing other implementations of cryptography on microcontrollers.

1 Motivation

Due to its wide deployment in RKE systems, the KEELOQ cipher has come to the attention of cryptographers in 2007 [1]. Several improved cryptanalytical attacks followed, but still, their complexity and other requirements make them impractical for real-world products.

This situation extremely changed with the first differential power analysis (DPA) of KEELOQ as presented on CRYPTO 2008 [5]. The paper describes how secret keys can be revealed in practice from the power consumption of KEELOQ implementations in hardware and software. In Sect. 3.3 we reflect, how especially knowing his master key allows for devastating attacks on all systems of a manufacturer. Unfortunately - from the attacker's point of view - the extraction of the master key remains difficult and requires some efforts, because the software implementations programmed into the receivers are very hard to analyze using DPA, as discussed in Sect. 4.

We illustrate in the following, that in some cases performing a key recovery by SPA is much easier and much more efficient than by DPA, and demonstrate that SPA constitutes a remedy for the open problem of extracting the master key from KEELOQ software implementations. Starting from a specific unprotected

---

* The work described in this paper has been supported in part by the European Commission through the ICT programme under contract ICT-2007-216676 ECRYPT II.

B. Preneel (Ed.): AFRICACRYPT 2009, LNCS 5580, pp. 403–420, 2009.

software implementation of the algorithm - as recommended by Microchip - we develop a highly effective SPA attack in Sect. 5. Usually, an SPA is performed based on tedious visual inspection, as detailed in Sect. 5.2, or by massive profiling of a similar device, which takes a lot of efforts and time. In Sect. 5.3, a non-heuristic method to avoid the visual inspection in some types of SPA attacks is presented, enabling a full key recovery from just a single measurement of the power consumption. We practically verify our findings by attacking some commercial KEELOQ implementations on PIC 8-bit microcontrollers and proof the effectiveness of our methods, even in the presence of a simple countermeasure. Removing the effect of reoccurring disturbing patterns in the traces, that hinder DPA and SPA in the first place, is detailed in Sect. 6. Before developing our new attack, we give some necessary background information about power analysis in Sect. 2 and briefly introduce KEELOQ RKE systems in Sect. 3. Finally, the effectiveness of DPA and SPA in the case of KEELOQ is discussed in Sect. 7.

This article meliorates the CRYPTO 2008 attacks in terms of a great reduction of the required time and computations to recover secret master keys of different manufacturers and hence allows to completely circumvent many KEELOQ systems in the field with almost no effort.

2   Power Analysis in a Nutshell

In contrast to a mathematical cryptanalysis which requires pairs of plain- and ciphertexts, in the context of power analysis knowing either the input or the output of the cipher is sufficient to mount a key-recovery attack. By measuring and evaluating the power consumption of a cryptographic device, information-dependent leakage is exploited and combined with the knowledge about the plaintext or ciphertext in order to extract, e.g., a secret key. Since intermediate results of the computations can be derived from the leakage, e.g., from the Hamming weight of the data processed in a software implementation, a divide-and-conquer strategy becomes possible, i.e., the secret key could be recovered bit by bit.

2.1   Preprocessing

For unknown implementations, it is often difficult to find an appropriate trigger point for starting the oscilloscope, e.g., a special feature in the traces, that reoccurs at the same instant in each measurement. Accordingly, the alignment of the measurements typically needs to be improved as a first preprocessing step after the acquisition. Furthermore, traces can be very large or too noisy for an effective evaluation – thus they might need to be compressed or averaged prior to statistical analysis.

Peak Extraction. The dynamic power consumption is the dominant factor disclosing the processed data of complementary metal oxide semiconductor (CMOS) circuits. The corresponding peaks appearing in the measurements on each edge of the clock hence play a prominent role for power analysis. Processing only

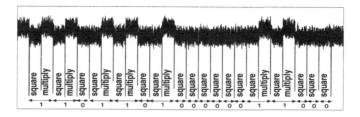

Fig. 1. SPA of an implementation of RSA

the amplitudes of these peaks - instead of all acquired data points - allows for a great reduction of computations and memory required during the analysis. Moreover, misalignments arising from a clock jitter due to an unstable oscillator in the cryptographic device are eliminated by peak extraction.

Averaging. In case of a bad quality of the acquired power consumption, e.g., due to a noisy environment, bad measurement setup or cheap equipment, averaging can be applied by decrypting the same ciphertext repeatedly and calculating the mean of the corresponding traces. This method reduces the noise floor and can enormously increase the signal-to-noise ratio of the power traces. As exactly the same input data is processed, the measurements can be accurately aligned using a cross-correlation between two full traces. Comparing (averaged) traces for different ciphertexts can help to find the time-window in the traces, where a data-dependent behavior occurs and hence the decryption takes place.

2.2   Simple Power Analysis

An SPA attack, as introduced in [7], relies on visual inspection of power traces, e.g., measured from an embedded microcontroller of a smartcard. The aim of an SPA is to reveal details about the execution path of a software implementation, like the detection of conditional branches depending on secret information. At first, implementations of RSA were in the focus of the attackers, because an SPA of them is rather straightforward. A typical modular exponentiation comprises two main function calls, i.e., "square" and "multiply". The execution time for processing a zero or one can often be distinguished visually from the power traces, as illustrated for an 8051-based microprocessor in Fig. 1. Obviously, an attacker can directly recover the secret exponent of the RSA encryption from the sequence of instructions visible in the measurements.

2.3   Differential Power Analysis

Contrary to SPA, DPA takes many traces with often uniformly distributed known plaintexts or known ciphertexts into account and evaluates them with statistical methods. A DPA requires no knowledge about the concrete implementation of the cipher and can hence be applied to any unprotected black box implementation. The points in time where secret information leaks during the execution of the cipher are an outcome of a DPA [7]. The traces are divided into sets according to intermediate values depending on key hypotheses and then statistically evaluated, e.g., by calculating the mean for each point in time of all traces of each set. The probability for a zero or one being processed should be uniformly distributed in each set, and thus the difference of the means will vanish, except for the set belonging to the correct key hypothesis.

In a correlation power analysis (CPA), each point in time for all measurements is compared to a theoretical model of the implementation by calculating the correlation coefficient. A maximum correlation between the hypothetical power consumption and actually measured power values indicates the correct key hypothesis [2].

3   KeeLoq RKE Systems

An RKE system consists of one receiver in the secured object and one or more remote controls that can send transmissions to the receiver and thereby control the access to the object. The early fixed-code or multi-code systems[1] were developed soon after digital circuitry became available. They rely on sending a fixed sequence of binary data when pressing the remote, and permit access in case the code is correctly identified. The obvious need for a protection against replay attacks, with only an unidirectional channel available, brought the invention and wide deployment of so-called hopping code systems. KeeLoq RKE systems typically employ hardware implementations of the cipher, such as HCSXXX [10], for generating hopping codes in the remote controls and a software implementation running on an 8-Bit PIC microcontroller [11] in the receiver to decrypt the transmissions.

3.1   Hopping Code Scheme

The remote control possesses an internal counter that is increased each time one of its buttons is pressed. The increased value is then encrypted and transmitted as a hopping code. For each remote, the receiver stores the counter value of the last valid transmission and updates the counter only upon decryption of a valid hopping code with a moderately increased counter value. The receiver is thus capable of rejecting repetitive codes and can thereby prevent replay attacks (except if combined with jamming, see Sect. 3.3). Extra remotes can usually be made known to the receiver by putting it into a learning mode in which the key of the extra remote is derived and stored.

Key Management. Microchip suggests several key derivation schemes for generating a unique device key K_{dev} for each remote control. All of these schemes involve a secret manufacturer key K_{man} that is used once in the factory for a freshly produced remote control, and later in the receiver when the key derivation

---

[1] Note that even these outdated systems are still available on the market.

takes place. This global master key for the RKE system is of course stored securely in the memory of the microcontroller.

For the most widespread key derivation in practice, the device key is a function f of the identifier ID (serial-number) of the remote control. The ID is no secret, because it is transmitted unencryptedly with every hopping code. Any party knowing the manufacturer key K_{man}, e.g., the receiver, is hence capable of calculating $K_{dev} = f(K_{man}, ID)$. For the key-derivation function f, Microchip proposes KEELOQ decryptions with K_{man} [9], as described in Sect. 3.2. Even if a K_{dev} and the corresponding ID are known, a straightforward inversion of f is impossible. The described scheme for the key-derivation enables different business models, as the receivers of one manufacturer will only cooperate with remotes of the same manufacturer and thus prohibit a competitor from selling spare remotes.

3.2 KEELOQ Decryption

The decryption algorithm described in the following is used both for deciphering the hopping codes and during the key-learning phase – note that the software in a receiver never encrypts data. Prior to a decryption employing the KEELOQ block cipher, a 32-bit state $Y = \{y_0, \ldots, y_{31}\}$ is initialized with the ciphertext C. After 528 rounds of the decryption involving a secret key $K = \{k_0, \ldots, k_{63}\}$ of length 64 bits, Y contains the plaintext P.

Details of the Cipher. In each round i, one key bit $k_{(15-i) \bmod 64}$ is XORed with two bits of the state and the output bit of a non-linear function (NLF) that combines five other bits of the state. Afterwards, the state is shifted left, such that the most significant bit (MSB) y_{31} is dropped, and the output of the XOR becomes the new least significant bit (LSB) y_0. The details of the cipher are given in Alg. 1, where $\oplus$ denotes a bitwise XOR. Note that each key bit is reused at least eight times, i.e., every 64 rounds of the decryption.

Algorithm 1. KEELOQ Decryption (Pseudo Code)

Input: ciphertext $C = \{c_0, \ldots, c_{31}\}$, key $K = \{k_0, \ldots, k_{63}\}$
Output: plaintext $P = dec_K(C)$, where dec denotes KEELOQ decryption with K

1. Load ciphertext: $Y = C$
2. For i = 0 to 527 do
    2.1. Output bit of NLF: $OUT = \mathrm{NLF}(y_{30}, y_{25}, y_{19}, y_8, y_0)$
    2.2. Output bit of XOR: $XOR = k_{(15-i) \bmod 64} \oplus y_{31} \oplus y_{15} \oplus OUT$
    2.3. Update state
        2.3.1. left-shift state: $Y = (Y << 1)$
        2.3.2. assign LSB: $y_0 = XOR$
3. RETURN Y

The Non-linear Function. While the NLF could also be realized by means of Boolean functions, performing table-look-ups as described in the following is common practice. Defining a look-up table by the hexadecimal constant $LUT =$ 0x3A5C742E, its j-th bit is equivalent to one output bit OUT of the non-linear function NLF$(x_4, x_3, x_2, x_1, x_0)$. The index $j \in \{0, 1, \ldots, 31\}$ thereby equals the decimal representation of the input bits x_4 to x_0, i.e., $j = 2^4 \cdot x_4 + 2^3 \cdot x_3 + 2^2 \cdot x_2 + 2^1 \cdot x_1 + 2^0 \cdot x_0$. The implementation of the NLF can be crucial for the susceptibility to SPA, as will be shown in Sect. 5.

3.3   History of Attacks on KeeLoq

A common method for electronically breaking into cars secured with hopping code systems is a combined eavesdropping-and-jamming attack: While the legitimate owner tries to lock his car with a remote control, the transmission is monitored and at the same time the frequency of the transmission is jammed, with the effect that the car won't be locked and the attacker possesses a temporarily valid hopping code. There are devices that automatically perform the described process, but in practice they are rather unreliable. One successful transmission of a new hopping code from the original remote to the car invalidates all previously eavesdropped hopping codes.

Mathematical Analysis. Recently, several cryptanalytic attacks on the KeeLoq cipher have been published [3,4,6]. Without taking precomputed tables into account, the most efficient attack has a complexity of 2^{48} and requires 2^{16} plain- and ciphertext pairs - hence KeeLoq has to be regarded as insecure from the cryptographic point of view. Still, for a practical RKE system using hopping codes the plaintext remains secret in the remote control, rendering the mathematical attacks impractical.

Power Analysis and Eavesdropping Attack. On CRYPTO 2008, a paper demonstrates the *Power of Power Analysis* [5] by describing how the K_{dev} and the master key K_{man} of commercial RKE systems based on KeeLoq can be extracted by means of DPA.

Hardware implementations of KeeLoq, such as HCS301 [10] application-specific integrated circuits (ASICs), are an ideal platform for conducting DPA attacks. The timing behavior of the chip can be foreseen very precisely, as it always performs exactly the same digital operations independent of the secret key. This implies that the power consumption at each point in time of the acquired traces is always related to the same step of the KeeLoq cipher, and extracting device keys K_{dev} with DPA is relatively straightforward. The authors of [5] report a full key recovery of K_{dev} from less than ten measurements, in the best case.

Extracting the manufacturer key K_{man} from software implementations turned out to be orders of magnitude harder, as explained below in Sect. 4. When the secret master key K_{man} gets into the hands of an attacker, two main implications arise. Firstly, the attacker can produce fake products that are compatible with

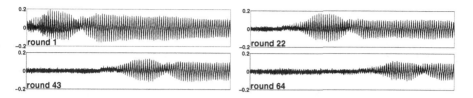

Fig. 2. Correlation coefficient of the correct key in a CPA attack on the software implementation of the KEELOQ decryption running in a PIC microcontroller

those of that manufacturer - the monopoly of the manufacturer, e.g., him being the only supplier of spare remote controls, collapses. Secondly, a remote control of this manufacturer - including its secret device key K_{dev} - can be cloned by monitoring a transmission from a distance, even without ever seeing the original. With this powerful eavesdropping approach, even a low-skilled intruder can spoof a KEELOQ receiver with technical equipment for less than US$ 50 and take over control of an RKE system, or deactivate an alarm system, leaving no physical traces.

4 Open Problem

The extraction of K_{man} from a software implementation of the KEELOQ decryption during the key-derivation mode of the receiver with DPA is much harder than a DPA attack on a hardware implementation of the cipher - mainly for two reasons. Firstly, lack of a suitable trigger point in the power consumption of the microcontroller leads to extra steps required for a proper alignment when preprocessing the traces. Secondly, as shown in Fig. 2, the correlation coefficient of the correct key continuously decreases with an increasing number of rounds, such that roughly 10 000 power traces need to be evaluated in order to fully recover the 64-bit K_{man} - a huge effort compared to 5-30 traces for extracting K_{dev} from hardware implementations. The authors of [5] predict that the cause is a data-dependent execution time for each round of a KEELOQ decryption in the program code.

4.1 Software Implementations of KEELOQ

Meanwhile, source code as proposed by Microchip for a PIC 8-bit microcontroller has become available on the Internet [12]. Appendix A shows an excerpt of the program code, revealing that the execution time of each round in the code example varies depending on the processed data. In fact, most of the program code takes the same amount of clock cycles, except for the specific implementation of the look-up table to build the NLF (compare with Sect. 3.2). As a result, the execution time of a decryption varies for different ciphertexts - a typical indicator for a susceptibility towards an SPA.

5   SPA-Attacking KEELOQ

In this section first the mathematical aspects of our proposed SPA on KEELOQ are illustrated; then, the effectiveness of visual inspection in practice is investigated for different platforms. Finally, a new method for performing an SPA devoid of visual inspection, and empirical results from attacking commercial products, are presented.

5.1   Mathematical Background

Let us denote the content of the state register during a KEELOQ decryption by a bitstream $S = \{s_i \ ; -31 \leq i \leq 528\}$. When the first 32 bits $\{s_{-31}, s_{-30}, \ldots, s_0\}$ of the bitstream are initialized with the corresponding ciphertext bits $\{c_{31}, c_{30}, \ldots, c_0\}$, the bits with indices $1 \leq i \leq 528$ can be computed according to step 2 of Alg. 1 using the iterative equation

$$s_{j+1} = k_{(15-j) \bmod 64} \oplus s_{j-31} \oplus s_{j-15} \oplus \text{NLF}\left(s_{j-30}, s_{j-25}, s_{j-19}, s_{j-8}, s_j\right). \quad (1)$$

According to Eq. (1), one bit of the secret key $k_{(15-j) \bmod 64}$ can be revealed from the knowledge of eight bits of the stream S. For extracting all 64 bits of the key a consecutive section of the stream with $32 + 64 = 96$ bits is sufficient to recover all keybits. Note that in a typical known-plaintext or known-ciphertext scenario up to 32 bit of the required stream might be known a priori. The following sections will describe how to determine the required consecutive bitstream by SPA.

5.2   Visual Inspection

Visual inspection and its utilization in an SPA attack on RSA are presented in Sect. 2.2. However, the KEELOQ algorithm is extremely different from RSA and there are no distinguishable functions called during the encryption or decryption routines. As illustrated in Sect. 4.1, there are conditional branches depending on the values of the state register in the software implementation of the KEELOQ decryption recommended by Microchip. Typically, no difference in the power patterns for taking or not taking these branches can be observed for a PIC microcontroller, as it mostly leaks the operands being processed, not the operations. Hence, pinpointing the small variations of two or three instructions between two rounds of the algorithm by visual inspection is a very challenging and sometimes impossible task.

Fig. 3 shows power traces measured from commercial implementations of the KEELOQ decryption on different PIC microcontrollers. Spending a lot of time and efforts with manually analyzing the details of the power consumption, in few cases a distinguishable behavior can be spotted in the periodic power patterns of the microcontroller, as illustrated in Fig.3(a) and Fig.3(b). If the difference in these patterns would furthermore directly depend on the values of the status register, a key recovery according to Sect. 5.1 could be possible. However, Fig. 3(c) illustrates that in some cases no difference between the periodic patterns can

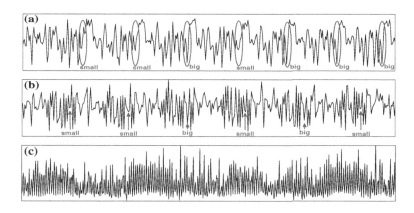

Fig. 3. Visual inspection of power traces of the KEELOQ cipher

be detected by means of heuristic methods, even by averaging the traces as detailed in Sect. 2.1. Note that Fig. 3 shows pure measurements without averaging, directly sampled by the oscilloscope.

5.3   A Non-heuristic Method for SPA

In the following, we will develop a non-heuristic method that allows for automatically identifying differences in power traces - as required for an SPA - even for those implementations in which a visual inspection is not effectual. First, we examine the time-variations occurring during a KEELOQ decryption more precisely, using the example of a program code proposed by Microchip. It will turn out that the conclusions drawn from analyzing this code can be applied to many different, unknown implementations of KEELOQ decryptions[2] on PIC microcontrollers.

Investigating the Code. The duration of all conditional branches that take place during the decryption is examined in Fig. 4 by comparing the number of clock-cycles required for each instruction of a PIC microcontroller [11]. While the decisions taken in the code excerpts (a), (c), (d), (e), (g), and (h) shown in Fig. 4 do not affect the execution time, the number of cycles required for (b), (f), and (i) varies with the respective condition being fulfilled or not - hence the time variations in different rounds are due to these three conditional branches. Table 1 summarizes the effect of the conditional branches on the difference in clock-cycles for the execution of one KEELOQ round. The duration of the program code in (f) and (i) of Fig. 4 can increase by a multiple of two cycles, depending on the state of the checked register, and (b) can likewise increase the length of a round by one cycle. As a consequence, taking the execution time of each round modulo 2 can reveal the result of the decision taken in (b), where the value of HOP2,0,

---

[2] In fact, the described method allows for extracting K_{man} from all implementations we are aware of.

1	0	= HOP3,3
1	2	0B: BTFSC HOP3,3
1	0	0C: MOVLW 10000B
2	**2**	sum

(a)

1	0	= HOP2,0
2	1	0E: BTFSS HOP2,0
0	2	0F: GOTO $+3
1	0	10: RLF MASK
1	0	11: RLF MASK
4	**3**	sum

(b)

1	0	= HOP1,0
1	2	12: BTFSC HOP1,0
1	0	13: RLF MASK
2	**2**	sum

(c)

1	0	= HOP4,1
1	2	15: BTFSC HOP4,1
1	0	16: IORLW 2
2	**2**	sum

(d)

1	0	= HOP4,6
1	2	17: BTFSC HOP4,6
1	0	18: IORLW 4
2	**2**	sum

(e)

6	4	2	0	= W Reg
2	2	2	2	19: ADDWF PC
0	0	0	1	1A: MOVLW 02EH
0	0	0	2	1B: GOTO T_END
0	0	1	0	1C: MOVLW 074H
0	0	2	0	1D: GOTO T_END
0	1	0	0	1E: MOVLW 05CH
0	2	0	0	1F: GOTO T_END
1	0	0	0	20: MOVLW 03AH
3	**5**	**5**	**5**	sum

(f)

1	0	= MASK
1	2	23: SKPZ
1	0	24: MOVLW 80H
2	**2**	sum

(g)

1	0	= KEY7,7
1	2	2F: BTFSC KEY7,7
1	0	30: SETC
2	**2**	sum

(h)

$\neq 1$	1	= CNT0
1	2	39: DECFSZ CNT0
2	0	3A: GOTO INLOOP
0	1	3B: DECFSZ CNT1
0	2	3C: GOTO OUTLOOP
0	1	02: MOVLW 48
0	1	03: MOVWF CNT0
3	**7**	sum

(i)

Fig. 4. Number of cycles required for the execution of an exemplary implementation of the KEELOQ decryption, depending on conditional branches

i.e., the 9th bit of the status register y_8, is tested. It is hence possible to deduce one bit of the status register from the duration of each round and, as described in Sect. 5.1, recover the whole 64-bit secret key from the execution time of at least 96 consecutive rounds[3].

As shown in Sect. 5.2, visual inspection is not feasible for some implementations - even less can the length of each round be precisely detected. In the following, we thus introduce a non-heuristic technique for determining the number of cycles in each round.

Power Leakage of PIC microcontrollers. Each execution cycle of a PIC microcontroller lasts four clock cycles [11], hence four peaks in a power trace relate to one execution cycle. Fig. 5 shows peaks extracted from power traces of a PIC microcontroller running a KEELOQ decryption. Extracting all four peaks of an execution cycle, as illustrated in Fig. 5(a), does not allow to locate the rounds of the decryption algorithm. In an attempt to facilitate the round-detection, only the first, second, third or fourth peak of each execution cycle are taken into account to yield Fig. 5(b), (c), (d) and (e), respectively. While focusing on the second or third peak does not improve the noticeability of the KEELOQ rounds, confining the analysis to the first or the fourth peak of each execution cycle, as shown in Fig. 5(b) and Fig. 5(e), allows for accurately distinguishing the successive rounds.

---

[3] 64 consecutive rounds may suffice if the ciphertext or the plaintext is known prior to the attack.

Table 1. Difference of the number of cycles depending on the conditional branches

CNT0=1	HOP4,1=1 and HOP4,6=1	HOP2,0=1	Diff. of no. of cycles	
			#	# mod 2
✗	✗	✗	2	0
✗	✗	✓	3	1
✗	✓	✗	0	0
✗	✓	✓	1	1
✓	✗	✗	6	0
✓	✗	✓	7	1
✓	✓	✗	4	0
✓	✓	✓	5	1

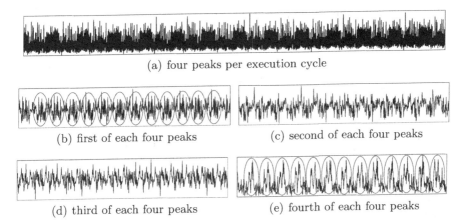

(a) four peaks per execution cycle

(b) first of each four peaks   (c) second of each four peaks

(d) third of each four peaks   (e) fourth of each four peaks

Fig. 5. Peaks of a power consumption trace of a PIC microcontroller running KEELOQ

Scrutinizing the Timing. In order to pinpoint the duration of each round of the algorithm, the cross-correlation between periodic patterns in the traces and a reference pattern is computed similarly to [8]. Suppose the reference pattern $\mathcal{R} = (r_1, r_2, \ldots, r_l)$ with a length of l which consists of the power-peaks of one particular round of the KEELOQ decryption. Furthermore, the vector containing the power-peaks of a whole KEELOQ decryption, with a length of n, is denoted by $\mathcal{P} = (p_1, p_2, \ldots, p_n)$. Then, a vector $\mathcal{C}$ showing the linear dependency between $\mathcal{R}$ and each section of $\mathcal{P}$ can be computed as

$$C = (c_1, c_2, \ldots, c_{n-l+1}) \quad ; \quad c_i = \mathrm{Correlation}\left(\mathcal{R}, (p_i, p_{i+1}, \ldots, p_{i+l-1})\right). \quad (2)$$

As illustrated in Fig. 6, the rounds can be clearly identified by consecutive maxima of $\mathcal{C}$. The locations of these maxima reveal the exact length of each round and hence, taking the length of each round modulo 2 discloses the content of the state register and consequently the bits of the secret key, as described in Sect. 5.1.

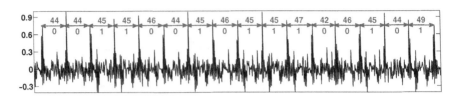

Fig. 6. An example for the correlation coefficients in vector C

5.4   Attack

Following the above described approach all 64 bits of the secret key can be recovered, but still there are three remaining problems:

i) Due to noise in the traces of the power-consumption the detection of the length of individual rounds may fail, leading to an incorrect detection of bits. Thus, a method to verify the recovered key bits would be convenient.

ii) The efficiency of the illustrated method depends strongly on the accuracy and correctness of the reference pattern $\mathcal{R}$. This demands for an in-depth study on choosing an accurate reference pattern.

iii) Suppose that all 64 bits of the key are recovered correctly. Since the key-bits are used periodically during the KEELOQ decryption and the attack can be started at any point in time with respect to the beginning of the decryption, the correct position of the recovered key bits in the secret key is not clear. Thus, the correct order of the bits needs to found out of 64 different alternatives.

Error Correction. Suppose all bits of the secret key are deduced with the described attack and let $\widehat{S} = \{\widehat{s}_i\ ;\ 1 \leq i \leq 528\}$ be the resulting bitstream, containing a part of the stream S of a decryption. The corresponding $\widehat{K} = \{\widehat{k}_i\ ;\ 0 \leq i \leq 527\}$ contains the key-bits computed from $\widehat{S}$. As each key bit is used at least eight times - every 64 rounds of the decryption - the correct key bits reappear in stream $\widehat{K}$. Let $\widehat{K}_i = (\widehat{k}_i, \widehat{k}_{i+1}, \ldots, \widehat{k}_{i+63})$, $0 \leq i \leq 464$, be a part of $\widehat{K}$ with a length of 64 bits, then

$$\exists i, j\ ;\ i \neq j\ ,\ i = j\ (\mathrm{mod}\ 64)\ ,\ \widetilde{K}_i = \widetilde{K}_j.$$

Errors in the detection of the correct key-bits due to noise can hence be corrected by a majority decision.

Generation of a Reference Pattern. Since the characteristics of the power consumption strongly depend on the device under test (DUT), the best basis for the reference pattern is a part of the power peaks produced by the DUT itself. As Fig. 5(b) illustrates, the durations of the rounds can be estimated by visual inspection of a decryption. Comparing with the source code described in Sect. 4.1 one can estimate that each round takes between 42 and 49 execution cycles - a reference pattern with a length of approximately 30 cycles is hence

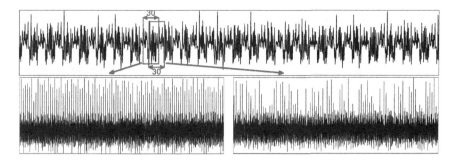

Fig. 7. Two exemplary reference patterns resulting in different correlation vectors C

adequate. However, as the beginning and the end of a round can only be guessed, the best position of the reference pattern in the power-peaks has to be found by moving the window until C contains regular maximums with a similar amplitude. In Fig. 7, two C vectors of the same power-peaks are plotted for two different reference patterns - the vector on the left-hand side is more appropriate.

As detailed in Sect. 3.1, a device key K_{dev} is obtained by a KEELOQ decryption of the corresponding ID of a remote control. Suppose that K_{dev} is already known from performing a DPA attack on the remote [5] and that $\widetilde{K}_k$ contains the correct bits of the secret master key. With $\widetilde{K}_k^{(i)}$ denoting a rotation of the bits[4] of $\widetilde{K}_k$ by i times, where $0 \leq i < 64$, the correct secret key is found if

$$\exists i \; ; \; f\left(\widetilde{K}_k^{(i)}, ID\right) = K_{dev},$$

where f denotes the key derivation function as detailed in Sect. 3.1. Hence, a known device key K_{dev} can be used to verify the correctness of the revealed key bits and furthermore simplifies to find the correct number of rotations i of $\widetilde{K}_k$.

Attack Results. The power traces of several PIC microcontrollers, such as PIC16C56 and PIC16F84A, were acquired using an Agilent Infiniium 54832D digital oscilloscope with a sampling rate of 125 MS/s by measuring the differential voltage of a $100\,\Omega$ resistor inserted in the ground path. Using the presented techniques we are able to extract the secret master key K_{man} of commercial KEELOQ code hopping receivers from only one single power trace. The efficiency of our attack is due to a software implementation leaking various key dependent information, and due to the nature of the KEELOQ cipher, i.e., using the key bits more than once.

6   Dealing with Interrupts

Most real-world implementations of the KEELOQ decryption algorithm are running on microcontrollers that are also responsible for other controlling tasks. In

---

[4] The direction of the rotations is not important, as long as it remains the same.

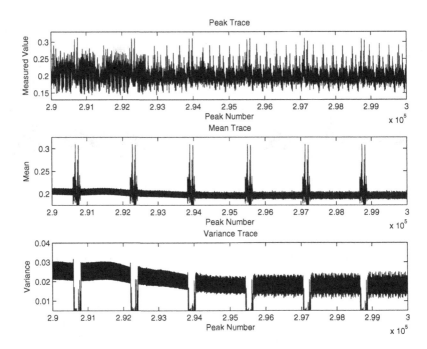

Fig. 8. A power trace with a periodic pattern (top), the mean of the aligned traces (middle), and their variance (bottom)

garage door systems this could be controlling the motor of the garage door or safety algorithms protecting users from injuries. These co-existing tasks of access control and other functionality may interfere by means of interrupt calls leading to unforeseen program flows. The resulting power traces prohibit averaging over multiple measurements and hinder straightforward CPA and SPA of the implementation. In this section we describe how power traces can be preprocessed in order to remove the power consumption of irrelevant program code inserted during the execution of the algorithm to still ensure the feasibility of side channel attacks.

Profiling. A recent implementation of the KEELOQ decryption, running on the 8-bit PIC microcontroller of a commercial product, proved to be resistant to both CPA and the SPA attack detailed above. Further investigations confirmed

---

Algorithm 2. Profiling of Interrupted Traces

---

1. Measure a reasonable amount of power traces (100 traces were sufficient)
2. Identify prominent parts of the pattern to be removed by visual inspection and select one occurrence as a template
3. Align all power traces on the first match of the template since the beginning of the decryption, e.g., using least square comparison
4. Calculate mean and variance of each data point over all aligned traces

---

Algorithm 3. Preprocessing of Interrupted Traces

---

1. Find the first occurrence of the pattern using least squares.
2. Jump to the end of the pattern, whose relative position is known from the profiling.
3. Save its absolute position in the data point index to *Start*.
4. From the beginning of the trace to its end calculate for each data point index:
   $$RelPos = CurrentDataPointIndex - Start \bmod PeriodLength$$
5. For each point decide:
   if $RelPos \leq (PeriodLength - PatternLength)$, append data point to *NewTrace*
   if $RelPos > (PeriodLength - PatternLength)$, discard data point

The *PeriodLength* and the *PatternLength* denote the least separation between identical points of different instances of the pattern and the length of the pattern, respectively.

---

the existence of a periodic pattern in the power consumption that appeared at unpredictable positions, independent of the start of the decryption algorithm. In order to remove the pattern, it was necessary to identify its exact length and position. Alg. 2 allows to extract the required information.

Practical results for the profiling are depicted in Fig. 8. The given power-, mean- and variance traces show the end of the KEELOQ decryption, which can be identified as a fast changing pattern on the left of the top picture. The right parts of the traces illustrate the situation after the microcontroller has finished the decryption. The mean and variance traces reveal the pattern contained in all traces that is independent of the KEELOQ algorithm. The variance allows to identify the extent and the position of the pattern, while the mean trace shows an averaged instance of the pattern that can be used as template to identify it. Note that the pattern occurs even after the KEELOQ algorithm has finished, indicating its independency from the execution of the cipher.

Preprocessing. For periodically occurring patterns, Alg. 3 provides a method to clean the traces. A similar preprocessing can be applied in case of non-periodic patterns in the power consumption, as long as they can be identified and characterized during profiling. The exact position and length of the unwanted pattern can again be found via the variance of adequately aligned traces.

Practical Results. While an improved CPA on the clean traces now succeeds with around 5000 power traces, we are again able to extract the master key from a single trace using SPA. The methods described in this section can generally be used to remove the effect of timer-based interrupts and inserted dummy operations from power traces, as long as their patterns are prominent enough to allow identification of their rough position.

7   Comparison of DPA and SPA

The efforts for performing an SPA are significantly smaller than those for a DPA, because the latter naturally requires acquiring many traces and a lot of memory for storing and evaluating them. Analyzing commercial black-box implementations with DPA moreover poses the in practice sometimes difficult tasks of triggering the oscilloscope and aligning the measurements accurately. The SPA described in Sect. 5 requires neither alignment nor memory, as one measurement starting at an almost arbitrary point[5] during a decryption is sufficient for a full key recovery. Furthermore, our proposed SPA requires no knowledge about neither the plaintext, nor the ciphertext of the attacked decryption, as all necessary parameters for the SPA can be derived solely from the power measurements. A DPA is clearly impossible under these premises.

The outcome, that conducting a DPA is difficult for an unknown implementation does not imply that the implementation is more secure. In the contrary, it may turn out - as demonstrated in this paper - that an even simpler and much more effective attack is applicable, due to data-dependent execution times in the algorithm.

Implementing the cipher such that the duration of a table look-up takes equally long for any input will most likely prevent from a key recovery with the SPA as described in this paper. However, this approach cannot be recommended, because it would simultaneously facilitate an extraction of the secret key via DPA of the - now well aligned - measurements.

8   Conclusion

Obtaining the device key K_{dev} of a remote control by DPA of the hardware implementation of KEELOQ is straightforward. However, recovering the manufacturer key K_{man} from a software implementation of the cipher was still a challenging task. In this paper, we developed an SPA targeting KEELOQ software implementations on 8-bit PIC microcontrollers, making an extraction of K_{man} from commercial KEELOQ systems much more feasible: where thousands of power traces were originally required to mount a successful DPA, now one single measurement suffices to recover the secret key.

After an in-depth analysis of a reference implementation of KEELOQ, we pinpointed a fatal vulnerability to SPA and exploited it to develop a very efficient key-recovery attack that requires no prior knowledge about neither the plaintext nor the ciphertext. The described approach includes a non-heuristic method for automatically extracting the parameters required for the SPA from power traces, and thus avoids tedious visual inspection. Our attack neither requires a sophisticated measurement setup, nor any preprocessing steps to align or average traces. We further detailed techniques for correcting errors, e.g., due to noisy measurements, and how irrelevant program code inserted during the execution of an algorithm can be removed a priori.

The feasibility of our attacks was demonstrated by successfully attacking several commercial products based on different PIC microcontrollers. In all cases, the efforts for extracting the correct K_{man} were reduced to evaluating one measurement. To our knowledge, and without naming any manufacturers, the

---

[5] Any starting point that captures ≥ 96 rounds of KEELOQ is appropriate.

described SPA can be applied to the vast majority of KEELOQ receivers in the field. Therefore, it becomes practical for criminals to extract and collect master keys of many manufacturers, and perform devastating attacks on KEELOQ RKE systems.

The assumption that extracting the manufacturer key from the software running in a receiver is very demanding and it thus could be regarded as being stored more securely than a device key of a remote control, does no longer hold. With the developed SPA attack, the manufacturer key can be extracted even much simpler than the device keys - a tragedy for the security of all owners of KEELOQ-based RKE systems.

References

1. Bogdanov, A.: Attacks on the KeeLoq Block Cipher and Authentication Systems. In: RFIDSec 2007 (2007),
   http://rfidsec07.etsit.uma.es/slides/papers/paper-22.pdf
2. Brier, E., Clavier, C., Olivier, F.: Correlation Power Analysis with a Leakage Model. In: Joye, M., Quisquater, J.-J. (eds.) CHES 2004. LNCS, vol. 3156, pp. 16–29. Springer, Heidelberg (2004)
3. Courtois, N.T., Bard, G.V., Bogdanov, A.: Periodic ciphers with small blocks and cryptanalysis of keeloq. Tatra Mountains Mathematical Publications (2008)
4. Courtois, N.T., Bard, G.V., Wagner, D.: Algebraic and Slide Attacks on KeeLoq. In: Nyberg, K. (ed.) FSE 2008. LNCS, vol. 5086, pp. 97–115. Springer, Heidelberg (2008)
5. Eisenbarth, T., Kasper, T., Moradi, A., Paar, C., Salmasizadeh, M., Shalmani, M.T.M.: On the Power of Power Analysis in the Real World: A Complete Break of the KeeLoq Code Hopping Scheme. In: Wagner, D. (ed.) CRYPTO 2008. LNCS, vol. 5157, pp. 203–220. Springer, Heidelberg (2008)
6. Indesteege, S., Keller, N., Dunkelman, O., Biham, E., Preneel, B.: A Practical Attack on KeeLoq. In: Smart, N.P. (ed.) EUROCRYPT 2008. LNCS, vol. 4965, pp. 1–18. Springer, Heidelberg (2008)
7. Kocher, P.C., Jaffe, J., Jun, B.: Differential Power Analysis. In: Wiener, M. (ed.) CRYPTO 1999. LNCS, vol. 1666, pp. 388–397. Springer, Heidelberg (1999)
8. Messerges, T.S., Dabbish, E.A., Sloan, R.H.: Power Analysis Attacks of Modular Exponentiation in Smartcards. In: Koç, Ç.K., Paar, C. (eds.) CHES 1999. LNCS, vol. 1717, pp. 144–157. Springer, Heidelberg (1999)
9. Microchip. AN642: Code Hopping Decoder using a PIC16C56,
   http://www.keeloq.boom.ru/decryption.pdf
10. Microchip. HCS301 KEELOQ Code Hopping Encoder Data sheet,
    http://ww1.microchip.com/downloads/en/DeviceDoc/21143b.pdf
11. Microchip. PIC16C5X Data Sheet,
    http://ww1.microchip.com/downloads/en/DeviceDoc/30453d.pdf
12. Webpage. Program Code for KeeLoq Decryption,
    http://www.pic16.com/bbs/dispbbs.asp?boardID=27&ID=19437

Appendix A: The KEELOQ Decryption Program Code

```
; DECRYPT   using [Key7 . . . Key0]
;   | HOP4 | HOP3 | HOP2 | HOP1 |<-- Feed

DECRYPT
00: MOVLW   11+1      ; OUTLOOP COUNTER
01: MOVWF   CNT1      ; 11+1 TIMES

OUTLOOP
02: MOVLW   48        ; INLOOP COUNTER
03: MOVWF   CNT0      ; 48 TIMES

INLOOP
04: CLRWDT            ;
05: MOVFW   CNT1      ;
06: XORLW   1         ;
07: SKPNZ             ; LAST 48 LOOPS
08: GOTO    ROT_KEY   ; RESTORE THE KEY

09: CLRC              ; CLEAR CARRY
0A: MOVLW   1         ; MASK = 1
0B: BTFSC   HOP3,3    ; SHIFT MASK 4X
0C: MOVLW   10000B    ; IF BIT 2 SET
0D: MOVWF   MASK      ;

0E: BTFSS   HOP2,0    ; SHIFT MASK
0F: GOTO    $+3       ; ANOTHER 2X
10: RLF     MASK      ; IF BIT 1 SET
11: RLF     MASK      ;

12: BTFSC   HOP1,0    ; SHIFT MASK
13: RLF     MASK      ; 1X MORE IF BIT 0

14: MOVLW   0         ; TABLE INDEX = 0
15: BTFSC   HOP4,1    ; IF BIT 3 SET
16: IORLW   2         ; TABLE INDEX += 2
17: BTFSC   HOP4,6    ; IF BIT 4 SET
18: IORLW   4         ; TABLE INDEX += 4

19: ADDWF   PC        ; PC += TABLE INDEX

TABLE
1A: MOVLW   02EH      ; BITS 4:3 WERE 00
1B: GOTO    T_END     ; END OF TABLE

1C: MOVLW   074H      ; BITS 4:3 WERE 01
1D: GOTO    T_END     ; END OF TABLE

1E: MOVLW   05CH      ; BITS 4:3 WERE 10
1F: GOTO    T_END     ; END OF TABLE

20: MOVLW   03AH      ; BITS 4:3 WERE 11

T_END
21: ANDWF   MASK      ; ISOLATE THE
22: MOVLW   0         ; CORRECT BIT
23: SKPZ              ;
24: MOVLW   80H       ; W = NLF OUTPUT

25: XORWF   HOP2,W    ; W XOR= HOP2,7
26: XORWF   HOP4,W    ; W XOR= HOP4,7
27: XORWF   KEY1,W    ; W XOR= KEYREG1,7

28: MOVWF   MASK      ; FEEDBACK = BIT 7
29: RLF     MASK      ; CARRY = BIT 7

2A: RLF     HOP1      ; SHIFT IN
2B: RLF     HOP2      ; THE NEW BIT
2C: RLF     HOP3      ;
2D: RLF     HOP4      ;

ROT_KEY
2E: CLRC              ; CLEAR CARRY
2F: BTFSC   KEY7,7    ; IF BIT 7 SET
30: SETC              ; SET CARRY

31: RLF     KEY0      ; LEFT-ROTATE
32: RLF     KEY1      ; THE 64-BIT KEY
33: RLF     KEY2      ;
34: RLF     KEY3      ;
35: RLF     KEY4      ;
36: RLF     KEY5      ;
37: RLF     KEY6      ;
38: RLF     KEY7      ;

39: DECFSZ  CNT0      ;
3A: GOTO    INLOOP    ; INLOOP 48 TIMES

3B: DECFSZ  CNT1      ;
3C: GOTO    OUTLOOP   ; OUTLOOP 12 TIMES
3D: RETLW   0         ; RETURN
```

An Improved Fault Based Attack of the Advanced Encryption Standard

Debdeep Mukhopadhyay

Computer Sc. and Engg, IIT Kharagpur, India
debdeep@cse.iitkgp.ernet.in

Abstract. In the present paper a new fault based attack has been proposed against AES-Rijndael. The paper shows that inducing a single random byte fault at the input of the eighth round of the AES algorithm the block cipher key can be deduced. Simulations show that when two faulty ciphertext pairs are generated, the key can be exactly deduced without any brute-force search. Further results show that with one single faulty ciphertext pair, the AES key can be ascertained with a brute-force search of 2^{32}.

1 Introduction

In order to satisfy the security requirements of various information disciplines e.g. networking, telecommunications, data base systems and mobile applications, applied cryptography has gained immense importance now-a-days. To satisfy the high throughput requirements of such applications, the complex cryptographic systems are implemented by means of either VLSI devices (crypto-accelerators) or highly optimized software routines (crypto-libraries). The high complexity of such implementations raises concerns regarding their reliability. Hence in this scenario it is imperative that the crypto-algorithms should not only prevent conventional cryptanalysis but also should prevent the deduction of the keys from accidental faults or intentional intrusions. Such attacks are known as fault attacks and were first conceived in September 1996 by Boneh, DeMillo and Lipton [1,2] from Bellcore. The fault attack was applicable to public key cryptosystems and was extended to various secret key ciphers like DES, the technique being known as Differential Fault Analysis (DFA)[3]. On 2nd October 2000, the US National Institute of Standards and Technology (NIST) selected Rijndael [4] as the Advanced Encryption Standard (AES) and thus replaced DES as a world-wide standard for symmetric key encryption. Thus smart cards and secure micro-controllers are designed using AES to protect both the confidentiality and the integrity of sensitive information. With the work on optical fault induction reported in [5], research in the field of fault-based side channel cryptanalysis of AES has gained considerable attention. Less costly methods for fault injection include variation of supply voltages, clock frequency, clock glitches or temperature variations. DFA on AES was reported in [6] by inducing faults at byte level to the input of 9th round of AES using 250 faulty ciphertexts. In the fault based

B. Preneel (Ed.): AFRICACRYPT 2009, LNCS 5580, pp. 421–434, 2009.

attack on AES reported in [7] around 128 to 256 faulty ciphertexts are required to discover the key. Dusart et. al. [8] performs a Differential Fault Analysis on AES and shows that using a byte level fault induction anywhere between the eighth round and ninth round the attacker is able to break the key with 40 faulty ciphertexts. Finally, [9] shows that using byte level faults at the input of the eighth round or the input of the ninth round of a ten round AES-128 algorithm, an attacker can retrieve the whole AES-128 key with two faulty ciphertexts. Recently fault attacks on AES exploiting the key-scheduling algorithms [10,11] have been developed which compute the value of the AES key with a minimum of two faulty ciphertexts at the cost of a brute force search of 48 and 40 bits respectively.

In the present work we present a fault based side-channel attack on AES using a single byte level fault. In the proposed attack the attacker induces a random non-zero byte level fault at the input of the 8th round by affecting a single byte of the data. Extensive experimentations have been performed on a PC and it has been found that the key can be obtained using only two faulty ciphertexts. The attack has a workload of around 2^{16} and does not require any brute-force to obtain the final key. It takes a few seconds on a PC for completion. The idea of using algebraic equations have also been adopted in [8]. But the equations proposed in the present paper lead to much simpler analysis and reduce the number of faulty ciphertexts required from 40 to 2. Unlike [10,11] the present work does not require any brute force search. The present paper shows that although the present attack, like [9] requires two faulty ciphertexts, it requires a fault induction at a single byte location, as opposed to two required in [9]. Further we also suggest in the paper a modification to the basic attack, so that with one faulty ciphertext, the present attack can obtain the key with a brute force search of 2^{32}, which can be performed in less than 15 minutes using a PC. In short the present paper demonstrates the fact that AES can be broken using a single byte fault for one single instance. This increases the probability of making a fault attack practical using simple and less costly methods, like clock glitches and voltage fluctuations.

The paper is organised as follows: *section 2* describes the AES-Rijndael algorithm. The fault model and the attack environment is stated in *section 3*. The working principle of the attack is described in *section 4*, while the proposed attack is described in *section 5*. Finally the results of the work are compared to existing research in *section 6*. The work is concluded in *section 7*.

2   The Description of AES-Rijndael Algorithm

The description of the AES-Rijndael algorithm may be found in [4]. The typical round is described in the current subsection. The 128 bit message and key sizes have been considered, but the discussion can be extended to other specifications of the Rijndael block cipher.

The 128 bit input block to AES is arranged as a 4 × 4 array of bytes, known as the state matrix, refer to *figure 1*. The elements of the matrix are represented

b_{00}	b_{01}	b_{02}	b_{03}
b_{10}	b_{11}	b_{12}	b_{13}
b_{20}	b_{21}	b_{22}	b_{23}
b_{30}	b_{31}	b_{32}	b_{33}

Fig. 1. The State Matrix of AES-Rijndael

by the variable, b_{ij}, where $0 \leq i, j \leq 3$ and i, j refers to the row and column positions.

The algorithm has ten rounds and the keys of each round are generated by a key scheduling algorithm. The design of the key scheduling algorithm of AES is such that the knowledge regarding any round key reveals the original input key (named as the master key) from which the round keys are derived. The input state matrix (plaintext) is transformed by the various round transforms. The state matrix evolves as it passes through the various steps of the cipher and finally emerges in the form of ciphertext.

The rounds of AES use the following steps (*figure 2*):

1. The Byte Sub Step: The Byte Sub is the only non-linear step of the cipher. It is a bricklayer permutation consisting of an S-box applied to the bytes of the state. Each byte of the state matrix is replaced by its multiplicative inverse, followed by an affine mapping. Thus the input byte x is related to the output y of the S-Box by the relation, $y = A.x^{-1} + B$, where A and B are constant matrices[4].

2. The ShiftRows Step: Each row of the state matrix is rotated by a certain number of byte positions. This is a byte transposition step.

3. The MixColumn Step: The MixColumn is a bricklayer permutation operating on the state column by column. Each column of the state matrix is considered as a 4-dimensional vector where each element belongs to $GF(2^8)$. A 4×4 matrix M whose elements are also in $GF(2^8)$ is used to map this column into a new vector. This operation is applied on all the 4 columns of the state matrix [4]. Here M is defined as follows:

$$M = \begin{pmatrix} 2 & 3 & 1 & 1 \\ 1 & 2 & 3 & 1 \\ 1 & 1 & 2 & 3 \\ 3 & 1 & 1 & 2 \end{pmatrix}$$

4. AddRoundKey: Each byte of the array is exclusive-ored with a byte from a corresponding array of round subkeys.

The first 9 rounds of AES-Rijndael are identical - only the last round is not because the MixColumn step does not exist.

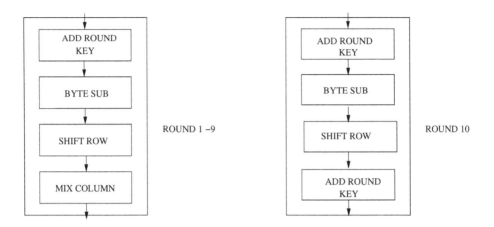

Fig. 2. The Round Transforms of AES-Rijndael

3   Fault Model Used and the Attack Environment

In this work, the fault assumed is a single byte fault. Single byte fault means the fault f_{ij} is injected in one particular byte b_{ij}, where i refers to the row position and j refers to the column position in the state matrix ($0 \leq i, j \leq 3$). The number of bits in the byte which are affected by the fault is indicated by $w(f_{ij})$, where $1 \leq w(f_{ij}) \leq 8$.

The attacker injects fault at the input of the eighth round in a single byte. The fault value can be arbitrary but non-zero. Before stating the fault attack techniques we outline two practical scenarios where the attack may be carried out. The scenarios show that depending upon the implementation of cryptographic hardware there are two different requirements on the attacker in order to inject the fault at any precise round location.

- Scenario 1: Certain implementations of AES-Rijndael requires pipelining at all stages (unrolled rounds), due to the requirement of throughput. Thus each key requires access to a key memory which cannot be shared among the rounds. Hence, the entire key has to be stored in a key register or memory. In such a case the attacker wishes to cause faults in the value that is being read from the memory while leaving the value stored in the memory unaffected. This does not hamper the normal functionality of the device and is thus undetectable. Further, in such an implementation large number of faulty ciphertexts can be obtained, since the key stored in the memory is unaffected. Thus the requirement on the attacker in such a case is *Control on Fault Location*.
- Scenario 2: The other way in which block ciphers like AES-Rijndael are implemented is through iterative structures (rolled rounds) or a combination of unrolled and rolled rounds. In such a case the key is not stored in the memory and thus the requirement on the attacker is *Control on Fault Timing*.

Imprecise control over fault location or fault timing hinders the attacker to be able to inject a fault at the intended round. In our present attack, we assume that the attacker intends to inject fault in a byte at the input of the 8th round. In the following sections we present strategies through which if the attacker induces only a single byte fault for one time he is able to discover the key. First we explain the principle of the attack.

4   Working Principle of the Proposed Attack in the Ninth Round

The proposed attack is based on the induction of a single byte fault at the input of the eighth round. In order to explain the attack, first let us consider the scenario when there is a single byte disturbance at the input of the ninth round.

The 9th round has a diffusion step, so a disturbance in one byte affects 4 bytes at the output. The last round does not have a diffusion step and so the disturbances remain in 4 bytes of the state matrix. If one traces the disturbance in the state matrix through the last two rounds the following properties can be identified. These properties can be utilized to develop an attack against the block cipher.

4.1   Property of the State Matrix

Figure 3 shows the propagation of the fault, when it is induced in a byte at the input of the ninth round Byte Sub. In the figure we consider the case when there is an arbitrary but non-zero disturbance at the 00th byte of the state matrix. The disturbance or fault value is say f, and after the Byte Sub, then it is transformed to a value f'. The byte fault is propagated to four byte positions. As the figure suggests, the faults at the input of the tenth round Byte Sub have values of

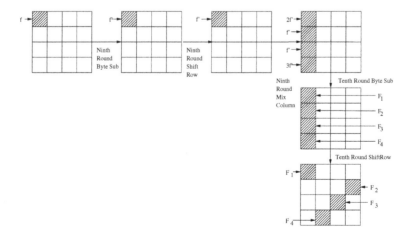

Fig. 3. Propagation of Fault Induced in the input of the ninth round of AES

$2f'$, f', f' and $3f'$. The faulty values after the tenth round Byte Sub gets transformed into F_1, F_2, F_3 and F_4. The attacker obtains a pair of ciphertext, (CT, CT'). The ciphertext CT is a fault free ciphertext and the ciphertext CT' is the ciphertext, when a fault is induced in a byte. When a bytewise fault is induced at the input of the ninth round Byte Sub, the difference of the ciphertexts CT and CT' has the pattern as shown in *figure 3* after the tenth round ShiftRow.

The fault pattern as shown in *figure 3* depicts the difference between the fault free ciphertext CT and the faulty ciphertext CT'. Let the values of the bytes shaded in *figure 3* after the tenth round Shift Row in the ciphertext CT be denoted by x_1, x_2, x_3 and x_4. Then the corresponding values for the faulty ciphertext CT' is denoted by $x_1 + F_1$, $x_2 + F_2$, $x_3 + F_3$ and $x_4 + F_4$. Here the sign $+$ stands for the bit-wise exclusive-or operation of two bytes. The corresponding key bytes are K_1, K_2, K_3 and K_4.

The fault pattern gives the following set of equations:

$$ISB(x_1 + K_1) + ISB(x_1 + F_1 + K_1) = 2[ISB(x_2 + K_2) + ISB(x_2 + F_2 + K_2)]$$
$$ISB(x_2 + K_2) + ISB(x_2 + F_2 + K_2) = \ ISB(x_3 + K_3) \ + \ ISB(x_3 + F_3 + K_3)$$
$$ISB(x_4 + K_4) + ISB(x_4 + F_4 + K_4) = 3[ISB(x_2 + K_2) + ISB(x_2 + F_2 + K_2)]$$

In the above set of equations ISB stands for the Inverse Byte Sub operation, which is defined as the inverse of the Byte Sub step. The values of (x_1, x_2, x_3, x_4) and $(x_1 + F_1, x_2 + F_2, x_3 + F_3, x_4 + F_4)$ are known to the adversary. The attacker intends to compute the values of K_1, K_2, K_3 and K_4 from the equations. The attacker evaluates the keys as follows: The attacker guesses the bytes K_1 and K_2 and checks whether they satisfy the first equation. The solution sets for the second and third equations are searched in parallel. Thus the time complexity of the key conjuring is 2^{16}. Finally since the variable K_2 is in all the three equations, the solution set of K_2 from each equation is intersected to arrive at a reduced solution space for K_2. The reduced space of K_2 is then used to find a reduced set of K_1, K_3 and K_4 from the three equations. It may be noted that the above attack does not depend upon the value of the induced fault, which may be random but non-zero. This completes a single pass of the algorithm with one (CT, CT') pair. In order to ascertain the key bytes, further passes of the algorithm are run with other (CT, CT') pairs. We have experimentally verified that the number of passes of the algorithm is two with a probability of around 0.99, thus revealing the key within two faulty encryptions. In the other few cases, a third faulty ciphertext is required to uniquely ascertain the key.

It may be observed that the attack proposed in [8] also use the properties of the MixColumn matrix, similar to the proposed attack. However one of the key differences between the present attack and the one proposed by [8] lies in the mechanism of solving the equations. [8] requires around $10 - 40$ faulty pairs to determine 4 key bytes of the 10th round, while the proposed attack requires only 2. The difference comes from the fact that we use the Inverse SubByte, compared to SubByte used in [8]. This helps to obtain a better filter for the wrong key bytes.

4.2   A Working Example

In the present section we outline the working of the attack through an example. Let the plaintext be:

$$PT_1 = \begin{pmatrix} 00000001 \ 11111110 \ 10000001 \ 11111100 \\ 10100110 \ 01101011 \ 11100001 \ 10100011 \\ 11110100 \ 10100010 \ 11110111 \ 01111000 \\ 01000000 \ 10100101 \ 10001110 \ 11110000 \end{pmatrix}$$

The plaintext is encrypted using AES-Rijndael encryption algorithm. The key matrix is:

$$K0 = \begin{pmatrix} 11100111 \ 00101000 \ 10010101 \ 01100001 \\ 01110110 \ 10101110 \ 11110111 \ 11001111 \\ 00010101 \ 11011010 \ 00110101 \ 01011111 \\ 00111110 \ 10000010 \ 10100100 \ 01001100 \end{pmatrix}$$

and the tenth round key is:

$$K10 = \begin{pmatrix} \mathbf{00101101} \ 00010100 \ 00011101 \ 11000101 \\ 11110000 \ 11100010 \ 11010111 \ \mathbf{01000001} \\ 11100010 \ 00000111 \ \mathbf{10100010} \ 10110010 \\ 01010101 \ \mathbf{11101010} \ 01110000 \ 00111100 \end{pmatrix}$$

The corresponding ciphertext is:

$$CT_1 = \begin{pmatrix} 11101110 \ 01111111 \ 11110100 \ 01100101 \\ 01011000 \ 01001101 \ 10110101 \ 10110101 \\ 11111001 \ 00101001 \ 11010010 \ 11100010 \\ 10000101 \ 00111011 \ 11111100 \ 11110111 \end{pmatrix}$$

The faulty ciphertext for a random fault induced in the 00th position at the input of the ninth round leads to the following faulty ciphertexts:

$$CT_1' = \begin{pmatrix} \mathbf{00101111} \ 01111111 \ 11110100 \ 01100101 \\ 01011000 \ 01001101 \ 10110101 \ \mathbf{11111111} \\ 11111001 \ 00101001 \ \mathbf{01111000} \ 11100010 \\ 10000101 \ \mathbf{10010101} \ 11111100 \ 11110111 \end{pmatrix}$$

The bytes in the faulty ciphertexts which are bolded show how the fault has propagated. The equations developed in the previous attack are applied to obtain the key bytes. From one faulty ciphertext the key space is reduced to around 32 possibilities per key byte.

The actual key can be ascertained if we consider another faulty encryption. Let another plaintext be:

$$PT_2 = \begin{pmatrix} 10101001 \ 01010110 \ 10001101 \ 11001100 \\ 11110110 \ 01001011 \ 10100011 \ 10000011 \\ 11110100 \ 00000010 \ 10100110 \ 11110000 \\ 11110000 \ 11100101 \ 00111100 \ 11110001 \end{pmatrix}$$

The corresponding ciphertext is:

$$\mathbf{CT_1} = \begin{pmatrix} 00001101 \ 11111101 \ 00111011 \ 10001101 \\ 11000110 \ 00100011 \ 11110101 \ 01110001 \\ 11001111 \ 00101110 \ 10100101 \ 11011010 \\ 01110011 \ 00001111 \ 10101101 \ 11000100 \end{pmatrix}$$

and the faulty ciphertext is

$$\mathbf{CT_1'} = \begin{pmatrix} \mathbf{01011100} \ 11111101 \ 00111011 \ 10001101 \\ 11000110 \ 00100011 \ 11110101 \ \mathbf{10100001} \\ 11001111 \ 00101110 \ \mathbf{11111011} \ 11011010 \\ 01110011 \ \mathbf{00011011} \ 10101101 \ 11000100 \end{pmatrix}$$

Intersection of the two solution sets leaves only one element, which is the correct solution. In this example we arrive at the key bytes: K_1=00101101, K_2=01000001, K_3=10100010 and K_4=11101010. The bold elements in the matrix of $K10$ show that the guesses are correct.

The above discussion shows that one byte fault reveals four bytes of the key. Thus for all the 16 bytes of AES it is necessary to induce faults at four bytes. However, often it may not be possible to induce faults at four byte positions. In the next section, we outline a modification to perform the fault based cryptanalysis of AES with one fault induction.

5   The Proposed Attack Strategy in the Eighth Round

In this attack we assume that the adversary has induced fault in a byte of the input to the eighth round. If the fault is induced in a byte of the state matrix, which is input to the eighth round, the disturbance spreads to the entire state matrix when it emerges out after the tenth round. In this case, a single byte fault creates four byte faults at the input of the ninth round. An attack similar to the previous attack can thus be used to compute the AES key.

5.1   Property of the State Matrix

Figure 4 shows the diffusion of a byte fault induced at the input of the eighth round. Similar to the previous section the various round operations transform the initial value of the fault f. The attacker observes, like in the previous cases two ciphers - one fault free and the other faulty. The difference of the state matrices of the two ciphers are depicted in *figure 4*.

The attacker knows the value of CT and CT' from the two ciphertexts that he obtains. Let, the two ciphertexts be represented by:

$$\mathbf{CT} = \begin{pmatrix} x_1 & x_2 & x_3 & x_4 \\ x_5 & x_6 & x_7 & x_8 \\ x_9 & x_{10} & x_{11} & x_{12} \\ x_{13} & x_{14} & x_{15} & x_{16} \end{pmatrix}$$

and

$$\mathbf{CT'} = \begin{pmatrix} x_1 + A_1 & x_2 + A_2 & x_3 + A_3 & x_4 + A_4 \\ x_5 + A_6 & x_6 + A_7 & x_7 + A_8 & x_8 + A_5 \\ x_9 + A_{11} & x_{10} + A_{12} & x_{11} + A_9 & x_{12} + A_{10} \\ x_{13} + A_{16} & x_{14} + A_{13} & x_{15} + A_{14} & x_{16} + A_{15} \end{pmatrix}$$

Here x_i and A_i $(1 \leq i \leq 16)$ are each one byte.
The corresponding key matrix for the tenth round is:

$$\mathbf{K_{10}} = \begin{pmatrix} K_{00} & K_{01} & K_{02} & K_{03} \\ K_{10} & K_{11} & K_{12} & K_{13} \\ K_{20} & K_{21} & K_{22} & K_{23} \\ K_{30} & K_{31} & K_{32} & K_{33} \end{pmatrix},$$

where each term k_{ij} $(0 \leq i, j \leq 3)$ is also a byte value.

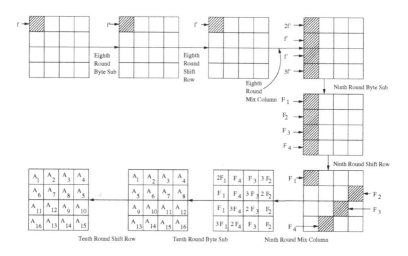

Fig. 4. Propagation of Fault Induced in the input of eighth round of AES

We note the state of the differences after the ninth round shift row from *figure 4*. Combining the above facts we obtain the following set of equations to evaluate the values of the key bytes K_{00}, K_{13}, K_{22} and K_{31}, thus revealing 32 bits of the AES key.

$$ISB(x_1 + K_{00}) + ISB(x_1 + A_1 + K_{00}) = 2[ISB(x_8 + K_{13}) + ISB(x_8 + A_5 + K_{13})]$$
$$ISB(x_8 + K_{13}) + ISB(x_8 + A_5 + K_{13}) = ISB(x_{11} + K_{22}) + ISB(x_{11} + A_9 + K_{22})$$
$$ISB(x_{14} + K_{31}) + ISB(x_{14} + A_{13} + K_{31}) = 3[ISB(x_8 + K_{13}) + ISB(x_8 + A_5 + K_{13})]$$

The unknowns in the above set of equations is the value of the key bytes K_{00}, K_{13}, K_{22} and K_{31}. The attacker similar to the previous strategy obtains reduced solution spaces for the bytes K_{00}, K_{13}, K_{22} and K_{31} from the three equations. The worst case complexity for one pass of the algorithm is 2^{16} and is again independent of the value of the fault induced. Another solution set for the key bytes is obtained with another CT and CT' pair. The two solution sets are intersected to arrive at the correct key bytes, as the intersection set leaves only one element.

The above system of equation is used to reduce the possibilities of 32 bits of the key. In order to obtain the remaining three 32 bits of the AES key the attacker uses three more similar systems of equations. We briefly state the three other system of equations as follows:

In order to obtain $(K_{01}, K_{10}, K_{23}, K_{32})$ the attacker uses the following equations:

$$ISB(x_{15} + K_{32}) + ISB(x_{15} + A_{14} + K_{32}) = 2[ISB(x_2 + K_{01}) + ISB(x_2 + A_2 + K_{01})]$$
$$ISB(x_2 + K_{01}) + ISB(x_2 + A_2 + K_{01}) = [ISB(x_5 + K_{10}) + ISB(x_5 + A_6 + K_{10})]$$
$$ISB(x_{12} + K_{23}) + ISB(x_{12} + A_{10} + K_{23}) = 3[ISB(x_2 + K_{01}) + ISB(x_2 + A_2 + K_{01})]$$

In order to obtain $(K_{02}, K_{11}, K_{20}, K_{33})$ the attacker uses the following equations:

$$ISB(x_9 + K_{20}) + ISB(x_9 + A_9 + K_{20}) = 2[ISB(x_3 + K_{02}) + ISB(x_3 + A_3 + K_{02})]$$
$$ISB(x_3 + K_{02}) + ISB(x_3 + A_3 + K_{02}) = [ISB(x_{16} + K_{33}) + ISB(x_{16} + A_{15} + K_{33})]$$
$$ISB(x_6 + K_{11}) + ISB(x_6 + A_7 + K_{11}) = 3[ISB(x_3 + K_{02}) + ISB(x_3 + A_3 + K_{02})]$$

In order to obtain $(K_{03}, K_{12}, K_{21}, K_{30})$ the attacker uses the following equations:

$$ISB(x_7 + K_{12}) + ISB(x_7 + A_8 + K_{12}) = 2[ISB(x_{10} + K_{21}) + ISB(x_{10} + A_{12} + K_{21})]$$
$$ISB(x_{10} + K_{21}) + ISB(x_{10} + A_{12} + K_{21}) = [ISB(x_{13} + K_{30}) + ISB(x_{13} + A_{16} + K_{30})]$$
$$ISB(x_4 + K_{03}) + ISB(x_4 + A_4 + K_{03}) = 3[ISB(x_{10} + K_{21}) + ISB(x_{10} + A_{12} + K_{21})]$$

It may be noted that the equations are identical to that obtained in the previous section and thus the solutions are of similar nature. If only one faulty pair is used for the guessing of the key, the number of possible key values for each of the bytes of 32 bits of the key is around 32. Thus, one pass of the algorithm leaves on the average around $32^4 \approx 2^{20}$ possible candidate 32 bits of the key. Thus after one pass of the attack, number of possible 128 bit 10th round AES keys is $(2^{20})^4 = 2^{80}$, as there are four 32 bit keys. Hence, brute force is not possible. However using another faulty ciphertext identifies the key uniquely. In the next section, we present an improvement to the above attack to reduce the brute force complexity to find the complete 128 bits AES key with one faulty ciphertext to only 2^{32}.

6   The Proposed Fault Attack with One Faulty Ciphertext

As discussed in the last section, the proposed attack computes four bytes of the AES key by using a faulty ciphertext and a fault free ciphertext and solving a system of equations. Each system of equation reduces the possible values of 32 bits of the AES key. There are four system of equations, each giving possible candidates for 32 bits of the key and the solutions are independent. Since, the system of equations are similar in nature, their pruning capability of the key values are also similar. As we have discussed that the number of values for each of the bytes is 32, the total number of AES key values is as large as 2^{80}. In the following we propose an improvement which reduces the possibility of AES key values to as low as 2^{32} without using any other faulty ciphertext.

The improvement is based on the following observation: The fault induced at the input of the eighth round gets spread to the entire column after the MixColumn operation. Due to the next round shift row, this disturbance gets spread to each column. More specifically, each column has one byte of disturbance. The other three bytes in each column are undisturbed. We exploit this property for the further pruning of the key space.

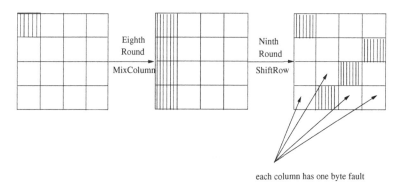

each column has one byte fault

Fig. 5. Improvement to perform the fault attack with one faulty ciphertext

Assuming that the fault has been induced at the 00th byte position of the input state matrix of the eighth round, the propagation of the fault through the MixColumn and Shift Row is as depicted in *figure 5*. It may be noted that each column at the output of the ninth round ShiftRow has a one byte disturbance. After the attack proposed in section 5, there are four reduced set of 32 bit keys. To further reduce the key size, the keys are used to check the fault pattern at the output of the ninth round ShiftRow. It is inspected whether only one specific byte position (depending on the initial fault location) is disturbed. For example, let us assume that after one pass of the previous algorithm we have a reduced set for 32 bits of the key, denoted by $(K_{00}, K_{13}, K_{22}, K_{31})$. We denote the fault free ciphertext by CT and faulty ciphertext by CT'. First we take the four ciphertext bytes, $(x_1, x_8, x_{11}, x_{14})$ and decrypt with the 32 bits of the key $(K_{00}, K_{13}, K_{22}, K_{31})$. Then we perform inverse Byte Sub. The same procedure is repeated over the faulty ciphertext, CT'. This reveals to us the fault pattern after AddRoundKey at the input of the tenth round. Since the fault pattern is not disturbed by the AddRoundKey after the ninth round, we have the same at the output of the ninth round MixColumn. Then we perform Inverse Mix-Column to find out the fault pattern of the first column after the ninth round ShiftRow. We check whether the left byte is non-zero, while the rest are zero. We eliminate all key bytes which do not satisfy the above property. Using exhaustive experimentations, we observe that after the above pruning there are around 240 possible key values for the four key bytes.

In order to obtain the entire key, the same pruning is applied on the reduced set of the other three 32 bits of the key. Hence, the total number of keys remaining after this additional pruning is $240^4 \approx 2^{32}$. This may be explored with todays computation power using a brute force search.

It may be noted that the proposed attack assumed that the fault was an arbitrary non-zero byte value at a known byte position (say the 00th byte position of the state matrix). However the attack is easily extended to an attack which does not require the knowledge of the fault location, that is which byte position. This is based on the observation that depending on the fault location (the byte value where the fault is induced) we have four sets of equations. This cluster of equations varies if the byte position where the fault is located changes. Since there can be 16 byte positions, there are 16 clusters of such equations. It may be easily noted that the cluster of equations are identical and have the same pruning power. Thus, if the attacker induces a byte fault of arbitrary but non-zero value at the input of the eighth round, but is not sure about the exact byte location, he assumes the byte location and runs the above attack. Thus for each possible byte location, the proposed attack gives 2^{32} possible AES keys. Now varying the fault position, the number of keys is reduced to $16 \times 2^{32} = 2^{36}$, which is also within practical limits.

Next we present comparisons of our work with existing research in this area.

7   Comparison with Existing Works and Experimental Results

There have been considerable number of works on the subject. In this section we compare the existing fault based attack on AES with the help of *table 1*. The comparisions show that the current fault attack requires the minimum of faulty encryptions in order to derive the key like [9]. Recently, there have been some related fault attacks on AES, using the properties of the key scheduling algorithm. We compare our result with these attacks in *table 2*.

If the work reported in [9] be compared with the present attack based on the result with one faulty encryption, then both the attacks reduce the number of possible 32 bits of the key. The present attack reduces the number of candidate keys to an average of about 240 compared to 1036 required in the previous attack. Often the second fault induction may not be possible, in such a case the present attack can lead to a brute force attack after reducing the key space by using one faulty ciphertext. Thus if only faulty ciphertext is used, the present attack requires a brute force search of 2^{32} if the byte position of the fault in the state matrix is known and 2^{36} if not known. In comparison the work of Piret et. al. [9] requires around $(2^{10})^4 = 2^{40}$, which is higher than that of the present attack.

Table 1. Comparison of Existing Fault Attacks on AES exploiting properties of the encryption function

Reference	Fault Model	Fault Location	No. of Faulty Encryptions
[7]	Force 1 bit to 0	Chosen	128
[7]	Implementation Dependent	Chosen	256
[6]	Switch 1 bit	Any bit of chosen bytes	≈ 50
[6]	Disturb 1 byte	Anywhere among 4 bytes	≈ 250
[8]	Disturb 1 byte	Anywhere between last two MixColumn	≈ 40
[9]	Disturb 1 byte	Anywhere between 7th round and 8th round MixColumn	2
This Paper	Disturb 1 byte	Anywhere between 7th round MixColumn and 8th round MixColumn	2

Table 2. Comparison with Existing Fault Attacks on AES exploiting key scheduling

Reference	No of fault Injection Points	No. of Faulty Encryptions	Brute-force Search
[10]	1	2	2^{48}
	2	4	2^{16}
	3	7	0
[11]	1	2	2^{40}
	3	7	0
Our Attack	1	2	0
	1	1	2^{32}

8   Conclusions

The paper proposes an improved fault based cryptanalysis of the AES algorithm. The work shows that using only one arbitrary but non-zero byte fault at the input of the eighth round MixColumn, the 128 bit AES key can be deduced. Results have been furnished to show that the proposed attack leads to the evaluation of the exact key without any requirement for brute force search if two faulty ciphertexts are available. The attack has been further improved to show that even if only one faulty ciphertext is available, the AES key can be ascertained with a brute force of only 2^{32}, thus significantly improving existing fault based cryptanalysis of AES.

References

1. Boneh, D., DeMillo, R.A., Lipton, R.J.: On the Importance of checking cryptographic Protocols for Faults. In: Fumy, W. (ed.) EUROCRYPT 1997. LNCS, vol. 1233, pp. 37–51. Springer, Heidelberg (1997)
2. Boneh, D., DeMillo, R.A., Lipton, R.J.: On the Importance of Eliminating Errors in Cryptographic Computations. Journal of Cryptology, 101–120 (2001)
3. Biham, E., Shamir, A.: Differential Fault Analysis of Secret Key Cryptosystems. In: Kaliski Jr., B.S. (ed.) CRYPTO 1997. LNCS, vol. 1294, pp. 513–525. Springer, Heidelberg (1997)
4. Daemen, J., Rijmen, V.: The Design of Rijndael. Springer, Heidelberg (2002)
5. Skorobogatov, S., Anderson, R.: Optical Fault Induction Attacks. In: Kaliski Jr., B.S., Koç, Ç.K., Paar, C. (eds.) CHES 2002. LNCS, vol. 2523, pp. 2–12. Springer, Heidelberg (2003)
6. Giraud, C.: DFA on AES. Cryptology ePrint Archive, Report 2003/008 (2003)
7. Blomer, J., Seifert, J.P.: Fault Based Cryptanalysis of the Advanced Encryption Standard (AES). In: Wright, R.N. (ed.) FC 2003. LNCS, vol. 2742, pp. 162–181. Springer, Heidelberg (2003)
8. Dusart, P., Letourneux, G., Vivolo, O.: Differential Fault Analysis on A.E.S. (2003), http://eprint.iacr.org/2003/010
9. Piret, G., Quisquater, J.J.: A Differential Fault Attack Technique against SPN Structures, with Application to the AES and Khazad. In: Walter, C.D., Koç, Ç.K., Paar, C. (eds.) CHES 2003. LNCS, vol. 2779, pp. 77–88. Springer, Heidelberg (2003)
10. Takahashi, J., Fukunaga, T., Yamakoshi, K.: DFA mechanism on the AES schedule. In: Proceedings of 4th International Workshop on Fault Detection and Tolerance in Cryptography, FDTC, pp. 62–72 (2007)
11. Takahashi, J., Fukunaga, T.: Differential Fault Analysis on the AES Key Schedule (2007), http://eprint.iacr.org/2007/480

Author Index